The

Metals Databook

The Metals Databook

Alok Nayar

Escorts Research Centre
Faridabad

McGraw-Hill

New York San Francisco Washington, D.C. Auckland Bogotá Caracas Lisbon
London Madrid Mexico City Milan Montreal New Delhi San Juan
Singapore Sydney Tokyo Toronto

McGraw-Hill

A Division of The ***McGraw-Hill*** *Companies*

First published © 1997, Tata McGraw-Hill Publishing Company Limited

1 2 3 4 5 6 7 8 9 0 DOC/DOC 9 0 2 1 0 9 8 7

ISBN 0-07-46088-4

Printed and bound by R. R. Donnelley & Sons Company.

This book is printed on recycled, acid-free paper containing a minimum of 50% recycled, de-inked fiber.

To

My wife, Payal,

Our Children, Mohit and Radhika,

and our Parents

PREFACE

The Metals Databook contains extensive metallurgical information on the grades of:

- Cast irons
- Steels
- Aluminium and its alloys
- Copper and its alloys
- Zinc and its alloys
- Sintered materials

It provides valuable data on various aspects including the chemical composition, mechanical and physical properties, applications, heat treatment temperatures, etc. from Indian Standards and some selected international and foreign standards. This book also provides information on the physical properties of pure elements. Numeric data have been expressed in SI units only.

The Databook is intended for designers, manufacturing engineers, maintenance engineers, quality assurance engineers, purchasing staff and others associated with the metal producing and metal using industries. For easy reference and comparability of properties, most of the information has been presented in a tabular format. The material has been collected and collated from standards, reference books, and publications of various technical societies and companies and then edited with the aim of maximising its value to the users. The readers are requested to always make use of the individual standards alongside the databook for updated and current information. Any feedback on the quality of information provided by way of suggestions is welcome.

The work on this Databook has involved a lot of effort. While I am obliged to many people and organizations for the strengths of the book, the weaknesses are entirely my responsibility.

ALOK NAYAR

ACKNOWLEDGEMENTS

I am indebted to the following individuals and organizations, both in India and abroad, who have contributed material for *The Metals Databook*:

Aluminium Federation Ltd., Broadway House Calthorpe Road, Five Ways, Birmingham B15 ITN, United Kingdom.

American Foundrymen's Society, Inc., 505 State Street, Des Plaines, Illinois 60016–8399, United States of America.

American National Standards Institute, 11 West, 42nd Street, 13th Floor, New York, N.Y. 10036 United States of America.

ASM International, Materials Park, Ohio 44073, United States of America.

American Society for Testing and Materials, 1916 Race Street, Philadelphia, Pennsylvania 19103–1187, United States of America.

Mr. Avtar Singh, Director, Bureau of Indian Standards, Manak Bhavan, 9 Bahadur Shah Zafar Marg, New Delhi 110 002, India.

Dr. R. J. Barnhurst, Noranda Technology Centre, 240 Hymus Boulevard, Pointe–Claire, Quebec H9R IG5, Canada.

BCIRA, Alvechurch, Birmingham B48 7QB, United Kingdom.

British Standards Institution, 389 Chiswick High Road, London W4 4AL, United Kingdom.

Bureau of Indian Standards, Manak Bhavan, 9 Bahadur Shah Zafar Marg, New Delhi 110 002, India.

Butterworth–Heinemann Ltd., Linacre House, Jordan Hill, Oxford OX2 8DP, United Kingdom.

Copper Development Association, Orchard House, Mutton Lane, Potters Bar, Herts EN6 3AP, United Kingdom.

Copper Development Association, Inc., 2 Greenwich Office, P.O. Box 1840, Connecticut 06836–1840, United States of America.

DIN Deutsches Institut fur Normung e.V., Burggrafenstrasse 6, D–10787 Berlin, Germany.

Eastern Alloys, Inc., P.O. Box Q, Maybrook, New York 12543–0316, United States of America.

Dr. R. Elliot, Materials Science Centre, University of Manchester, Manchester, United Kingdom.

Mr. Sandeep Handa, House No.150, Sector 15, Faridabad 121 007, India.

India Lead Zinc Information Centre, Jawahar Dhatu Bhawan, 39 Tughlaqabad Institutional Area, Mehrauli–Badarpur Road, New Delhi 110 062, India.

International Lead Zinc Research Organization Inc., P.O. Box 12036, Research Triangle Park, North Carolina 27709–2036, United States of America.

International Organization for Standardization, 1 rue de Varembe, Case postale 56, CH–1211 Geneva 20, Switzerland.

Japanese Industrial Standards Committee, c/o Standards Department, Agency of Industrial Science and Technology, Ministry of International Trade and Industry, 1–3–1, Kasumigaseki, Chiyoda–ku, Tokyo 100, Japan.

Mr. Tom H. Lewinson, Noranda Sales Corporation Ltd., Suite 2700, 1 Adelaide Street East, Toronto, Ontario M5C 2Z6, Canada.

Machine Design, A Penton Publication, 1100 Superior Avenue, Cleveland, Ohio 44114–2543, United States of America.

Metal Powder Industries Federation, 105 College Road East, Princeton, New Jersey 08540–6692, United States of America.

I am grateful to the management of Escorts Ltd., for permission to publish this book. I would also like to thank my colleagues at Escorts Research Centre for their support during the preparation of this book.

I would also like to express my appreciation to Ms J. Crasta of the J. N. Tata Endowment Trust for her help during the initial stages of the project.

I would particularly like to thank Ms V. Deepa, Mr P.L. Pandita, and the editorial and production staff at Tata McGraw-Hill for making the publication of this book possible.

Finally, and most importantly, I am indebted to my wife, Payal, and our children, Mohit and Radhika, for their patience, support, and understanding during the preparation of this book.

CONTENTS

Preface *vii*
Acknowledgements *ix*
Abbreviations and Symbols *xiii*

CHAPTER ONE The Elements and Their Properties 1.1–1.21

CHAPTER TWO Cast Irons 2.1–2.84

2.1 Introduction to cast irons *2.5*
2.2 Designation system for cast irons *2.12*
2.3 Designation of microstructure of graphite in cast irons *2.15*
2.4 Grey cast iron *2.22*
2.5 Spheroidal graphite cast iron *2.33*
2.6 Austempered ductile iron *2.44*
2.7 Compacted graphite cast iron *2.49*
2.8 Malleable cast iron *2.53*
2.9 Abrasion-resistant cast iron *2.60*
2.10 Austenitic cast iron *2.67*
2.11 Corrosion-resistant high-silicon cast iron *2.82*

CHAPTER THREE Steels 3.1–3.261

3.1 Classification of steels *3.5*
3.2 Designation system for wrought steels *3.13*
3.3 General data on steels *3.17*
3.4 Heat treatment of steel *3.29*
3.5 Low-carbon steel sheet and strip *3.37*
3.6 Steel sheet and strip for porcelain enamelling *3.46*
3.7 Metal coated steel sheet and strip *3.50*
3.8 Structural steels *3.61*
3.9 Steel tubes for structural and mechanical purposes *3.71*
3.10 Free-cutting steels *3.81*
3.11 Bright bars *3.86*
3.12 Steels for cold heading and cold extruding *3.94*
3.13 Microalloyed ferritic-pearlitic forging steels *3.112*
3.14 Steels for hardening and tempering *3.115*
3.15 Case hardening steels *3.132*

3.16 Nitriding steels *3.141*
3.17 Steels for flame and induction hardening *3.146*
3.18 Spring steels *3.159*
3.19 Wrought stainless steels *3.185*
3.20 Valve steels for internal combustion engines *3.202*
3.21 Bearing steels *3.213*
3.22 Tool steels *3.221*
3.23 Cast steels *3.238*

CHAPTER FOUR **Aluminium and Its Alloys** 4.1–4.86

4.1 Aluminium *4.5*
4.2 Designation system for aluminium and aluminium alloys *4.8*
4.3 Temper designation system for aluminium and aluminium alloys *4.12*
4.4 Cast aluminium and aluminium alloys *4.16*
4.5 Wrought aluminium and aluminium alloys *4.35*
4.6 Heat treatment of aluminium and aluminium alloys *4.77*

CHAPTER FIVE **Copper and Its Alloys** 5.1–5.102

5.1 Copper *5.5*
5.2 Designation system for copper and copper alloys *5.10*
5.3 Temper designation system for copper and copper alloys *5.12*
5.4 Cast copper and copper alloys *5.15*
5.5 Wrought copper and copper alloys *5.41*
5.6 Heat treatment of copper and copper alloys *5.98*

CHAPTER SIX **Zinc and Its Alloys** 6.1–6.35

6.1 Zinc *6.5*
6.2 Zinc casting alloys *6.8*
6.3 Wrought zinc and zinc alloys *6.31*

CHAPTER SEVEN **Powder Metallurgy** 7.1–7.36

CHAPTER EIGHT **Reference Data** 8.1–8.44

CHAPTER NINE **Units and Measures** 9.1–9.77

Annexure *A.1–A.6*
Index *I.1–I.6*

ABBREVIATIONS AND SYMBOLS

Standards organizations, and technical societies and associations

AA	Aluminium Association
AFS	American Foundrymen's Society
AISI	American Iron and Steel Institute
ANSI	American National Standards Institute
ASM	American Society for Metals (now ASM International)
ASME	American Society of Mechanical Engineers
ASTM	American Society for Testing and Materials
BCIRA	British Cast Iron Research Association (now BCIRA)
BIS	Bureau of Indian Standards
BSI	British Standards Institution
CDA	Copper Development Association
CEN	European Committee for Standardization
DIN	Deutsches Institut fur Normung e.V.
GOST	Committee of the Russian Federation for Standardization, Metrology and Certification (now GOST R)
ISO	International Organization for Standardization
JISC	Japanese Industrial Standards Committee
SAE	Society of Automotive Engineers

Standards

ASTM	refers to the standards published by the American Society for Testing and Materials (ASTM)
BS	refers to the national standards of the United Kingdom
DIN	refers to the national standards of Germany
EN	refers to the European Standards published by the European Committee for Standardization (CEN). European standards are adopted as BS EN, DIN EN, etc.
GOST	refers to the former national standards of the U.S.S.R. (now GOST R)
IS	refers to the national standards of India
ISO	refers to the International Standards published by the International Organization for Standardization (ISO)
JIS	refers to the national standards of Japan
SAE	refers to the standards published by the Society of Automotive Engineers

Chemical Composition

Ag	silver
Al	aluminium
As	arsenic
B	boron
Bi	bismuth
C	carbon
Ca	calcium
Cd	cadmium
Ce	cerium
Co	cobalt
Cr	chromium
Cu	copper
Fe	iron
Ga	gallium
H	hydrogen
Hg	mercury
In	indium
La	lanthanum
Mg	magnesium
Mn	manganese
Mo	molybdenum
N	nitrogen
Na	sodium
Nb	niobium
Ni	nickel
O	oxygen
P	phosphorus
Pb	lead
S	sulphur
Sb	antimony
Se	selenium
Si	silicon
Sn	tin
Ta	tantalum
Te	tellurium
Ti	titanium
Tl	thallium
V	vanadium
W	tungsten
Zn	zinc
Zr	zirconium

Mechanical properties

A	percentage elongation after fracture on original gauge length (L_o) of $5.65\sqrt{S_o}$
$A_{x\,mm}$	percentage elongation on original gauge length (L_o) of x mm, e.g. $A_{50\,mm}$
E	modulus of elasticity or Young's modulus
G	modulus of rigidity or shear modulus
HB, HBS, HBW	Brinell hardness
HK	Knoop hardness
HR	Rockwell hardness (should be followed by the scale designation, e.g. HRC, HRB, HRH, HRF, or HR3OT)
HV	Vickers hardness
KU	impact strength of longitudinal ISO U-notch test pieces
KV	impact strength of longitudinal ISO V-notch test pieces
L_o	original gauge length
R_{cm}	ultimate compressive strength
$R_{c0.1}$	0.1 % proof stress in compression
R_e	yield strength
R_{eH}	upper yield stress
R_{eL}	lower yield stress
R_m	tensile strength
$R_{p0.2}$	0.2 % proof stress
$R_{p0.5}$	0.5 % proof stress
R_{trm}	transverse rupture strength
S_o	original cross-sectional area of the parallel length
Z	percentage reduction of area
μ, ν	Poisson's ratio

Transformation temperatures

Ac_{cm}	in hypereutectoid steel, the temperature at which the solution of cementite in austenite is completed during heating
Ac_1	temperature at which austenite begins to form during heating
Ac_3	temperature at which transformation of ferrite to austenite is completed during heating
Ae_{cm}	equilibrium temperature defining the upper limit of existence of cementite in a hypereutectoid steel
Ae_1	equilibrium temperature defining the lower limit of existence of austenite
Ae_3	equilibrium temperature defining the upper limit of existence of ferrite
Ar_{cm}	in hypereutectoid steel, the temperature at which precipitation of cementite starts during cooling
Ar_1	temperature at which transformation of austenite to ferrite or to ferrite plus cementite is completed during cooling

Ar_3	temperature at which austenite begins to transform to ferrite during cooling
M_f	temperature at which transformation of austenite to martensite is completed during cooling
M_s	temperature at which transformation of austenite to martensite starts during cooling

Units

A	ampere	N	newton
Å	angstrom	nm	nanometer
cm	centimetre	s	second
g	gram	S	siemens
GPa	gigapascal	T	tesla
H	henry	W	watt
h	hour	°	degree
J	joule	°C	degree Celsius
K	kelvin	%	per cent
kA	kiloampere	% IACS	a unit of electrical conductivity
kg	kilogram	% (*m*/*m*)	per cent by mass
kJ	kilojoule	Ω	ohm
m	metre	μΩ	microhm
min	minute	nΩ	nanohm
mm	millimetre	μH	microhenry
MPa	megapascal	μm	micrometre
MS	megasiemens		

Mathematical symbols and operations

$a = b$	a is equal to b	$a - b$	a minus b
$a \neq b$	a is not equal to b	$a \pm b$	a plus or minus b
$a \approx b$	a is approximately equal to b	$a \times b$, ab	a multiplied by b
$a < b$	a is less than b	$\frac{a}{b}$, a/b, ab^{-1}	a divided by b
$a \leq b$	a is less than or equal to b	a^p	a to the power p
$a > b$	a is greater than b	$\sqrt{a}$	square root of a
$a \geq b$	a is greater than or equal to b	$\sqrt[n]{a}$	nth root of a
$a + b$	a plus b	π	pi (3.141592)

Roman numerals

I	1	VIII	8
II	2	IX	9
III	3	X	10
IV	4	L	50
V	5	C	100
VI	6	D	500
VII	7	M	1000

Other abbreviations

CE	carbon equivalent
max.	maximum
min.	minimum
SI	International System of Units

The

Metals Databook

1

THE ELEMENTS AND THEIR PROPERTIES

CHAPTER 1

The Elements and Their Properties

1	The elements and their properties	1.5

1

THE ELEMENTS AND THEIR PROPERTIES

1 PERIODIC TABLE OF THE ELEMENTS

← Metals → ← Nonmetals →

Key: Symbol → Al 13 ← Atomic number; Name → Aluminum

IA	IIA	IIIB	IVB	VB	VIB	VIIB	VIII	VIII	VIII	IB	IIB	IIIA	IVA	VA	VIA	VIIA	0
H 1 Hydrogen																	He 2 Helium
Li 3 Lithium	Be 4 Beryllium											B 5 Boron	C 6 Carbon	N 7 Nitrogen	O 8 Oxygen	F 9 Fluorine	Ne 10 Neon
Na 11 Sodium	Mg 12 Magnesium											Al 13 Aluminium	Si 14 Silicon	P 15 Phosphorus	S 16 Sulphur	Cl 17 Chlorine	Ar 18 Argon
K 19 Potassium	Ca 20 Calcium	Sc 21 Scandium	Ti 22 Titanium	V 23 Vanadium	Cr 24 Chromium	Mn 25 Manganese	Fe 26 Iron	Co 27 Cobalt	Ni 28 Nickel	Cu 29 Copper	Zn 30 Zinc	Ga 31 Gallium	Ge 32 Germanium	As 33 Arsenic	Se 34 Selenium	Br 35 Bromine	Kr 36 Krypton
Rb 37 Rubidium	Sr 38 Strontium	Y 39 Yttrium	Zr 40 Zirconium	Nb 41 Niobium (Columbium)	Mo 42 Molybdenum	Tc 43 Technetium	Ru 44 Ruthenium	Rh 45 Rhodium	Pd 46 Palladium	Ag 47 Silver	Cd 48 Cadmium	In 49 Indium	Sn 50 Tin	Sb 51 Antimony	Te 52 Tellurium	I 53 Iodine	Xe 54 Xenon
Cs 55 Caesium	Ba 56 Barium	La* 57 Lanthanum	Hf 72 Hafnium	Ta 73 Tantalum	W 74 Tungsten	Re 75 Rhenium	Os 76 Osmium	Ir 77 Iridium	Pt 78 Platinum	Au 79 Gold	Hg 80 Mercury	Tl 81 Thallium	Pb 82 Lead	Bi 83 Bismuth	Po 84 Polonium	At 85 Astatine	Rn 86 Radon
Fr 87 Francium	Ra 88 Radium	Ac† 89 Actinium	Rf 104 Rutherfordium	Ha 105 Hahnium													

*Lanthanide series	Ce 58 Cerium	Pr 59 Praseodymium	Nd 60 Neodymium	Pm 61 Promethium	Sm 62 Samarium	Eu 63 Europium	Gd 64 Gadolinium	Tb 65 Terbium	Dy 66 Dysprosium	Ho 67 Holmium	Er 68 Erbium	Tm 69 Thulium	Yb 70 Ytterbium	Lu 71 Lutetium
†Actinide series	Th 90 Thorium	Pa 91 Protactinium	U 92 Uranium	Np 93 Neptenium	Pu 94 Plutonium	Am 95 Americium	Cm 96 Curium	Bk 97 Berkelium	Cf 98 Californium	Es 99 Einsteinium	Fm 100 Fermium	Md 101 Mendelevium	No 102 Nobelium	Lr 103 Lawrencium

Source: Reference 1

2 PHYSICAL PROPERTIES OF THE ELEMENTS

TABLE 1.1 Physical properties of the elements

Element		Atomic number	Atomic weight[a)]	Density		Melting point	Boiling point	Specific heat capacity		Latent heat of fusion
Name	Symbol			kg/m^3	°C[b)]	°C	°C	J/(kg K)	°C[b)]	KJ/kg
Actinium	Ac	89	227.0277	10100[d)]	–	1430[e)]	3200 ± 300[e)]	–	–	–
Aluminium	Al	13	26.98154	2698.9	20	660.4	2494	900	25	397
Americium	Am	95	(243)	13610	–	1173	2067	–	–	–
Antimony	Sb	51	121.75	6697	26	630.7	1587	207	25	163.17
Argon	Ar	18	39.948	1.784[f)]	20	–189.4 ± 0.2	–185.8	523	20	28.03
Arsenic	As	33	74.9216	5778	26	816 at 3.91 MPa	615[g)]	328	–	370.3
Astatine	At	85	(210)	–	–	302[e)]	337[e)]	–	–	–
Barium	Ba	56	137.3	3500	20	729 ± 2	1637[d)]	285	0–100	56.4 ± 7.6
Berkelium	Bk	97	(247)	14780	–	1050	2900	–	–	–
Beryllium	Be	4	9.0122	1848	20	1283	2770	1886	20	1300
Bismuth	Bi	83	208.980	9808	25	271.4	1564	122	25	53.976
Boron	B	5	10.81	2300[h)]	< 800	≈ 2300	¢ 2550	1285	0–100	22000
Bromine	Br	35	79.904	3120	–	–7.2 ± 0.2	58	290	20	67.78
Cadmium	Cd	48	112.40	8642	26	321.1	767	230	20	55

(Contd.)

TABLE 1.1 Physical properties of the elements (continued)

Element		Coefficient of thermal expansion		Thermal conductivity		Electrical resistivity		Modulus of elasticity (in tension)		Poisson's ratio	Crystal structure[c)]	
Name	Symbol	μm/(m K)	°C[b)]	W/(m K)	°C[b)]	nΩ m	°C[b)]	GPa	°C[b)]			°C[b)]
Actinium	Ac	–	–	–	–	–	–	–	–	–	fcc	–
Aluminium	Al	23.6	20–100	247	25	26.2	20	62	–	–	fcc	25
Americium	Am	–	–	–	–	–	–	–	–	–	dhcp	25
Antimony	Sb	8–11	20	25.9	–	370	0	77.759	–	–	rhom	–
Argon	Ar	–	–	0.017	20	–	–	–	–	–	fcc	–233
Arsenic	As	5.6	20	–	–	α : 260	0	–	–	–	rhom	26
Astatine	At	4.7	20	1.7	25	333	20	–	–	–	–	–
Barium	Ba	18	0–100	18.4	–	600	0	12.8	–	0.28	bcc	26
Berkelium	Bk	–	–	–	–	–	–	–	–	–	dhcp	25
Beryllium	Be	11.6	25–100	210	–	40	20	303	–	0.07–0.075	α : hcp	k)
Bismuth	Bi	13.2	20	8.2	0	1050	0	32	–	–	rhom	25
Boron	B	1.1–8.3	20–750	–	–	6.5×10^3	27	0.440[h)]	–	–	h)	≤ 800
Bromine	Br	–	–	–	–	–	–	–	–	–	ortho	–150
Cadmium	Cd	31.3	≈ 20	96.8	27	72.7	22	55	–	0.33[k)]	hcp	26

(Contd.)

TABLE 1.1 Physical properties of the elements (continued)

Element		Atomic number	Atomic weight[a)]	Density		Melting point	Boiling point	Specific heat capacity		Latent heat of fusion
Name	Symbol			kg/m³	°C[b)]	°C	°C	J/(kg K)	°C[b)]	KJ/kg
Calcium	Ca	20	40.08	1550	25	842 at 1 atm	1495	631.5	25	213.1
Californium	Cf	98	(251)	15100	–	900	1745	–	–	–
Carbon, graphite	C	6	12.011	2250	–	3727[g)]	4830	691	20	–
Cerium	Ce	58	140.115	a : 6770	24	798	3443	192	25	38.97
Caesium	Cs	55	132.9054	1892	18	28.64 ± 0.17	670	201.6	20	16.38
Chlorine	Cl	17	35.4527	3.214[f)]	20	–100.99	–34.7	486	20	90.37
Chromium	Cr	24	51.996	7190	20	1875	2680	459.8	20	258–283
Cobalt	Co	27	58.9332	α : 8832	20	1495	≈ 2900	414	–	292
Copper	Cu	29	63.54	8930	20	1084.88	2595	386	20	205
Curium	Cm	96	(247)	13530	–	1347	3110	–	–	–
Dysprosium	Dy	66	162.50	α : 8551	24	1412	2567	170.5	25	68.1
Einsteinium	Es	99	(252)	8840	–	860	996[d)]	–	–	–
Erbium	Er	68	167.26	9066	24	1529	2868	168.0	25	119.0
Europium	Eu	63	151.96	5244	24	822	1529	182.3	–	60.6
Fermium	Fm	100	(257)	8800[e)]	–	857[e)]	1077[d)]	–	–	–

(Contd.)

TABLE 1.1 Physical properties of the elements (continued)

Element		Coefficient of thermal expansion		Thermal conductivity		Electrical resistivity		Modulus of elasticity (in tension)		Poisson's ratio	Crystal structure[c)]	
Name	Symbol	μm/(m K)	°C[b)]	W/(m K)	°C[b)]	nΩ m	°C[b)]	GPa	°C[b)]			°C[b)]
Calcium	Ca	22.3	0–400	–	–	α : 31.6	0	19.6	–	0.31	α : fcc	26.6
Californium	Cf	–	–	–	–	–	–	–	–	–	dhcp	25
Carbon, graphite	C	0.6–4.3	20–100	23.8	≈ 20	13750	0	4.83	–	–	hex	20
Cerium	Ce	γ : 6.3	24	γ : 11.3	25	γ : 744	25	γ : 33.6	27	γ : 0.24	γ : fcc	24
Caesium	Cs	–	–	18.42[i)]	28.64	200	20	–	–	–	bcc	–10
Chlorine	Cl	–	–	0.0072	20	–	–	–	–	–	tet	–185
Chromium	Cr	6.2	–	67	20	130	20	248	20	–	bcc	20
Cobalt	Co	13.8	k)	69.04	20	52.5	20	211	–	0.32	α : hcp	–
Copper	Cu	16.5	20	398	27	16.730	20	128	–	0.308[d)]	fcc	25
Curium	Cm	–	–	–	–	–	–	–	–	–	dhcp	–
Dysprosium	Dy	9.9	24	10.7	25	926	25	61.4	27	0.237	α : hcp	24
Einsteinium	Es	–	–	–	–	–	–	–	–	–	fcc	–
Erbium	Er	12.2	24	14.5	25	860	25	69.9	27	0.237	hcp	24
Europium	Eu	35.0	24	13.9[e)]	25	900	25	18.2	–	0.152	bcc	24
Fermium	Fm	–	–	–	–	–	–	–	–	–	fcc	–

(Contd.)

TABLE 1.1 Physical properties of the elements (continued)

Element		Atomic number	Atomic weight[a]	Density		Melting point	Boiling point	Specific heat capacity		Latent heat of fusion
Name	Symbol			kg/m^3	°C[b]	°C	°C	J/(kg K)	°C[b]	KJ/kg
Fluorine	F	9	18.9984032	1.696[f]	20	–219.6	–188.2	750	20	42.26
Francium	Fr	87	(223)	–	–	27	677	–	–	–
Gadolinium	Gd	64	157.25	α : 7901	24	1313	3273	235.9	25	63.6
Gallium	Ga	31	69.723 ± 0.001	5907	20	29.78	2204	373.8	25	80.16
Germanium	Ge	32	72.59	5323	25	937.4	2830	321.7	25	466.5
Gold	Au	79	196.9665	19302	25	1064.43	2857	128	25	62.762
Hafnium	Hf	72	178.49	13310	20	2222 ± 30	5400	147	20	–
Helium	He	2	4.0020602	0.1785[f]	20	–269.7	–268.9	5230	20	–
Holmium	Ho	67	164.93032	8795	24	1474	2700	164.9	25	103[e]
Hydrogen	H	1	1.00794	0.0899[f]	20	–259.19	–252.7	14400	20	62.76
Indium	In	49	114.82	7300	20	156.61	2080	233	25	28.47
Iodine	I	53	126.90447	4940	–	113.7	183	220	20	59.41
Iridium	Ir	77	192.9	22650	20	2447	4500	130	–	–
Iron	Fe	26	55.847	α : 7870	20	1538	2870	–	–	α : 247 ± 7
Krypton	Kr	36	83.80	3.743[f]	20	–157.3	–152	–	–	–

(Contd.)

TABLE 1.1 Physical properties of the elements (continued)

Element		Coefficient of thermal expansion		Thermal conductivity		Electrical resistivity		Modulus of elasticity (in tension)		Poisson's ratio	Crystal structure[c)]	
Name	Symbol	µm/(m K)	°C[b)]	W/(m K)	°C[b)]	nΩ m	°C[b)]	GPa	°C[b)]			°C[b)]
Fluorine	F	–	–	–	–	–	–	–	–	–	–	–
Francium	Fr	–	–	–	–	–	–	–	–	–	–	–
Gadolinium	Gd	9.4	100	10.5	25	1310	25	54.8	27	0.259	α : hcp	24
Gallium	Ga	–	–	33.49[l)]	29.8	150.5[l)]	20	–	–	–	ortho	24
Germanium	Ge	6.0	27	59.9	27	53×10^7	25	–	–	–	fcc (diam)	25
Gold	Au	14.2	20	317.9	0	23.5	20	78	–	–	fcc	–
Hafnium	Hf	519	20–200	93.3	20	,351	25	–	–	–	hcp	20
Helium	He	–	–	0.14	20	–	–	–	–	–	hcp	–271.5
Holmium	Ho	11.2	24	16.2	25	814	25	64.8	27	0.231	hcp	24
Hydrogen	H	–	–	0.17	20	–	–	–	–	–	hex	–271
Indium	In	24.8	20	83.7	0	84	20	12.74	20	0.4498	tet	26
Iodine	I	93	–	0.44	–	1.3×10^{16}	20	–	–	–	ortho	20
Iridium	Ir	6.8	20	147	0–100	53	20	517	–	0.26	fcc	20
Iron	Fe	α : 11.8	20	80.4	25	–	–	208.2	k)	0.291	α : bcc	20
Krypton	Kr	–	–	0.0088	20	–	–	–	–	–	fcc	–191

(Contd.)

TABLE 1.1 Physical properties of the elements (continued)

Element		Atomic number	Atomic weight[a)]	Density		Melting point	Boiling point	Specific heat capacity		Latent heat of fusion
Name	Symbol			kg/m^3	°C[b)]	°C	°C	J/(kg K)	°C[b)]	KJ/kg
Lanthanum	La	57	138.9055	α : 6146	24	918	3464	195.1	25	44.6
Lawrencium	Lr	103	(260)	–	–	–	–	–	–	–
Lead	Pb	82	207.19	11340	16	327.4	1750	128.7	25	22.98–23.38
Lithium	Li	3	6.939	533.4	20	180.7	1336	3305.4	20	433.9
Lutetium	Lu	71	174.967	9841	24	1663	3402	150.3	25	126[e)]
Magnesium	Mg	12	24.312	1738	20	650	1107 ± 10	1025	20	360–377
Manganese	Mn	25	54.938	α : 7430 at 1 atm	20	1246	2065	α : 475	20	219
Mendelevium	Md	101	(258)	–	–	–	–	–	–	–
Mercury	Hg	80	200.59	13546[i)]	20	–38.87[j)]	356.58 at 1 atm	139.6[i)]	25	11.8
Molybdenum	Mo	42	95.94	10220	20	2610	5560	276	20	270[e)]
Neodymium	Nd	60	144.24	α : 7008	24	1021	3074	190.0	25	49.5
Neon	Ne	10	20.1797	0.8999[f)]	20	–248.6 ± 0.3	–246.0	–	–	–
Neptunium	Np	93	237.0482	α : 20480	25	637	≈ 3902	–	–	–
Nickel	Ni	28	58.71	8902	25	1453	≈ 2730	471	100	–

(Contd.)

TABLE 1.1 Physical properties of the elements (continued)

Element		Coefficient of thermal expansion		Thermal conductivity		Electrical resistivity		Modulus of elasticity (in tension)		Poisson's ratio	Crystal structure[c]	
Name	Symbol	μm/(m K)	°C[b]	W/(m K)	°C[b]	nΩ m	°C[b]	GPa	°C[b]			°C[b]
Lanthanum	La	12.1	24	α : 13.4	25	α : 615	25	36.6	27	0.280	a : dhcp	24
Lawrencium	Lr	–	–	–	–	–	–	–	–	–	–	–
Lead	Pb	29.3	17–100	–	–	206.43	20	–	–	–	fcc	–
Lithium	Li	56	20	44.0	180.7	93.5	20	–	–	–	α : hcp	–
Lutetium	Lu	9.9	24	16.4	25	582	25	68.6	27	0.261	hcp	24
Magnesium	Mg	25.2[1)]	20	418	20	44.5[1)]	20	40	20	–	hcp	25
Manganese	Mn	21.7	20	7.82	27	α : 1440	22	191	–	–	α : cc	20
Mendelevium	Md	–	–	–	–	–	–	–	–	–	–	–
Mercury	Hg	–	–	8.21	0	958	20	–	–	–	rhom	–39
Molybdenum	Mo	–	–	142	20	52	0	–	–	–	bcc	25
Neodymium	Nd	9.6	24	16.5	25	643	25	41.4	27	0.281	α : dhcp	24
Neon	Ne	–	–	0.046	20	–	–	–	–	–	fcc	–268
Neptunium	Np	α : 27.5	40–240	–	–	1164	37	–	–	–	α : ortho	25
Nickel	Ni	13.3	0–100	82.9	100	68.44	20	207	–	0.31	fcc	20

(Contd.)

TABLE 1.1 Physical properties of the elements (continued)

Element		Atomic number	Atomic weight[a)]	Density		Melting point	Boiling point	Specific heat capacity		Latent heat of fusion
Name	Symbol			kg/m³	°C[b)]	°C	°C	J/(kg K)	°C[b)]	KJ/kg
Niobium	Nb	41	92.9064	8570	20	2468	4927	–	–	290
Nitrogen	N	7	14.00674	1.250[f)]	20	–209.97	–195.8	1030	20	25.94
Nobelium	No	102	(259)	–	–	–	–	–	–	–
Osmium	Os	76	190.2	22583[d)]	26	≈ 2700	≈ 5500	129.73	0	–
Oxygen	O	8	15.9994	1.429[f)]	20	–218.83	–183.0	913	20	13.81
Palladium	Pd	46	106.4	12020	20	1552	≈ 3980	245	0	–
Phosphorus, white	P	15	30.97362	1830	–	44.25	280	741	20	20.92
Platinum	Pt	78	195.09	21450[d)]	20	1769	3800	132	0	113
Plutonium	Pu	94	239.052	α : 19860	21	640	3235	α : 33900	25	–
Polonium	Po	84	(209)	9400	–	254	962	–	–	–
Potassium	K	19	39.09	855	20	63.2	756.5	770	20	59.45
Praseodymium	Pr	59	140.90765	α : 6773	24	931	3520	194.6	25	48.9
Promethium	Pm	61	145	α : 7264	24	1042	3000[e)]	188[e)]	25	53[e)]
Protactinium	Pa	91	231.0359	15430	27	1572 ± 20	4027	–	–	–

(Contd.)

TABLE 1.1 Physical properties of the elements (continued)

Element		Coefficient of thermal expansion		Thermal conductivity		Electrical resistivity		Modulus of elasticity (in tension)		Poisson's ratio	Crystal structure[c]	
Name	Symbol	μm/(m K)	°C[b]	W/(m K)	°C[b]	nΩ m	°C[b]	GPa	°C[b]			°C[b]
Niobium	Nb	7.31	18–300	54.4	100	–	–	103	25	0.38	bcc	–
Nitrogen	N	–	–	0.025	20	–	–	–	–	–	hex	–234
Nobelium	No	–	–	–	–	–	–	–	–	–	–	–
Osmium	Os	2.6	50	–	–	≈ 95	20	560[e]	–	–	hcp	26
Oxygen	O	–	–	0.025	20	–	–	–	–	–	sc	–225
Palladium	Pd	11.76	20	70	18	108	20	–	–	–	fcc	20
Phosphorus, white	P	125	20	–	–	1×10^{18}	11	–	–	–	sc	–35
Platinum	Pt	9.1	20–100	71.1	0	106	20	–	–	–	fcc	25
Plutonium	Pu	56	21–104	6.5	[k]	1414	107	107	–	0.15–0.21	α : mono	21
Polonium	Po	–	–	–	–	–	–	–	–	–	mono	20
Potassium	K	83	0–95	108.3	20	72	20	–	–	–	bcc	20
Praseodymium	Pr	α : 6.7	24	12.5	25	700	25	37.3	27	0.281	α : dhcp	24
Promethium	Pm	11[e]	24	15[e]	27	750[e]	25	46[e]	27	0.28[e]	α : dhcp	24
Protactinium	Pa	9.9	30–700	–	–	150	27	–	–	–	α : bct	27

(Contd.)

TABLE 1.1 Physical properties of the elements (continued)

Element		Atomic number	Atomic weight[a)]	Density		Melting point	Boiling point	Specific heat capacity		Latent heat of fusion
Name	Symbol			kg/m^3	°C[b)]	°C	°C	J/(kg K)	°C[b)]	KJ/kg
Radium	Ra	88	226.025	5000	–	700	1140	–	–	–
Radon	Rn	86	(222)	9.960[f)]	20	–71[e)]	–61.8	–	–	–
Rhenium	Re	75	186.2	21020	20	3180	5627	25700	25	178
Rhodium	Rh	45	102.905	12410	–	1963	≈ 3700	247	–	–
Rubidium	Rb	37	85.467	1532	20	38.89	688	334.89	0	25.535
Ruthenium	Ru	44	101.07	12450	20	2310 ± 20	≈ 3900	240	0	–
Samarium	Sm	62	150.4	α : 7520	24	1074	1794	196.2	25	57.3
Scandium	Sc	21	44.95591	α : 2989	24	1541	2836	567.4	25	313.6
Selenium	Se	34	78.96	γ : 4809	25	γ : 217	684.9	γ : 317	25	84.93
Silicon	Si	14	28.08	2329.0	25	1414	3145 at 1 atm	713	27	1807.9
Silver	Ag	47	107.868	10490	20	961.9	2163	235	25	104.2
Sodium	Na	11	22.9898	967.4	25	97.82	881.4	1222.0	25	113
Strontium	Sr	38	87.62	α : 2600	20	768	1370[d)]	54.4	–253	104.7[d)]
Sulphur, yellow	S	16	32.066	α : 2070	–	119.0 ± 0.5	444.6	733	20	38.91
Tantalum	Ta	73	180.948	16600	20	2996	5427	139.1	0	145–174
Technetium	Tc	43	99.0000	11500	25	2204	4265	–	–	–

(Contd.)

TABLE 1.1 Physical properties of the elements (continued)

Element		Coefficient of thermal expansion		Thermal conductivity		Electrical resistivity		Modulus of elasticity (in tension)		Poisson's ratio	Crystal structure[c)]	
Name	Symbol	μm/(m K)	°C[b)]	W/(m K)	°C[b)]	nΩ m	°C[b)]	GPa	°C[b)]			°C[b)]
Radium	Ra	–	–	–	–	–	–	–	–	–	–	–
Radon	Rn	–	–	–	–	–	–	–	–	–	–	–
Rhenium	Re	6.6	20–100	71.2	20	193	20	460	20	0.49	hcp	–
Rhodium	Rh	8.3	20–100	150	0–100	45.1	20	–	–	–	fcc	20
Rubidium	Rb	90	20	58.2	25	128.4	20	2.35	–	–	bcc	0
Ruthenium	Ru	5.05	20	–	–	76	0	–	–	–	hcp	–
Samarium	Sm	12.7	24	13.3	25	940	25	49.7	27	0.274	α : rhombo	24
Scandium	Sc	10.2	24	15.8	25	562	25	74.4	27	0.279	α : hcp	24
Selenium	Se	γ : 49	20	γ : 2.48	25	γ : 1×10^{15}	25	53.82	–	–	γ : hex	20
Silicon	Si	2.616	27	156	27	–	–	–	–	–	dc (0–12.5 GPa)	25
Silver	Ag	19.0	20	428	20	14.7	0	71.0	–	0.37	fcc	25
Sodium	Na	68.93	–	131.4	25	47.7	20	–	–	–	β : bcc	20
Strontium	Sr	–	–	–	–	–	–	–	–	–	α : fcc	25
Sulphur, yellow	S	64	20	0.264	20	2×10^{24}	20	–	–	–	ortho	20
Tantalum	Ta	6.5	20	54.4	20	135.0	20	186	20	0.35	bcc	20
Technetium	Tc	7.05[l)]	−123 to 25	50.2	25	185.0	25	322	–	0.31	hcp	25

TABLE 1.1 Physical properties of the elements (continued)

Element		Atomic number	Atomic weight[a)]	Density		Melting point	Boiling point	Specific heat capacity		Latent heat of fusion
Name	Symbol			kg/m³	°C[b)]	°C	°C	J/(kg K)	°C[b)]	KJ/kg
Tellurium	Te	52	127.60	6237	25	449.5	988	201	25	86.113
Terbium	Tb	65	158.92534	α : 8230	24	1356	3230	181.8	25	67.9
Thallium	Tl	81	204.37	11872	20	303	1473	130	–	20.27
Thorium	Th	90	232.038	11720	25	1755	≈ 4800	113.08	25	59.50
Thulium	Tm	69	168.93421	9321	24	1545	1950	159.8	25	99.4
Tin	Sn	50	118.69	β : 7298.4	15	231.9	2270	β : 222	25	59.5
Titanium	Ti	22	47.9	α : 4507	20	1668 ± 10	3260[e)]	522.3	25	440[e)]
Tungsten	W	74	183.85	19254	–	3410 ± 20	5700 ± 200	–	–	220 ± 36
Uranium	U	92	238.029	α : 19050	25	1133	3818	α : 117	27	38.72
Vanadium	V	23	50.941	6160	20	1910	3350–3400	498	0–100	314
Xenon	Xe	54	131.29	5.896[f)]	–	–111.9	–108.0	–	–	–
Ytterbium	Yb	70	173.04	α : 6903	23	819	1196	154.3	25	44.3
Yttrium	Y	39	88.90585	α : 4469	24	1522	3345	298.1	25	128.2
Zinc	Zn	30	65.38	7133	25	420	906	382	20	100.9
Zirconium	Zr	40	91.22	α : 6505–6574	–	1852 ± 2	4377	–	–	25

(Contd.)

TABLE 1.1 Physical properties of the elements (continued)

Element		Coefficient of thermal expansion		Thermal conductivity		Electrical resistivity		Modulus of elasticity (in tension)		Poisson's ratio	Crystal structure[c)]	
Name	Symbol	μm/(m K)	°C[b)]	W/(m K)	°C[b)]	nΩ m	°C[b)]	GPa	°C[b)]			°C[b)]
Tellurium	Te	18.2	20	5.98–6.02[l)]	20–28	–	–	–	–	–	hex	25
Terbium	Tb	10.3	24	11.1	25	1150	25	55.7	27	0.261	α : hcp	24
Thallium	Tl	28	20	47	0	150	0	–	–	–	hcp	< 230
Thorium	Th	11.4	25	77	25	157	25	72.4	–	0.27	α : fcc	25
Thulium	Tm	13.3	24	16.9	25	676	25	74.0	27	0.213	hcp	25
Tin	Sn	α : 23.8	100	β : 60.7	100	155	100	41.6	k)	0.33	β : bct	–
Titanium	Ti	8.41	20	11.4	–240	420	20	–	–	–	α : hcp	–
Tungsten	W	3.7	–93	–	–	53	27	–	–	–	α : bcc	25
Uranium	U	α : 12	25	α : 27.6	27	α : 300	27	203	–	0.22	α : ortho	25
Vanadium	V	8.3	23–100	31.0	100	248–260	20	124–137	–	0.36	bcc	–
Xenon	Xe	–	–	0.052	20	–	–	–	–	–	fcc	–185
Ytterbium	Yb	26.3	24	38.5	25	250	25	23.9	27	0.207	α : hcp	24
Yttrium	Y	10.6	–	17.2	25	596	25	63.5	–	0.243	α : hcp	25
Zinc	Zn	39.7[l)]	20–250	113	25	59.16[l)]	20	–	–	–	hcp	–
Zirconium	Zr	5.85[l)]	20	21.1	25	450	–	–	–	0.35	α : hcp	20

(Contd.)

TABLE 1.1 Physical properties of the elements (continued)

Key:

a) The number in parentheses is the mass number of the most stable isotope of that element.

b) Temperature at which the property has been measured.

c) Abbreviations used for crystal structure:

bcc	body-centered cubic;
bct	body-centered tetragonal;
cc	complex cubic;
dc	diamond cubic;
dhcp	double-hexagonal close packed;
fcc	face-centered cubic;
fcc (diam)	face-centered cubic (diamond);
fct	face-centered tetragonal;
hex	hexagonal;
hcp	hexagonal close-packed;
mono	monoclinic;
ortho	orthorhombic;
rhom	rhombohedral;
sc	simple cubic;
tet	tetragonal.

d) Calculated.

e) Estimated.

f) Gas (in grams per litre at 760 mm).

g) Sublimes.

h) Amorphous.

i) Liqunid.

j) Freezing temperature.

k) Room temperature.

l) Polycrystalline.

Source: References 1–4.

(Contd.)

3 REFERENCES

Books

1. *ASM Handbook,* vol. 2, ASM International, Materials Park, Ohio, U.S.A., 1990.
2. *Metals Handbook,* Desk Edition, American Society for Metals, Metals Park, Ohio, U.S.A., 1985.
3. *Metals Handbook,* vol. 1, 8th ed., American Society for Metals, Metals Park, Ohio, U.S.A., 1961.
4. D.R. Lide (Ed.), *CRC Handbook of Chemistry and Physics,* 73rd ed., CRC Press, Boca Raton, Florida, U.S.A., 1992.

2

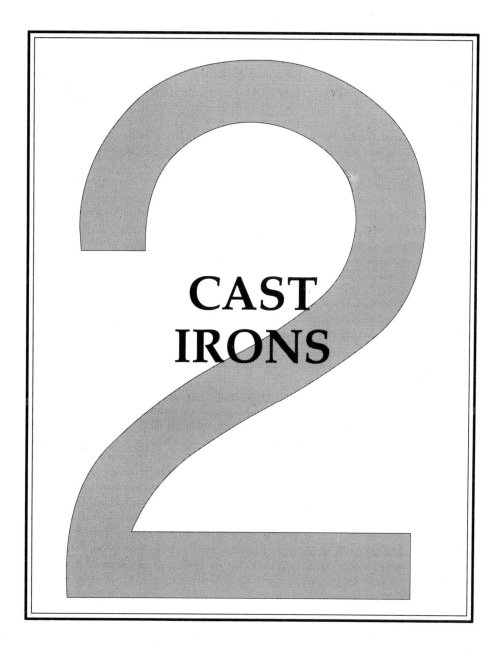

CAST IRONS

CHAPTER **2**

Cast Irons

2.1	Introduction to cast irons	2.5
2.2	Designation system for cast irons	2.12
2.3	Designation of microstructure of graphite in cast irons	2.15
2.4	Grey cast iron	2.22
2.5	Spheroidal graphite cast iron	2.33
2.6	Austempered ductile iron	2.44
2.7	Compacted graphite cast iron	2.49
2.8	Malleable cast iron	2.53
2.9	Abrasion-resistant cast iron	2.60
2.10	Austenitic cast iron	2.67
2.11	Corrosion-resistant high-silicon cast iron	2.82

2.1

INTRODUCTION TO CAST IRONS

1 INTRODUCTION

Cast irons are iron-carbon-silicon alloys (see Fig. 2.1) which generally contain more than 2 % carbon and from 1 to 3 % silicon. They also contain minor (< 0.1 %) and often alloying (> 0.1 %) elements. Cast irons are used in the as-cast condition or after heat treatment. They can be cast singly or repetitively into a wide variety of simple or complex shapes at relatively low cost and to a wide range of properties. Wide variations in properties can be achieved by controlling the chemical composition, and by varying melting, casting and heat treating practices.

2 CLASSIFICATION OF CAST IRONS

Cast irons are commonly classified on the basis of microstruture and/or application into:

a) General purpose cast irons; and
b) Special purpose cast irons.

General purpose cast irons are unalloyed and low-alloy cast irons which are used for general engineering applications. These cast irons are classified on the basis of the graphite form (see Table 2.1) into grey cast iron, spheroidal graphite cast iron, compacted graphite cast iron and malleable cast iron. Each of these types may be heat treated without changing its basic classification.

Special purpose cast irons are white and high-alloy (alloying additions greater than 3 %) cast irons which are used for abrasion-resistant, corrosion-resistant, heat-resistant and other special applications.

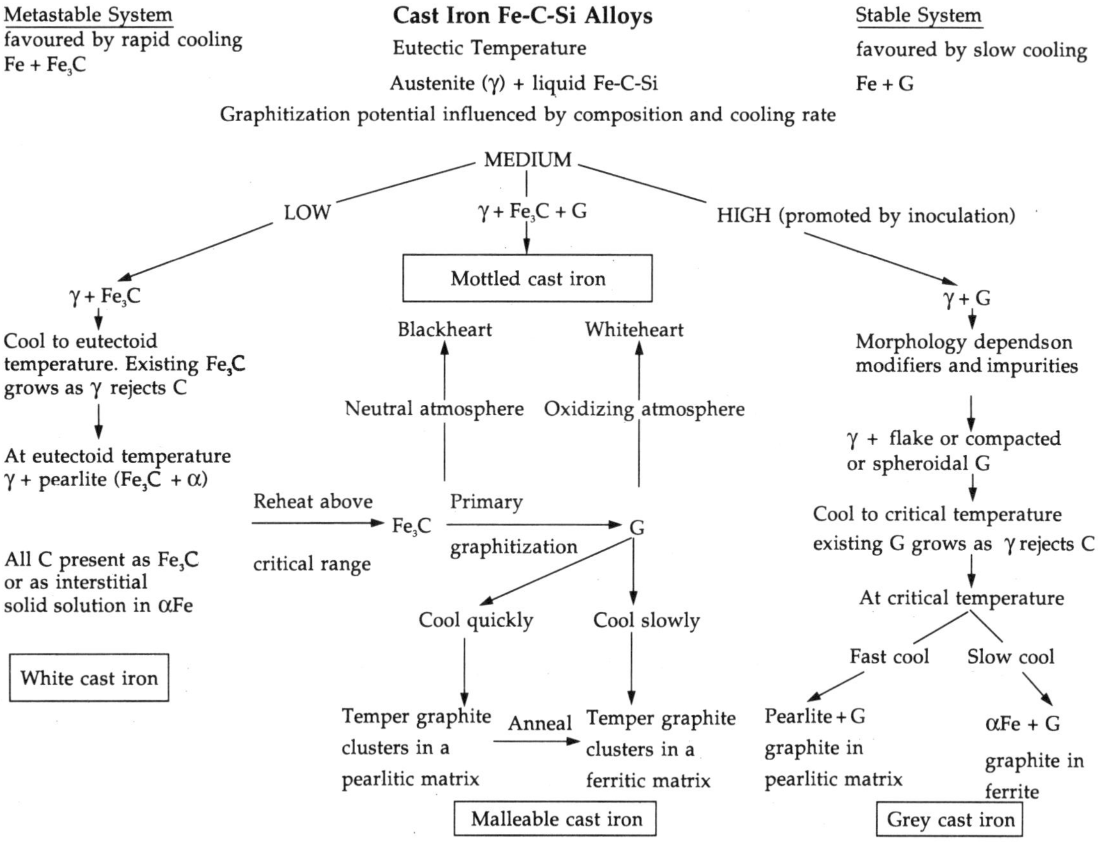

Fig. 2.1 Cast iron family (Source: Reference 3)

TABLE 2.1 Classification of general purpose cast irons on the basis of graphite form

Type of cast iron	Graphite form	Fracture
Grey cast iron	Flake graphite	Grey
Spheroidal graphite cast iron	Spheroidal graphite	Silver-grey
Compacted graphite cast iron	Compacted vermicular graphite	Grey
Malleable cast iron	Temper graphite	Silver-grey
Source: Reference 1		

3 EFFECT OF ELEMENTS ON THE STRUCTURE OF CAST IRONS

TABLE 2.2 Effect of elements on the structure of cast irons

Element		Grey cast iron	Spheroidal graphite cast iron	Malleable cast iron
Name	Symbol			
Carbon	C	Present as either graphite or carbide. Excessive amounts (CE > 4.3 %), can lead to kish graphite. Low levels tend to be carbidic.	Present as either graphite or carbide. High levels, especially in heavy sections, can lead to graphite floatation.	Lower levels of carbon are difficult to anneal, have strong white iron tendency, and low fluidity. Higher levels are easier to anneal, have better fluidity, but show a tendency towards mottle.
Cerium	Ce	Stabilizes carbides. Used to reduce the effect of lead.	Promotes nodule formation and quality. Promotes chunky graphite in austenitic grades. Reduces or eliminates the effect of subversive elements (e.g., Te and Ti).	Retards mottling.
Chromium	Cr	Greatly affects chill. Promotes pearlite in matrix but not without tendency to promote primary carbides.	Very powerful carbide former. The carbides resist annealing. Must be below 0.04 % for as-cast ferritic grades and generally no higher than 0.10% for pearlitic grades.	Strongly retards FSG and SSG.
Copper	Cu	Used primarily as a means of ensuring pearlite.	No effect on nodule quality or on primary carbides. Has a significant effect on pearlite forming tendency — about 0.50 % yields fully pearlitic matrix, depending on section size and other alloys present.	Increases the rates of FSG and SSG. Slightly increases the nodule number.

(Contd.)

TABLE 2.2 Effect of elements on the structure of cast irons (continued)

Element Name	Element Symbol	Grey cast iron	Spheroidal graphite cast iron	Malleable cast iron
Magnesium	Mg	Stabilizes carbides.	Causes graphite to solidify in the nodular form. Lowers sulphur and oxygen levels during the nodularizing process.	Promotes carbides and retards mottling.
Manganese	Mn	Controls the detrimental effect of sulphur. % Mn = 1.7 × % S + 0.15 % produces a ferritic grade with minimum pearlite. % Mn = 3 × % S + 0.35 % used for pearlitic grades.	Intercellular carbides formed at % Mn > 0.70. Generally less than 0.20 % for ferritic grades. From 0.50 % to 0.90 % added for as-cast pearlitic grades.	Removes the detrimental effect of sulphur. % Mn > 1.7 × % S stabilizes carbides with little influence on solidification. Weakly retards FSG and SSG.
Molybdenum	Mo	Increases tensile strength, hardness, and hardenability.	Mild carbide forming tendency. Used to promote pearlite and/or bainite.	Increases depth of chill, and chill plus mottle. Prevents galvanic embrittlement. Slightly retards FSG and SSG.
Nickel	Ni	Used as a pearlite stabilizer. Increases strength and hardness without increasing chill.	Promotes pearlite and/or bainite formation.	Promotes FSG and SSG while slightly increasing nodule count. Increases strength by increasing hardenability. A nickel plus copper addition can be used to produce an austenitic structure.
Phosphorus	P	Present as steadite, but also dissolved in ferrite to strengthen and harden the matrix. High levels increase fluidity but also increases brittleness.	Forms phosphide network. Generally kept as low as possible. Increases the nil ductility transition temperature.	Improves fluidity and promotes mottling. Should be less than 0.15 % because it increases the nil ductility transition temperature.

(Contd.)

TABLE 2.2 Effect of elements on the structure of cast irons (continued)

Element		Grey cast iron	Spheriodal graphite cast iron	Malleable cast iron
Name	Symbol			
Silicon	Si	Promotes graphitization. Hardens and promotes ferrite.	Promotes graphitization. Hardens ferrite. Increases nil ductility transition temperature.	Acts as a graphitizer. Hardens ferrite and raises the nil ductility transition temperature. Reduces shrinkage during solidification. Promotes ferrite during annealing.
Sulphur	S	Present as sulphides — type depends on Mn/S ratio. Minimum amount appears necessary to control graphitization (0.04–0.06 %). Increased sulphur over 0.15 % reduces fluidity somewhat, increases cell count and shrinkage, and may promote gas holes.	Reacts preferentially with Mg and/or rare earths, limiting the economical production of SG iron, if too much is present in the base iron.	If amount present gives a Mn/S ratio greater than 1.7, it retards FSG and SSG, and decreases the nodule number. Stabilizes carbides.
Tin	Sn	About 0.03–0.08 % used to promote a pearlitic matrix.	Powerful effect on pearlite formation. More than 0.10 % can result in non-spheroidal graphite in cellular boundaries.	High levels retard FSG and SSG.
Titanium	Ti	Increases chill, reduces cell count and promotes type D graphite. Renders iron much more responsive to inoculation; however, 0.02–0.03 % residual is desirable. Ties up nitrogen.	Considered a subversive element. Excessive amounts (depending on magnesium content) result in compacted graphite.	Increases mottling tendency.
Vanadium	V	Forms very stable carbides.	Forms very stable carbides which are highly resistant to annealing.	Forms very stable carbides which are highly resistant to annealing.

(Contd.)

TABLE 2.2 Effect of elements on the structure of cast irons (continued)

Key: CE (carbon equivalent) = % C + % Si/3; FSG = First stage graphitization; SSG = Second stage graphitization.

Source: Reference 4

4 COMPARISON RATINGS OF UNALLOYED CAST IRONS

TABLE 2.3 Comparison of the characteristics of various types of cast iron with those of 0.3 % C cast steel

Characteristic	Grey cast iron	Spheroidal graphite cast iron	Malleable cast iron	White cast iron	0.3 % C cast steel
Castability	1	1	2	3	4
Machinability	1	2	2	–	3
Reliability	5	1	3	4	2
Vibration damping	1	2	2	4	4
Surface hardenability	1	1	1	–	3
Modulus of elasticity	3	1	2	–	1
Impact resistance	5	2	3	–	1
Wear resistance	3	2	4	1	5
Corrosion resistance	1	1	2	2	4
Strength/weight ratio	5	1	4	–	3
Cost of manufacture	1	2	3	3	4

Key: BEST 1 2 3 4 5 WORST

Source: Reference 3

5 REFERENCES

Books

1. *ASM Handbook,* vol. 1, ASM International, Materials Park, Ohio, U.S.A., 1990.
2. C.F. Walton and T.J. Opar (Eds.), *Iron Castings Handbook,* 3rd ed., Iron Castings Society, Des Plaines, Illinois, U.S.A., 1981.
3. R. Elliot, *Cast Iron Technology,* Butterworths, London, U.K., 1988.

Datasheet

4. "Datasheet: How Master and Tramp Elements Affect the Structures of Cast Irons", 115(5), 64–67, *Metal Progress,* 1978.

2.2

DESIGNATION SYSTEM FOR CAST IRONS

1 DESIGNATION SYSTEM

The grades of cast iron covered by the Indian Standards are designated either on the basis of their mechanical properties or on the basis of their chemical composition.

1.1 Cast irons designated on the basis of mechanical properties

1.1.1 Grey cast iron: The grades of grey cast iron are designated as follows, in the order given:

a) by capital letters FG;
b) by a space; and
c) by three digits indicating the specified minimum tensile strength, in MPa, of a test piece machined from a separately cast 30 mm diameter test bar.

EXAMPLE FG 260

1.1.2 Spheroidal graphite cast iron: The grades of spheroidal graphite cast iron are designated as follows, in the order given:

a) by capital letters SG;
b) by a space;
c) by three digits indicating the specified minimum tensile strength, in MPa, of a test piece machined from a separately cast test sample;
d) by a solidus (/);
e) by one or two digits indicating the specified minimum percentage elongation after fracture ($L_o = 5d$) of a test piece machined from a separately cast test sample;
f) by the capital letter A to indicate that the properties are obtained on cast-on test samples, to distinguish them from those obtained on separately cast test samples; and
g) by the capital letter L to indicate that the corresponding grade has a specified impact strength at low temperature.

EXAMPLES SG 600/3, SG 400/18A, SG 350/22AL

1.1.3 Malleable cast iron: The grades of malleable cast iron are designated as follows, in the order given:

a) by capital letters representing the type of malleable cast iron:
  i) WM for whiteheart malleable cast iron,
  ii) BM for blackheart malleable cast iron, and
  iii) PM for pearlitic malleable cast iron;
b) by a space; and
c) by three digits indicating the specified minimum tensile strength, in MPa, of a 15 mm diameter separately cast, unmachined test bar.

EXAMPLE PM 690

1.2 Cast irons designated on the basis of chemical composition

1.2.1 Austenitic cast iron: The grades of austenitic cast iron are designated as follows, in the order given:

a) by capital letters representing the type of austenitic cast iron:
  i) AFG for flake graphite austenitic cast iron,
  ii) ASG for spheroidal graphite austenitic cast iron;
b) by a space; and
c) by the chemical symbol(s) for the alloying element(s), each followed by a number indicating its approximate mean percentage content, rounded to the nearest whole number. The sequence of symbols should be in decreasing order of the value of their content. The symbol for the alloying element is separated from the number representing the approximate mean percentage content by a space.

EXAMPLE AFG Ni 13 Mn 7

1.2.2 Abrasion-resistant cast iron: The grades of abrasion-resistant cast iron are designated as follows, in the order given:

a) by letters representing the alloy group:
  i) NiLCr for nickel containing low-chromium white cast iron,
  ii) NiHCr for nickel containing high-chromium white cast iron,
  iii) CrMoHC for high-carbon, chromium-molybdenum white cast iron,
  iv) CrMoLC for low-carbon, chromium-molybdenum white cast iron,
  v) HCrNi for nickel containing high-chromium white cast iron, and
  vi) HCr for nickel-free, high-chromium white cast iron;
b) by a space;
c) by a number indicating ten times the specified mean percentage carbon content, rounded to the nearest whole number;
d) by a solidus (/); and
e) by a number indicating the specified minimum Brinell hardness, in HBW

EXAMPLE NiLCr 30/500

2 REFERENCES

Standards

1. IS 210 – 1993: *Grey Iron Castings—Specification.*
2. IS 1865 – 1991: *Iron Castings with Spheroidal or Nodular Graphite—Specification.*
3. IS 2107 – 1977: *Specification for Whiteheart Malleable Iron Castings.*
4. IS 2108 – 1977: *Specification for Blackheart Malleable Iron Castings.*
5. IS 2640 – 1977: *Specification for Pearlitic Malleable Iron Castings.*
6. IS 2749 – 1974: *Specification for Austenitic Iron Castings.*
7. IS 4771 – 1985: *Specification for Abrasion-Resistant Iron Castings.*

2.3

DESIGNATION OF MICROSTRUCTURE OF GRAPHITE IN CAST IRONS

1 DESIGNATION OF MICROSTRUCTURE OF GRAPHITE IN CAST IRONS

The graphite occurring in cast irons is designated by:

a) its form (designated by Roman numerals I to VI, see Fig. 2.2);
b) its distribution (designated by capital letters A to E, see Fig. 2.3); and
c) its size (designated by Arabic numerals 1 to 8, see Fig. 2.4 to 2.7, and Table 2.4).

EXAMPLES

1. Type I A 4 indicates graphite particles of form I, distribution A and size 4 (linear dimension 12 to 25 mm when observed at × 100 magnification).
2. If the graphite observed lies between two sizes, reference to both is possible (for example, size 3/4). In a given case the predominating size may be emphasized by underlining (for example, size 3/4). This method can be extended to cover structures where more than two sizes are present.
3. Mixed structures with different types of graphite may be defined by indicating the percentage proportion of the different types of graphite: for example,

60 % I A 4 + 40 % I D 7

indicates 60 % graphite of the form I, distribution A and size 4, and 40 % graphite of the form I, distribution D and size 7.

FORM

Form I Flake

Form II Star

Form III Worm

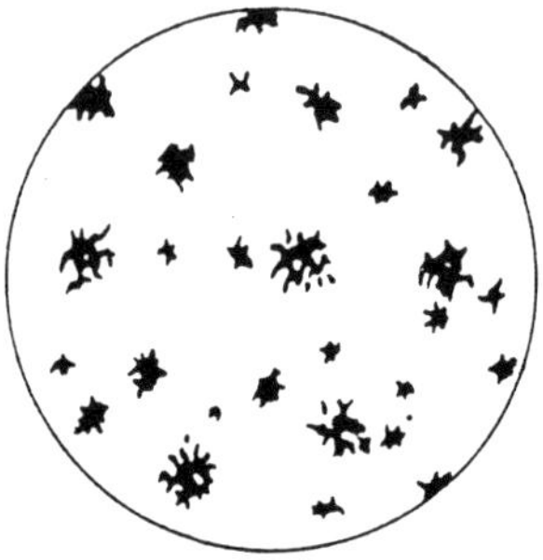

Form IV Chinese Script

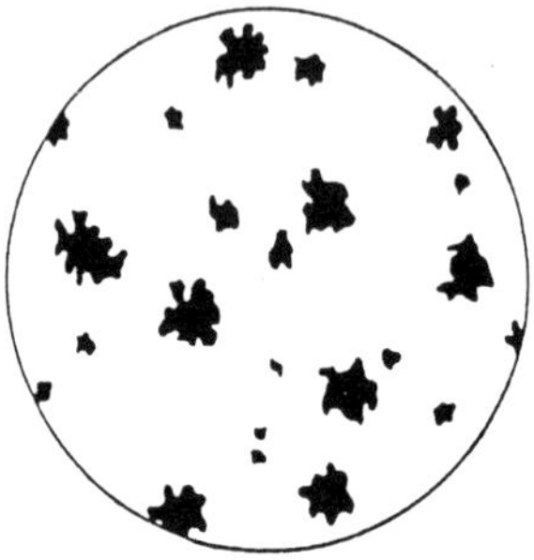

Form V Amoeba

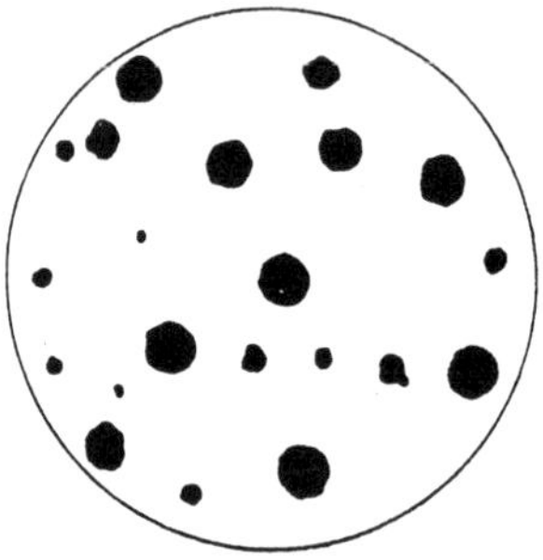

Form VI Round

Fig. 2.2 Reference diagrams for graphite form (distribution A) (Source: Reference 3)

DISTRIBUTION

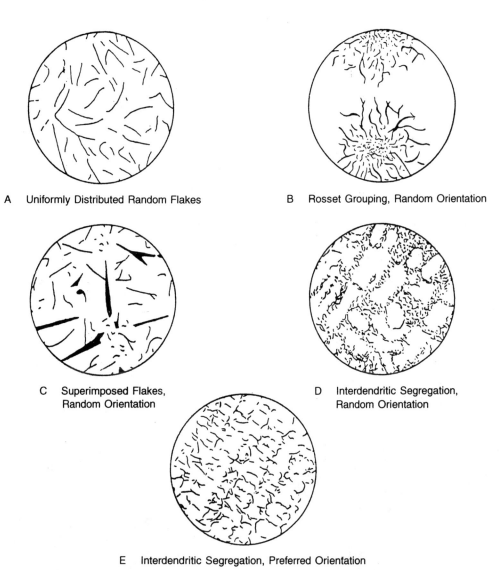

A Uniformly Distributed Random Flakes

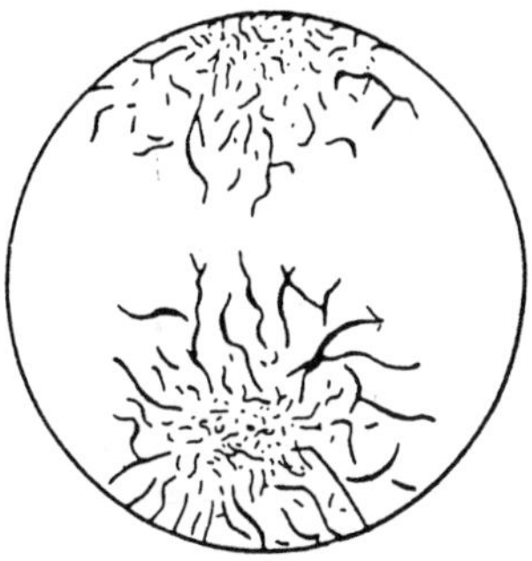

B Rosset Grouping, Random Orientation

C Superimposed Flakes, Random Orientation

D Interdendritic Segregation, Random Orientation

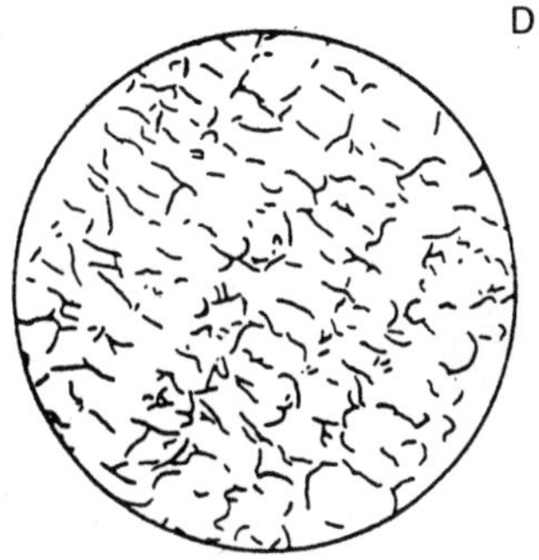

E Interdendritic Segregation, Preferred Orientation

Fig. 2.3 Reference diagrams for graphite distribution (form I) (Source: Reference 3)

SIZE

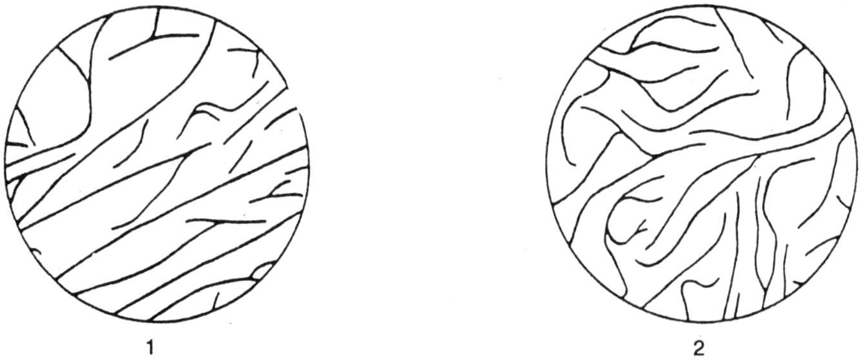

Fig. 2.4 Reference diagrams for graphite size (form I and distribution A) (magnification × 100)—Reference Nos. 1 and 2 (Source: Reference 3)

SIZE

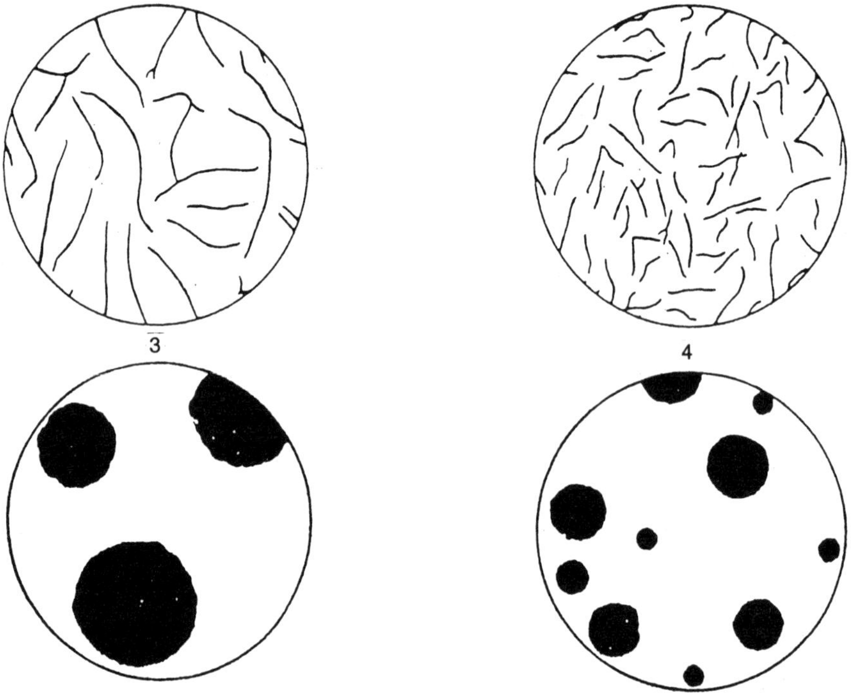

Fig. 2.5 Reference diagrams for graphite size (forms I and VI, and distribution A) (magnification × 100)—Reference Nos. 3 and 4 (Source: Reference 3)

SIZE

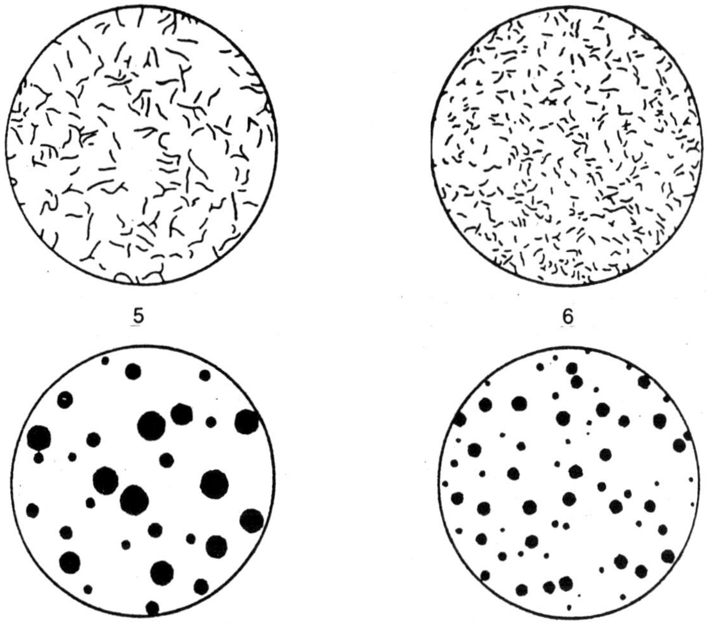

Fig. 2.6 Reference diagrams for graphite size (forms I and VI, and distribution A) (magnification × 100)—Reference Nos. 5 and 6 (Source: Reference 3)

SIZE

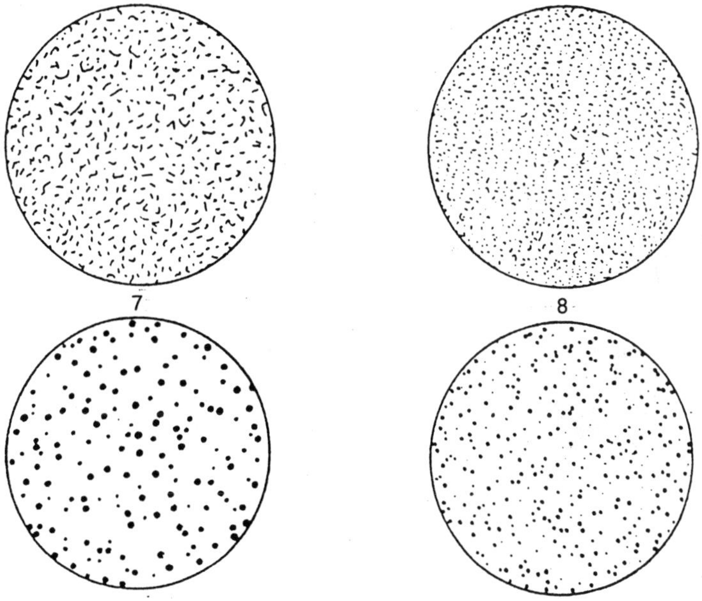

Fig. 2.7 Reference diagrams for graphite size (forms I and VI, and distribution A) (magnification × 100)—Reference Nos. 7 and 8 (Source: Reference 3)

TABLE 2.4 Dimension of the graphite particles—forms I to VI

Reference number	Dimension of the particles observed at × 100 magnification mm	True dimension mm
1	> 100	> 1
2	50 to 100	0.5 to 1
3	25 to 50	0.25 to 0.5
4	12 to 25	0.12 to 0.25
5	6 to 12	0.06 to 0.12
6	3 to 6	0.03 to 0.06
7	1.5 to 3	0.015 to 0.03
8	< 1.5	< 0.015

Source: Reference 3

2 EQUIVALENT GRAPHITE FORMS

TABLE 2.5 Comparison of IS/ISO and equivalent ASTM classification of graphite forms in cast iron

Graphite form		
Designation		Description
IS[1)]/ISO[2)]	ASTM[3)]	
I[4)]	VII	Flake graphite
II	V	Crab-form graphite
III	IV	Quasi flake graphite
IV	III	Aggregate, or temper carbon
V	VI	Irregular or "open" type nodules

(Contd.)

TABLE 2.5 Comparison of IS/ISO and equivalent ASTM classification of graphite forms in cast iron (continued)

Graphite form		
Designation		Description
IS[1)]/ISO[2)]	ASTM[3)]	
VI	I II	Spheroidal or nodular graphite

[1)] As defined in IS 7754 – 1975.
[2)] As defined in ISO 945 – 1975.
[3)] As defined in ASTM A 247 – 1967.
[4)] Subdivided into five types (see Fig. 2.3).

Source: Reference 1

3 REFERENCES

Book

1. *Metals Handbook*, vol. 1, 9th ed., American Society for Metals, Metals Park, Ohio, U.S.A., 1978.

Standards

2. IS 7754 – 1975: *Method for Designation of the Microstructure of Graphite in Cast Iron.*
3. ISO 945 – 1975: *Cast Iron—Designation of Microstructure of Graphite.*
4. ASTM A 247 – 1967: *Test Method for Evaluating the Microstructure of Graphite in Iron Castings.*

2.4

GREY CAST IRON

1 INTRODUCTION

Grey cast iron, also known as flake graphite cast iron, is a type of cast iron in which most of the carbon is present as flake graphite corresponding to form I of IS 7754.

The properties of grey cast iron depend on the distribution, size and amount of the graphite flakes, and the matrix structure. These factors are, in turn, influenced mainly by the manufacturing conditions, chemical composition, solidification time and rate of cooling in the mould.

Grey cast irons exhibit low to moderate strength, low ductility and toughness, low modulus of elasticity, low notch sensitivity, high resistance to wear and seizure, excellent vibration damping capacity, excellent machinability, high thermal conductivity, moderate resistance to thermal shock, and outstanding castability.

In general, the following properties of grey cast iron increase with increasing tensile strength:

a) all strengths, including strength at elevated temperature;
b) ability to produce a fine machined finish;
c) modulus of elasticity; and
d) wear resistance.

On the other hand, the following properties decrease with increasing tensile strength:

a) machinability;
b) resistance to thermal shock;
c) damping capacity; and
d) ability to be cast into thin sections.

2 SPECIFICATION

TABLE 2.6 Mechanical properties[1] of grey cast iron, conforming to IS 210 – 1993

Grade	R_m min. MPa	Brinell hardness HB
FG 150	150	130–180
FG 200	200	160–220
FG 220	220	180–220
FG 260	260	180–230
FG 300	300	180–230
FG 350	350	207–241
FG 400	400	207–270

1) Measured on test pieces machined from separately cast 30 mm diameter test bars.

3 TYPICAL MECHANICAL PROPERTIES

TABLE 2.7 Typical mechanical properties[1] of grey cast iron, conforming to IS 210 – 1993

Property	Unit	FG 150	FG 200	FG 220	FG 260	FG 300	FG 350	FG 400
Tensile strength	MPa	150	200	220	260	300	350	400
0.01 % proof stress	MPa	42	56	62	73	84	98	112
0.1 % proof stress	MPa	98	130	143	169	195	228	260
Total strain at failure	%	0.60–0.75[2]	0.48–0.67[2]	0.39–0.63[2]	0.57	0.50	0.50	0.50
Elastic strain at failure	%	0.15	0.17	0.18	0.20	0.22	0.25	0.28
Total minus elastic strain at failure	%	0.45–0.60[2]	0.31–0.50[2]	0.21–0.45[2]	0.37	0.28	0.25	0.22

(Contd.)

TABLE 2.7 Typical mechanical properties[1] of grey cast iron, conforming to IS 210 – 1993 (continued)

Property	Unit	FG 150	FG 200	FG 220	FG 260	FG 300	FG 350	FG 400
Notched tensile strength[3]								
Circumferential 45° V-notch: root radius 0.25 mm (*notch depth 2.5 mm, notch diameter 20 mm or notch depth 3.3 mm, notch diameter 7.6 mm*)	MPa	120	160	176	208	240	280	320
Circumferential notch: radius 9.5 mm (*notch depth 2.5 mm, notch diameter 20 mm*)	MPa	150	200	220	260	300	350	400
Compressive strength	MPa	600	720	768	864	960	1080	1200
0.01 % proof stress	MPa	84	112	123	146	168	196	224
0.1 % proof stress	MPa	195	260	286	338	390	455	520
Shear strength	MPa	173	230	253	299	345	403	460
Torsional strength	MPa	173	230	253	299	345	403	460
Shear strain at failure	%	>4	>4	>4	>4	≤4	≤4	≤4
Modulus of elasticity								
Tension	GPa	100	114	120	128	135	140	145
Compression	GPa	100	114	120	128	135	140	145
Modulus of rigidity	GPa	40	46	48	51	54	56	58
Poisson's ratio	–	0.26	0.26	0.26	0.26	0.26	0.26	0.26
Fatigue limit (Wöhler)								
Unnotched (8.4 mm diameter)	MPa	68	90	99	117	135	149	152
V-notched (*Circumferential 45° V-notch with 0.25 mm root radius. Diameter at notch 8.4 mm, depth of notch 3.4 mm*)	MPa	68	87	94	108	122	129	127
Hardness	–	4)						

(Contd.)

TABLE 2.7 Typical mechanical properties[1)] of grey cast iron, conforming to IS 210 – 1993 (continued)

1) Determined on a separately cast 30 mm diameter test bar or in a casting section correctly represented by this size of test bar.
2) Value depends on the composition of iron.
3) The notched tensile strength increases slightly as the notch severity ratio (ratio of notch radius to notch diameter) increases above 0.47.
4) Hardness is not simply related to tensile strength and varies with casting section thickness and material composition.

Source: Reference 8

4 TYPICAL PHYSICAL PROPERTIES

TABLE 2.8 Typical physical properties[1)] of grey cast iron, conforming to IS 210 – 1993

Property	Unit	FG 150	FG 200	FG 220	FG 260	FG 300	FG 350	FG 400
Density	kg/m^3	7050	7100	7150	7200	7250	7300	7300
Coefficient of thermal expansion								
–100 to 20 °C	μm/(m K)	10.0	10.0	10.0	10.0	10.0	10.0	10.0
20 to 200 °C	μm/(m K)	11.0	11.0	11.0	11.0	11.0	11.0	11.0[2)] (15.0)[3)]
20 to 400 °C	μm/(m K)	12.5	12.5	12.5	12.5	12.5	12.5	12.5[2)] (16.5)[3)]
Specific heat capacity								
20 to 200 °C	J/(kg K)	265	375	420	460	460	460	460
20 to 300 °C	J/(kg K)	355	430	455	495	495	495	495
20 to 400 °C	J/(kg K)	400	450	465	505	505	505	505
20 to 500 °C	J/(kg K)	425	470	475	515	515	515	515
20 to 600 °C	J/(kg K)	445	490	495	535	535	535	535
20 to 700 °C	J/(kg K)	490	545	560	605	605	605	605

(Contd.)

TABLE 2.8 Typical physical properties[1] of grey cast iron, conforming to IS 210 – 1993 (continued)

Property	Unit	FG 150	FG 200	FG 220	FG 260	FG 300	FG 350	FG 400
Thermal conductivity								
100 °C	W/(m K)	52.5	50.8	50.1	48.8	47.4	45.7	44.0
200 °C	W/(m K)	51.5	49.8	49.1	47.8	46.4	44.7	43.0
300 °C	W/(m K)	50.5	48.8	48.1	46.8	45.4	43.7	42.0
400 °C	W/(m K)	49.5	47.8	47.1	45.8	44.4	42.7	41.0
500 °C	W/(m K)	48.5	46.8	46.1	44.8	43.4	41.7	40.0
Magnetic and electrical properties								
Maximum magnetic permeability	μH/m	310–380	310–380	310–380	310–380	310–380	310–380	310–380
Remanent magnetism	T	0.4–0.5	0.4–0.5	0.4–0.5	0.4–0.5	0.4–0.5	0.4–0.5	0.4–0.5
Coercive force	A/m	560–720	560–720	560–720	560–720	560–720	560–720	560–720
Hysteresis loss (B = 1T)	J/m^3	2500–3000	2500–3000	2500–3000	2500–3000	2500–3000	2500–3000	2500–3000
	W/kg at 50 Hz	17.6–20.9	17.6–20.9	17.6–20.9	17.6–20.9	17.6–20.9	17.6–20.9	17.6–20.9
Electrical resistivity	μΩ m	0.800	0.770	0.760	0.730	0.700	0.670	0.640

1) Determined on a separately cast 30 mm diameter test bar or in a casting section correctly represented by this size of test bar.
2) For pearlitic irons.
3) For acicular irons.

Source: References 3 and 8

5 DESIGN STRESSES

TABLE 2.9 Maximum design stresses for grey cast iron, conforming to IS 210 – 1993, at room temperature

Stress condition	Maximum design stress, in MPa, for grades						
	FG 150	FG 200	FG 220	FG 260	FG 300	FG 350	FG 400
Tension[1)]	38	50	55	65	75	88	100
Compression[2)]	156	208	229	270	312	364	416
Fatigue (unnotched)[3)]	23	30	33	39	45	50	51

[1)] Maximum design stress = 0.25 × tensile strength.
[2)] Maximum design stress = 0.8 × 0.1 % proof stress in compression.
[3)] Maximum design stress = $\frac{1}{3}$ × unnotched fatigue limit.
Source: Reference 3

6 HEAT TREATMENT

6.1 Stress relieving

Purpose: To relieve residual stresses resulting from casting, machining, welding, or heat treatment.

Recommended heat treatment cycle: Heat uniformly to:

a) 510–565 °C for unalloyed grey cast irons; and
b) 595–650 °C for low-alloy grey cast irons.

Hold at this temperature for:

Section thickness mm	Holding time h
< 50	2 h
≥ 50 ≤ 100	1.5 h per 25 mm of maximum section thickness
> 100	6 h

Furnace cool to 315 °C or lower (up to 95 °C for castings of intricate design), and continue cooling to ambient temperature in still air.

6.2 Ferritizing annealing

Purpose: To convert pearlitic carbide to ferrite and graphite in unalloyed or low-alloy cast irons for improved machinability.

Recommended heat treatment cycle: Heat to 700–760 °C. Hold at this temperature for 1 h per 25 mm of maximum section thickness. Furnace cool to 315 °C at a rate not exceeding 55 °C/h, and continue cooling to ambient temperature in still air.

6.3 Medium ("full") annealing

Purpose:

a) To convert pearlitic carbide to ferrite and graphite in alloyed irons unresponsive to low temperature ferritizing annealing; and
b) To eliminate minor amounts of well dispersed carbides in unalloyed irons.

Recommended heat treatment cycle: Heat to 790–900 °C. Hold at this temperature for 1 h per 25mm of maximum section thickness. Furnace cool to 315 °C and continue cooling to ambient temperature in still air.

6.4 Graphitizing annealing

Purpose:

a) To convert massive carbides to pearlite and graphite in mottled or chilled irons; and
b) To convert pearlitic carbide to ferrite and graphite for maximum machinability.

Recommended heat treatment cycle: Heat to 900–925 °C in a controlled atmosphere. Hold at this temperature for 0.5–5 h. For maximum machinability, furnace cool to 540 °C. If a stronger, wear resistant pearlitic matrix is required, air cool to 540 °C. In both instances, cool from 540 °C to 290 °C at a rate not exceeding 110 °C/h, before finally cooling to ambient temperature in still air.

6.5 Normalizing

Purpose:

a) To obtain higher strength and hardness than can be achieved in either the as-cast or annealed condition;
b) To improve the response to flame or induction hardening; and
c) To restore as-cast properties that have been modified by another heating process, such as graphitizing, or the preheating and post heating associated with repair welding.

Recommended heat treatment cycle: Heat to 885–925 °C. Hold at this temperature for 1 h per 25 mm of maximum section thickness or 1 h minimum. Cool to ambient temperature in still air.

6.6 Hardening and tempering

Purpose: To produce higher strength and hardness (usually to improve wear resistance) than can be achieved by normalizing.

Recommended heat treatment cycle: Preheat slowly to 600–650 °C, before heating rapidly to the austenitizing temperature of 840–880 °C. Hold at this temperature for 20 min per 25 mm of maximum section thickness. Quench in agitated oil. Withdraw castings from oil bath at a temperature of approximately 150–200 °C, and transfer immediately to the tempering furnace. Temper for 1-2 h per 25 mm of maximum section thickness at:

a) 150–200 °C for maximum hardness; and
b) 370–510 °C for optimum strength and toughness.

6.7 Martempering

Purpose: To replace conventional hardening for one or more of the following reasons:

a) To minimize distortion;
b) To reduce or eliminate susceptibility to cracking; and
c) To reduce the level of residual stresses.

Recommended heat treatment cycle: Preheat slowly to 600–650 °C, before rapidly heating to the austenitizing temperature of 840–880 °C. Hold at this temperature for 20 min per 25 mm of maximum section thickness Quench in a salt, oil or lead bath maintained at a temperature just above the temperature at which martensite begins to form (usually in the range of 205-260 °C for unalloyed grey cast iron), at a rate fast enough to prevent the formation of pearlite, ferrite, or bainite. Hold at the bath temperature for a sufficient length of time to equalize the temperature throughout the casting — but not long enough to permit bainite to form. Cool to ambient temperature at a moderate rate to achieve a fully martensitic structure. Temper in the same manner as if the part had been conventionally hardened.

6.8 Austempering

Purpose: To replace conventional hardening and tempering, for one or more of the following reasons:

a) To obtain higher ductility or notch toughness at a given hardness;
b) To reduce the likelihood of cracking; and
c) To reduce distortion.

Recommended heat treatment cycle: Preheat slowly to 600–650 °C, before heating rapidly to the austenitizing temperature of 840–880 °C. Hold at this temperature for 20 min per 25 mm of maximum section thickness. Quench in a salt, oil or lead bath maintained at a temperature of 230–425 °C (for high hardness and wear resistance maintain bath at a temperature of 230–290 °C), at a rate fast enough to prevent the formation of pearlite and ferrite. Hold at the bath temperature for a sufficient length of time to ensure complete isothermal transformation of austenite to bainitic ferrite and high-carbon austenite. Once the transformation is complete cool to ambient temperature in still air.

6.9 Surface hardening (flame or induction hardening)

Purpose:

a) To produce a hard, wear-resistant surface; and
b) To improve fatigue strength.

Recommended heat treatment cycle: Heat the surface to be hardened with an oxy-acetylene flame (as in flame hardening) or by electromagnetic induction (as in induction hardening) to 870–925 °C. Quench the surface with water, soluble-oil mixture or solution of polyvinyl alcohol in water. Temper at 150–200 °C for a minimum of 1 h.

Notes:

1) A minimum combined carbon content of 0.40 % (as pearlite) is recommended for grey cast irons to be flame or induction hardened.
2) Parts designed for flame hardening should preferably have phosphorus contents below 0.40 % to reduce the risk of pitting or cracking.

7 TYPICAL APPLICATIONS

TABLE 2.10 Typical applications of grey cast iron, conforming to IS 210 – 1993

Grade	Typical applications
FG 150	Andirons, exhaust manifolds, frames, grates, housings, machine bases, manhole covers, and traffic signs.
FG 200	Air-cooled cylinders, clutch housings, clutch plates, compressor frames, cylinder heads, flywheels, gearboxes, grate bars, impellers, light-duty brake drums, oil pump bodies, pipe and fittings, pistons, rams, small cylinder blocks, and transmission cases.
FG 220	Air-cooled cylinders, clutch housings, clutch plates, compressor frames, cylinder heads, flywheels, gearboxes, grate bars, impellers, light-duty brake drums, oil pump bodies, pipe and fittings, pistons, rams, small cylinder blocks, and transmission cases.
FG 260	Anvils, automobile and diesel cylinder blocks, compressors, cylinder heads, cylinder liners, face plates, heavy flywheels, heavy machine beds, impellers, medium-duty brake drums and clutch plates, pistons, pumps, rams, steam pressure castings, valves, and wheels.
FG 300	Anvils, brake drums and clutch plates for heavy-duty service, compressors, diesel engine blocks, differential carrier castings, face plates, heavy flywheels, heavy gearboxes, heavy machine beds, impellers, pumps, rams, steam pressure castings, tractor transmission cases, truck and tractor cylinder blocks and heads, valves, and wheels.

(Contd.)

TABLE 2.10 Typical applications of grey cast iron, conforming to IS 210 – 1993 (continued)

Grade	Typical applications
FG 350	Anvils, camshafts, compressors, cylinders, cylinder liners, diesel engine castings, face plates, heavy machine beds, impellers, light crankshafts, pistons, pumps, rams, steam pressure castings, valves, and wheels.
FG 400	Camshafts, connecting rods, crusher frames, cylinder heads, diesel liners, high pressure well pumps, hydraulic cylinders, pressure castings in the chemical industry, truck cylinder blocks, and water works sluice gate valves.

Source: References 7 and 9

8 EQUIVALENT GRADES

TABLE 2.11 Comparison of grades of grey cast iron specified in IS 210–1993 with nearest equivalent grades specified in other national and international standards

IS 210–93 IS 6331 – 87	ASTM A 48–94a	BS 1452 – 90	DIN 1691–85	GOST 1412–85	ISO 185–88	JIS G 5501–89	SAE J431–93 ASTM A 159–83
FG 150	25 B	150	GG-15	СЧ15	150	FC 150	G2500
FG 200	30 B	200	GG-20	СЧ20	200	FC 200	G3000
FG 220[1)]	35 B	220	–	СЧ24	–	–	G3500
FG 260	40 B	250	GG-25	СЧ25	250	FC 250	G4000
FG 300	45 B	300	GG-30	СЧ30	300	FC 300	–
FG 350	50 B	350	GG-35	СЧ35	350	FC 350	–
FG 400[1)]	60 B	–	–	–	–	–	–

[1)] Not covered in IS 6331 – 1987.

9 REFERENCES

Books

1. R. Elliot, *Cast Iron Technology*, Butterworths, London, U.K., 1988.
2. *ASM Handbook*, vol. 1, ASM International, Materials Park, Ohio, U.S.A., 1990.
3. G.N.J. Gilbert, *Engineering Data on Grey Cast Irons—SI Units*, BCIRA, Alvechurch, Birmingham, U.K., 1976.
4. H.T. Angus, *Cast Iron: Physical and Engineering Properties*, 2nd ed., Butterworths, London, U.K., 1976.
5. *ASM Handbook*, vol. 4, ASM International, Materials Park, Ohio, U.S.A., 1991.
6. C.F. Walton and T.J. Opar (Eds.), *Iron Castings Handbook*, 3rd ed., Iron Castings Society, Des Plaines, Illinois, U.S.A., 1981.
7. S.L. Hoyt (Ed.), *ASME Handbook of Metals Properties*, 1st ed., McGraw-Hill, New York, U.S.A., 1954.

Standards

8. IS 210 – 1993: *Grey Iron Castings—Specification.*
9. IS 6331 – 1987: *Specification for Automotive Grey Iron Castings.*

2.5

SPHEROIDAL GRAPHITE CAST IRON

1 INTRODUCTION

Spheroidal graphite cast iron (SG iron), also known as ductile or nodular iron, is a cast iron in which most of the carbon is present as spheroidal graphite corresponding to form VI of IS 7754. It is produced by treating low-sulphur molten iron of suitable composition with a small but definite amount of magnesium and/or cerium, followed by or combined with inoculation with a silicon containing material to control the size and distribution of the graphite spheroids and also to ensure a graphitic structure free from carbides.

Spheroidal graphite cast iron possesses higher tensile strength and higher ductility than grey cast iron. It also exhibits a high modulus of elasticity and a linear stress-strain relationship for most of the region below the yield point. Properties such as machinability and corrosion resistance are comparable to those of grey cast iron, except that spheroidal graphite cast iron has better resistance to elevated temperature oxidation than grey cast iron. The damping capacity of spheroidal graphite cast iron is, however, inferior to that of grey cast iron. Like grey cast iron, spheroidal graphite cast iron can be heat treated or alloyed to enhance certain properties, especially wear resistance.

2 SPECIFICATION

TABLE 2.12 Mechanical properties of spheroidal graphite cast iron, conforming to IS 1865 – 1991, measured on test pieces machined from separately cast test samples

Grade	R_m min. MPa	$R_{p0.2}$ min. MPa	*A* min. %	For information only: Brinell hardness HB	For information only: Predominant constituent of structure
SG 900/2	900	600	2	280–360	Bainite or tempered martensite
SG 800/2	800	480	2	245–335	Pearlite or tempered structure
SG 700/2	700	420	2	225–305	Pearlite
SG 600/3	600	370	3	190–270	Pearlite + ferrite
SG 500/7	500	320	7	160–240	Ferrite + pearlite
SG 450/10	450	310	10	160–210	Ferrite
SG 400/15	400	250	15	130–180	Ferrite
SG 400/18	400	250	18	130–180	Ferrite
SG 350/22	350	220	22	≤ 150	Ferrite

TABLE 2.13 Minimum impact values of spheroidal graphite cast iron, conforming to IS 1865 –1991, measured on test pieces machined from separately cast test samples

Grade	*KV* min., in J, at a temperature of					
	23 ± 5 °C		–20 ± 2 °C		–40 ± 2 °C	
	Mean value from 3 tests	Individual value	Mean value from 3 tests	Individual value	Mean value from 3 tests	Individual value
SG 400/18	14	11	–	–	–	–
SG 400/18L	–	–	12	9	–	–
SG 350/22	17	14	–	–	–	–
SG 350/22L	–	–	–	–	12	9

TABLE 2.14 Mechanical properties of spheroidal graphite cast iron, conforming to IS 1865–1991, measured on test pieces machined from cast-on test samples

Grade	Typical casting thickness	R_m min.	$R_{p0.2}$ min.	A min.	For information only	
					Brinell hardness	Predominant constituent of structure
	mm	MPa	MPa	%	HB	
SG 700/2A	> 30 ≤ 60	700	400	2	220–320	Pearlite
	> 60 ≤ 200	650	380	1		
SG 600/3A	> 30 ≤ 60	600	360	2	180–270	Pearlite + ferrite
	> 60 ≤ 200	550	340	1		
SG 500/7A	> 30 ≤ 60	450	300	7	170–240	Ferrite + pearlite
	> 60 ≤ 200	420	290	5		
SG 400/15A	> 30 ≤ 60	390	250	15	130–180	Ferrite
	> 60 ≤ 200	370	240	12		
SG 400/18A	> 30 ≤ 60	390	250	15	130–180	Ferrite
	> 60 ≤ 200	370	240	12		
SG 350/22A	> 30 ≤ 60	330	220	18	≤ 150	Ferrite
	> 60 ≤ 200	320	210	15		

TABLE 2.15 Minimum impact values of spheroidal graphite cast iron, conforming to IS 1865–1991, measured on test pieces machined from cast-on test samples

Grade	Typical casting thickness mm	*KV* min., in J, at a temperature of					
		23 ± 5 °C		–20 ± 2 °C		–40 ± 2 °C	
		Mean value from 3 tests	Individual value	Mean value from 3 tests	Individual value	Mean value from 3 tests	Individual value
SG 400/18A	> 30 ≤ 60	14	11	–	–	–	–
	> 60 ≤ 200	12	9	–	–	–	–
SG 400/18AL	> 30 ≤ 60	–	–	12	9	–	–
	> 60 ≤ 200	–	–	10	7	–	–
SG 350/22A	> 30 ≤ 60	17	14	–	–	–	–
	> 60 ≤ 200	15	12	–	–	–	–
SG 350/22AL	> 30 ≤ 60	–	–	–	–	12	9
	> 60 ≤ 200	–	–	–	–	10	7

3 TYPICAL MECHANICAL PROPERTIES

TABLE 2.16 Typical mechanical properties of spheroidal graphite cast iron, conforming to IS 1865 – 1991

Property	Unit	SG 350/22	SG 400/18	SG 400/15	SG 450/10	SG 500/7	SG 600/3	SG 700/2	SG 800/2	SG 900/2
Tensile strength	MPa	350	400	400	450	500	600	700	800	900
Limit of proportionality	MPa	153	186	186	203	194	208	231	264	405
0.1 % proof stress	MPa	203	247	247	293	323	346	385	440	675
0.2 % proof stress	MPa	215	259	259	305	339	372	416	471	710
0.5 % proof stress	MPa	229	273	273	319	356	409	462	517	744
Percentage elongation after fracture	%	22	18	15	10	7	3	2	2	2
Compressive strength										
Limit of proportionality	MPa	181	216	216	253	272	288	318	362	550
0.1 % proof stress	MPa	226	270	270	316	340	360	397	452	687
0.2 % proof stress	MPa	229	273	273	319	351	382	425	480	719
0.5 % proof stress	MPa	232	276	276	322	360	414	468	523	750
Shear strength	MPa	315	360	360	405	450	540	630	720	810
Torsional strength	MPa	315	360	360	405	450	540	630	720	810
Limit of proportionality	MPa	118	144	144	170	181	185	203	231	355
0.1 % proof stress	MPa	157	191	191	227	241	247	270	308	473
0.2 % proof stress	MPa	167	201	201	236	253	266	291	330	497
0.5 % proof stress	MPa	177	212	212	247	265	291	323	362	521
Plastic strain at failure	%	← Higher than elongation in tension →								
Modulus of elasticity										
Tension	GPa	169	169	169	169	169	174	176	176	172
Compression	GPa	169	169	169	169	169	174	176	176	172

(Contd.)

TABLE 2.16 Typical mechanical properties of spheroidal graphite cast iron, conforming to IS 1865 – 1991 (continued)

Property	Unit	SG 350/22	SG 400/18	SG 400/15	SG 450/10	SG 500/7	SG 600/3	SG 700/2	SG 800/2	SG 900/2
Modulus of rigidity	GPa	65.9	65.9	65.9	65.9	65.9	67.9	68.6	68.6	67.1
Poisson's ratio	–	0.275	0.275	0.275	0.275	0.275	0.275	0.275	0.275	0.275
Fatigue limit (Wöhler)										
Unnotched (10.6 mm diameter)	MPa	180	195	195	210	224	248	280	304	317
Notched (10.6 mm diameter at root) *Circumferential 45° V-notch with 0.25 mm root radius and notch depth of 3.6 mm.*	MPa	114	122	122	128	134	149	168	182	190
Source: Reference 4										

4 TYPICAL PHYSICAL PROPERTIES

TABLE 2.17 Typical physical properties of spheroidal graphite cast iron, conforming to IS 1865 – 1991

Property	Unit	SG 350/22	SG 400/18	SG 400/15	SG 450/10	SG 500/7	SG 600/3	SG 700/2	SG 800/2	SG 900/2
Density	kg/m^3	7100	7100	7100	7100	7100	7170	7200	7200	7150
Coefficient of thermal expansion										
–100 to 20 °C	µm/(m K)	10.0	10.0	10.0	10.0	10.0	10.0	10.0	10.0	10.0
20 to 200 °C	µm/(m K)	11.0	11.0	11.0	11.0	11.0	11.0	11.0	11.0	11.0
20 to 400 °C	µm/(m K)	12.5	12.5	12.5	12.5	12.5	12.5	12.5	12.5	12.5

(Contd.)

TABLE 2.17 Typical physical properties of spheroidal graphite cast iron, conforming to IS 1865–1991 (continued)

Property	Unit	SG 350/22	SG 400/18	SG 400/15	SG 450/10	SG 500/7	SG 600/3	SG 700/2	SG 800/2	SG 900/2
Specific heat capacity										
20 to 200 °C	J/(kg K)	461	461	461	461	461	461	461	461	461
20 to 300 °C	J/(kg K)	494	494	494	494	494	494	494	494	494
20 to 400 °C	J/(kg K)	507	507	507	507	507	507	507	507	507
20 to 500 °C	J/(kg K)	515	515	515	515	515	515	515	515	515
20 to 600 °C	J/(kg K)	536	536	536	536	536	536	536	536	536
20 to 700 °C	J/(kg K)	603	603	603	603	603	603	603	603	603
Thermal conductivity										
100 °C	W/(m K)	36.5	36.5	36.5	36.5	35.50	32.80	31.40	31.40	33.50
200 °C	W/(m K)	36.3	36.3	36.3	36.3	35.35	32.65	31.25	31.25	33.35
300 °C	W/(m K)	36.2	36.2	36.2	36.2	35.20	32.50	31.10	31.10	33.20
400 °C	W/(m K)	36.0	36.0	36.0	36.0	35.05	32.35	30.95	30.95	33.05
500 °C	W/(m K)	35.8	35.8	35.8	35.8	34.90	32.20	30.80	30.80	32.90
Magnetic and electrical properties										
Maximum magnetic permeability	μH/m	2136	2136	2136	2136	1596	866	501	501	< 501
Remanent magnetism	T	0.56	0.56	0.56	0.56	0.58	0.608	0.62	0.62	> 0.62
Coercive force	A/m	159	159	159	159	468	740	875	875	> 875
Hysteresis loss (B = 1T)	J/m^3	600	600	600	600	1345	2248	2700	2700	> 2700
	W/kg at 50 Hz	4.41	4.41	4.41	4.41	9.57	16.11	19.4	19.4	> 19.4
Electrical resistivity	μΩ. m	0.500	0.500	0.500	0.500	0.510	0.530	0.540	0.540	> 0.540

Source: Reference 4

5 DESIGN STRESSES

TABLE 2.18 Maximum design stresses for spheroidal graphite cast iron, conforming to IS 1865 – 1991

Stress condition	Maximum design stress, in MPa, for grades								
	SG 350/22	SG 400/18	SG 400/15	SG 450/10	SG 500/7	SG 600/3	SG 700/2	SG 800/2	SG 900/2
Tension[1)]	114	138	138	152	145	156	173	198	304
Compression[2)]	136	162	162	190	204	216	238	271	412
Alternating fatigue (Wöhler)[3)]									
Unnotched	60	65	65	70	75	83	93	101	106
Notched	38	41	41	43	45	50	56	61	63

1) Maximum design stress in tension:
(a) For grades SG 350/22, SG 400/18 and SG 400/15: 0.56 × 0.1 % proof stress in tension;
(b) For grades SG 450/10: 0.52 × 0.1 % proof stress in tension; and
(c) For grades SG 500/7, SG 600/3, SG 700/2, SG 800/2 and SG 900/2: 0.45 × 0.1 % proof stress in tension;
2) Maximum design stress in compression = 0.6 × 0.1 % proof stress in compression;
3) Maximum design stress in alternating fatigue (Wöhler):
(a) Unnotched: $\frac{1}{3}$ × unnotched fatigue limit;
(b) Notched : $\frac{1}{3}$ × notched fatigue limit;

Source: Reference 4

6 HEAT TREATMENT

6.1 Stress relieving

Purpose: To relieve residual stresses resulting from casting, machining, welding or heat treatment.

Recommended heat treatment cycle: Heat to 510–595 °C. Hold at this temperature for 1 h plus 1 h per 25 mm of maximum section thickness. Furnace cool to about 200 °C and continue cooling to ambient temperature in still air.

6.2 Ferritizing annealing

Purpose: To achieve maximum ductility and good machinability.

Recommended heat treatment cycle:

a) Heat to 900–955 °C. Hold at this temperature for 1 h plus 1 h per 25 mm of maximum section thickness. Cool uniformly to 690 °C. Hold at this temperature for 5 h plus 1 h per 25 mm of maximum section thickness. Cool to ambient temperature in still air; or

b) Heat to 900–955 °C. Hold at this temperature for 1 h plus 1 h per 25 mm of maximum section thickness. Furnace cool to 650 °C at a rate not exceeding 20 °C/h in the temperature range 790–650 °C, and continue cooling to ambient temperature in still air.

6.3 Normalizing

Purpose:

a) To obtain higher strength and hardness than can be achieved in either the as-cast or annealed condition; and

b) To improve the response to flame or induction hardening.

Recommended heat treatment cycle: Heat to 870–940 °C. Hold at this temperature for 1 h per 25 mm of maximum section thickness or 1 h minimum. Cool to ambient temperature in still air.

Normalizing is commonly followed by tempering:

a) To attain the desired hardness;

b) To relieve residual stresses; and

c) To obtain high toughness along with high tensile properties.

Tempering cycle: Heat to 500–625 °C. Hold at this temperature for 1 h per 25 mm of maximum section thickness. Furnace cool to 300 °C, and continue cooling to ambient temperature in still air.

6.4 Hardening and tempering

Purpose: To produce higher strength and hardness (usually to improve wear resistance) than can be achieved by normalizing.

Recommended heat treatment cycle: Heat to 845–925 °C. Hold at this temperature for 1 h per 25 mm of maximum section thickness. Quench in agitated oil (complicated castings may have to be quenched in oil at 80–100 °C to avoid cracks). Temper immediately after quenching at 200–600 °C for 1 h plus 1 h per 25 mm of maximum section thickness, and cool to ambient temperature in still air.

6.5 Surface hardening (flame or induction hardening)

Purpose:

a) To produce a hard, wear-resistant surface; and
b) To improve fatigue strength.

Recommended heat treatment cycle: Heat the surface to be hardened with an oxy-acetylene flame (as in flame hardening) or by electromagnetic induction (as in induction hardening) to 900–925 °C. Quench the surface with water, soluble-oil mixture or solution of polyvinyl alcohol in water. Temper to 150–200 °C for a minimum of 1 h.

7 TYPICAL APPLICATIONS

Automotive and agricultural: Axle housings, brake calipers, brake cylinders, camshafts, connecting rods, crankshafts, cylinder liners, exhaust manifolds, gears and pinions, piston rings, wheel hubs, and yokes.

Other transportation modes: Bulldozer parts, conveyor frames, couplers, crawler sprockets, elevator buckets, furnace skids, hoist drums, railway wheels, rollers, and track crossovers.

General engineering: Boiler segments, briquetting rams, coal crusher gears, compressor bodies, crusher hammers, damper frames, die blocks, frames and jigs, furnace grates, machine frames, nuclear fuel containers, pipe forming dies, rolls, shafts, tank covers, tunnel segments, and turret heads.

8 EQUIVALENT GRADES

TABLE 2.19 Comparison of grades of spheroidal graphite cast iron specified in IS 1865 – 1991 with nearest equivalent grades specified in other national and international standards

IS 1865–91	ASTM A 536–84	BS 2789–85	DIN 1693 (1)–73 1693 (2)–77	GOST 7293–85	ISO 1083–87	JIS G5502–89	SAE J434–86
SG 900/2	–	900/2	–	–	900-2	–	–
SG 800/2	120-90-02	800/2	GGG-80	**ВЧ80**	800-2	FCD 800	–
SG 700/2	100-70-03	700/2	GGG-70	**ВЧ70**	700-2	FCD 700	D7003
SG 600/3	80-55-06	600/3	GGG-60	**ВЧ60**	600-3	FCD 600	D5506
SG 500/7	–	500/7	GGG-50	**ВЧ50**	500-7	FCD 500	–
SG 450/10	65-45-12	450/10	–	**ВЧ45**	450-10	FCD 450	D4512

(Contd.)

TABLE 2.19 Comparison of grades of spheroidal graphite cast iron specified in IS 1865 – 1991 with nearest equivalent grades specified in other national and international standards (continued)

IS 1865–91	ASTM A 536–84	BS 2789–85	DIN 1693 (1)–73 1693 (2)–77	GOST 7293–85	ISO 1083–87	JIS G5502–89	SAE J434–86
SG 400/15	60-40-18	–	GGG-40	ВЧ40	400-15	FCD 400	D4018
SG 400/18	–	400/18	GGG-40.3	–	400-18	FCD 370	–
SG 350/22	–	350/22	GGG-35.3	ВЧ35	350-22	–	–

9 REFERENCES

Books

1. *ASM Handbook,* vol. 1, ASM International, Materials Park, Ohio, U.S.A., 1990.
2. *ASM Handbook,* vol. 4, ASM International, Materials Park, Ohio, U.S.A., 1991.
3. C.F. Walton and T.J. Opar (Eds.), *Iron Castings Handbook,* 3rd ed., Iron Castings Society, Des Plaines, Illinois, U.S.A., 1981.
4. G.N.J. Gilbert, *Engineering Data on Nodular Cast Irons—SI Units,* BCIRA, Alvechurch, Birmingham, U.K., 1986.
5. *A Design Engineer's Digest of Ductile Iron,* 7th ed., QIT-Feret et Titane Inc., Montreal, Quebec, Canada, 1990.

Standards

6. IS 1865 – 1991: *Iron Castings with Spheroidal or Nodular Graphite—Specification.*
7. BS 2789 – 1985: *Specification for Spheroidal Graphite or Nodular Graphite Cast Iron.*

2.6

AUSTEMPERED DUCTILE IRON

1 INTRODUCTION

Austempered ductile iron (ADI) is a recent addition to the cast iron family and represents a new group of ductile irons with an outstanding combination of high strength, toughness and wear resistance previously unattainable in ductile irons (see Fig. 2.8).

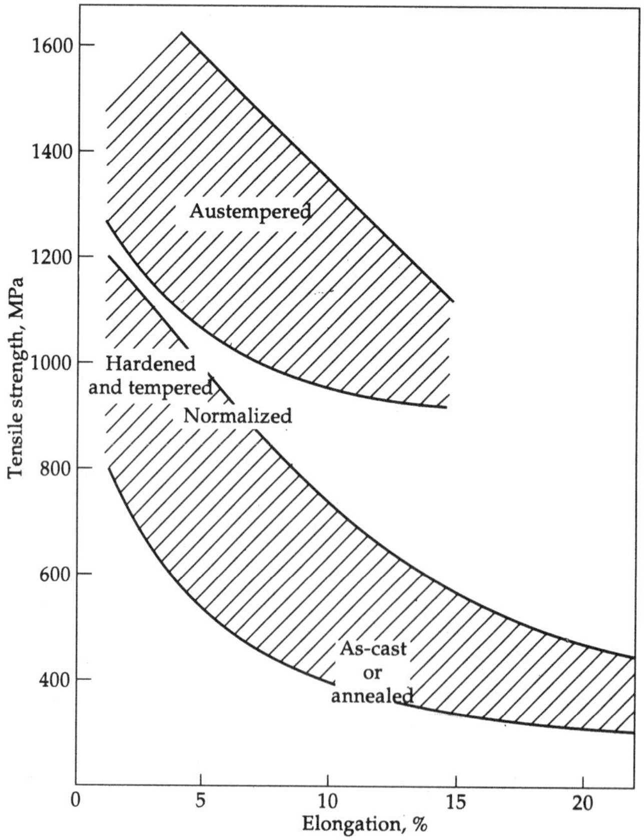

Fig. 2.8 Tensile strength and elongation in austempered ductile iron compared with other ductile irons (Source: Reference 2)

The remarkable properties of ADI are developed by a closely controlled heat treatment operation (austempering) which develops a unique matrix structure of bainitic ferrite and retained high-carbon austenite. The retained high-carbon austenite is thermally stable to extremely low temperatures but is work hardenable and will locally transform to martensite under suitable conditions of stress.

Austempered ductile iron can be produced with a wide range of mechanical properties by varying the heat treatment cycle (principally the austempering temperature — see Fig. 2.9). They have impact values that are much higher than those of pearlitic ductile irons and are comparable with those of ferritic ductile irons. Fracture toughness values are also higher than those for other ductile cast irons and are comparable with the values obtained on engineering steels.

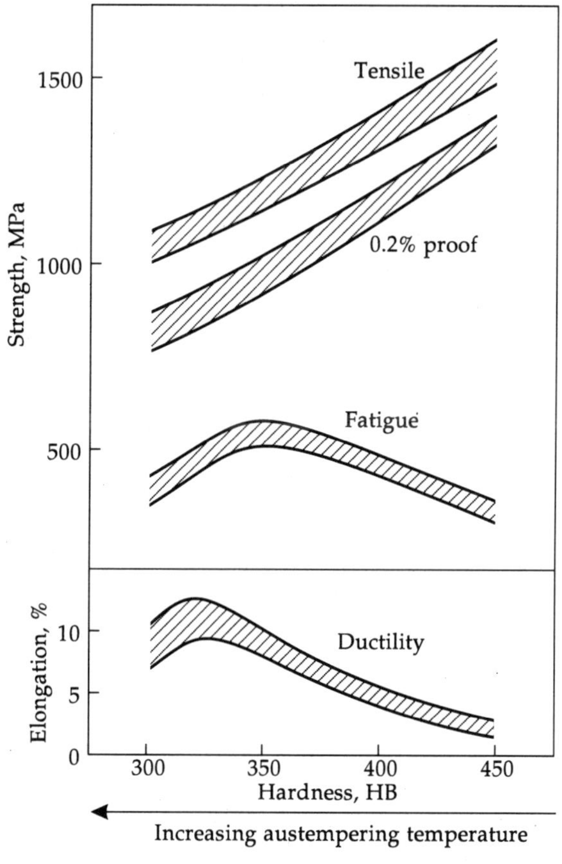

Fig. 2.9 Properties of austempered ductile iron (Source: Reference 3)

Machinability of austempered ductile iron is generally good, and (in the ductile grades) comparable with that of other ductile irons or low-alloy steels. It may, however, be an advantage in the higher hardness materials to do most of the machining prior to the austempering heat treatment

Another important property of these materials is wear resistance which because of the bainitic structure, is greater than that of the other ductile irons and some steels, though not in the same class as white irons.

Austempering heat treatment

Heat the ductile iron castings in a controlled atmosphere to an austenitizing temperature between 815 °C and 925 °C. Hold them at the austenitizing temperature for a sufficient length of time (about 1–4 h) to saturate the austenite with carbon. Then quench the castings rapidly in a molten salt or in a hot oil (use of hot oil is usually restricted to 260 °C) bath maintained at a constant temperature above the M_s temperature (this may vary from 235–400 °C depending upon the strength and hardness required) at a rate fast enough to prevent the formation of pearlite or ferrite. Hold the castings at this temperature for a sufficient length of time (for a period of up to 4 h) to allow the austenite to isothermally transform to bainitic ferrite and high-carbon austenite. Once the transformation is complete, cool the castings to ambient temperature either in still air or by quenching with water.

2 SPECIFICATIONS

TABLE 2.20 Mechanical properties of austempered ductile iron

Grade	R_m min. MPa	$R_{p0.2}$ min. MPa	A %	Minimum Charpy impact strength[1)] J	Brinell hardness HB
		ASTM A 897M – 1990			
850/550/10	850	550	≥ 10[2)]	100[3)]	269-321[4)]
1050/700/7	1050	700	≥ 7[2)]	80[3)]	302-363[4)]
1200/850/4	1200	850	≥ 4[2)]	60[3)]	341-444[4)]
1400/1100/1	1400	1100	≥ 1[2)]	35[3)]	388-477[4)]
1600/1300/-	1600	1300	–	–	444-555[4)]

(Contd.)

TABLE 2.20 Mechanical properties of austempered ductile iron (continued)

Grade	R_m min. MPa	$R_{p0.2}$ min. MPa	A %	Minimum Charpy impact strength[1)] J	Brinell hardness HB
JIS G 5503 – 1989					
FCD 900 A	900	600	≥ 8[2)]	100[4),5)] 80[4),6)]	–
FCD 1000 A	1000	700	≥ 5[2)]	–	–
FCD 1200 A	1200	900	≥ 2[2)]	–	≥ 340

1) Measured on unnotched, 10 mm × 10 mm test pieces.
2) L_o = 50 mm
3) At 22 ± 4 °C. These values are a minimum for the average of the three highest of the four tested samples.
4) For information only.
5) Mean value from three tests.
6) Individual value.

3 TYPICAL APPLICATIONS

Engineering applications

Agriculture: Undercarriage parts and constructional equipment.

Automotive: Camshafts, crankshafts, helical gears, hypoid pinion and ring gears, internal gears, steering knuckles, and suspension components.

Pumps and compressors: Bodies, crankshafts and drive shafts.

Railway: Couplings.

Wear-resistant applications

Agriculture and forestry: Plough shares.

Machinery: Conveyor rollers, guides, blades and shredders.

Railway: Friction blocks and locomotive wheels.

Sludge handling: Impellers and pump casings.

4 REFERENCES

Book/articles

1. *A Design Engineer's Digest of Ductile Iron*, 7th ed., QIT-Fer et Titane Inc., Montreal, Quebec, Canada, 1990.
2. I.C.H. Hughes: "Austempered Ductile Irons—Their Properties and Significance", *Materials & Design*, 6(3), 124–126, June/July 1985.
3. A.G. Fuller, "Austempered Ductile Irons—Present Applications", *Materials & Design*, 6(3), 127–130, June/July 1985.
4. R.A. Harding, "Austempered Ductile Irons—Gears", *Materials & Design*, 6(4), 177-184, August/Septemer 1985.
5. R. Elliot, *Cast Iron Technology*, Butterworths, London, U.K., 1988.

Standards

6. ASTM A 897M – 1990: *Specification for Austempered Ductile Iron Castings (Metric).*
7. JIS G 5503 – 1989: *Austempered Spheroidal Graphite Iron Castings.*

2.7

COMPACTED GRAPHITE CAST IRON

1 INTRODUCTION

Compacted graphite cast iron (CG iron), frequently referred to as vermicular iron, is a cast iron in which the graphite is present as short, thick flakes with rounded or blunted ends corresponding to form III of IS 7754. In general, an acceptable compacted graphite cast iron is one in which at least 80 % of the graphite is compacted graphite with not more than 20 % spheroidal graphite and no flake graphite.

Compacted graphite cast iron can be produced by inoculating low-sulphur molten iron of suitable composition with a controlled amount of magnesium and/or cerium, in the presence of titanium. Under-treatment with magnesium will result in a grey cast iron structure, while over-treatment with magnesium will result in a spheroidal graphite cast iron structure. Foundry practices used in casting grey cast iron can be used for compacted graphite cast iron with only a slight modification. The casting yield for compacted graphite cast iron is compararble to that for grey cast iron and substantially greater than that for spheroidal graphite cast iron.

The properties of compacted graphite cast iron are intermediate between those of grey cast iron and spheroidal graphite cast iron (see Table 2.22). Compacted graphite cast iron possesses higher tensile strength, modulus of elasticity, ductility, fatigue life, impact strength and elevated temperature properties compared to grey cast iron with a similar matrix structure. At the same time its machinability and thermal conductivity are superior to those of spheroidal graphite cast iron. It is this combination of mechanical and physical properties that makes compacted graphite cast irons an ideal choice for several applications where neither grey cast iron nor spheroidal graphite cast iron are entirely satisfactory.

TABLE 2.21 Comparison of the characteristics of compacted graphite cast iron with those of other ferrous castings

Property	Compacted graphite cast iron	Grey cast iron	Spheroidal graphite cast iron	Malleable cast iron	White cast iron	Cast steel
Strength	M	L	H	H	L	H
Ductility	M	L	H	H	L	H
Hardness	L-M	L-M	L-M	L-M	H	L-M
Machinability	M-H	H	M	M	L	L-M
Castability	M-H	H	M	M-L	M-L	L
Thermal conductivity	M-H	H	M	M	L	M-L
Damping capacity	M-H	H	M	M	L	L-M

Key: H = high, M = moderate, L = low.

Source: Reference 5

2 SPECIFICATION

TABLE 2.22 Mechanical properties[1] of compacted graphite cast iron, conforming to ASTM A 842 – 1985

Grade	R_m min. MPa	$R_{p0.2}$ min. MPa	A min. %	Brinell hardness[2] HB
250[3]	250	175	3.0	≤179
300	300	210	1.5	143–207
350	350	245	1.0	163–229
400	400	280	1.0	197–255
450[4]	450	315	1.0	207–269

1) Measured on test pieces machined from separately cast 30 mm diameter test bars or separately cast keel blocks.
2) For information only.
3) Ferritic grade.
4) Pearlitic grade.

3 TYPICAL MECHANICAL AND PHYSICAL PROPERTIES

TABLE 2.23 Typical mechanical and physical properties of compacted graphite cast iron

Property	Unit	Value
Limit of proportionality		
Ferritic CG iron	MPa	125
Pearlitic CG iron	MPa	125
Compressive strength		
Ferritic CG iron (annealed)	MPa	< 1400
Pearlitic CG iron	MPa	3 × Tensile strength
Modulus of elasticity	MPa	138–165
Poisson's ratio	–	0.27–0.28

(Contd.)

TABLE 2.23 Typical mechanical and physical properties of compacted graphite cast iron (continued)

Property	Unit	Value
Unnotched fatigue limit (Wöhler) **Fatigue strength reduction factor**	MPa –	0.45 × Tensile strength 1.7–1.8
Density Ferritic CG iron	kg/m^3	7000–7200
Coefficient of thermal expansion at 20–200°C (ferritic CG iron)	μm/(m K)	12–14
Thermal conductivity at 100°C	W/(m K)	36.0–43.0
Electrical resistivity (ferritic CG iron)	μΩ m	0.70–0.80
Source: References 1, 2 and 4		

4 TYPICAL APPLICATIONS

Brake components: Brake drums and discs.

Engine components: Cylinder heads, exhaust manifolds, flywheels and piston rings.

Hydraulic equipment: Housings for industrial power-steering units and gear-plate castings for heavy-duty gear pumps.

Ingot moulds.

5 REFERENCES

Books/Articles

1. *ASM Handbook*, vol. 1, ASM International, Materials Park, Ohio, U.S.A., 1990.
2. C.F. Walton and T.J. Opar (Eds.), *Iron Castings Handbook,* 3rd ed., Iron Castings Society, Des Plaines, Illinois, U.S.A., 1981.
3. R. Elliot, *Cast Iron Technology,* Butterworths, U.K., 1988.
4. J. Powell, "A review of Some Recent Work on Compacted Graphite Irons", *The British Foundryman, 77,* 472–483, 1984.
5. D.R. Dreger, "Compacted Graphite Iron Finds A Niche", *Machine Design, 54*(5), 177-180, 1982.

Standard

6. ASTM A 842 – 1985: *Specification for Compacted Graphite Iron Castings.*

2.8

MALLEABLE CAST IRON

1 INTRODUCTION

Malleable cast iron is a type of cast iron in which most of the carbon is present as irregularly shaped nodules of graphite. This form of graphite is called temper carbon because it is formed in the solid state during heat treatment. Malleable cast iron is produced by first casting iron of suitable composition as white iron and then heat treating the white iron to convert the iron carbide into irregularly shaped nodules of graphite. By varying the chemical composition of the white iron and the heat treatment process, malleable cast iron can be produced with different matrix microstructures (ferrite, pearlite, tempered martensite or bainite).

Malleable cast iron, like spheroidal graphite cast iron, possesses considerable ductility and toughness. It also exhibits high modulus of elasticity, good fatigue strength, good damping capacity, high corrosion resistance, excellent machinability, good magnetic permeability and low magnetic retentivity.

Malleable cast iron is classified into whiteheart, blackheart and pearlitic types, differentiated by the chemical composition, temperature and time cycle of the annealing process, the annealing atmosphere, and the properties and microstructure resulting therefrom.

1.1 Whiteheart malleable cast iron

Whiteheart malleable cast iron is produced by annealing white iron of suitable composition in a decarburizing atmosphere. This annealing treatment causes a carbon gradient from the surface to the centre of the casting and produces a microstructure which is dependent as follows:

a) Small section size: ferrite (+ pearlite + temper carbon);
b) Large section size:
   i) surface zone: ferrite;
   ii) intermediate zone: pearlite + ferrite + temper carbon; and
   iii) core zone: pearlite (+ ferrite) + temper carbon

1.2 Blackheart malleable cast iron

Blackheart malleable cast iron is produced by annealing white iron of lower total carbon and slightly higher silicon than that used for whiteheart malleable cast iron, in an inert atmosphere. This annealing treatment results in an almost homogeneous structure of temper carbon in a matrix of ferrite.

1.3 Pearlitic malleable cast iron

The microstructure of pearlitic malleable cast iron consists of temper carbon in a matrix of pearlite or some other transformation product of austenite. They have been developed for applications where higher strength are required than those associated with whiteheart and blackheart malleable cast iron. Pearlitic malleable cast iron is produced by any one of the following methods:

a) By alloying normal blackheart compositions with carbide stabilizers so that the required structure can be obtained after a normal annealing cycle;
b) By interrupting a normal blackheart anneal and air or oil quenching; or
c) By a suitable reheating treatment, e.g., air or oil quenching and tempering, of the normal blackheart product.

2 SPECIFICATIONS

TABLE 2.24 Mechanical properties[1] of malleable cast iron specified in Indian Standards

Grade	Sectional thickness represented mm	Diameter of test bar mm	R_m min. MPa	$R_{p0.5}$ min. MPa	A min. ($L_o = 3d$) %	Brinell hardness HB
Whiteheart malleable cast iron (IS 2107 – 1977)						
WM 410	≤ 8	9	350	200	10	≤ 220[2]
	> 8 ≤ 13	12	390	230	6	≤ 220[2]
	> 13	15	410	250	4	≤ 220[2]
WM 340	≤ 8	9	270	–	7	≤ 220[2]
	> 8 ≤ 13	12	310	–	4	≤ 220[2]
	> 13	15	340	–	3	≤ 220[2]
Blackheart malleable cast iron (IS 2108 – 1977)						
BM 340	All sizes	15	340	200	12	≤ 150[2]
BM 310	All sizes	15	310	190	10	≤ 150[2]
BM 290	All sizes	15	290	–	6	≤ 160[2]

(Contd.)

TABLE 2.24 Mechanical properties[1] of malleable cast iron specified in Indian Standards (continued)

Grade	Sectional thickness represented mm	Diameter of test bar mm	R_m min. MPa	$R_{p0.5}$ min. MPa	A min. ($L_o = 3d$) %	Brinell hardness HB
Pearlitic malleable cast iron (IS 2640 – 1977)						
PM 690	All sizes	15	690	540	2	241–285
PM 590	All sizes	15	590	420	3	200–248
PM 540	All sizes	15	540	350	4	180–230
PM 490	All sizes	15	490	310	5	170–229
PM 440	All sizes	15	440	270	7	149–201

[1] Measured on separately cast, unmachined test bars.
[2] For information only.

3 TYPICAL MECHANICAL PROPERTIES

TABLE 2.25 Typical mechanical properties of malleable cast iron specified in Indian Standards

Property	Unit	Whiteheart malleable cast iron		Blackheart malleable cast iron			Pearlitic malleable cast iron				
		WM 340	WM 410	BM 290	BM 310	BM 340	PM 440	PM 490	PM 540	PM 590	PM 690
Tensile strength	MPa	340	410	290	310	340	440	490	540	590	690
Compressive strength[1]	MPa	← 1) →									
Shear strength	MPa	306	369	261	279	306	396	441	486	531	621
Torsional strength[2]	MPa	306	369	261	279	306	396	441	486	531	621

(Contd.)

TABLE 2.25 Typical mechanical properties of malleable cast iron specified in Indian Standards (continued)

Property	Unit	Whiteheart malleable cast iron		Blackheart malleable cast iron			Pearlitic malleable cast iron				
		WM 340	WM 410	BM 290	BM 310	BM 340	PM 440	PM 490	PM 540	PM 590	PM 690
Modulus of elasticity											
Tension	GPa	175.8	175.8	168.9	168.9	172.4	172.4	172.4	172.4	172.4	172.4
Compression	GPa	175.8	175.8	168.9	168.9	172.4	172.4	172.4	172.4	172.4	172.4
Modulus of rigidity	GPa	70.3	70.3	67.6	67.6	67.6	68.9	68.9	68.9	68.9	68.9
Poisson's ratio	–	0.26	0.26	0.26	0.26	0.26	0.26	0.26	0.26	0.26	0.26
Fatigue limit (Wöhler)											
Unnotched (10 mm diameter)	MPa	← 0.45 × R_m →		← 0.45 × R_m →			0.45 × R_m for low strength grades, and decreasing with increasing tensile strength to 0.40 × R_m for the high strength grades				
Notched (10 mm diameter at root) *Circumferential 45° V-notch with a root radius of 0.25 mm and a notch depth of 2 mm*	MPa	← 0.60 × unnotched fatigue limit →		← 0.60 × unnotched fatigue limit →			← 0.60 × unnotched fatigue limit →				

[1] The stress-strain properties in compression are similar to those in tension. The proof stress values in compression are, however, slightly higher than in tension because the onset of plastic deformation is delayed under compressive stress.

[2] The proof stress values in torsion are:
(i) for whiteheart malleable cast iron: 0.75 times the value in tension;
(ii) for blackheart malleable cast iron: 0.775 times the value in tension; and
(iii) for pearlitic malleable cast iron: 0.75 times the value in tension for the low strength grades and decreasing progressively to about 0.70 times the value in tension for the high strength grades.

Source: References 7–9

4 TYPICAL PHYSICAL PROPERTIES

TABLE 2.26 Typical physical properties of malleable cast iron specified in Indian Standards

Property	Unit	Whiteheart malleable cast iron		Blackheart malleable cast iron			Pearlitic malleable cast iron				
		WM 340	WM 410	BM 290	BM 310	BM 340	PM 440	PM 490	PM 540	PM 590	PM 690
Density	kg/m³	7400	7400	7350	7350	7350	7300	7300	7300	7300	7300
Coefficient of thermal expansion[1] (up to 400 °C)	µm/(m K)	10.0–12.5		10.0–12.5			10.0–12.5				
Specific heat capacity	J/(kg K)	460		500–690			–	–	–	–	–
Thermal conductivity (up to 500 °C)[2]	W/(m K)	41.9–45.2		45.6–49.0			38.5–45.2				
Magnetic and electrical properties											
Maximum magnetic permeability	µH/m	1822–917		2513			1005–251				
Remanent magnetism	T	0.75–0.74		0.5			0.6–0.9				
Coercive force	A/m	200–360		120			480–1750				
Hysteresis loss (B=1T)	J/m³	1490–840		600			1000–3000				
Electrical resistivity	µΩ m	0.24–0.26		0.30			0.34				

[1] The higher value being obtained at higher temperature.
[2] The value falls with increasing temperature.
Source: References 7–9

5 HEAT TREATMENT (SURFACE HARDENING OF PEARLITIC MALLEABLE CAST IRON)

Purpose:

a) To produce a hard, wear-resistant surface; and
b) To improve fatigue strength.

Recommended heat treatment cycle: Heat the surface to be hardened with an oxy-acetylene flame (as in flame hardening) or by electromagnetic induction (as in induction hardening) to 870–925 °C. Quench the surface with water, soluble-oil mixture or solution of polyvinyl alcohol in water. Temper at 150–200 °C for a minimum of 1 h.

Note: A surface hardness of 55–60 HRC can be expected.

6 TYPICAL APPLICATIONS

Agricultural implements: Hand implements, harrows, plows, rakes, reapers, and spreaders.

Armaments: Automatic rifle parts, gun mounts, machine gun parts, mortar parts, rifle grenades, shell parts, and bomb parts.

Automotive: Automotive transmission parts, bearing caps, brake and clutch pedals, differential carriers, differential cases, disc brake calipers, rocker arms, spring hangers, steering-gear housings, steering knuckle, and wheel hubs.

Hand tools:

Railway: Parts for the construction of locomotives, freight and passenger cars, and track accessories.

7 EQUIVALENT GRADES

TABLE 2.27 Comparison of grades of whiteheart malleable cast iron specified in IS 2107 – 1977 with nearest equivalent grades specified in other national and international standards

IS 2107–77	ASTM	BS 6681–86	DIN 1692–82	GOST	ISO 5922–81	JIS G 5703–88	SAE
WM 410	–	W 40–05	GTW-40-05	–	W 40–05	FCMW 370	–
WM 340	–	W 35–04	GTW-35-04	–	W 35–04	FCMW 330	–

TABLE 2.28 Comparison of grades of blackheart malleable cast iron specified in IS 2108 – 1977 with nearest equivalent grades specified in other national and international standards

IS 2108–77	ASTM A 47M–90	BS 6681–86	DIN 1692–82	GOST 1215–79	ISO 5922–81	JIS G 5703–88	SAE J158–86 ASTM A 602–70
BM 340	22010	B 35–12	GTS-35-10	КЧ35–10	B 35–10	–	M3210
BM 310	–	B 32–10	–	КЧ33–8	B 32–12	FCMB 310	–
BM 290	–	B 30–06	–	КЧ30–6	B 30–06	FCMB 270	–

TABLE 2.29 Comparison of grades of pearlitic malleable cast iron specified in IS 2640 – 1977 with nearest equivalent grades specified in other national and international standards

IS 2640–77	ASTM A 220M–88	BS 6681–86	DIN 1692–82	GOST 1215–79	ISO 5922–81	JIS G 5703–88	SAE J158–86 ASTM A 602–70
PM 690	550M2	P 70–02	GTS-70-02	КЧ70–2	P 70–02	FCMP 690	–
PM 590	410M4	P 60–03	–	КЧ60–3	P 60–03	FCMP 590	–
PM 540	–	P 55–04	GTS-55-04	КЧ55–4	P 55–04	FCMP 540	M5503
PM 490	340M5	P 50–05	–	КЧ50–5	P 50–05	FCMP 490	M5003
PM 440	310M6 310M8	P 45–06	GTS-45-06	КЧ45–7	P 45–06	FCMP 440	M4504

8 REFERENCES

Books

1. ASM Handbook, vol. 1, ASM International, Materials Park, Ohio, U.S.A., 1990.
2. ASM Handbook, vol. 4, ASM International, Materials Park, Ohio, U.S.A., 1991.
3. C.F. Walton and T.J. Opar (Eds.), *Iron Castings Handbook,* 3rd ed., Iron Castings Society, Des Plaines. Illinois, U.S.A., 1981.
4. H.T. Angus, *Cast Iron: Physical and Engineering Properties,* 2nd ed., Butterworths, London, U.K., 1976.
5. G.N.J. Gilbert, *Engineering Data on Malleable Cast Irons—SI Units,* BCIRA, Alvechurch, Birmingham, U.K., 1983.
6. Cast Metals Handbook, 3rd ed., American Foundrymen's Association, Chicago, Illinois, U.S.A., 1944.

Standards

7. IS 2107 – 1977: *Specification for Whiteheart Malleable Iron Castings.*
8. IS 2108 – 1977: *Specification for Blackheart Malleable Iron Castings.*
9. IS 2640 – 1977: *Specification for Pearlitic Malleable Iron Castings.*
10. ISO 5922 – 1981: *Malleable Cast Iron.*

2.9

ABRASION-RESISTANT CAST IRON

1 INTRODUCTION

Abrasion-resistant cast irons are unalloyed, low-alloy and high-alloy white irons in which most of the carbon is present in the combined form as carbides. In the microstructure, these carbides provide the high hardness necessary for crushing and grinding other materials without degradation. The supporting matrix structure may be modified by alloy content and/or heat treatment to develop the most cost effective balance between resistance to abrasive wear and the toughness required to withstand repeated impact loading. The white irons are readily cast to the shapes required for crushing and grinding, or handling of abrasive materials.

The unalloyed and low-alloy (containing up to 2 % chromium) white irons are used in a variety of applications where the abrading material is not fine or where replacement is not frequent. These white irons develop a hardnesses in the range of 350 to 550 HBW.

The high-alloy white irons are nickel-chromium, chromium-molybdenum and high-chromium white irons which display higher abrasion resistance and toughness than the unalloyed and low-alloy grades. These high-alloy white irons develop a hardness in the range of 500 to 700 HBW.

2 SPECIFICATIONS

2.1 Chemical composition

TABLE 2.30 Chemical composition of abrasion-resistant cast iron specified in Indian Standards

Type	Grade	Chemical composition [% (m/m)]							
		C	Si	Mn	Ni	Cr	Mo	P max.	S max.
Low-alloy abrasion-resistant cast iron (IS 7925 – 1976)									
–	1	2.4–3.4	0.5–1.5	0.2–0.8	–	≤2.0	–	0.15	–

(Contd.)

TABLE 2.30 Chemical composition of abrasion-resistant cast iron specified in Indian Standards (continued)

Type	Grade	Chemical composition [% (m/m)]							
		C	Si	Mn	Ni	Cr	Mo	P max.	S max.
Low-alloy abrasion-resistant cast iron (IS 7925 – 1976)									
–	2	2.4–3.4	0.5–1.5	0.2–0.8	–	≤2.0	–	0.50	–
–	3	2.4–3.0	0.5–1.5	0.2–0.8	–	≤2.0	–	0.15	–
Nickel-chromium abrasion-resistant cast iron (IS 4771 – 1985)									
1A	NiLCr 30/500	2.7–3.3	0.3–0.6	0.3–0.6	3.0–5.5	1.5–2.5	≤0.5	0.3	0.15
1A	NiLCr 34/550	3.2–3.6	0.3–0.6	0.3–0.6	3.0–5.5	1.5–2.5	≤0.5	0.3	0.15
1B	NiHCr 27/500	2.5–2.9	1.5–2.2	0.3–0.6	4.0–6.0	8.0–10.0	≤0.5	0.3	0.15
1B	NiHCr 30/550	2.8–3.3	1.5–2.2	0.3–0.6	4.0–6.0	8.0–10.0	≤0.5	0.3	0.15
1B	NiHCr 34/600	3.2–3.6	1.5–2.2	0.3–0.6	4.0–6.0	7.5–9.5	≤0.5	0.3	0.15
Chromium-molybdenum abrasion-resistant cast iron (IS 4771 – 1985)									
2	CrMoHC 34/500	3.1–3.6	0.3–0.8	0.4–0.9	≤0.5	14.0–18.0	2.5–3.5	0.3	0.15
2	CrMoLC 34/500	2.4–3.1	0.3–0.8	0.4–0.9	≤0.5	14.0–18.0	2.5–3.5	0.3	0.15
High-chromium abrasion-resistant cast iron (IS 4771 – 1985)									
3	HCrNi 27/400	2.3–3.0	0.2–1.5	≤1.5	≤1.2	24.0–28.0	≤0.6	0.3	0.15
3	HCr 27/400	2.3–3.0	0.2–1.5	≤1.5	≤0.5	24.0–28.0	≤0.6	0.3	0.15

Notes:
1) For NiLCr (type 1A) and NiHCr (type 1B) grades:
   (a) low-nickel, high-silicon and low-chromium contents are preferred for the smaller sections; and
   (b) high-nickel, low-silicon and high-chromium contents are preferred for the heavier sections.
2) Toughness and resistance to repeated shock increases as carbon content decreases.
3) Resistance to abrasive wear increases as carbon content increases.

2.2 Hardness

TABLE 2.31 Hardness requirements for abrasion-resistant cast iron, specified in Indian Standards, in the sand-cast, hardened and annealed conditions

Type	Grade	Minimum Brinell hardness[1] in the condition		
		Sand-cast HBW	Hardened HBW	Annealed HBW
Low-alloy abrasion-resistant cast iron (IS 7925 – 1976)				
–	1	400	–	–
–	2	400	–	–
–	3		–	≥ 250
Nickel-chromium abrasion-resistant cast iron (IS 4771–1985)				
1A	NiLCr 30/500	500	–	–
1A	NiLCr 34/550	550	–	–
1b	NiHCr 27/500	500	–	–
1B	NiHCr 30/550	550	–	–
1B	NiHCr 34/600	600	–	–
Chromium-molybdenum abrasion-resistant cast iron (IS 4771–1985)				
2	CrMoHC 34/500	500	600	≤ 380
2	CrMoLC 28/500	500	550	≤ 380
High-chromium abrasion-resistant cast iron (IS 4771–1985)				
3	HCrNi 27/400	400	550	–
3	HCr 27/400	400	550	≤ 380

[1] Minimum hardness values up to 50 HBW lower than those specified in this table should be permitted in castings with section thickness greater than 125 mm.

3 TYPICAL PHYSICAL PROPERTIES

TABLE 2.32 Typical physical properties of selected abrasion-resistant white cast irons

Property	Unit	Low-carbon white cast iron	Martensitic nickel-chromium white cast iron
Density	kg/m^3	7600–7800	7600–7800
Coefficient of thermal expansion (at 10–260 °C)	µm/(m K)	12	8–9
Thermal conductivity	W/(m K)	22	30
Electrical resistivity	µΩ m	0.53	0.80
Source: Reference 1			

4 HEAT TREATMENT

4.1 Stress-relief heat treatment of nickel-chromium (type 1A) abrasion-resistant white cast iron

Purpose

a) To relieve transformation stresses;
b) To transform much of the retained austenite to martensite;
c) To increase strength and impact resistance; and
d) To temper the martensite.

Note: This heat treatment does not reduce hardness or abrasion resistance.

Recommended heat treatment cycle

a) Heat to 200–275 °C. Hold at this temperature for 12–16 h. Air or furnace cool to ambient temperature; or
b) Heat to 425-475 °C. Hold at this tempperature for 4–8 h. Cool to ambient temperature in still air. Reheat to 200-275 °C. Hold at this temperature for 12–16 h. Air or furnace cool to ambient temperature.

4.2 High temperature heat treatment of nickel-chromium (type 1B) abrasion-resistant white cast iron

Purpose: To transform a substantial portion of the retained austenite to martensite for maximum wear resistance.

Recommended heat treatment cycle

a) Heat slowly to 775–825 °C. Hold at this temperature for 6-12 h. Air or furnace cool to ambient temperature; or
b) Heat slowly to 775–825 °C. Hold at this temperature for 6-12 h. Air or furnace cool to ambient temperature. Temper at 200-350 °C for 4 h and air or furnace cool to ambient temperature

Note: This heat treatment cycle is recommended where a maximum resistance to repeated impact is required.

4.3 Annealing of chromium-molybdenum (type 2) and high-chromium (type 3) abrasion-resistant cast irons

Purpose: To improve machinability.

Recommended heat treatment cycle: Heat slowly to 920–975 °C. Hold at this temperature for a minimum of 1 h. Furnace cool to 810–815 °C and then down to 590-645 °C at a rate not exceeding 40 °C/h. Subsequent cooling to ambient temperature may be in the furnace or in still air.

4.4 Hardening and tempering of chromium-molybdenum (type 2) and high-chromium (type 3) abrasion-resistant white cast irons

Purpose: To improve abrasion resistance and toughness.

Recommended heat treatment cycle

a) **Hardening**: Heat slowly to 950-1010 °C for type 2 irons and 980-1120 °C for type 3 irons. Hold at this temperature for 1 h per 25 mm of maximum section thickness, or a minimum of 1 h. Air cool to ambient temperature or quench in oil.
b) **Tempering after hardening**: Heat slowly to 200–550 °C. Hold at this temperature for a sufficient length of time, appropriate to the section size. Air or furnace cool to ambient temperature.

5 TYPICAL APPLICATIONS

Unalloyed and low-alloy abrasion-resistant white irons: Breaker bars, grinding balls, liner plates, pulleys, and sand or slurry pumps.

High-alloy abrasion-resistant white irons: Used widely in the power, cement, mining, paint and other industries for pulverizing, crushing and grinding, and for various other uses including earth-moving equipment, gravel and slurry pumps, shot- blasting machines, chutes and sheaves for pulleys involving low- and high-stress abrasion. Typical applications include grinding balls, metal-working rolls, mill liners, pipe and elbows, pulverizer rings and roll heads, slurry pump parts, and wearbacks.

6 EQUIVALENT GRADES

TABLE 2.33 Comparison of grades of low-alloy abrasion-resistant white cast iron specified in IS 7925 – 1976 with nearest equivalent grades specified in other national standards

IS 7925–76	BS 4844–86
1	1A
2	1B
3	1C

TABLE 2.34 Comparison of grades of nickel-chromium, chromium-molybdenum and high-chromium abrasion-resistant white cast iron specified in IS 4771–1985 with nearest equivalent grades specified in other national standards

IS 4771–85	ASTM A 532/A 532M–93a	BS 4844–86	DIN 1695–81	Trade name
Type 1A, NiLCr 30/500	Class I Type B Ni-Cr-Lc	2A	G-X 260 NiCr 4 2	Ni-Hard 2
Type 1A, NiLCr 34/550	Class I Type A Ni-Cr-Hc	2B	G-X 330 NiCr 4 2	Ni-Hard 1
Type 1B, NiHCr 27/500	Class I Type D Ni-HiCr	2C	G-X 300 CrNiSi 9 5 2	Ni-Hard 4
Type 1B, NiHCr 30/550	Class I Type D Ni-HiCr	2D	G-X 300 CrNiSi 9 5 2	Ni-Hard 4
Type 1B, NiHCr 34/600	Class I Type D Ni-HiCr	2E	G-X 300 CrNiSi 9 5 2	Ni-Hard 4
Type 2, CrMoHC 34/500	Class II Type B 15% Cr-Mo	3B	G-X 300 CrMo 15 3	Alloy 15-3
Type 2, CrMoLC 28/500	Class II Type B 15% Cr-Mo	3A	G-X 300 CrMo 15 3	Alloy 15-3
Type 3, HCrNi 27/400	Class III Type A 25% Cr	3D	G-X 260 Cr 27	–
Type 3, HCr 27/400	–	–	–	–

7 REFERENCES

Books

1. *ASM Handbook*, vol. 1, ASM International, Materials Park, Ohio, U.S.A., 1990.

2. C.F. Walton and T.J. Opar (Eds.), *Iron Castings Handbook,* 3rd ed., Iron Castings Society, Des Plaines, Illinois, U.S.A., 1981.
3. R. Elliot, *Iron Casting Technology,* Butterworths, London, U.K., 1988.

Standards

4. IS 7925–1976: *Specification for Low Alloy Types of Abrasion-Resistant Iron Castings.*
5. IS 4771–1985: *Specification for Abrasion-Resistant Iron Castings.*

2.10

AUSTENITIC CAST IRON

1 INTRODUCTION

Austenitic cast irons are high-alloy cast irons in which the metallic matrix is rendered austenitic at ambient temperature by the addition of alloying elements, principally nickel, and in which the carbon is present predominantly as either flake graphite or spheroidal graphite. These cast irons exhibit outstanding resistance to many types of corrosion and erosion, and to the effects of high temperature. By varying the chemical composition austenitic cast irons can be produced with a low coefficient of thermal expansion, non-magnetic properties and good low temperature impact properties. When compared with the corrosion- and heat-resistant steels, austenitic cast irons have superior castability and machinability.

2 SPECIFICATION

2.1 *Flake graphite austenitic cast iron*

TABLE 2.35 Chemical composition and mechanical properties of flake graphite austenitic cast iron, conforming to IS 2749 – 1974

Grade	Chemical composition[1)] [% (m/m)]						Mechanical property[2)]	
	C	Si	Mn	Ni	Cr	Cu	R_m min. MPa	Maximum Brinell hardness HB
AFG Ni 13 Mn 7	≤3.0	1.5–3.0	6.0–7.0	12.0–14.0	â 0.2	≤0.5	140	150
AFG Ni 15 Cu 6 Cr 2	≤3.0	1.0–2.8	0.5–1.5	13.5–17.5	1.0–2.5	5.5–7.5	170	200
AFG Ni 15 Cu 6 Cr 3	≤3.0	1.0–2.8	0.5–1.5	13.5–17.5	2.5–3.5	5.5–7.5	190	250
AFG Ni 20 Cr 2	≤3.0	1.0–2.8	0.5–1.5	18.0–22.0	1.0–2.5	≤0.5	170	215

(Contd.)

TABLE 2.35 Chemical composition and mechanical properties of flake graphite austenitic cast iron, conforming to IS 2749–1974 (continued)

Grade	Chemical composition[1)] [% (m/m)]						Mechanical property[2)]	
	C	Si	Mn	Ni	Cr	Cu	R_m min. MPa	Maximum Brinell hardness HB
AFG Ni 20 Cr 3	≤3.0	1.0–2.8	0.5–1.5	18.0–22.0	2.5–3.5	≤0.5	190	250
AFG Ni 20 Si 5 Cr 3	≤2.5	4.5–5.5	0.5–1.5	18.0–22.0	1.5–4.5	≤0.5	190	250
AFG Ni 30 Cr 3	≤2.5	1.0–2.0	0.5–1.5	28.0–32.0	2.5–3.5	≤0.5	190	215
AFG Ni 30 Si 5 Cr 5	≤2.5	5.0–6.0	0.5–1.5	29.0–32.0	4.5–5.5	≤0.5	170	210
AFG Ni 35	≤2.4	1.0–2.0	0.5–1.5	34.0–36.0	≤0.2	≤0.5	120	140

1) Unless otherwise specified, other elements may be present at the discretion of the manufacturer, provided they do not alter the microstructure substantially, or affect the properties adversely.
2) Measured on test pieces machined from separately cast test samples.

2.2 Spheroidal graphite austenitic cast iron

TABLE 2.36 Chemical composition of spheroidal graphite austenitic cast iron, conforming to IS 2749 – 1974

Grade	Chemical composition[1)] [% (m/m)]						
	C max.	Si	Mn	Ni	Cr	P max.	Cu max.
ASG Ni 13 Mn 7	3.0	2.0–3.0	6.0–7.0	12.0–14.0	≤0.2	0.080	0.5
ASG Ni 20 Cr 2	3.0	1.5–3.0	0.5–1.5	18.0–22.0	1.0–2.5	0.080	0.5
ASG Ni 20 Cr 3	3.0	1.5–3.0	0.5–1.5	18.0–22.0	2.5–3.5	0.080	0.5
ASG Ni 20 Si 5 Cr 2	3.0	4.5–5.5	0.5–1.5	18.0–22.0	1.0–2.5	0.080	0.5
ASG Ni 22	3.0	1.0–3.0	1.5–2.5	21.0–24.0	≤0.5	0.080	0.5

(Contd.)

TABLE 2.36 Chemical composition of spheroidal graphite austenitic cast iron, conforming to IS 2749 – 1974 (continued)

Grade	Chemical composition[1)] [% (*m*/*m*)]						
	C max.	Si	Mn	Ni	Cr	P max.	Cu max.
ASG Ni 23 Mn 4	2.6	1.5–2.5	4.0–4.5	22.0–24.0	≤0.2	0.080	0.5
ASG Ni 30 Cr 1	2.6	1.5–3.0	0.5–1.5	28.0–32.0	1.0–1.5	0.080	0.5
ASG Ni 30 Cr 3	2.6	1.5–3.0	0.5–1.5	28.0–32.0	2.5–3.5	0.080	0.5
ASG Ni 30 Si 5 Cr 5	2.6	5.0–6.0	0.5–1.5	28.0–32.0	4.5–5.5	0.080	0.5
ASG Ni 35	2.4	1.5–3.0	0.5–1.5	34.0–36.0	≤0.2	0.080	0.5
ASG Ni 35 Cr 3	2.4	1.5–3.0	0.5–1.5	34.0–36.0	2.0–3.0	0.080	0.5

[1)] Unless otherwise specified, other elements may be present at the discretion of the manufacturer, provided they do not alter the microstructure substantially, or affect the properties adversely.

TABLE 2.37 Mechanical properties[1)] of spheroidal graphite austenitic cast iron, conforming to IS 2749 – 1974

Grade	R_m min.	$R_{p0.2}$ min.	*A* min.	Minimum mean impact value from 3 tests		Maximum Brinell hardness
				V-notch (Charpy)	U-notch (Mesnager)	
	MPa	MPa	%	J	J	HB
ASG Ni 13 Mn 7	390	210	15	16	not indicated	170
ASG Ni 20 Cr 2	370	210	7	13	16	200
ASG Ni 20 Cr 3	390	210	7	not indicated	not indicated	255
ASG Ni 20 Si 5 Cr 2	370	210	10	not indicated	not indicated	230
ASG Ni 22	370	170	20	20	24	170

(Contd.)

TABLE 2.37 Mechanical properties[1)] of spheroidal graphite austenitic cast iron, conforming to IS 2749 – 1974 (continued)

Grade	R_m min.	$R_{p0.2}$ min.	A min.	Minimum mean impact value from 3 tests		Maximum Brinell hardness
				V-notch (Charpy)	U-notch (Mesnager)	
	MPa	MPa	%	J	J	HB
ASG Ni 23 Mn 4	440	210	25	24	28	180
ASG Ni 30 Cr 1	370	210	13	not indicated	not indicated	190
ASG Ni 30 Cr 3	370	210	7	not indicated	not indicated	200
ASG Ni 30 Si 5 Cr 5	390	240	not indicated	not indicated	not indicated	250
ASG Ni 35	370	210	20	not indicated	not indicated	180
ASG Ni 35 Cr 3	370	210	7	not indicated	not indicated	190

1) Measured on test pieces machined from separately cast test samples.

3 SUPPLEMENTARY DATA ON MECHANICAL AND PHYSICAL PROPERTIES, AND TYPICAL APPLICATIONS

3.1 Flake graphite austenitic cast iron

TABLE 2.38 Mechanical properties of flake graphite austenitic cast iron, conforming to IS 2749–1974

Grade	R_m	R_{cm}	A	E	Brinell hardness
	MPa	MPa	%	GPa	HB
AFG Ni 13 Mn 7	140–220	630–840	not indicated	70–90	120–150
AFG Ni 15 Cu 6 Cr 2	170–210	700–840	2	85–105	140–200
AFG Ni 15 Cu 6 Cr 3	190–240	860–1100	1–2	98–113	150–250

(Contd.)

TABLE 2.38 Mechanical properties of flake graphite austenitic cast iron, conforming to IS 2749 – 1974 (continued)

Grade	R_m MPa	R_{cm} MPa	A %	E GPa	Brinell hardness HB
AFG Ni 20 Cr 2	170–210	700–840	2–3	85–105	120–215
AFG Ni 20 Cr 3	190–240	860–1100	1–2	98–113	160–250
AFG Ni 20 Si 5 Cr 3	190–280	860–1100	2–3	110	140–250
AFG Ni 30 Cr 3	190–240	700–910	1–3	98–113	120–215
AFG Ni 30 Si 5 Cr 5	170–240	560	not indicated	105	150–210
AFG Ni 35	120–180	560–700	1–3	74	120–140

TABLE 2.39 Physical properties of flake graphite austenitic cast iron, conforming to IS 2749 – 1974

Grade	Nominal density kg/m^3	Coefficient of thermal expansion between 20 °C and 200 °C µm/(m K)	Thermal Conductivity W/(m K)	Specific heat capacity J/(kg K)	Specific electrical resistance Ω mm^2/m	Relative permeability µ (where H = 8 kA/m)
AFG Ni 13 Mn 7	7300	17.7	37.7–41.9	460–500	1.4	1.02
AFG Ni 15 Cu 6 Cr 2	7300	18.7	37.7–41.9	460–500	1.6	1.03
AFG Ni 15 Cu 6 Cr 3	7300	18.7	37.7–41.9	460–500	1.1	1.05
AFG Ni 20 Cr 2	7300	18.7	37.7–41.9	460–500	1.4	1.04
AFG Ni 20 Cr 3	7300	18.7	37.7–41.9	460–500	1.2	1.04
AFG Ni 20 Si 5 Cr 3	7300	18.0	37.7–41.9	460–500	1.6	1.1
AFG Ni 30 Cr 3	7300	12.4	37.7–41.9	460–500	not indicated	not indicated
AFG Ni 30 Si 5 Cr 5	7300	14.6	37.7–41.9	460–500	1.6	2
AFG Ni 35	7300	5.0	37.7–41.9	460–500	not indicated	not indicated

TABLE 2.40 Properties and typical applications of flake graphite austenitic cast iron, conforming to IS 2749 – 1974

Grade	Properties	Typical applications
AFG Ni 13 Mn 7	Non-magnetic.	Pressure covers for turbine generator sets, housings for switchgear, insulator flanges, terminals, and ducts.
AFG Ni 15 Cu 6 Cr 2	Good resistance to corrosion, particularly in alkalis, dilute acids, sea water and salt solutions. Good heat resistance, good bearing properties, high thermal expansion, non-magnetic at low chromium contents.	Pumps, valves, furnace components, bushings, and piston ring carriers for light alloy metal pistons.
AFG Ni 15 Cu 6 Cr 3	Better corrosion and erosion resistance than AFG Ni 15 Cu 6 Cr 2.	
AFG Ni 20 Cr 2	Similar to AFG Ni 15 Cu 6 Cr 2, but more corrosion resistant to alkalis. High coefficient of thermal expansion.	As for AFG Ni 15 Cu 6 Cr 2, but preferred for pumps handling alkalis, vessels for caustic alkalis, uses in the soap, food, artificial silk and plastic industries. Suitable where copper-free materials are required.
AFG Ni 20 Cr 3	Similar to AFG Ni 20 Cr 2, but more resistant to erosion, heat and growth.	As for AFG Ni 20 Cr 2, but preferred also for high temperature applications.
AFG Ni 20 Si 5 Cr 3	Good resistance to corrosion, even to dilute sulphuric acid. More heat resistant than AFG Ni 20 Cr 2 and AFG Ni 20 Cr 3. This grade is not suitable for use in the temperature range 500 to 600 °C.	Pump components, valves, castings for industrial furnaces.
AFG Ni 30 Cr 3	Resistant to heat and thermal shock up to 800°C. Good corrosion resistance at high temperatures; excellent erosion resistance in wet steam and salt slurry; average thermal expansion.	Pumps, pressure vessels, valves, filter parts, exhaust gas manifolds, turbocharger housings.
AFG Ni 30 Si 5 Cr 5	Particularly resistant to corrosion, erosion and heat; average thermal expansion.	Pump components, valves, castings for industrial furnaces.
AFG Ni 35	Resistant to thermal shock; low thermal expansion.	Such applications as parts with dimensional stability (for example, machine tools), scientific instruments, glass moulds.

3.2 Spheroidal graphite austenitic cast iron

TABLE 2.41 Mechanical properties of spheroidal graphite austenitic cast iron, conforming to IS 2749 – 1974

Grade	R_m MPa	$R_{p0.2}$ MPa	A %	E GPa	KV J	Brinell hardness HB
ASG Ni 13 Mn 7	390–460	210–260	15–25	140–150	15.0–27.5	130–170
ASG Ni 20 Cr 2	370–470	210–250	7–20	112–130	13.5–27.5	140–200
ASG Ni 20 Cr 3	390–490	210–260	7–15	112–133	12.0	150–255
ASG Ni 20 Si 5 Cr 2	370–430	210–260	10–18	112–133	14.9	180–230
ASG Ni 22	370–440	170–250	20–40	85–112	20.0–33.0	130–170
ASG Ni 23 Mn 4	440–470	210–240	25–45	120–140	24.0	150–180
ASG Ni 30 Cr 1	370–440	210–270	13–18	112–130	17.0	130–190
ASG Ni 30 Cr 3	370–470	210–260	7–18	92–105	8.5	140–200
ASG Ni 30 Si 5 Cr 5	390–490	240–310	1–4	91	3.9–5.9	170–250
ASG Ni 35	370–410	210–240	20–40	112–140	20.5	130–180
ASG Ni 35 Cr 3	370–440	210–290	7–10	112–123	7.0	140–190

TABLE 2.42 Physical properties of spheroidal graphite austenitic cast iron, conforming to IS 2749 – 1974

Grade	Nominal density kg/m^3	Coefficient of thermal expansion between 20 °C and 200 °C μm/(m K)	Thermal conductivity W/(m K)	Specific electrical resistance $\Omega . mm^2/m$	Relative permeability μ (where H = 8 kA/m)
ASG Ni 13 Mn 7	7300	18.2	12.6	1.0	1.02
ASG Ni 20 Cr 2	7400	18.7	12.6	1.0	1.04

(Contd.)

TABLE 2.42 Physical properties of spheroidal graphite austenitic cast iron, conforming to IS 2749–1974 (continued)

Grade	Nominal density kg/m^3	Coefficient of thermal expansion between 20 °C and 200 °C μm/(m K)	Thermal conductivity W/(m K)	Specific electrical resistance Ω . mm^2/m	Relative permeability μ (where H = 8 kA/m)
ASG Ni 20 Cr 3	7400	18.7	12.6	1.0	1.05
ASG Ni 20 Si 5 Cr 2	7400	18.0	12.6	not indicated	not indicated
ASG Ni 22	7400	18.4	12.6	1.0	1.02
ASG Ni 23 Mn 4	7400	14.7	12.6	not indicated	not indicated
ASG Ni 30 Cr 1	7400	12.6	12.6	not indicated	not indicated
ASG Ni 30 Cr 3	7400	12.6	12.6	not indicated	not indicated
ASG Ni 30 Si 5 Cr 5	7400	14.4	12.6	not indicated	not indicated
ASG Ni 35	7600	5	12.6	not indicated	not indicated
ASG Ni 35 Cr 3	7600	5	12.6	not indicated	not indicated

TABLE 2.43 Properties and typical applications of spheroidal graphite austenitic cast iron, conforming to IS 2749 – 1974

Grade	Properties	Typical applications
ASG Ni 13 Mn 7	Non-magnetic.	Pressure covers for turbine generator sets, housings for switchgear, insulator flanges, terminals, and ducts.
ASG Ni 20 Cr 2	Similar to AFG Ni 20 Cr 2 in relation to composition, corrosion and heat resistance.	Pumps, valves, compressors, bushings, turbosupercharger housings, exhaust gas manifolds.
ASG Ni 20 Cr 3	Similar to ASG Ni 20 Cr 2, but better erosion and heat resistance.	

(Contd.)

TABLE 2.43 Properties and typical applications of spheroidal graphite austenitic cast iron, conforming to IS 2749–1974 (continued)

Grade	Properties	Typical applications
ASG Ni 20 Si 5 Cr 2	Good resistance to corrosion even to dilute sulphuric acid. Good heat resistance. This grade is not suitable for use in the temperature range 500 to 600 °C.	Pump components, valves, castings for industrial furnaces subject to high mechanical stress.
ASG Ni 22	High coefficient of thermal expansion; lower corrosion and heat resistance than ASG Ni 20 Cr 2. Good impact properties down to –100 °C. Non-magnetic.	Pumps, valves, compressors, bushings, turbosupercharger housings, exhaust gas manifolds.
ASG Ni 23 Mn 4	Good impact properties down to –196 °C. Non-magnetic.	Castings for refrigeration engineering for use down to –196 °C (see Table 2.45).
ASG Ni 30 Cr 1	Similar to ASG Ni 30 Cr 3. Good bearing properties.	Pumps, boilers, filter parts, exhaust gas manifolds, valves, turbosupercharger housings.
ASG Ni 30 Cr 3	Similar to AFG Ni 30 Cr 3.	Pumps, boilers, valves, filter parts, exhaust gas manifolds, turbosupercharger housings.
ASG Ni 30 Si 5 Cr 5	Properties similar to AFG Ni 30 Si 5 Cr 5.	Pump components, valves, castings for industrial furnaces subject to high mechanical stress.
ASG Ni 35	Lower thermal expansion similar to AFG Ni 35, but more resistant to thermal shock.	Parts with dimensional stability (for example, machine tools), scientific instruments, glass moulds.
ASG Ni 35 Cr 3	Similar to ASG Ni 35.	Parts of gas turbine housings, glass moulds.

TABLE 2.44 Typical mechanical properties of ASG Ni 23 Mn 4, conforming to IS 2749 – 1974, at low temperatures

Temperature °C	R_m MPa	$R_{p0.2}$ MPa	A %	Z %	KV J
+20	450	220	35	32	29
0	450	240	35	32	31
–50	460	260	38	35	32
–100	490	300	40	37	34
–150	530	350	38	35	33
–183	580	430	33	27	29
–196	620	450	27	25	27

4 DESIGN STRESSES

TABLE 2.45 Recommended maximum tensile design stresses for austenitic cast iron, conforming to IS 2749–1974, in the temperature range 20 to 675 °C

Grade	Maximum tensile design stress, in MPa, at a temperature, in °C, of																	
	20	50	100	150	200	250	300	350	400	450	500	525	550	575	600	625	650	675
Flake graphite austenitic cast iron																		
AFG Ni 15 Cu 6 Cr 2	54	53	49	48	46	43	42	40	37	36	–	–	–	–	–	–	–	–
AFG Ni 20 Cr 2 (Cr = 1%)	49	48	45	43	42	39	37	36	32	31	–	–	–	–	–	–	–	–
AFG Ni 20 Cr 2 (Cr = 2%)	57	56	54	51	49	48	45	43	42	39	–	–	–	–	–	–	–	–
AFG Ni 20 Cr 3 (Cr = 3%)	69	68	65	63	62	59	57	56	53	51	–	–	–	–	–	–	–	–
AFG Ni 20 Si 5 Cr 3	73	69	68	66	63	62	60	57	56	54	–	–	–	–	–	–	–	–
AFG Ni 30 Cr 3	66	63	62	60	57	56	54	51	49	48	–	–	–	–	–	–	–	–

(Contd.)

TABLE 2.45 Recommended maximum tensile design stresses for austenitic cast iron, conforming to IS 2749 – 1974, in the temperature range 20 to 675 °C (continued)

Grade	Maximum tensile design stress, in MPa, at a temperature, in °C, of																	
	20	50	100	150	200	250	300	350	400	450	500	525	550	575	600	625	650	675
Spheroidal graphite austenitic cast iron																		
ASG Ni 20 Cr 2 (Cr = 1%)	80	77	74	74	71	69	68	65	63	62	–	–	–	–	–	–	–	–
ASG Ni 20 Cr 2 (Cr = 2%)	86	85	83	80	79	62	74	73	69	68	–	–	–	–	–	–	–	–
ASG Ni 20 Cr 3 (Cr = 3%)	91	90	88	85	83	82	79	77	76	73	49	40	32	25	20	17	12	11
Spheroidal graphite austenitic cast iron																		
ASG Ni 20 Si 5 Cr 2	91	90	86	85	83	80	79	77	74	73	36	29	23	19	14	11	9	8
ASG Ni 22	80	77	76	74	71	69	68	65	63	62	36	29	23	19	14	11	9	8
ASG Ni 23 Mn 4	77	76	74	71	69	68	66	63	62	59	–	–	–	–	–	–	–	–
ASG Ni 30 Cr 3	–	85	83	82	79	77	76	73	71	69	49	40	32	25	20	17	12	11
ASG Ni 30 Si 5 Cr 5	No results. Take as for ASG Ni 20 Si 5 Cr 2.										36	29	23	19	14	11	9	8
Basis for design stress	Based on 0.45 × 0.1 % proof stress value.										Based on one-third stress required to produce rupture in 100 000 h.							
Source: Reference 4																		

5 HEAT TREATMENT

5.1 Stress relieving

Purpose: To relieve residual stresses resulting from casting, machining, or both.

Recommended heat treatment cycle: Heat to 625–650 °C at a rate not exceeding 150 °C/h. Hold at this temperature for 2 h plus 1 h per 25 mm of maximum section thickness. Furnace cool to 200 °C at a rate not exceeding 100 °C/h, and continue cooling to ambient temperature in still air.

5.2 Annealing

Purpose: To soften castings whose hardness is higher than desired, as may occur through excessive carbide formation (particularly in thin sections and in rapidly cooled sections).

Recommended heat treatment cycle: Heat to 980–1040 °C. Hold at this temperature for 0.5–5 h. Furnace cool to ambient temperature in still air.

5.3 Heat treatment for high temperature stability

Purpose: To stabilize the microstructure and minimize growth and distortion in austenitic iron castings intended for either static or cyclic elevated temperature service at 500 °C or above.

Note: It is usually advisable to stabilize castings prior to final machining.

Recommended heat treatment cycle: Heat to 875–900 °C at a rate not exceeding 150 °C/h. Hold at this temperature for 2 h plus 1 h per 25 mm of maximum section thickness. Furnace cool to 500 °C at a rate not exceeding 50 °C/h, and continue cooling to ambient temperature in still air.

Note: For certain critical components, this heat treatment can be followed by a stress relief after rough machining. It should be noted that copper containing austenitic cast iron grade AFG Ni 15 Cu 6 Cr 2 is not amenable to high temperature structural stabilizing treatment.

5.4 Heat treatment for dimensional stability

Purpose: To ensure dimensional stability in austenitic iron castings, such as those used in precision machinery or scientific instruments.

Recommended heat treatment cycle: Heat to 870 °C. Hold at this temperature for 2 h plus 1 h per 25 mm of maximum section thickness. Furnace cool to 540 °C at a rate not exceeding 50 °C/h. Hold at 540 °C for 1 h per 25 mm of maximum section thickness, and cool to ambient temperature in still air.

After rough machining reheat to 455–480 °C. Hold at this temperature for 1 h per 25 mm of maximum section thickness, and cool to ambient temperature in still air. Finish machine and once again reheat to 260–315 °C, and cool uniformly to ambient temperature in still air.

6 EQUIVALENT GRADES

TABLE 2.46 Comparison of grades of flake graphite austenitic cast iron specified in IS 2749 – 1974 with nearest equivalent grades specified in other national and international standards

IS 2749–74	ASTM A 436–84	BS 3468–86	DIN 1694–81	ISO 2892–73	JIS G 5510–87	Trade name
AFG Ni 13 Mn 7	–	–	GGL-NiMn 13 7	L - Ni Mn 13 7	FCA-NiMn 13 7	–
AFG Ni 15 Cu 6 Cr 2	Type 1	F1	GGL-NiCuCr 15 6 2	L - Ni Cu Cr 15 6 2	FCA-NiCuCr 15 6 2	Ni-Resist 1
AFG Ni 15 Cu 6 Cr 3	Type 1b	–	GGL-NiCuCr 15 6 3	L - Ni Cu Cr 15 6 3	FCA-NiCuCr 15 6 3	Ni-Resist 1b
AFG Ni 20 Cr 2	Type 2	F2	GGL-NiCr 20 2	L - Ni Cr 20 2	FCA-NiCr 20 2	Ni-Resist 2
AFG Ni 20 Cr 3	Type 2b	–	GGL-NiCr 20 3	L - Ni Cr 20 3	FCA-NiCr 20 3	Ni-Resist 2b
AFG Ni 20 Si 5 Cr 3	–	–	GGL-NiSiCr 20 5 3	L - Ni Si Cr 20 5 3	FCA-NiSiCr 20 5 3	Nicrosilal
AFG Ni 30 Cr 3	Type 3	F3	GGL-NiCr 30 3	L - Ni Cr 30 3	FCA-NiCr 30 3	Ni-Resist 3
AFG Ni 30 Si 5 Cr 5	Type 4	–	GGL-NiSiCr 30 5 5	L - Ni Si Cr 30 5 5	FCA-NiSiCr 30 5 5	Ni-Resist 4
AFG Ni 35	Type 5	–	–	L - Ni 35	FCA-Ni 35	Ni-Resist 5

TABLE 2.47 Comparison of grades of spheroidal graphite austenitic cast iron specified in IS 2749 – 1974 with nearest equivalent grades specified in other national and international standards

IS 2749–74	ASTM A 439–83 ASTM A 571M–84	BS 3468–86	DIN 1694–81	ISO 2892–73	JIS G 5510–87	Trade name
ASG Ni 13 Mn 7	–	S6	GGG-NiMn 13 7	S - Ni Mn 13 7	FCDA-NiMn 13 7	–
ASG Ni 20 Cr 2	Type D-2	S2	GGG-NiCr 20 2	S - Ni Cr 20 2	FCDA-NiCr 20 2	Ni-Resist D-2

(Contd.)

TABLE 2.47 Comparison of grades of spheroidal graphite austenitic cast iron specified in IS 2749 – 1974 with nearest equivalent grades specified in other national and international standards (continued)

IS 2749–74	ASTM A 439–83 ASTM A 571M–84	BS 3468–86	DIN 1694–81	ISO 2892–73	JIS G 5510–87	Trade name
ASG Ni 20 Cr 3	Type D-2B	S2B	GGG-NiCr 20 3	S - Ni Cr 20 3	FCDA-NiCr 20 3	Ni-Resist D-2B
ASG Ni 20 Si 5 Cr 2	–	–	GGG-NiSiCr 20 5 2	S - Ni Si Cr 20 5 2	FCDA-NiSiCr 20 5 2	Nicrosilal Spheronic
ASG Ni 22	Type D-2C	S2C	GGG-Ni 22	S - Ni 22	FCDA-Ni 22	Ni-Resist D-2C
ASG Ni 23 Mn 4	Type D-2M	S2M	GGG-NiMn 23 4	S - Ni Mn 23 4	FCDA-NiMn 23 4	Ni-Resist D-2M
ASG Ni 30 Cr 1	Type D-3A	–	GGG-NiCr 30 1	S - Ni Cr 30 1	FCDA-NiCr 30 1	Ni-Resist D-3A
ASG Ni 30 Cr 3	Type D-3	S3	GGG-NiCr 30 3	S - Ni Cr 30 3	FCDA-NiCr 30 3	Ni-Resist D-3
ASG Ni 30 Si 5 Cr 5	Type D-4	–	GGG-NiSiCr 30 5 5	S - Ni Si Cr 30 5 5	FCDA-NiSiCr 30 5 5	Ni-Resist D-4
ASG Ni 35	Type D-5	–	GGG-Ni 35	S - Ni 35	FCDA-Ni 35	Ni-Resist D-5
ASG Ni 35 Cr 3	Type D-5B	–	GGG-NiCr 35 3	S - Ni Cr 35 3	FCDA-NiCr 35 3	Ni-Resist D-5B

7 REFERENCES

Books

1. *ASM Handbook,* vol. 1, ASM International, Materials Park, Ohio, U.S.A., 1990.
2. *Properties and Applications of Ni-Resist Austenitic Cast Irons,* Inco Europe, London, U.K., 1965.
3. *ASM Handbook,* vol. 4, ASM International, Materials Park, Ohio, U.S.A., 1991.
4. H.T. Angus, *Cast Iron: Physical and Engineering Properties,* 2nd ed., Butterworths, London, U.K., 1976.

Standards

5. IS 2749 – 1974: *Specification for Austenitic Iron Castings.*
6. ISO 2892 – 1973: *Austenitic Cast Iron.*
7. BS 3468 – 1986: *Specification for Austenitic Cast Iron.*
8. DIN 1694 – 1981: *Austenitic Cast Iron.*

2.11

CORROSION-RESISTANT HIGH-SILICON CAST IRON

1 INTRODUCTION

High-silicon cast irons containing between 10 % and 18 % silicon are corrosion-resistant alloys which are extensively used in the chemical industry for processing and transporting highly corrosive fluids even under abrasive conditions. These cast irons are extremely resistant to attack by sulphuric and nitric acids and mixtures of the two at all concentrations and temperatures, and by phosphoric acid at room temperature. However, they have limited usefulness in handling hydrochloric acid and have no useful resistance to hydrofluoric or sulphurous acids. Their resistance to strong hot caustics is also not satisfactory.

High-silicon cast irons have low strength, are extremely hard and brittle, and possess a low resistance to thermal shock. They are difficult to cast and are virtually unmachinable, except by grinding.

2 SPECIFICATION

TABLE 2.48 Chemical composition of corrosion-resistant high-silicon cast iron, conforming to IS 7520 – 1974

Grade	Chemical composition [% (*m*/*m*)]				
	C max.	Si	Mn max.	P max.	S max.
1	1.2	10.00–12.00	0.5	0.25	0.1
2	1.0	14.00–16.00	0.5	0.25	0.1

(Contd.)

TABLE 2.48 Chemical composition of corrosion-resistant high-silicon cast iron, conforming to IS 7520 – 1974 (continued)

Grade	Chemical composition [% (*m*/*m*)]				
	C max.	Si	Mn max.	P max.	S max.
3	0.8	16.00–18.00	0.5	0.25	0.1

Note: Grade 2 is recommended for general applications involving corrosion resistance. Grade 1 has greater tensile strength than Grade 2 but a reduced corrosion resistance. Grade 3 is recommended where greater corrosion resistance is required at the expense of tensile strength.

3 TYPICAL MECHANICAL AND PHYSICAL PROPERTIES

TABLE 2.49 Typical mechanical and physical properties of corrosion-resistant high-silicon cast iron

Tensile strength MPa	Modulus of elasticity GPa	Brinell hardness HB	Impact strength[1)] J	Density kg/m^3	Coefficient of thermal expansion µm/(m K)	Specific heat capacity J/(kg K)	Electrical resistivity µΩ m
93–154	124	450–520	5–8	7000	12.0–15.9	544	0.50

[1)] Measured on 20 mm diameter unnotched test pieces.

Source: Reference 3

4 HEAT TREATMENT (STRESS RELIEVING)

Purpose: To relieve residual stresses in castings.

Recommended heat treatment procedure: Strip the castings from the mould immediately after solidification and, without delay, charge the hot castings into a furnace maintained at a temperature of 600 °C. When the furnace is fully charged, heat the castings uniformly to 750–850 °C. Hold at this temperature for a period ranging from 2 h for small castings of simple form and a maximum thickness of 18 mm, to 8 h for heavy castings of intricate design. Furnace cool to 300 °C and continue cooling to ambient temperature in still air.

5 TYPICAL APPLICATIONS

Pump rotors, agitators, kettles, evaporators, separator towers and Rachid rings, tank outlets, crucibles, insoluble anodes, and pipe and fittings for plumbing in chemical laboratories of hospitals, colleges and industry.

6 EQUIVALENT GRADES

TABLE 2.50 Comparison of grades of corrosion-resistant high-silicon cast iron specified in IS 7520 – 1974 with nearest equivalent grades specified in other national standards

IS 7520–74	ASTM A 518M–92	BS 1591–75	GOST 11849–76
1	–	Si 10	–
2	1	Si 14	ЧС15
3	–	Si 16	ЧС17

7 REFERENCES

Books

1. *ASM Handbook,* vol. 1, ASM International, Materials Park, Ohio, U.S.A. 1990.
2. C.F. Walton and T.J. Opar (Eds.), *Iron Castings Handbook,* 3rd ed., Iron Castings Society, Des Plaines, Illinois, U.S.A., 1981.
3. H.T. Angus, *Cast Iron: Physical and Engineering Properties,* 2nd ed., Butterworths, London, U.K., 1976.
4. *ASM Handbook,* vol. 4, ASM International, Materials Park, Ohio, U.S.A., 1991.

Standard

5. IS 7520 – 1974: *Specification for Corrosion-Resistant High-Silicon Iron Castings.*

3
STEELS

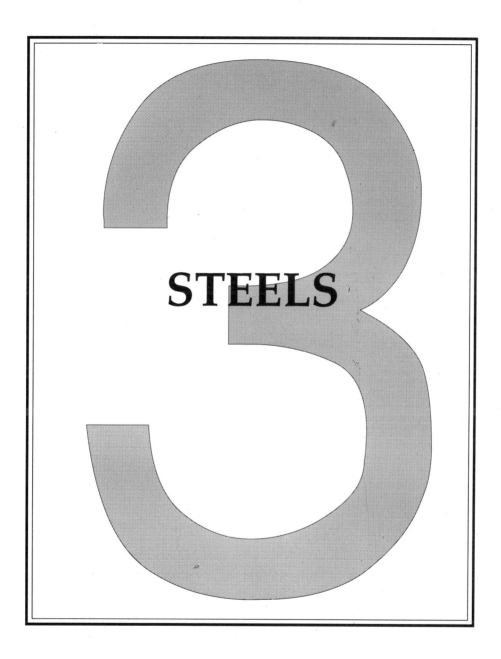

CHAPTER 3

Steels

3.1	Classification of steels	3.5
3.2	Designation system for wrought steels	3.13
3.3	General data on steels	3.17
3.4	Heat treatment of steel	3.29
3.5	Low-carbon steel sheet and strip	3.37
3.6	Steel sheet and strip for porcelain enamelling	3.46
3.7	Metal coated steel sheet and strip	3.50
3.8	Structural steels	3.61
3.9	Steel tubes for structural and mechanical purposes	3.71
3.10	Free-cutting steels	3.81
3.11	Bright bars	3.86
3.12	Steels for cold heading and cold extruding	3.94
3.13	Microalloyed ferritic-pearlitic forging steels	3.112
3.14	Steels for hardening and tempering	3.115
3.15	Case hardening steels	3.132
3.16	Nitriding steels	3.141

3.17	Steels for flame and induction hardening	3.146
3.18	Spring steels	3.159
3.19	Wrought stainless steels	3.185
3.20	Valve steels for internal combustion engines	3.202
3.21	Bearing steels	3.213
3.22	Tool steels	3.221
3.23	Cast steels	3.238

3.1

CLASSIFICATION OF STEELS

1 DEFINITION OF STEEL

Steel is a material with iron as the predominant element, having a carbon content generally less than 2 % and containing other elements. A limited number of chromium steels may have more than 2 % carbon, but 2 % is the usual dividing line between steel and cast iron.

2 CLASSIFICATION OF STEELS ON THE BASIS OF CHEMICAL COMPOSITION

Steels are classified as:

a) unalloyed steels; and
b) alloy steels.

From the different values given for the chemical composition of the steel, the following are taken into account, for classifying it as unalloyed or alloyed:

a) Where a minimum value or range is specified for the cast analysis of the elements given in Table 3.1, the minimum value should be taken for classification;
b) Where the manganese content of the cast analysis is specified as a maximum value only, this maximum value should be taken for classification;
c) Where, for elements other than manganese, only a maximum value is specified for the cast analysis, a value of 0.7 times this maximum value should be taken for classification;
d) Where there is no standard or specification or ordered composition, the cast analysis reported by the manufacturer should be taken for classification; and
e) The results of the product analysis may deviate from those of the cast analysis to an extent permitted by the appropriate product standard. Where the product analysis indicates a value which would place the steel in a class other than that standard, then its inclusion in the class originally intended should, if necessary, be separately and reliably substantiated.

TABLE 3.1 Boundary between unalloyed and alloy steels

Element	Limiting value % (m/m)
Al	0.10
B	0.0008
Bi	0.10
Cr	0.30
Co	0.10
Cu	0.40
Mn	1.65[1)]
Mo	0.08
Ni	0.30
Nb	0.06
Pb	0.40
Se	0.10
Si	0.50
Te	0.10
Ti	0.05
W	0.10
V	0.10
Zr	0.05
Lanthanides (each)	0.05
Other specified elements (except C, P, S and N)	0.05

[1)] If only a maximum is specified for the Mn content of the steel, the boundary should be at 1.80 %.
Source: Reference 2

1.1 Unalloyed steels

Unalloyed steels are those steels in which, for all elements listed in Table 3.1, the percentage of each element taken in accordance with the guidelines given above is less than the boundary value given for the relevant element in Table 3.1.

1.2 Alloy steels

Alloy steels are those steels in which, for any element listed in Table 3.1, the percentage of any element taken in accordance with the guidelines given above is equal to or greater than the value given for the relevant element in Table 3.1.

Depending on the alloy content (exclusive of C, P, S and N), alloy steels are subdivided as shown below:

Subdivision	Total alloying content % (*m*/*m*)
Low-alloy steels	≤ 5
Medium-alloy steels	> 5 ≤ 10
High-alloy steels	> 10

3 CLASSIFICATION OF UNALLOYED AND ALLOY STEELS ACCORDING TO MAIN QUALITY CLASSES AND MAIN PROPERTY OR APPLICATION CHARACTERISTICS

The main classes of steels are characterized by:

a) the main quality class; and
b) the main characteristic of the steel.

Steels are subdivided into the following main quality classes:

a) base steel (applicable to unalloyed steels only);
b) quality steel; and
c) special steel.

Main characteristics are considered to be those characteristics which are applied with a certain priority, for example, in designation systems or for classification of steels.

3.1 Main classes of unalloyed steels

3.1.1 Main quality classes of unalloyed steels

Base steel: The term base steel (also known as regular steel, commercial steel or merchant steel) applies to all steels for which no quality requirement, which would necessitate special care during steel production, is specified. These steels should simultaneously meet the following four conditions:

a) The steel is unalloyed;
b) No heat treatment is specified (annealing or normalizing is not considered as heat treatment);
c) The characteristics, if specified in product standards or specifications, are as follows:

Minimum tensile strength	≤ 690 MPa
Minimum yield strength	≤ 360 MPa
Minimum elongation on $L_0 = 5d$	≤ 26 %
Minimum diameter of bending mandrel	≥ 1 × thickness of test piece
Minimum energy absorbed at 20 °C (on ISO V-notch test piece, taken longitudinally)	≤ 27 J
Maximum Rockwell hardness	≥ 60 HRB
Maximum carbon content	≥ 0.10 %
Maximum phosphorus content	≥ 0.050 %
Maximum sulphur content	≥ 0.050 %
Maximum nitrogen content	≥ 0.007 %

Note: The indicated mechanical characteristics correspond to the range of thicknesses from 3 to 16 mm and apply to test pieces taken in the longitudinal or transverse direction in accordance with the requirements of the relevant standard or specification.

d) No other quality requirement is specified.

Unalloyed quality steel: The term unalloyed quality steel applies to those unalloyed steels which require special care during production (for example, grain size control, decrease of phosphorus and sulphur content, improvement of surface finish or increased production control, etc.) to achieve, in comparison with base steels, special quality characteristics such as improved resistance against brittle fracture, improved cold-forming properties, etc. However, requirements concerning careful production of these steels are less stringent than those for classical unalloyed special steels, i.e., steels with controlled hardenability.

Unalloyed special steel The term unalloyed special steel applies to those steels whose production requires special care comparable in extent to the care necessary for the production of classical special steels, i.e., unalloyed steels with controlled (special) hardenability requirements.

In view of their special manufacturing conditions, special steels are generally cleaner— especially from the point of view of inclusions—than quality steels.

The following unalloyed steels are special steels:

a) All unalloyed steels (including unalloyed free-cutting steels and tool steels) destined for heat treatment for which specific requirements, for at least one of the following characteristics, are to be observed:
   i) Requirements concerning the impact properties in the hardened and tempered or simulated case hardened condition,
   ii) Requirements concerning the hardening depth or surface hardness after hardening, or after hardening and tempering,
   iii) Requirements concerning limitation of surface discontinuities, and

iv) Requirements concerning limitation of the non-metallic inclusion content and/or the internal homogeneity;

b) All unalloyed steels not destined for heat treatment for which at least one of the following requirements is to be observed;

i) Requirements concerning limitation of the non-metallic inclusion content and/ or the internal homogeneity, for example, plates resistant to lamellar tearing,

ii) The maximum phosphorus and/or sulphur content is limited as follows:

- for cast analysis ≤ 0.020 %,
- for product analysis ≤ 0.025 %;

EXAMPLES Certain steels for welding wire, steel for wire in tyres.

iii) The contents of the following residual elements are simultaneously restricted as follows:
Cu max., cast ≤ 0.10 %,
Co max., cast ≤ 0.05 %,
V max., cast ≤ 0.05 %;

iv) The requirements for surface quality are more stringent than those specified in IS 11169 (Parts 1 and 2) for cold-heading and cold-extruding steels.

EXAMPLES Certain cold forging, cold drawing and plating qualities.

c) Steels with a specified electrical conductivity that is greater than or equal to 9 S/m or with specified magnetic properties, except in the case of magnetic sheets and strips for which only the maximum magnetic losses and the minimum magnetic induction, and not, for example, the permeability, is specified.

3.1.2 Main characteristics of unalloyed steels

For unalloyed steels, the classification according to main characteristics used is as follows:

a) Unalloyed steels with maximum values of R_m, R_e, or HB (or maximum diameter of bending mandrel, etc.) as the main characteristic;

EXAMPLE Soft sheet for cold forming.

b) Unalloyed steels with minimum values of R_m or R_e as the main characteristic;

EXAMPLES Structural steels including steels for ships, pipelines and pressure purposes as well as unalloyed steels with improved weather resistance.

c) Unalloyed steels with carbon content as the main characteristic, with the exception of the steels indicated under items d) and e) below;

EXAMPLES Steels for wire rod, steels for hardening and tempering, etc.

d) Unalloyed free-cutting steels (% S min., cast ≥ 0.070 and/or additions of Pb, Bi, Te, Se or P);

e) Unalloyed tool steels;

f) Unalloyed steels with particular specifications for magnetic or electrical properties; and

EXAMPLES Magnetic sheet and strip, steel with permeability requirements for transmitters, telephone wire, etc.

g) Other unalloyed steels.

3.2 Main classes of alloy steels

3.2.1 Main quality classes of alloyed steels

Alloyed quality steel: The term alloyed quality steel applies to steels with low alloy contents which are manufactured in relatively large quantities and according to quality requirements which are, in comparison with those for alloyed special steels, relatively easy to fulfil.

The following alloy steels are quality steels:

a) Structural weldable fine grain steels with high yield strength, which simultaneously meet the following conditions:

i) The specified minimum yield strength is less than 420 MPa (for thicknesses less than or equal to 16 mm),

TABLE 3.2 High yield strength alloyed steels—Limiting contents of alloying elements for alloyed quality steels

Alloying element	Limiting value % (*m*/*m*)
Cr[1)]	0.50
Cu[1)]	0.50
Lanthanides	0.06
Mn	1.80
Mo[1)]	0.10
Ni[1)]	0.50
Nb[2)]	0.08
Ti[2)]	0.12
V[2)]	0.12
Zr[2)]	0.12

(Contd.)

TABLE 3.2 High yield strength alloyed steels—Limiting contents of alloying elements for alloyed quality steels (continued)

1) When two, three or four of these elements are specified together for the steel under consideration, it is necessary to consider simultaneously (a) the limiting contents for each one of these elements, and (b) the limiting content for all these elements which should be taken as equal to 70% of the sum of the limiting contents indicated for each one of the two, three or four elements in question.
2) The rule in note[1)] is also applicable to these elements.
Source: Reference 2

ii) The alloy contents, defined by a minimum value or the lower value of a range, are less than the values given in Table 3.2. If the alloying element is defined by a maximum value only, the class to which it belongs is given by the value corresponding to 70 % of this maximum value;

b) Steels which are alloyed only with copper and have a specified minimum copper content that is greater than or equal to 0.40 % but less than 0.50 %, or, if no minimum value is specified, a specified maximum copper content that is greater than or equal to 0.57 % but less than 0.70 %;

c) Alloy steels for rails;

d) Silico-manganese steels for springs or parts resistant to abrasion with the elements P and S greater than 0.035 %; and

e) Steels for sheets and strips containing only Si and/or Al as an alloying element, and with requirements for magnetic losses as well as for the minimum values of magnetic induction (which means, for example, no requirements of permeability).

Alloyed special steel: All alloy steels, excluding those indicated in the section on alloyed quality steels are special steels. This class includes stainless steels, heat and creep resisting steels, tool steels, bearing steels, engineering steels, special structural steels, and steels with special physical properties.

3.2.2 Main characteristics of alloyed steels

The criteria used for the main characteristics of alloy steels are characteristic applications, properties and/or alloy contents.

4 REFERENCES

Book

1. ASM Handbook, vol. 1, ASM International, Materials Park, Ohio, U.S.A., 1990.

Standard

2. IS 7598 – 1990: *Classification of Steels.*

3.2

DESIGNATION SYSTEM FOR WROUGHT STEELS

1 DESIGNATION SYSTEM

The grades of wrought steel covered by the Indian Standards are designated either on the basis of their mechanical properties or on the basis of their chemical composition.

1.1 Steels designated on the basis of mechanical properties

Unalloyed and low-alloy steels which are primarily characterized by their minimum tensile strength or yield strength are designated as follows, in the order given:

a) by the symbol:
   i) Fe, for steels specified on the basis of the minimum tensile strength, or
   ii) FeE, for steels specified on the basis of the minimum yield strength;
b) by a space; and
c) By a number indicating the minimum tensile strength, in MPa, or the minimum yield strength, in MPa. If the minimum tensile strength or yield strength is not specified the number should be 00.

EXAMPLE Fe 410

1.2 Steels designated on the basis of chemical composition

1.2.1 Unalloyed steels (except free-cutting steels)

The grades of unalloyed steels (except free-cutting steels) with a mean manganese content less than 1%, or greater than or equal to 1% are designated as follows, in the order given:

a) by a number indicating 100 times the specified mean percentage carbon content, rounded off to the nearest whole number. Where the carbon content is not specified by a range, a realistic mean value should be used;
b) by the capital letter C; and
c) by a number indicating 10 times the specified mean percentage manganese content, rounded off to the nearest whole number. Where the manganese content is not specified by a range, a realistic mean value should be used.

EXAMPLE 45C8

1.2.2 Unalloyed tool steels

The grades of unalloyed tool steels are designated as follows, in the order given:

a) by a number indicating 100 times the specified mean percentage carbon content, rounded off to the nearest whole number;
b) by the capital letter T; and
c) by a number indicating 10 times the specified mean percentage manganese content, rounded off to the nearest whole number.

EXAMPLE 80T6

1.2.3 Unalloyed free-cutting steels

The grades of unalloyed free-cutting steels are designated as follows, in the order given:

a) by a number indicating 100 times the specified mean carbon content, rounded off to the nearest whole number. Where the carbon content is not specified by a range, a realistic mean value should be used;
b) by the capital letter C;
c) by a number indicating 10 times the specified mean percentage manganese content, rounded off to the nearest whole number; and
d) by the chemical symbol(s) for the alloying elements each followed by a number indicating its specified mean percentage content, multiplied by the factor given in Table 3.3 and rounded off to the nearest whole number. The sequence of symbols should be in decreasing order of the value of their content.

EXAMPLE 40C10S18

TABLE 3.3 Multiplication factors to be used with the alloy content of low-alloy and medium-alloy steels[1)]

Element	Multiplication factor
Cr, Co, Mn, Ni, Si and W	4
Al, Be, Cu, Mo, Nb, Pb, Ta, Ti, V and Zr	10
Ce, N, P and S	100
B	1000

1) The mean content of the alloying element is multiplied by the factor and rounded off to the nearest whole number.

1.2.4 Low-alloy and medium-alloy steels (total alloying content less than or equal to 10 %)

The grades of low-alloy and medium-alloy steels are designated as follows, in the order given:

a) by a number indicating 10 times the specified mean percentage carbon content, rounded off to the nearest whole number; and
b) by the chemical symbol(s) for the alloying element(s), each followed by a number indicating its specified mean percentage content, multiplied by the factor given in Table 3.3 and rounded off to the nearest whole number. The sequence of symbols should be in decreasing order of the value of their content; where the values of contents are the same for two or more elements, the corresponding symbols should be indicated in alphabetical order.

EXAMPLE 16Mn5Cr4

1.2.5 High-alloy steels (total alloying content greater than 10 %)

The grades of high-alloy steels are designated as follows, in the order given:

a) by the capital letter X;
b) by a number indicating 100 times the specified mean percentage carbon content, rounded off to the nearest whole number. Where the carbon content is not specified by a range, a realistic mean value should be used. A nought (0) should be the first digit if the number is less than 10; and
c) by the chemical symbol(s) for the alloying element(s), each followed by a whole number indicating its specified mean percentage content. The sequence of symbols should be in decreasing order of the value of their content; where the values of contents are the same for two or more elements, the corresponding symbols should be indicated in alphabetical order.

EXAMPLE X02Cr19Ni10

1.2.6 Alloy tool steels

The grades of alloy tool steels are designated as follows, in the order given:

a) by the capital letter(s) T for low-alloy and medium-alloy tool steels, or XT for high-alloy tool steels;
b) by a number indicating 100 times the specified mean percentage carbon content, rounded off to the nearest whole number; and
c) by the chemical symbol(s) for the alloying element(s) each followed by:
    i) For low-alloy and medium alloy steels: A number indicating the specified mean percentage alloy content, multiplied by the factor given in Table 3.3 and rounded off to the nearest whole number,
    ii) For high-alloy steels: A whole number indicating the specified mean percentage alloy content, rounded off to the nearest whole number.

The sequence of symbols should be in decreasing order of the value of their content; where the values of contents are the same for two or more elements, the corresponding symbols should be indicated in alphabetical order.

EXAMPLE T110W6Cr4

2 REFERENCES

Standards

1. IS 1762 (I) – 1974: *Code for Designation of Steels: Part I—Based on Letter Symbols.*

3.3

GENERAL DATA ON STEELS

1 IRON-CARBON EQUILIBRIUM DIAGRAM

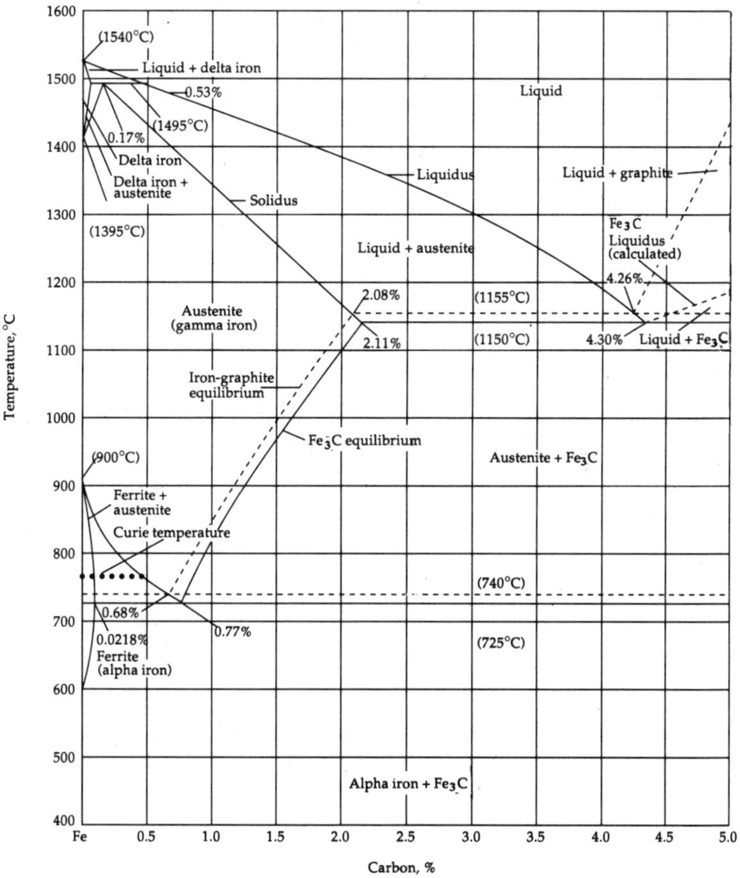

Fig. 3.1 Iron-carbon equilibrium diagram (Source: Reference 1)

2 TERMINOLOGY

2.1 *Chemical analysis*

Cast analysis (ladle analysis): Cast analysis represents the chemical analysis of a cast/heat of steel as reported to the purchaser. It is the analysis for all the specified elements and is determined by analyzing one or more test samples obtained during the pouring of the steel.

Product analysis (check analysis): Product analysis is the chemical analysis of the steel in the semi-finished or finished product form. It is carried out either for the purpose of verifying the average composition of a cast or lot as represented by the cast analysis, or to determine variations in the composition of a cast or lot.

Residual elements (incidental elements): Residual elements are small quantities of certain elements, not quoted in the relevant specification, which are retained from the raw materials used in steelmaking. Through careful steelmaking practices, the amounts of these residual elements are generally held to acceptable levels.

2.2 *Types of steel based on deoxidation practice*

Fully killed (killed) steels: Fully killed steels are fully deoxidized (by silicon, aluminium, or both, or by vacuum treatment of the molten steel) steels in which the evolution of gas during solidification is minimized. These steels are characterized by minimum chemical segregation and uniform mechanical properties. In general, fully killed steels are specified when a homogeneous structure and internal soundness are required, such as in forgings, carburizing steels, or when improved low temperature impact properties are desired. Fully killed steels can be produced either fine grained or coarse grained. Fully killed steels are usually the most expensive to produce.

Semi-killed steels: Semi-killed steels are partially deoxidized steels which do not exhibit the same degree of chemical uniformity or freedom from surface imperfections as fully killed steels. These steels are the most economical to produce and find wide application in plates, bars and structural shapes. Semi-killed steels are usually made in carbon ranges below 0.50 %. Semi-killed steels are not produced to a controlled grain size.

Rimmed steels: Rimmed steels are produced by casting steel that has not been treated for oxygen removal, thus allowing considerable evolution of gas during solidification. Rimming is limited to steels having a carbon content of less than 0.25 %. This type of steel, when properly made, has a minimum of pipe and a smooth, clean, low-carbon, homogeneous surface or "rim" particularly suited for applications where the surface is of prime importance. However, rimmed steels are characterized by pronounced lack of chemical homogeneity, which persists through all rolling operations into the final product. Rimmed steels are more expensive because of the careful control required of the slag composition, pouring temperature, oxygen levels and other variables. However, the amount of pipe is quite small and a high ingot yield is obtained.

2.3 *Fine grain and coarse grain steels*

The austenitic grain size in fully killed steels is classifed as either coarse (grain size of 1 to 5) or fine (grain size of 5 and finer). The grain structure is considered to be satisfactory when 70 % of the grains examined are within the specified size limits.

2.4 Equivalent diameter

The equivalent diameter of any product, or part of a product, is the diameter at the time of heat treatment of a hypothetical very long bar (effectively of infinite length) of uniform circular cross-section which, if subjected to the same cooling conditions as the product (i.e., the same initial and final temperatures, and same cooling medium), would have a cooling rate at its axis equivalent to that at the slowest cooling position in the product or relevant part.

2.5 Ruling section

Ruling section is the equivalent diameter of that portion of the product at the time of heat treatment that is most important in relation to the mechanical properties.

2.6 Limiting ruling section

The limiting ruling section is the largest diameter in which certain specified mechanical properties are achieved after a specified heat treatment.

3 EFFECT OF ELEMENTS IN STEEL

Aluminium (Al)

1. Used as a deoxidizer.
2. Restricts austenite grain growth.
3. Used as an alloying element in nitriding steels.

Boron (B)

1. Significantly increases hardenability.

Carbon (C)

1. Principal hardening element in steel.
2. Ductility and weldability decrease with increasing carbon content.
3. Hardenability increases with increasing carbon content.

Chromium (Cr)

1. Increases hardenability.
2. Increases resistance to corrosion and oxidation.
3. Increases high-temperature strength.
4. Increases abrasion resistance in high-carbon compositions.

Cobalt (Co)

1. Contributes to red hardness by hardening ferrite.
2. Used as an alloying element in certain high-speed steels.

Copper (Cu)

1. Improves resistance to atmospheric corrosion.
2. Detrimental to surface quality and hot-working behaviour.

Lead (Pb)

1. Improves machinability.

Manganese (Mn)

1. Reduces susceptibility to hot shortness.
2. Slightly increases the strength of ferrite.
3. Increases hardenability moderately.

Molybdenum (Mo)

1. Raises the grain coarsening temperature of austenite.
2. Increases hardenability.
3. Minimizes susceptibility to temper embrittlement.
4. Increases resistance to softening in tempering.
5. Increases high-temperature tensile and creep strengths.
6. Enhances corrosion resistance in stainless steel, especially in chloride solutions.

Nickel (Ni)

1. Strengthens unhardened steels by solid solution effects.
2. Promotes toughness in pearlitic-ferritic steels.
3. Produces steels of moderate or high hardenability (depending on other elements).
4. Renders high-chromium ferrous alloys austenitic.

Niobium (Nb)

1. Imparts a fine grain size.
2. Retards tempering.
3. Increases the elevated temperature strength.
4. Produces precipitation hardening in ferritic-pearlitic microalloyed steels.

Phosphorus (P)

1. Increases strength and hardness, but decreases ductility and notch impact toughness.
2. Increases resistance to atmospheric corrosion in low-alloy structural steels.
3. Improves machinability in free-cutting steels.
4. Increases susceptibility of medium-carbon alloy steels to temper embrittlement.

Silicon (Si)

1. Used as a deoxidizer.
2. Slightly increases the strength of ferrite, without causing serious loss of ductility.
3. Used as an alloying element in electrical and magnetic sheet steels.
4. Improves oxidation resistance in several heat-resisting compositions.

Sulphur (S)

1. Lowers transverse ductility and notch impact toughness, but has only a slight effect on longitudinal mechanical properties.
2. Impairs surface quality and weldability.
3. Improves machinability.

Titanium (Ti)

1. Fixes carbon as inert particles.
2. Reduces martensitic hardness and hardenability in medium-carbon steels.
3. Prevents formation of austenite in high-chromium steels.
4. Prevents localized depletion of chromium in stainless steels during long heating periods.

Tungsten (W)

1. Forms hard, abrasion resisting carbide particles in tool steels.
2. Develops high temperature ("red") hardness in hardened and tempered steels.
3. Contributes to creep strength in some high temperature alloys.

Vanadium (V)

1. Elevates coarsening temperature of austenite.
2. Increases hardenability when dissolved.
3. Resists softening in tempering and causes marked secondary hardening.
4. Produces precipitation hardening in ferritic-pearlitic microalloyed steels.

4 TYPICAL PHYSICAL PROPERTIES

TABLE 3.4 Typical physical properties of unalloyed and alloy steels

Steel designation	Treatment condition[1)]	Density at 20 °C	Specific heat capacity between 20 °C and – – – °C					Coefficient of thermal expansion between 20 °C and – – – °C					Thermal conductivity at – – – °C						Electrical resistivity at – – – °C		
			100	200	300	400	500	100	200	300	400	500	20	100	200	300	400	500	20	100	200
		kg/m³	J/(kg K)					μm/(m K)					W/(m K)						μΩ cm		
4C2	A.930	7871	477	498	515	532	557	12.2	13.2	13.5	13.9	14.3	–	60.3	55.7	51.1	46.5	41.0	13.0	17.8	25.2
5C4	A.930	7856	481	498	519	536	557	12.4	13.2	13.7	14.0	14.4	–	57.8	53.6	49.4	44.8	40.2	14.2	19.0	26.3
10C4	–	7840	461	486	507	523	544	–	–	–	–	–	–	–	–	–	–	–	17.3	22.6	29.6
15C8	N	7870	477	–	–	–	–	12	–	13	–	14	–	–	–	–	–	–	–	–	–
20C8	A.930	7860	486	498	515	536	557	12.4	12.8	13.2	13.6	14.0	51.9	51.1	48.6	44.4	42.7	39.4	16.9	21.9	29.2
30C8	N.880	7830	–	502	523	548	569	12.1	12.4	13.2	13.7	14.2	–	–	–	–	–	–	19.5	23.7	31.0
40C8	A.860	7854	481	498	511	532	553	11.3	12.2	13.1	13.7	14.2	51.9	50.7	48.1	45.6	41.9	38.1	17.1	22.1	29.6
55C8	N	7850	–	–	–	–	–	12.7	–	–	–	–	46.1	–	–	–	–	–	17.7	–	–
14C14S14	N.950	7800	–	502	507	519	540	12.45	13.15	13.65	14.05	14.50	–	–	–	–	–	–	21.1	24.2	30.4
11C10S25	As-forged	7820	–	473	502	532	561	12.60	13.05	13.40	13.95	14.45	–	54.4	51.5	47.3	43.1	40.2	15.9	20.6	27.2
40Si7	–	← Similar to that given for 55Si7 →																			
55Si7	A.900	7720	498	507	523	540	561	11.3	12.3	13.1	13.5	13.9	26.0	28.5	30.1	31.0	31.0	31.0	42.9	47.0	52.9
20C15	As-rolled	7840	–	494	507	519	548	12.40	13.05	13.70	14.15	14.55	–	–	–	–	–	–	23.5	27.9	34.8

(Contd.)

TABLE 3.4 Typical physical properties of unalloyed and alloy steels (continued)

Steel design-ation	Treatment condition[1)]	Density at 20°C	Specific heat capacity between 20 °C and - - - °C					Coefficient of thermal expansion between 20 °C and - - - °C					Thermal conductivity at - - - °C						Electrical resistivity at - - - °C		
			100	200	300	400	500	100	200	300	400	500	20	100	200	300	400	500	20	100	200
		kg/m^3	J/(kg K)					µm/(m K)					W/(m K)						µΩ cm		
20C15	As-rolled	7840	–	494	507	519	548	12.40	13.05	13.70	14.15	14.55	–	–	–	–	–	–	23.5	27.9	34.8
27C15	Softened	7860	–	–	–	–	–	–	–	–	–	–	–	36.4	–	–	–	–	21.5	–	–
37C15	OQ.850 + T.600	7880	–	494	507	515	532	12.00	12.95	13.50	13.90	14.3	–	46.1	44.4	42.7	39.8	36.8	23.1	28.2	35.4
35Mn6Mo3	OQ.850 + T.650	7820	–	461	490	519	548	12.3	12.8	13.35	13.8	14.3	–	48.1	45.6	42.3	39.4	36.8	21.6	26.4	33.3
35Mn6Mo4	OQ.850+ T.650	≈7830	–	486	498	515	540	12.70	13.20	13.70	14.15	14.55	–	45.6	44.0	42.3	39.8	36.8	23.5	28.2	35.0
15Cr3	Softened	7860	481	–	–	–	–	12	–	–	–	–	54.4	–	–	–	–	–	16.5	–	–
55Cr3	OQ.830+ T.650	7840	–	465	494	519	548	12.10	12.80	13.45	13.90	14.35	–	48.6	46.1	43.1	39.8	36.8	20.8	25.4	32.6
40Cr4	OQ.870+ T.600	7830	–	473	494	519	540	12.2	12.75	13.3	13.9	14.35	–	44.8	43.5	41.4	37.7	35.2	23.0	28.0	35.2

(Contd.)

TABLE 3.4 Typical physical properties of unalloyed and alloy steels (continued)

Steel design-ation	Treatment condition[1)]	Density at 20°C	Specific heat capacity between 20 °C and – – – °C					Coefficient of thermal expansion between 20 °C and – – – °C					Thermal conductivity at – – – °C						Electrical resistivity at – – – °C		
			100	200	300	400	500	100	200	300	400	500	20	100	200	300	400	500	20	100	200
		kg/m^3	J/(kg K)					μm/(m–K)					W/(m K)						μΩ cm		
50Cr4	OQ.850+ T.600	7820	–	–	–	–	–	–	–	–	–	–	–	–	–	–	–	–	–	–	–
103Cr6	WQ.810	7810	–	–	–	–	–	12.2	–	–	–	–	–	–	–	–	–	–	–	–	–
42Cr4Mo2	OQ.870+ T.600	7830	–	473	507	536	557	11.85	12.4	13.3	13.75	14.0		42.3	41.9	40.2	36.8	34.8	23.6	27.4	33.8
15Cr13 Mo6	Softened	7860	–					12.2	–	12.9	–	13.6	–	–	–	–	–	–	–	–	–
25Cr13 Mo6	Softened	7860	–	–	–	–	–	12.2	–	12.9	–	13.6	–	–	–	–	–	–	30	–	–
50Cr4V2	Softened	7830	481	–	–	–	–	12.0	–	–	–	–	50.2	–	–	–	–	–	23	–	–
40Cr13 Mo10V2	–	7840	486	–	–	–	–	11.6	–	12.8	–	13.6	–	–	–	–	–	–	–	–	–
40Cr7Al10 Mo2	Q + T	7690	477	–	–	–	–	12.7	–	13.4	–	14.0	29.3	–	–	–	–	–	–	–	–
40Ni14	–	7830	–	498	519	544	569	11.6	12.3	12.9	13.4	13.9	–	351.7	360.1	364.3	360.1	339.1	28.0	32.9	40.0
16Ni3Cr2	Softened	7850	494	–	–	–	–	12	–	13	–	14	35.6	–	–	–	–	–	25	–	–
16Ni4Cr3	Softened	7850	494	–	–	–	–	12	–	13	–	14	–	–	–	–	–	–	26	–	–

(Contd.)

TABLE 3.4 Typical physical properties of unalloyed and alloy steels (continued)

Steel designation	Treatment condition[1)]	Density at 20°C	Specific heat capacity between 20 °C and – – – °C					Coefficient of thermal expansion between 20 °C and – – – °C					Thermal conductivity at – – – °C						Electrical resistivity at – – – °C		
			100	200	300	400	500	100	200	300	400	500	20	100	200	300	400	500	20	100	200
		kg/m³	J/(kg K)					μm/(m–K)					W/(m K)						μΩ cm		
13Ni13Cr3	Softened	7870	494	–	–	–	–	12	–	13	–	14	37.7	–	–	–	–	–	–	–	–
15Ni16Cr5	Softened	7880	494	–	–	–	–	12	–	13	–	13	33.5	–	–	–	–	–	–	–	–
35Ni5Cr2	Softened	7800	481	–	–	–	–	11	–	13	–	14	43.5	–	–	–	–	–	–	–	–
30Ni16Cr5	Softened	7870	502	–	–	–	–	11	–	13	–	14	–	–	–	–	–	–	–	–	–
15Ni5Cr4 Mo1	Softened	7850	494	–	–	–	–	12	–	13	–	14	–	–	–	–	–	–	26.0	–	–
15Ni7Cr4 Mo2	Softened	7850	494	–	–	–	–	–	–	–	–	–	–	–	–	–	–	–	–	–	–
16Ni8Cr6 Mo2	Softened	7850	494	–	–	–	–	12	–	13	–	14	–	–	–	–	–	–	–	–	–
40Ni6Cr4 Mo2	Softened	7860	494	–	–	–	–	11	–	13	–	14	–	–	–	–	–	–	–	–	–
40Ni6Cr4 Mo3	OQ830+ T.600	7840	–	–	–	–	–	–	12.4	13.1	13.6	14.0	–	–	–	–	–	–	24.8	29.8	36.7
31Ni10Cr3 Mo6	Softened	7860	–	–	–	–	–	–	–	–	–	–	–	–	–	–	–	–	–	–	–
40Ni10Cr3 Mo6	OQ830+ T.650	7830	–	–	–	–	–	11.5	12.3	13.0	13.6	14.1	–	–	–	–	–	–	27.1	30.6	35.3

1) A = annealed; N = normalized; T = tempered; Q= quenched; OQ = oil quenched.
The numbers following the abbreviations indicate the treatment temperature, in °C.

Source: References 5 – 7

5 MACHINABILITY RATING

TABLE 3.5 Machinability rating of unalloyed and alloy steels

Steel designation	Treatment condition[1)]	Machinability rating %	Steel designation	Treatment condition[1)]	Machinability rating %
20C8	–	100	55Cr3	N	45
30C8	N	80		Q+T	35
	Q+T	70	40Cr4	A	50–55
40C8	N	72		Q+T	40
	Q+T	68	50Cr4	Softened	45
55C8	N	50–60	103Cr6	Softened	40
	Q+T	35–45	42Cr4Mo2	Softened	50
14C14S14	–	150		Q+T	40
11C10S25	–	200	15Cr13Mo6	Softened	35
40Si7	Softened	45	25Cr13Mo6	Softened	35
55Si7	Softened	45	50Cr4V2	Softened	45
20C15	N	70	40Cr13Mo10V2	Softened	30
	Q+T	65	40Cr7Al10Mo2	Softened	45
37C15	N	65		Q+T	35–40
	Q+T	65	40Ni14	Softened	60
35Mn6Mo3	Q+T	40–50		Q+T	45–55
35Mn6Mo4	Q+T	40–50	16Ni3Cr2	Softened	80

(Contd.)

TABLE 3.5 Machinability rating of unalloyed and alloy steels (continued)

Steel designation	Treatment condition[1)]	Machinability rating %
16Ni4Cr3	Softened	75
13Ni13Cr3	Softened	65
15Ni16Cr5	Softened	60
35NiCr2	A	60
	Q+T	40–50
30Ni16Cr5	Softened	40
15Ni5Cr4Mo1	Softened	75
15Ni7Cr4Mo2	Softened	70

Steel designation	Treatment condition[1)]	Machinability rating %
16Ni8Cr6Mo2	Softened	70
40Ni6Cr4Mo2	Softened	55
	Q+T	40
40Ni6Cr4Mo3	Softened	53
	Q+T	35–40
31Ni10Cr3Mo6	A	50
	Q+T	40
40Ni10Cr3Mo6	A	50
	Q+T	20–35

[1)] A = annealed; Q + T = quenched and tempered; N = normalized.

Source: Reference 5 – 7

6 REFERENCES

Books

1. P.M. Unterweiser, H.E. Boyer and J.J. Kubbs (Eds.), *Heat Treater's Guide: Standard Practices and Procedures for Steels*, American Society for Metals, Metals Park, Ohio, U.S.A., 1982.
2. *Metals Handbook*, vol. 1, 8th ed., American Society for Metals, Metals Park, Ohio, U.S.A., 1961.
3. *ASM Handbook*, vol. 1, ASM International, Materials Park, Ohio, U.S.A., 1990.
4. E.C. Bain and H.W. Paxton, *Alloying Elements in Steel*, 2nd ed., American Society for Metals, Metals Park, Ohio, U.S.A., 1966.
5. J. Woolman and R.A. Mottram, *The Mechanical and Physical Properties of the British Standard En Steels*, vol.1, 1st ed., British Iron and Steel Research Association, Pergamon Press, Oxford, U.K., 1964.
6. J. Woolman and R.A. Mottram, *The Mechanical and Physical Properties of the British Standard En Steels*, vol. 2, 1st ed., British Iron and Steel Research Association, Pergamon Press, Oxford, U.K., 1966.

7. J. Woolman and R.A. Mottram, *The Mechanical and Physical Properties of the British Standard En Steels*, vol. 3, 1st ed., British Iron ad Steel Research Association, Pergamon Press, Oxford, U.K., 1968.
8. C.W. Wegst, *Stahlsclüssel*, 16th ed., Verlag Stahlschlüssel Wegst, Marbach, Germany, 1992.

Standards

9. BS 970 (1) – 1991: *Specification for Wrought Steels for Mechanical and Allied Engineering Purposes —* Part 1. *General Inspèction and Testing Procedures and Specific Requirements for Carbon, Carbon Manganese, Alloy and Stainless Steels.*
10. BS 5046 – 1974: *Method for the Estimation of Equivalent Diameters in the Heat Treatment of Steel.*
11. SAE J408 – 1983: *Methods of Sampling Steel for Chemical Analysis.*
12. SAE J411 – 1989: *Carbon and Alloy Steels.*

3.4

HEAT TREATMENT OF STEEL

1 STRESS RELIEVING

Purpose: To relieve stresses that remain locked in a structure as a consequence of a manufacturing sequence.

Recommended heat treatment cycle: Heat uniformly to a suitable temperature below the transformation range (Ac_1 for ferritic steels), namely:

a) 550–650 °C for unalloyed and low-alloy steels; and
b) 600–700 °C for hot-work and high-speed tool steels.

Hold at this temperature for a sufficient length of time to achieve the desired reduction in residual stresses. Cool slowly to ambient temperature to avoid creation of new residual stresses.

2 ANNEALING

2.1 Full annealing

Purpose:

a) To lower strength and hardness, and increase dutility; and
b) To break up continuous carbide networks in high-carbon steels by agglomeration into separated, spherical carbide particles.

Recommended heat treatment cycle: Heat to a temperature of about 30–50 °C above Ac_3 for hypoeutectoid steels, and above Ac_1 for hypereutectoid steels. Soak thoroughly (holding time ≈ 1 h/25 mm of cross-section). Furnace cool through the critical transformation range at a controlled rate and continue cooling to ambient temperature in still air.

2.2 Isothermal annealing

Purpose: To replace full annealing for the following reasons:

a) To reduce the total duration of the annealing operation; and
b) To achieve a more homogeneous structure.

Recommended heat treatment cycle: Heat to a temperature of about 30–50 °C above Ac_3 for hypoeutectoid steels, and above Ac_1 for hypereutectoid steels. Soak thoroughly (holding time ≈

1 h/25 mm of cross-section). Cool rapidly to a temperature below Ar_1 (usually 50–100 °C below Ar_1). Hold at this temperature for a sufficient length of time until transformation of austenite to a relatively soft ferrite-pearlite aggregate is complete. Cool to ambient temperature (the rate of cooling need not be controlled after transformation is complete).

2.3 Spheroidizing

Purpose: To produce a microstructure consisting of spherical carbide particles uniformly dispersed in a ferrite matrix. The spheroidized structure is desirable for low-carbon and medium-carbon steels that are cold formed, and for high-carbon steels that undergo extensive machining prior to final machining.

Recommended heat treatment cycle: Steels may be spheroidized by any one of the following methods:

a) Heat to a temperature just below Ae_1. Hold at this temperature for a sufficient length of time to achieve the desired microstructure;
b) Heat and cool alternately between temperatures that are just above Ac_1 and just below Ar_1; or
c) Heat to a temperature above Ac_1, and then either furnace cool to a temperature just below Ar_1 or cool rapidly to a temperature just below Ar_1 and hold for a sufficient length of time.

2.4 Process annealing

Purpose:

a) To restore ductility to cold worked steel products for subsequent cold deformation; and
b) To soften hot-worked high-carbon and alloy steels for shearing, turning or straightening.

Recommended heat treatment cycle: Heat to a temperature below Ae_1 (usually 10–20 °C below Ae_1). Hold at this temperature for an appropriate length of time. Cool to ambient temperature in still air.

2.5 Softening (soft annealing)

Purpose: To reduce the hardness of a steel workpiece to a given level to facilitate machining.

Recommended heat treatment cycle: Heat to about 650–750 °C. Hold at this temperature for an appropriate length of time. Cool slowly to ambient temperature.

2.6 Homogenizing

Purpose: To eliminate or decrease the chemical segregation existing in steel ingots or castings, by diffusion.

Recommended heat treatment cycle: Heat rapidly to 1100–1200 °C. Hold at this temperature for 8–16 h. Furnace cool to 800–850 °C and continue cooling to ambient temperature in still air.

3 NORMALIZING

Purpose:

a) To provide the desired mechanical properties (for a given steel composition, a normalized steel will have higher strength and hardness, and slightly lower ductility than a fully annealed steel);
b) To modify and refine coarse as-rolled, forged or as-cast dendritic structures;
c) To improve the response in hardening by refining the grain size and homogenizing the microstructure;
d) To improve machining characteristics. This treatment is especially beneficial for carbon steels containing 0.15–0.40 % carbon; and
e) To eliminate carbide networks in hypereutectoid steels.

Recommended heat treatment cycle: Heat to a temperature of about 55 °C above Ac_3 for hypoeutectoid steels and above Ac_{cm} for hypereutectoid steels. Soak thoroughly (holding time ≈ 1 h/25 mm of cross-section). Cool to ambient temperature in still air.

4 HARDENING AND TEMPERING

Purpose:

a) To produce high strength and hardness; and
b) To develop a combination of strength, ductility and toughness superior to that which can be achieved after normalizing.

Recommended heat treatment cycle: Heat to a temperature about 30–50 °C above Ac_3 for hypoeutectoid steels, and above Ac_1 for hypereutectoid steels. Soak thoroughly. Quench in oil or water, with or without additives, at a rate fast enough to develop the desired microstructure and hardness. Temper immediately after quenching at a temperature below Ac_1 to achieve the desired properties.

5 MARTEMPERING (MARQUENCHING)

Purpose: To replace conventional hardening for one or more of the following reasons:

a) To minimize distortion;
b) To reduce or eliminate susceptibility to cracking; and
c) To reduce the level of residual stresses.

Recommended heat treatment cycle: Heat to 815–870 °C. Soak thoroughly. Quench in a hot fluid medium (usually hot oil or molten salt) maintained at a temperature generally above the M_S point (usually in the range of 160–400 °C), at a rate fast enough to prevent the formation of pearlite, ferrite or bainite. Hold in the quenching medium for a sufficient length of time to equalize the temperature throughout the part—but not long enough to permit bainite to form. Cool to ambient temperature at a moderate rate to achieve a fully martensitic structure. Temper in the same manner as if the part had been conventionally hardened.

6 TEMPERING

Purpose: To obtain the desired mechanical properties, such as increased toughness and ductility, lower hardness, and dimensional stability, in previously hardened or normalized parts.

Recommended heat treatment cycle: Heat to a temperature below Ac_1:

a) When it is desired to reduce internal stresses and to increase the toughness without any appreciable loss in hardness (e.g., as in case-hardened and surface-hardened parts), heat to 150–200 °C;
b) When it is desired to achieve the highest attainable elastic limit together with ample toughness (e.g., as in springs), heat to 350–500 °C; and
c) When it is desired to achieve an optimum combination of strength and toughness, heat to 500–700 °C.

Hold at this temperature for a sufficient length of time (≈ 1 h minimum) to achieve the desired properties. Cool to ambient temperature in still air.

Note: When tempering steels which contain Mn, Mn and Cr, Cr and V, or Cr and Ni, the temperature range between 350 and 550 °C should be avoided, and if tempering is carried out at higher temperatures, the steel should be cooled rapidly through this temperature range (e.g., by quenching in oil) in order to avoid temper embrittlement.

7 AUSTEMPERING

Purpose: To replace hardening and tempering for either or both of the following reasons:

a) To obtain higher ductility or notch toughness at a given hardness; and
b) To reduce the likelihood of cracking and distortion.

Recommended heat treatment cycle: Heat to 790–870 °C. Soak thoroughly. Quench in molten salt maintained at a temperature just above the M_s point (usually 260–400 °C), at a rate fast enough to prevent the formation of pearlite and ferrite. Hold at the bath temperature for a sufficient length of time to ensure complete isothermal transformation of austenite to bainite. Cool to ambient temperature in still air.

8 CASE HARDENING

8.1 Carburizing

Carburizing is a case hardening process in which a solid ferrous alloy is heated to a temperature usually above Ac_3 in a solid, gaseous or liquid carburizing atmosphere of such composition as to cause absorption of carbon by the surface and, by diffusion, to create a concentration gradient. Hardening completes the process.

Purpose:

a) To produce a hard, wear resistant case; and
b) To improve fatigue strength.

Recommended heat treatment cycle:

Carburizing: Heat to 880–980 °C (preferably between 900–950 °C) in a carburizing atmosphere composed of:

a) For pack carburizing—a solid mixture consisting of charcoal/coke and catalyzing agents (e.g., barium carbonate);
b) For gas carburizing—hydrocarbon gases or vaporized liquid hydrocarbons, and carrier gases; and
c) For liquid carburizing—a molten salt bath consisting of activating alkaline earth chlorides with additives of alkaline chloride and alkaline cyanide, and reaction products of alkaline cyanate and carbonate.

Hold at this temperature for a sufficient length of time to achieve the desired case depth.

Hardening of carburized parts:

a) Direct hardening (adapatable to fine grain steels only): Quench in oil or water either directly from the carburizing temperature, or if there is a danger of distortion air cool to a temperature between the core hardening and case hardening temperatures before quenching;
b) Single hardening: Cool slowly to ambient temperature. Reheat to the core hardening temperature. Quench in oil or water; and
c) Double hardening: Cool slowly to ambient temperature. Reheat to the core hardening temperature. Quench in oil. Once again reheat to the case hardening temperature (780–820 °C). Quench in oil or water.

Tempering: Heat to 150–200 °C. Hold at this temperature for a sufficient length of time (≈ 1 h minimum). Cool to ambient temperature in still air.

8.2 Carbonitriding

Carbonitriding is a case hardening process in which a suitable ferrous alloy is heated above the lower transformation temperature in a gaseous atmosphere of such composition as to cause simultaneous absorption of carbon and nitrogen by the surface and, by diffusion, to create a concentration gradient. The process is completed by cooling at a rate that produces the desired properties in the workpiece.

Purpose:

a) To produce a hard, wear resistant case; and
b) To improve fatigue strength.

Recommended heat treatment cycle: Heat to 750–930 °C (preferably between 850–900 °C) in a carburizing atmosphere containing ammonia. Hold at this temperature for a sufficient length of time to acheive the desired case depth. Quench in oil or water, with or without additives, either directly from the carbonitriding temperature, or after slow cooling to ambient temperature and reheating to the hardening temperature. Temper at 150–200 °C (minimum holding time ≈ 1h).

8.3 Cyaniding (liquid carbonitriding)

Cyaniding is a case hardening process in which a ferrous alloy is heated above the lower transformation temperature in a molten salt containing cyanide to cause simultaneous absorption of carbon and nitrogen by the surface and, by diffusion, to create a concentration gradient. Hardening completes the process.

Purpose:

a) To produce a hard, wear resistant case; and
b) To improve fatigue strength.

Recommended heat treatment cycle: Heat to 700–870 °C in a molten salt bath containing alkali cyanides and cyanates. Hold at this temperature for a sufficient length of time (usually 0.5–1.0 h) to develop the desired case depth. Quench in oil or water, with or without additives, directly from the cyaniding temperature. Temper at 150–200 °C (minimum holding time ≈ 1h).

8.4 Nitriding

Nitriding is a case hardening process in which a solid ferrous alloy is heated to a suitable temperature (below Ac_1 for ferritic steels) in a nitrogen providing gaseous atmosphere or fused salt bath to cause absorption of nitrogen by the surface and, by diffusion, to create a concentration gradient. Quenching is not required to produce a hard case.

Purpose:

a) To obtain high surface hardness;
b) To increase wear resistance and provide anti-galling properties;
c) To improve fatigue strength;
d) To improve corrosion resistance of non-stainless steels; and
e) To obtain a surface that is resistant to the softening effect of heat at temperatures up to the nitriding temperature.

Notes: Compared to other methods of case hardening, nitriding offers the following advantages:
a) Nitriding can be accomplished with a minimum of distortion and excellent dimensional control; and
b) Nitrided cases are harder than those produced by carburizing and are quite stable in service up to the temperature of the nitriding process.

Recommended heat treatment cycle: Heat 490–565 °C in any of the following atmospheres:

a) For gas nitriding—in an ammonia enriched atmosphere; and
b) For liquid nitriding—in a molten nitrogen bearing, fused salt bath primarily containing cyanides and cyanates.

Hold at this temperature for a sufficient length of time to develop the desired case depth. Cool to ambient temperature.

8.5 Gaseous ferritic nitrocarburizing

Gaseous ferritic nitrocarburizing is a case hardening process in which a ferrous alloy is heated to a temperature below the lower critical temperature in a gaseous atmosphere of such composition as to cause simultaneous absorption of nitrogen and carbon by the surface layer and, by diffusion, to create a concentration gradient.

Purpose:

a) To provide an antiscuffing surface layer;
b) To improve fatigue strength; and
c) To enhance corrosion resistance.

Recommended heat treatment cycle: Heat to 570–580 °C in a gaseous nitrocarburizing atmosphere containing ammonia and a carbon rich gas. Hold at this temperature for about 1–3 h till the desired case depth is achieved. Quench in oil.

9 SURFACE HARDENING

9.1 Induction hardening

Induction hardening is a surface hardening process in which only the surface layer of a suitable ferrous workpiece is heated by electromagnetic induction to a temperature above the upper critical temperature and quenched immediately.

Purpose:

a) To develop a hard, wear resistant case; and
b) To improve fatigue strength.

Recommended heat treatment cycle: Rapidly heat the surface of the steel to be hardened by electromagnetic induction to the temperatures indicated below for unalloyed and alloy steel:

0.30 % C	900–925	°C
0.35 % C	900	°C
0.40 % C	870–900	°C
0.45 % C	870–900	°C
0.50 % C	870	°C
0.60 % C	845–870	°C

Note: Alloy steels containing carbide forming elements (Cr, Mo, V, W) should be heated 50–110 °C above the temperatures indicated.

After the grain structure of the case has become austenitic quickly quench in water, with or without additives. Temper at 150–200 °C for a minimum of 1 h.

9.2 Flame hardening

Flame hardening is a surface hardening process in which the surface layer of a suitable ferrous workpiece is heated by an intense flame to a temperature above the upper critical temperature and quenched immediately.

Purpose:

a) To develop a hard, wear resistant case; and
b) To improve fatigue strength.

Recommended heat treatment cycle: Rapidly heat the surface of the steel to be hardened by an oxyacetylene or hydrogen torch to the temperatures indicated below for unalloyed and alloy steel:

0.30 % C	900–925	°C
0.35 % C	900	°C
0.40 % C	870–900	°C
0.45 % C	870–900	°C
0.50 % C	870	°C
0.60 % C	845–870	°C

Note: Alloy steels containing carbide forming elements (Cr, Mo, V, W) should be heated 50–110 °C above the temperatures indicated.

After the grain structure of the case has become austenitic quickly quench in water, with or without additives. Temper at 150–200 °C for a minimum of 1 h.

10 TERMINOLOGY

Austenitizing

The process of forming austenite by heating a ferrous alloy into the transformation range (partial austenitizing) or above the transformation range (complete austenitizing). When used without qualification the term implies complete austenitizing.

Soaking

The process of prolonged holding at a selected temperature to effect homogenization of structure or composition.

11 REFERENCES

Books

1. *ASM Handbook*, vol. 4, ASM International, Materials Park, Ohio, U.S.A., 1991.
2. *Metals Handbook*, Desk Edition, American Society for Metals, Metals Park, Ohio, U.S.A., 1985.
3. G. Krauss, *Principles of Heat Treatment of Steel*, American Society for Metals, Metals Park, Ohio, U.S.A., 1980.
4. K.E. Thelning, *Steel and its Heat Treatment: Bofors Handbook*, Butterworths, London, U.K., 1984.
5. Y. Lakhtin, *Engineering Physical Metallurgy and Heat Treatment*, MIR Publishers, Moscow, Russia, 1974.

Standards

6. DIN 17014 (1)–1988: *Heat Treatment of Ferrous Materials—Terminology.*
7. DIN 17022 (3)–1989: *Heat Treatment of Ferrous Materials—Heat Treatment Methods—Case Hardening.*

3.5

Low-Carbon Steel Sheet and Strip

1 INTRODUCTION

Low-carbon steel sheets and strips are hot-rolled and cold-rolled flat products which possess in various degrees and combinations good formability, adequate strength, good weldability and good finishing characteristics. These products are available in three basic qualities:

a) Ordinary quality: Intended for general fabricating purposes where sheets and strips are used in the flat or for bending, or moderate forming;
b) Drawing quality: Intended for drawing or severe forming; and
c) Structural quality: Intended for structural purposes where particular mechanical properties are required.

Low-carbon steel sheets and strips may be ordered in the rimmed, semi-killed or fully killed condition. Rimmed steels are usually preferred because of their superior surface finish and formability characteristics. However, both rimmed and semi-killed steels tend to strain age. Hence, when strain ageing is to be avoided and/or when exceptionally deep draws are employed, steel stabilized (fully killed) with aluminium is recommended.

Hot-rolled sheets and strips are suitable for applications where the presence of oxide or scale, or normal surface imperfections disclosed after removal of oxide or scale, are not objectionable. They are not suitable for applications where the surface is of prime importance.

Cold-rolled sheets and strips, on the other hand are suitable for applications where surface is of prime importance. They are produced to closer size tolerances and with an improved surface finish than hot-rolled sheets and strips. They are available in thicknesses lower than what can be achieved by hot rolling.

2 SPECIFICATIONS

2.1 Hot-rolled low-carbon steel sheet and strip for cold forming

2.1.1 Chemical composition

TABLE 3.6 Chemical composition (cast analysis) of hot-rolled low-carbon steel sheet and strip for cold forming, conforming to IS 1079 – 1994

Grade	Quality	Chemical composition [% (m/m)]			
		C max.	Mn max.	P max.	S max.
O	Ordinary	0.15	0.60	0.055	0.055
D	Drawing	0.12	0.50	0.040	0.040
DD	Deep drawing	0.10	0.40	0.035	0.035
EDD	Extra deep drawing	0.08	0.40	0.030	0.030

Notes:

1) Grades O, D, and DD may be supplied in the rimmed, semi-killed or fully killed condition.
2) Grade EDD should be supplied in the fully Al-killed condition.
3) Si-killed steels should contain ≥ 0.10 % Si.
4) Al-killed steels should contain ≥ 0.02 % total Al.
5) Si-Al-killed steels should contain ≥ 0.03 % Si and ≥ 0.01 % total Al.
6) The N content in rimmed, semi-killed or Si-killed steels should be ≤ 0.007 %, whereas in Al-killed or Si–Al-killed steels, it should be ≤ 0.012 %.
7) For improved atmospheric corrosion resistance the steels may be supplied with a Cu content of 0.20–0.35 % in the cast analysis, or 0.17–0.38 % in the product analysis.
8) The steels may be supplied with microalloying additions such as B, Nb, Ti and V. The total microalloying content, as defined by the cast analysis, should be ≤ 0.006 % in case of steels containing B, and ≤ 0.20 % in case of steels containing Nb, Ti and V.

2.1.2 Mechanical properties

TABLE 3.7 Mechanical properties of hot-rolled low-carbon steel sheet and strip, conforming to IS 1079 – 1994

Grade	R_m MPA	R_e min. MPa	A min. %	Bend mandrel diameter (180° bend)[1)]
O	–	–	–	2*a*
D	240–400	–	25	1*a*
DD	260–390	–	28	0*a*
EDD	260–380	–	32	0*a*

[1)] *a* = thickness of the bend test piece; 0*a* means flat on itself.

Notes:
1) The longitudinal axis of the tensile test piece should be:
   (a) transverse to the direction of rolling for widths ≥ 150 mm, and
   (b) parallel to the direction of rolling for widths < 150 mm.

2) The axis of the bend should be parallel to the direction of rolling.

2.2 Hot-rolled low-carbon steel sheet and strip produced to a specified minimum strength

2.2.1 Chemical composition

TABLE 3.8 Chemical composition (cast analysis) of hot-rolled low-carbon steel sheet and strip produced to a specified minimum strength, conforming to IS 5986 – 1992

Grade	Chemical composition [% (*m*/*m*)]				
	C max.	Mn max.	P max.	S max.	CE[1)] max.
Fe 330	0.17	1.00	0.045	0.045	–
Fe 360	0.17	1.20	0.045	0.045	–
Fe 410	0.20	1.30	0.045	0.045	0.42
Fe 510	0.20	1.50	0.045	0.045	0.45

(Contd.)

TABLE 3.8 Chemical composition (cast analysis) of hot-rolled low-carbon steel sheet and strip produced to a specified minimum strength, conforming to IS 5986 – 1992 (continued)

[1)] The carbon equivalent (CE) value of steel based on cast analysis, is calculated using the formula

$$CE = C + Mn/6 + (Cr + Mo + V)/5 + (Ni + Cu)/15.$$

Notes:
1) Grades Fe 330, Fe 360, Fe 410 and Fe 510 may be supplied in the semi-killed or fully killed condition.
2) Si-killed steels should contain ≥ 0.10 % Si.
3) Al-killed steels should contain ≥ 0.02 % total Al.
4) Si-Al-killed steels should contain ≥ 0.03 % Si and ≥ 0.01 % total Al.
5) The N content in semi-killed or Si-killed steels should be ≤ 0.009 %, whereas in Al-killed or Si-Al-killed steels, the N content should be ≤ 0.012 %.
6) For improved atmospheric corrosion resistance the steel may be supplied with a Cu content of 0.20–0.35 % in the cast analysis.
7) The steels may be supplied with microalloying additions such as B, Nb, Ti and V. The total microalloying content, as defined by the cast analysis should be < 0.001 % in case of steels containing B, and < 0.20 % in case of steels containing Nb, Ti and V.

2.2.1 Mechanical properties

TABLE 3.9 Mechanical properties of hot-rolled low-carbon steel sheet and strip produced to a specified minimum strength, conforming to IS 5986 – 1992

Grade	R_m	R_e min.	*A* min. for a thickness, in mm, of		Bend mandrel diameter (180° bend)[1)]
			≤ 3	> 3	
			$L_o = 80$ mm	$L_o = 5.65\sqrt{S_o}$	
	MPa	MPa	%	%	
Fe 330	330–440	205	18	25	1*a*
Fe 360	360–470	235	18	25	1*a*
Fe 410	410–520	255	17	23	2*a*
Fe 510	510–620	355	16	20	2*a*

[1)] *a* = thickness of the bend test piece; 0*a* means flat on itself.

Notes:
1) The longitudinal axis of the tensile test piece should be transverse to the direction of rolling.
2) The axis of the bend should be parallel to the direction of rolling.

2.3 Cold-rolled low-carbon steel sheet and strip for cold forming

2.2.1 Chemical composition

TABLE 3.10 Chemical composition (cast analysis) of cold-rolled low-carbon steel sheet and strip for cold forming, conforming to IS 513–1994

Grade	Quality	Chemical composition [% (*m*/*m*)]			
		C max.	Mn max.	P max.	S max.
O	Ordinary	0.15	0.60	0.055	0.055
D	Drawing	0.12	0.50	0.040	0.040
DD	Deep drawing	0.10	0.45	0.035	0.035
EDD	Extra deep drawing	0.08	0.40	0.030	0.030

Notes:

1) Grades O, D and DD may be supplied in the rimmed, semi-killed or fully killed condition.
2) Grade EDD should be supplied in the fully Al-killed condition.
3) Si-killed steels should contain ≥ 0.10 % Si.
4) Al-killed steels should contain ≥ 0.02 % total aluminium.
5) Si-Al-killed steels should contain ≥ 0.03 % Si and ≥ 0.01% total Al.
6) The nitrogen content in rimmed, semi-killed or Si-killed steels, should be ≤ 0.007%, whereas in Al-killed or Si-Al-killed steels, it should be ≤ 0.012 %.
7) For improved atmospheric corrosion resistance the steels may be supplied with a Cu content of 0.20–0.35 % in the cast analysis or 0.17–0.38 % in the product analysis.
8) The steels may be supplied with microalloying additions such as B, Nb, Ti and V. The total microalloying content, as defined by the cast analysis, should be ≤ 0.006 % in case of steels containing B, and ≤ 0.20 % in case of steels containing Nb, Ti and V.

2.2.2 Mechanical properties

TABLE 3.11 Mechanical properties of annealed/skin passed cold-rolled low-carbon steel sheet and strip for cold forming, conforming to IS 513 – 1994

Grade	R_m MPa	R_e min. MPa	A min. %	Maximum hardness HRB	Maximum hardness HR30T	Bend mandrel diameter[1] (180° bend)	Modified Erichsen cupping test
O	–	–	–	[2]	[2]	1*a*	See Fig. 3.2 for minimum values
D	270–410	280	23	65	60	0*a*	
DD	270–370	250	26	57	55	0*a*	
EDD	270–350	220	32	50	50	0*a*	

[1] *a* = thickness of the bend test piece; 0*a* means flat on itself.
[2] See Table 3.12.

Notes:
1) The longitudinal axis of the tensile test piece should be:
   (a) transverse to the direction of rolling for widths > 250 mm, and
   (b) parallel to the direction of rolling for widths ≤ 250 mm.
2) The axis of the bend should be parallel to the direction of rolling.
3) The modified Erichsen cupping test is applicable only to the formability grades D, DD and EDD having a nominal thickness of 0.5–2.0 mm.

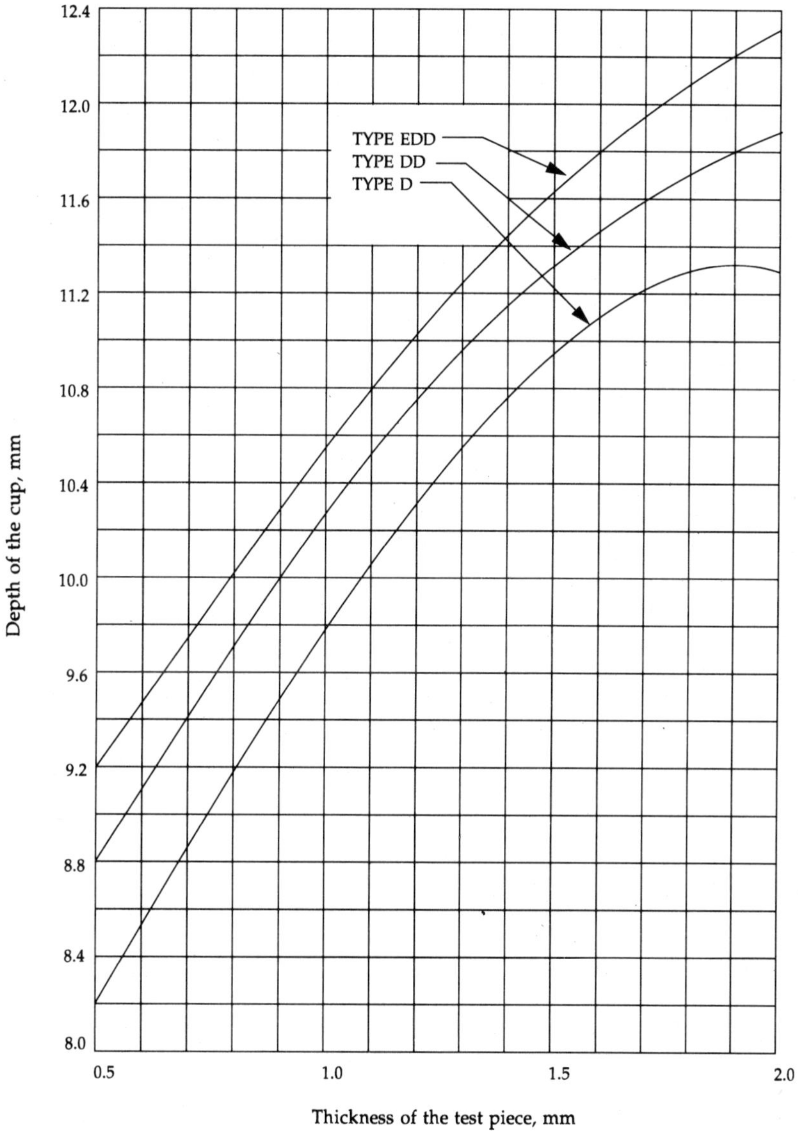

Fig. 3.2 Minimum modified Erichsen cupping test values (Source: Reference 2)

TABLE 3.12 Hardness and bend test requirements for IS 513 – Grade 0 sheet and strip in various tempers

Temper	Hardness	Bend test	
		Angle of bend	Bend mandrel diameter[1)]
	HRB	°	
Hard	≥ 85	–	–
Half hard	75–85	180	3*a*
Quarter hard	60–75	180	2*a*
Skin passed	≤ 70	180	1*a*
Annealed	≤ 60	180	1*a*

1) *a* = thickness of the bend test piece; 0*a* means flat on itself.

3 EQUIVALENT GRADES

TABLE 3.13 Comparison of grades of hot-rolled low-carbon steel sheet and strip for cold forming specified in IS 1079–1994 with nearest equivalent grades specified in other national and international standards

IS 1079–94	ASTM A 569/A 569M–91a A 621/A 621M–92 A 622/A 622M–92	BS 1449 (1.2)–91 BS 1449 (1.8)–91	DIN 1614 (2)–86	ISO 3573–86 ISO 6317–82	JIS G 3131–90	SAE J403–94
O	A 569/A 569M	HR14, HS14	–	HR1	SPHC	1012 1010 1009
D	–	HR4, HS4	StW 22	HR2	–	–
DD	A 621/A 621M	HR3, HS3	UStW 23	HR3	SPHD	1008
EDD	A 622/A 622M	HR1, HS1	StW 24	HR4	SPHE	1006

TABLE 3.14 Comparison of grades of hot-rolled low-carbon steel sheet and strip produced to a specified minimum strength, specified in IS 5986 – 1992 with nearest equivalent grades specified in other national and international standards

IS 5986–92	ASTM A 570/A 570M–92	BS 1449 (1.4)–91 BS 1449 (1.10)–91	DIN	ISO 4995–93 ISO 6316–93	JIS G 3101–87
Fe 330	30	HR34/20, HS34/20	–	HR235	SS 330
Fe 360	33	HR37/23, HS37/23	–	–	–
Fe 410	40	HR43/25, HS43/25	–	HR275	SS 400
Fe 510	55	HR50/35, HS50/35	–	HR355	–

TABLE 3.15 Comparison of grades of cold-rolled low-carbon steel sheet and strip for cold forming specified in IS 513 – 1994 with nearest equivalent grades specified in other national and international standards

IS 513–94	ASTM A 366/A 366M–91 ASTM A 619/A 619M–92 ASTM A 620/A 620M–92	BS 1449 (1.9)–91 EN 10130–91	DIN EN 10130–91	ISO 3574–86	JIS G 3141–90	SAE J403–94
O	A 366/A 366M	CS14	–	CR1	SPCC	1012 1010 1009
D	–	CS4, Fe P 01	Fe P 01	CR2	SPCCT	–
DD	A 619/A 619M	CS3, Fe P 03	Fe P 03	CR3	SPCD	1008
EDD	A 620/A 620M	CS1, Fe P 04	Fe P 04	CR4	SPCEN	1006

4 REFERENCES

Book

1. *ASM Handbook*, vol. 1, ASM International, Materials Park, Ohio, U.S.A., 1990.

Standards

2. IS 513 – 1994: *Cold-Rolled Low-Carbon Steel Sheets and Strips—Specification.*
3. IS 1079 – 1994: *Hot-Rolled Carbon Steel Sheets and Strips—Specification.*
4. IS 5986 – 1992: *Hot-Rolled Steel Plates, Sheets, Strips and Flats for Flanging and Forming Operation—Specification.*

3.6

STEEL SHEET AND STRIP FOR PORCELAIN ENAMELLING

1 INTRODUCTION

Steel sheets and strips for porcelain (vitreous) enamelling are low-carbon steel flat products which are characterized by good enamellability, freedom from surface defects, good formability, good sag resistance, adequate strength and good weldability. The relative importance of each of these factors depends on the requirement of the finished product.

The various grades of steel sheet and strip that are commonly used for porcelain enamelling are:

a) Cold rolled low-carbon ($C \leq 0.15$ %) steel sheet and strip of commercial and drawing qualities: Suitable for ground coat, or two-coat enamelling (i.e., cover coat applied over ground coat);
b) Cold rolled enamelling iron ($C \leq 0.03$ %) sheet and strip: Suitable for two-coat enamelling;
c) Cold rolled decarburized enamelling steel ($C \leq 0.008$ %) sheet and strip: Suitable for direct cover coat enamelling, and also for two-coat enamelling if distortion during enamelling is to be minimized; and
d) Hot rolled low-carbon steel sheet and strip: Generally not recommended for porcelain enamelling because of its high susceptibility to fishscale formation and carbon boiling. When the use of hot rolled steels cannot be avoided, it is recommended that porcelain enamelling be restricted to one surface only.

2 SPECIFICATION

2.1 Chemical composition

TABLE 3.16 Chemical composition (cast analysis) of cold-rolled and hot-rolled low-carbon steel sheet for porcelain enamelling, conforming to IS 9485 – 1980

Grade	Chemical composition [% (*m*/*m*)]				
	C max.	Mn max.	P max.	S max.	P+S max.
A	0.05	0.30	0.030	0.030	0.050
B	0.08	0.40	0.035	0.035	0.060
C	0.10	0.60	0.050	0.050	0.090

Notes:
1) The steels may be supplied in the fully killed, semi-killed or rimmed condition.
2) Si-killed steels should contain 0.15–0.35 % Si.
3) Si–Al-killed or Al-killed steels may be supplied with a Si content of ≤ 0.15 %.
4) Other elements such as Nb, Ti, etc., may be added for stabilization of C.

2.2 Mechanical properties

TABLE 3.17 Bend test and modified Erichsen cupping test requirements for cold-rolled and hot-rolled low-carbon steel sheet for porcelain enamelling, conforming to IS 9485 – 1980

Grade	Bend mandrel diameter[1),2)] (180° bend)	Modified Erichsen cupping test3)
A	1*a*	See Fig. 3.3 for minimum values
B	1*a*	
C	1*a*	

1) The axis of the bend should be parallel to the direction of rolling.
2) *a* = thickness of the bend test piece.
3) Applicable only to cold-rolled sheet of drawing quality.

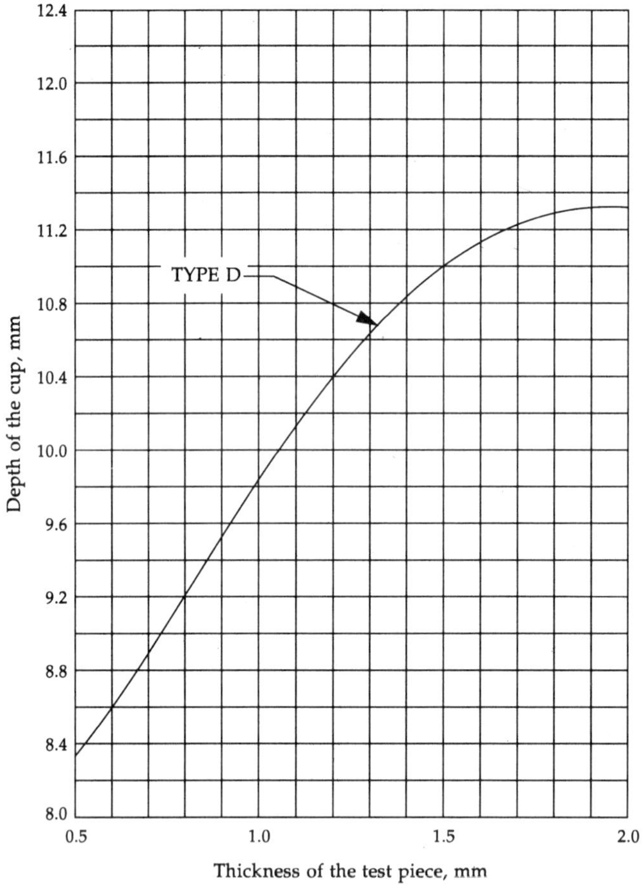

Fig. 3.3 Minimum modified Erichsen cupping test values (Source: Reference 3)

3 EQUIVALENT GRADES

TABLE 3.18 Comparison of grades of low-carbon steel sheet for porcelain enamelling specified in IS 9485–1980 with nearest equivalent grades specified in other national and international standards

IS 9485–80	ASTM A 424–92	BS 1449 (1.3)–91	DIN 1623 (3)–87	ISO 5001–93	JIS G 3133–88	SAE J403–94
A	Type II, Composition A	–	–	–	–	–
B	Type II, Composition B	CR1VE, CR2VE	EK2, EK4	Grade 2, VE04	–	1006
C	–	CR3VE	–	Grade 2, VE03	–	1008

4 REFERENCES

Books

1. *Metals Handbook*, vol. 1, 9th ed., American Society for Metals, Metals Park. Ohio, U.S.A., 1980.
2. *Metals Handbook*, vol. 5, 9th ed., American Society for Metals, Metals Park, Ohio, U.S.A., 1982.

Standard

3. IS 9485 – 1980: *Specification for Cold-Reduced and Hot-Rolled Carbon Steel Sheet for Porcelain Enamelling.*

3.7

METAL COATED STEEL SHEET AND STRIP

1 INTRODUCTION

1.1 Galvanized steel sheet and strip

Galvanized steel sheet and strip is hot-rolled or cold-rolled low-carbon steel sheet and strip coated with zinc either electrolytically or by dipping in a bath of molten zinc. The zinc coating may be applied on one surface only or on both surfaces with equal or differential coating thicknesses.

The zinc coating may have a normal spangle finish (that is a coating finish having a metallic lustre obtained when the zinc coating is left to solidify normally), minimized spangle (obtained by influencing the solidification process in a special way), or a regular finish (obtained by heat treatment in which iron diffuses through the zinc; the surface has a uniform matt grey appearance). They are also available with a smooth finish which is suitable for decorative painting.

After zinc coating, the surfaces may be passivated by chromating or phosphating.

Galvanized steel sheet and strip possess good formability, good resistance to atmospheric corrosion and to corrosion in underground or underwater environments, and good paintability.

1.2 Terne coated steel sheet and strip

Terne coated steel sheet and strip is cold-rolled low-carbon steel sheet and strip coated with terne metal (a lead-tin alloy generally containing 8–15 % Sn) either by dipping in a bath of molten alloy or electrolytically.

Terne coated steel sheet and strip possess good formability, good resistance to corrosion in petroleum products, water and air, as well as excellent solderability and paintability. They are extensively used for the manufacture of fuel tanks, radiator parts, valve rocker arm covers, and chassis for radios, televisions, and tape recorders.

1.3 Tinplate

Tinplate is cold-rolled low-carbon steel sheet and strip which has been single or double reduced and coated on both surfaces with tin either electrolytically or by dipping in a bath of molten tin. The tin coating can be produced with equal or differential coating weights on the two surfaces of the sheet or strip. Tinplate is generally produced with a matt or bright finish. It is normally supplied with a passivation treatment and a protective layer of oil and is suitable for varnishing (lacquering) or printing.

Tinplate is nontoxic, and has good formability, good solderability, good surface appearance and a degree of corrosion resistance. It is used extensively in food handling, packaging and dairy equipment.

1.4 Aluminium/aluminium-silicon alloy coated steel sheet and strip

Aluminium/aluminium-silicon alloy coated steel sheet and strip is cold-rolled low-carbon steel sheet and strip coated with commercially pure aluminium or an aluminium-silicon alloy (containing 5-11 % Si) by dipping in a molten bath.

Aluminium coated steel sheet and strip is used for applications requiring protection against atmospheric corrosion. Aluminium-silicon alloy coated steel sheet and strip is used principally for heat resisting applications and also for uses where resistance to corrosion is required. Both these types have a smooth satin-like surface, with excellent reflectivity

1.5 Aluminium-zinc alloy coated steel sheet and strip

Aluminium-zinc alloy coated steel sheet and strip is a cold-rolled low-carbon steel sheet and strip coated with an aluminium-zinc alloy (nominally 55 % Al, 1.6 % Si and the balance Zn) by dipping in a bath of molten alloy. It is intended for use in applications requiring corrosion resistance or heat resistance, or both. This coating has a corrosion resistance which is two to four times the corrosion resistance of a galvanized coating.

2 SPECIFICATIONS

2.1 Galvanized steel sheets

2.1.1 Chemical composition (base metal)

TABLE 3.19 Chemical composition (cast analysis) of hot-rolled (IS 1079 – 1994) and cold-rolled (IS 513 – 1994) low-carbon steel sheet for galvanizing

Grade	Chemical composition [% (*m*/*m*)]			
	C max.	Mn max.	P max.	S max.
Hot-rolled low-carbon steel sheet (IS 1079 – 1994)[1)]				
O	0.15	0.60	0.055	0.055
D	0.12	0.50	0.040	0.040
DD	0.10	0.40	0.035	0.035
EDD	0.08	0.40	0.030	0.030

(Contd.)

TABLE 3.19 Chemical composition (cast analysis) of hot-rolled (IS 1079 – 1994) and cold-rolled (IS 513 – 1994) low-carbon steel sheet for galvanizing (continued)

Grade	Chemical composition [% (*m*/*m*)]			
	C max.	Mn max.	P max.	S max.
Cold-rolled low-carbon steel sheet (IS 513 – 1994)[2)]				
O	0.15	0.60	0.055	0.055
D	0.12	0.50	0.040	0.040
DD	0.10	0.45	0.035	0.035
EDD	0.08	0.40	0.030	0.030

1) See notes in Table 3.6.
2) See notes in Table 3.10.

2.1.2 *Zinc coating mass*

TABLE 3.20 Zinc coating masses for galvanized steel sheets, conforming to IS 277 – 1992

Grade of coating	Minimum coating mass (total both surfaces)	
	Triple spot test g/m^2	Single spot test g/m2
600	600	510
450	450	380
350	350	300
275	275	235
220	220	190
200	200	170
180	180	155
120	120	100

2.1.3 Coating bend test requirements

TABLE 3.21 Coating bend test requirements for galvanized sheets, conforming to IS 277 – 1992

Grade of coating	Bend mandrel diameter (180° bend) for a thickness, in mm, of										
	> 3	> 2.3 ≤ 3	> 1.6 ≤ 2.3	> 1.25 ≤ 1.6	> 1.0 ≤ 1.25	> 0.8 ≤ 1.0	> 0.5 ≤ 0.8	> 0.4 ≤ 0.5	> 0.3 ≤ 0.4	> 0.22 ≤ 0.3	> 0.16 ≤ 0.22
750	6*a*	8*a*	10*a*	10*a*	11*a*	12*a*	14*a*	15*a*	–	–	–
600	4*a*	6*a*	8*a*	8*a*	9*a*	10*a*	11*a*	12*a*	–	–	–
450	3*a*	4*a*	6*a*	6*a*	7*a*	8*a*	8*a*	8*a*	9*a*	10*a*	11*a*
375	3*a*	4*a*	4*a*	4*a*	5*a*	6*a*	6*a*	7*a*	8*a*	8*a*	9*a*
300	3*a*	4*a*	4*a*	4*a*	5*a*	6*a*	6*a*	6*a*	7*a*	7*a*	8*a*
250	2*a*	3*a*	3*a*	3*a*	4*a*	4*a*	4*a*	4*a*	5*a*	5*a*	5*a*
200	2*a*	2*a*	2*a*	3*a*	3*a*	3*a*	3*a*	3*a*	4*a*	4*a*	4*a*
175	2*a*	2*a*	2*a*	3*a*	3*a*	3*a*	3*a*	3*a*	3*a*	4*a*	4*a*
120	2*a*	2*a*	2*a*	3*a*	3*a*	3*a*	3*a*	3*a*	3*a*	4*a*	4*a*

[1] *a* = thickness of the bend test piece.

2.2 Tinplate

2.2.1 Chemical composition (base metal)

TABLE 3.22 Chemical composition (cast analysis) of cold-reduced low-carbon steel sheet and strip for tin coating, conforming to IS 2385 – 1977

Chemical composition [% (*m*/*m*)]			
C max.	Mn max.	P max.	S max.
0.12	0.60	0.050	0.050

Notes:
1) The steel may be supplied in the fully killed, semi-killed or rimmed condition.
2) Al-killed steels should contain 0.02–0.08 % Al and ≤ 0.010 % N.

2.2.2 *Tin coating mass*

TABLE 3.23 Coating masses for electrolytic tinplate, equally coated, conforming to IS 1993 – 1993 and IS 9025 – 1993

Code	Nominal coating mass		Minimum average coating mass
	Per surface g/m^2	Total both surfaces g/m^2	Total both surfaces g/m2
E2.8/2.8	2.8	5.6	4.9
E5.6/5.6	5.6	11.2	10.5
E8.4/8.4	8.4	16.8	15.7
E11.2/11.2	11.2	22.4	20.2

TABLE 3.24 Coating masses for electrolytic tinplate, differentially coated, conforming to IS 1993 – 1993 and IS 9025 – 1993

Code	Nominal coating mass		Minimum average coating mass	
	Heavily coated surface g/m^2	Lightly coated surface g/m^2	Heavily coated surface g/m2	Lightly coated surface g/m^2
D5.6/2.8	5.6	2.8	4.75	2.25
D8.4/2.8	8.4	2.8	7.85	2.25
D8.4/5.6	8.4	5.6	7.85	4.75
D11.2/2.8	11.2	2.8	10.1	2.25
D11.2/5.6	11.2	5.6	10.1	4.75

TABLE 3.25 Coating masses for hot-dipped tinplate, conforming to IS 1993 – 1993

Code	Nominal coating mass (total both surfaces) g/m^2	Minimum average coating mass (total both surfaces) g/m2
H12/12	24.0	21.0
H14/14	28.0	24.6
H15/15	30.0	26.0
H17/17	33.6	28.0

2.2.3 Temper Classifications

TABLE 3.26 Rockwell HR30T hardness values normally associated with the temper classification of batch annealed and continuously annealed tinplate, conforming to IS 1993 – 1993 and IS 9025 – 1993

Temper classification	Rockwell HR30T hardness aim	
	Mean	Maximum deviation of sample average
T 50	≤ 52	
T 52	52	+4 –4
T 55	55	+4 –3
T 57	57	+4 –3
T 61	61	+4 –4
T 65	65	+4 –4
T 70	70	+3 –4

2.3 Hot-dip terne coated low-carbon steel sheet

2.3.1 Chemical composition (base metal)

TABLE 3.27 Chemical composition (cast analysis) of cold-rolled low-carbon steel sheet, conforming to IS 12313 – 1988

Quality		Chemical composition [% (m/m)]			
Designation	Name	C max.	Mn max.	P max.	S max.
T0 01	Ordinary	0.15	0.60	0.055	0-055
T0 02	Drawing	0.12	0.50	0.040	0.040
T0 03	Deep drawing	0.10	0.45	0.030	0.030
T0 04	Extra deep drawing	0.08	0.45	0.030	0.030

Notes:
1) Si-killed steels should contain ≥ 0.10 % Si.
2) Al-killed steels should contain 0.02–0.06 % total aluminium.
3) Si–Al-killed or Al-killed steels may be supplied with a Si content of < 0.10 %.
4) When the steel is rimmed, semi-killed or not fully killed by Al alone, the N content should be ≤ 0.007 %.
5) For improved atmospheric corrosion resistance the steels may be supplied with a Cu content of 0.20–0.35 % in the cast analysis or 0.17–0.38 % in the product analysis.

2.3.2 Mechanical properties

TABLE 3.28 Mechanical properties of cold-rolled low-carbon steel sheet for hot-dip terne coating, conforming to IS 12313 – 1988

Steel designation	R_m[1)] max	A[2)] min.		Bend mandrel diameter[3)] (180° bend)
		L_o = 50 mm	L_o = 80 mm	
	MPa	%	%	
T0 01	–	–	–	1*a*
T0 02	430	24	23	–
T0 03	410	26	25	–

(Contd.)

TABLE 3.28 Mechanical properties of cold-rolled low-carbon steel sheet for hot-dip terne coating, conforming to IS 12313 – 1988 (continued)

Steel designation	R_m[1)] max	A[2)] min.		Bend mandrel diameter[3)] (180° bend)
		L_o = 50 mm	L_o = 80 mm	
	MPa	%	%	
T0 04	410	29	28	–

1) A minimum tensile strength of 260 MPa can be assumed for qualities T0 02, T0 03 and T0 04.
2) For thicknesses ≤ 0.6 mm, the elongation values in the table should be reduced by 2.
3) a = thickness of the bend test piece.

TABLE 3.29 Coating bend test requirements for hot-dip terne coated steel sheet, conforming to IS 12313 – 1988

Bend mandrel diameter[1)] (180° bend) for all thicknesses and all coating designations	
Ordinary quality	Drawing qualities
1a	0a

1) a = thickness of the bend test piece; 0a means flat on itself.

2.3.3 Terne coating mass

TABLE 3.30 Terne coating masses for hot-dip terne coated steel sheet, conforming to IS 12313 – 1988

Coating designation	Minimum coating mass (total both surfaces)	
	Triple spot test	Single spot test
	g/m^2	g/m2
001	no minimum	no minimum
050	50	40

(Contd.)

TABLE 3.30 Terne coating masses for hot-dip terne coated steel sheet, conforming to IS 12313 – 1988 (continued)

Coating designation	Minimum coating mass (total both surfaces)	
	Triple spot test g/m^2	Single spot test g/m2
075	75	60
100	100	75
120	120	90

Note: "no minimum" means that there are no established minimum check limits for triple spot and single spot tests.

3 EQUIVALENT STANDARDS

TABLE 3.31 Comparison of national and international standards on coated steel sheets

Product	IS	ASTM	BS	DIN	ISO	JIS
Galvanized steel sheet	277–1992	A 446/A 446M–1993 A 525M–1991a A 526/A 526M–1990 A 528/A 528M–1990 A 591/A 591M–1990 A 642/A 642M–1990	EN 10147–1992 EN 10152–1993	EN 10142–1991 EN 10147–1992 EN 10152–1994	3575–1976 4998–1991 5002–1982	G 3302–1994 G 3313–1990
Zn-5Al coated steel sheet	–	A 875/A 875M–1988	–	–	–	G 3317–1994
Tinplate	1993–1993 9025–1993 13954–1994	A 599–1992 A 623M–1992 A 624M–1990 A 625/A 625M–1992 A 626M–1990 A 650/A 650M–1988	EN 10203–1991 –	EN 10203–1991 –	11949–1995 5950–1991	G 3303–1987
Terne coated steel sheet	12313–1988	A 308–1994	6582–1985	–	4999–1991	–
Al coated steel sheet	–	A 463/A 463M–1994	–	–	–	
Al-Si coated steel sheet	–	A 463/A 463M–1994	6536–1985	–	5000–1993	G 3314–1977
Al-Zn coated steel sheet	–	A 792/A 792M–1994	EN 10215–1995	–	9364–1991	–
$Cr\text{-}CrO_2$ coated steel sheet	–	–	EN 10202–1990	EN 10202–1990	11950–1995	–

4 REFERENCES

Books

1. *ASM Handbook*, vol. 1, ASM International, Materials Park, Ohio, U.S.A., 1990.
2. *Metals Handbook*, vol. 5, 9th ed., American Society for Metals, Metals Park, Ohio, U.S.A., 1982.

Standards

3. IS 277 – 1992: *Galvanized Steel Sheets (Plain and Corrugated)—Specification.*
4. IS 1993 – 1993: *Single Cold-Reduced Tinplate and Single Cold-Reduced Blackplate—Electrolytic and Hot-Dipped Tinplate Sheet and Blackplate Sheet.*
5. IS 9025 – 1993: *Single Cold-Reduced Tinplate and Single Cold-Reduced Blackplate—Electrolytic Tinplate Coil and Blackplate Coil for Subsequent Cutting into Sheet Form.*
6. IS 12313 – 1988: *Specification for Hot-Dip Terne Coated Carbon Steel Sheets.*
7. ISO 6929 – 1987: *Steel Products—Definitions and classification.*
8. SAE HS J447 – 1981: *Prevention of Corrosion of Motor Vehicle Body and Chassis Components.*

3.8

STRUCTURAL STEELS

1 INTRODUCTION

Structural steels are unalloyed and low-alloy steels which are used in the rivetted, bolted or welded construction of buildings and bridges, and for general structural purposes. These steels are available in the form of sheet, strip, plate, structural shapes, bar and special sections, and are produced to specified mechanical properties. Structural steels are broadly classified into four categories:

a) As-rolled carbon steels;
b) As-rolled high-strength (HSLA) steels: These steels have better mechanical properties than the as-rolled carbon steels;
c) Normalized or hardened and tempered carbon steels; and
d) Hardened and tempered low-alloy steels.

The mechanical properties of the structural steel products are influenced mainly by:

a) Chemical composition: Carbon together with manganese is used to produce the desired mechanical properties in the as-rolled condition. To achieve higher strength it is not always advisable to raise the carbon content beyond a certain level, because carbon impairs formability, weldability and notch toughness. Higher strength is, therefore, achieved by alloying and/or heat treatment;
b) Deoxidation practice: Structural steels are usually furnished in the semi-killed or fully-killed condition. For applications requiring guaranteed impact strength or those requiring hardening and tempering the steels should be supplied in the fully killed condition;
c) Rolling practice: Controlled rolling improves strength and toughness;
d) Section thickness: Strength tends to decrease with increasing section sizes; and
e) Heat treatment: Unalloyed, and low-alloy steels may be normalized, or hardened and tempered to improve mechanical properties.

2 SPECIFICATIONS

2.1 Structural steel (ordinary quality)

2.1.1 Chemical composition

TABLE 3.32 Chemical composition (cast analysis) of structural steel (ordinary quality), conforming to IS 1977 – 1975

Steel designation	Chemical composition [% (*m*/*m*)]		
	C max.	P max.	S max.
Fe 310-0	0.30	0.070	0.070
Fe 410-0	0.30	0.070	0.070

Notes:
1) The steels should be supplied in the semi-killed or fully-killed condition.

2) The steels may be supplied with a Cu content of ≤ 0.35 %. The copper bearing steels should be designated as Fe 310 Cu-O and Fe 410 Cu-O, respectively.

2.1.2 Mechanical properties

TABLE 3.33 Mechanical properties of structural steel (ordinary quality), conforming to IS 1977 – 1975

Steel designation	Product form	Nominal thickness or diameter mm	R_m MPa	R_e min. MPa	*A* min. %
Fe 310-0	1)	< 6	Bend test only is required		
		≥ 6	310–430	–	26
	2)	< 10	Bend test only is required		
		≥ 10	310–430	–	26

(Contd.)

TABLE 3.33 Mechanical properties of structural steel (ordinary quality), conforming to IS 1977 – 1975 (continued)

Steel designation	Product form	Nominal thickness or diameter mm	R_m MPa	R_e MPa	A min. %
Fe 410-0	1)	< 6	Bend test only is required		
		≥ 6	410–530	230	23
	2)	< 10	Bend test only is required		
		≥ 10	410–530	230	23

1) Plates, sections (for example, angle, beam, channel, tee, etc.) and flats.
2) Bars (round, square and hexagonal).

TABLE 3.34 Bend test requirements for structural steel (ordinary quality), conforming to IS 1977 – 1975

Product form	Bend mandrel diameter[1)] (180° bend)
Plates, sections, flats and bars[2)]	3*a*
Bars[3)]	2*a*

1) *a* = thickness or diameter of the bend test piece.
2) For product forms other than round bars of diameter ≤ 25 mm.
3) For round bars of diameter ≤ 25 mm.

2.2 Steel for general structural purposes

2.2.1 Chemical composition

TABLE 3.35 Chemical composition (cast analysis) of steel for general structural purposes, conforming to IS 2062 – 1992

Steel designation	Method of deoxidation	Chemical composition [% (*m*/*m*)]					
		C max.	Si max.	Mn max.	P max.	S max.	CE max.
Fe 410 W A	Semi-killed or fully-killed	0.23	–	1.50	0.050	0,050	0.42
Fe 410 W B	Fully killed or semi-killed	0.22	0.40	1.50	0.045	0.045	0.41
Fe 410 W C	Fully-killed	0.20	0.40	1.50	0.040	0.040	0.39

[1] The carbon equivalent (CE) value of steel, based on cast analysis, is calculated using the formula: $CE = C + Mn/6 + (Cr + Mo + V)/5 + (Ni + Cu)/15$.

Notes:
1) Si-killed steels should contain ≥ 0.10 % Si. Si–Al-killed and Al-killed steels may be supplied with < 0.10 % Si.
2) The steels may be supplied with microalloying additions, such as Nb, Ti, and V, added singly or in combination. The total microalloying content, however, should be ≤ 0.20 %.
3) For improved formability, the steels may be treated with rare earths.
4) For improved atmospheric corrosion resistance, the steels may be supplied with a copper content of 0.20–0.35 %. The Cu-bearing steel should be designated with the suffix Cu, for example Fe 410 Cu W A.
5) The N content of the steel should be ≤ 0.012 %.

2.2.2 Mechanical properties

TABLE 3.36 Mechanical properties of steel for general structural purposes, conforming to IS 2062 – 1992

Steel designation	Supply condition	R_m min.	R_e min. for a thickness or diameter, in mm, of			A min.	Bend mandrel diameter[1)] (180° bend)	KV[2)] min.
			< 20	≥ 20 ≤ 40	> 40			
		MPa	MPa	MPa	MPa	%		J
Fe 410 W A	As-rolled	410	250	240	230	23	3*a*	–
Fe 410 W B	As-rolled or normalized[5)]	410	250	240	230	23	2*a*[3)]	27[4)]
							3*a*[6)]	
Fe 410 W C	As-rolled or normalized[8)]	410	250	240	230	23	2*a*	27[7)]

1) *a* = thickness or diameter of the bend test piece.

2) Measured on standard 10 mm × 10 mm test pieces. When tested on subsidiary test pieces, the *KV* values should be as follows:

Test piece size mm × mm	*KV* min. J
10 × 7.5	22
10 × 5	19.5

3) For product thicknesses or diameters ≤ 25 mm.

4) To be guaranteed at 0 °C.

5) Plates > 12 mm thick may be supplied in the normalized condition, if agreed between the purchaser and the manufacturer.

6) For product thicknesses or diameters > 25 mm.

7) To be guaranteed at –20 °C or – 40 °C.

8) Plates > 12 mm thick should be supplied in the normalized condition.

2.3 Microalloyed structural steels

2.3.1 Chemical composition

TABLE 3.37 Chemical composition (cast analysis) of microalloyed structural steels, conforming to IS 8500 – 1991

Steel designation	Chemical composition [% (*m*/*m*)]					
	C	Si	Mn	P	S	CE[1)]
	max.	max.	max.	max.	max.	max.
Fe 440	0.20	0.45	1.30	2a)	2a)	0.40
Fe 440 B	0.20	0.45	1.30	2a)	2a)	0.40
Fe 490	0.20	0.45	1.50	2a)	2a)	0.42
Fe 490 B	0.20	0.45	1.50	2a)	2a)	0.42
Fe 540	0.20	0.45	1.60	2b)	2b)	0.44
Fe 540 B	0.20	0.45	1.60	2b)	2b)	0.44
Fe 570	0.22	0.45	1.60	2b)	2b)	0.46
Fe 570 B	0.22	0.45	1.60	2b)	2b)	0.46
Fe 590	0.22	0.45	1.80	2b)	2b)	0.48
Fe 590 B	0.22	0.45	1.80	2b)	2b)	0.48

1) The carbon equivalent (CE) value of steel, based on cast analysis is calculated using the formula

$$CE = C + Mn/6 + (Cr + Mo + V)/5 + (Ni + Cu)/15.$$

2) The steels may be ordered to the following limits of phosphorus and sulphur:

	a)		b)	
	% P max.	% S max.	% P max.	% S max.
Flat products	0.040	0.040	0.040	0.040
Other structurals	0.050	0.050	0.045	0.045

TABLE 3.37 Chemical composition (cast analysis) of microalloyed structural steels, conforming to IS 8500 – 1991 (continued)

Notes:
1) The steel should be supplied in the fully killed or semi-killed condition. Steels with guaranteed impact properties should be supplied in the fully killed condition.
2) Si-killed steels should contain ≥ 0.10 % Si in the product analysis.
3) Al-killed steels should contain ≥ 0.02 % total Al.
4) Si–Al-killed steels should contain ≥ 0.05 % Si and ≥ 0.01% total Al.
5) The steels should be supplied with microalloying additions, such as B, Nb, Ti and V. The total microalloying content, however, should be ≤ 0.25 %.
6) For improved formability, the steels may be treated with rare earths.
7) For improved atmospheric corrosion resistance, the steels may be ordered with a Cu content of 0.20–0.35 %.

2.3.2 Mechanical properties

TABLE 3.38 Mechanical properties of microalloyed structural steels, conforming to IS 8500–1991

Steel designation	R_m min.	R_e min. for a thickness or diameter, in mm, of				*A* min.	Bend mandrel diameter (180° bend) for a thickness or diameter, in mm of[1),2)]		*KV* min. at a temperature, in °C, of[3)]	
		< 6	≥ 16 ≤ 40	> 40 ≤ 63	> 63		< 12	≥ 12 ≤ 25	25 ± 2	– 20
	MPa	MPa	MPa	MPa	MPa	%			J	J
Fe 440	440	300	290	280	[4)]	22	2*a*	3*a*	–	–
Fe 440 B	440	300	290	280	[4)]	22	2*a*	3*a*	50	30
Fe 490	490	350	330	320	[4)]	22	2*a*	3*a*	–	–
Fe 490 B	490	350	330	320	[4)]	22	2*a*	3*a*	50	25
Fe 540	540	410	390	380	[4)]	20	2*a*	3*a*	–	–
Fe 540 B	540	410	390	380	[4)]	20	2*a*	3*a*	50	25
Fe 570	570	450	430	420	[4)]	20	2*a*	3*a*	–	–
Fe 570 B	570	450	430	420	[4)]	20	2*a*	3*a*	45	20
Fe 590	590	450	430	420	[4)]	20	2*a*	3*a*	–	–
Fe 590 B	590	450	430	420	[4)]	20	2*a*	3*a*	45	20

1) *a* = thickness or diameter of the bend test piece.
2) For thicknesses or diameters > 25 mm, the bend test is not required.
3) Average of three tests; no individual value should be less than 70 % of the specified minimum average value.
4) By agreement between the purchaser and the manufacturer.

Note: The steels may be supplied in the as-rolled or normalized condition.

3 EQUIVALENT GRADES

TABLE 3.39 Comparison of grades of structural steel (ordinary quality) specified in IS 1977 – 1975 with nearest equivalent grades specified in other national and international standards

IS 1977–75	ASTM A 283/A 283M–93a	BS EN 10025–93	DIN EN 10025–94	ISO 630–95	JIS G 3101–87
Fe 310-0	A	S185	S185	E 185 O	SS 330
Fe 410-0	–	–	–	–	SS 400

TABLE 3.40 Comparison of grades of steel for general structural purposes specified in IS 2062 – 1992 with nearest equivalent grades specified in other national and international standards

IS 2062–92	ASTM A 283/A 283M–93a ASTM A 284/A 284M–90	BS EN 10025–93	DIN EN 10025–94	ISO 630–95	JIS G 3106–92
Fe 410 W A	A 283/A 283M - D	S275	–	E 275 A	SM400A
Fe 410 W B	A 284/A 284M - D	S275J0	S275J0	E 275 C	SM400B
Fe 410 W C	–	S275J2G3 S275J2G4	S275J2G3 S275J2G4	E 275 D	–

TABLE 3.41 Comparison of grades of microalloyed structural steel specified in IS 8500 – 1991 with nearest equivalent steel grades specified in other national and international standards

IS 8500–91	ASTM A 572/A 572M–94b ASTM A 588/A 588M–94 ASTM A 633/A 633M–94a	BS EN 10025–93	DIN EN 10025–94	ISO 630–95	JIS G 3106–92
Fe 440	A 572/A 572M - 42 A 633/A 633M - A	–	–	–	–
Fe 440 B	–	–	–	–	–
Fe 490	A 572/A 572M - 50 A 588/A 588M A 633/A 633M - C, D	S355	S355	–	SM490A

(Contd.)

TABLE 3.41 Comparison of grades of microalloyed structural steel specified in IS 8500 – 1991 with nearest equivalent steel grades specified in other national and international standards (continued)

IS 8500–91	ASTM A 572/A 572M–94b ASTM A 588/A 588M–94 ASTM A 633/A 633M–94a	BS EN 10025–93	DIN EN 10025–94	ISO 630–95	JIS G 3106–92
Fe 490 B	–	S355J2G3 S355J2G4	S355J2G3 S355J2G4	E 355 D	–
Fe 540	A 572/A 572M - 60 A 633/A 633M - E	–	–	–	–
Fe 540 B	–	–	–	–	–
Fe 570	A 572/A 572M - 65	–	–	–	SM570A
Fe 570 B	–	–	–	–	–
Fe 590	–	–	–	–	–
Fe 590 B	–	–	–	–	–

5 REFERENCES

Book

1. *ASM Handbook*, vol. 1, ASM International, Materials Park, Ohio, U.S.A., 1990.

Standards

2. IS 1977 – 1975: *Specification for Structural Steel (Ordinary Quality).*
3. IS 2062 – 1992: *Steel for General Structural Purposes—Specification.*
4. IS 8500 – 1991: *Structural Steel—Microalloyed (Medium and High Strength Qualities)—Specification.*

3.9

Steel Tubes for Structural and Mechanical Purposes

1 INTRODUCTION

Steel tubes for structural and mechanical purposes are seamless and welded tubes which are available in the form of circular, oval, square, rectangular, and special shapes. These tubes are made from unalloyed and low-alloy steels and are produced to specified mechanical properties and are delivered in any one of the following conditions:

a) Hot finished;
b) Cold finished: Tubular products may be cold finished by cold drawing, turning, grinding, polishing, or by a combination of these processes for one or more of the following reasons:
   i) to increase or decrease the diameter;
   ii) to produce shapes other than rounds;
   iii) to improve surface finish;
   iv) to improve dimensional accuracy; and
   v) to increase strength.
c) Heat treated: Hot finished tubes may be annealed, normalized or stress relieved to provide a uniform grain structure or ductility for severe forming operations or cold-drawing. In addition, cold drawn tubes may be annealed or stress relieved to meet the desired specification.

Steel tubes for structural purposes are used for welded, riveted or bolted construction of bridges and buildings, and for general structural purposes. Steel tubes for mechanical purposes are used extensively in automobiles, agricultural machinery, electrical and household equipment.

2 SPECIFICATIONS

2.1 Steel tubes for structural purposes

2.1.1 Chemical composition

TABLE 3.42 Chemical composition (cast analysis) of unalloyed steel tubes for structural purposes, conforming to IS 1161 – 1979

Chemical composition [% (m/m)]				
C	Si	Mn	P	S
–	–	–	≤ 0.060	≤ 0.060

2.1.2 Mechanical properties

TABLE 3.43 Tensile properties of unalloyed steel tubes for structural purposes, conforming to IS 1161 – 1979

Grade	R_m min.	R_e min.	A min.	
			$L_o = 5.65\sqrt{S_o}$	$L_o = 4\sqrt{S_o}$
	MPa	MPa	%	%
YSt 210	330	210	$9500/R_m$[1)]	$11000/R_m$
YSt 240	410	240	$9500/R_m$	$11000/Rm$
YSt 310	540	310	$9500/R_m$	$11000/Rm$

1) $A \geq 12$ % is permitted for tubes with inside diameter ≤ 25 mm.

TABLE 3.44 Flattening test[1)] requirements for unalloyed steel tubes for structural purposes, conforming to IS 1161–1979

Grade	Distance between platens or outside surfaces[2)]
Hot finished welded (HFW) steel tubes	
YSt 210	$5t$ or 2/3 × inside diameter, whichever is smaller
YSt 240	$6t$ or 2/3 × inside diameter, whichever is smaller

(Contd.)

TABLE 3.44 Flattening test[1] requirements for unalloyed steel tubes for structural purposes, conforming to IS 1161 – 1979 (continued)

Grade	Distance between platens or outside surfaces[2]
Hot finished seamless (HFS) steel tubes	
YSt 210	5*t* or 2/3 × inside diameter, whichever is smaller
YSt 240	6*t* or 2/3 × inside diameter, whichever is smaller
YSt 310	8*t* or 7/8 × inside diameter, whichever is smaller
Electric resistance or induction butt welded (ERW) steel tubes	
YSt 210	5*t* or 2/3 × inside diameter, whichever is smaller
YSt 240	6*t* or 2/3 × inside diameter, whichever is smaller
YSt 310	8*t* or 7/8 × inside diameter, whichever is smaller

[1] Applicable to steel tubes with inside diameter:
(a) for HFW steel tubes: > 50 mm and ≤ 100 mm;
(b) for HFS steel tubes: ≤ 100 mm;
(c) for ERW steel tubes: > 50 mm.
[2] *t* = thickness of the tube.

TABLE 3.45 Bend test (whole tube)[1] requirements for unalloyed steel tubes for structural purposes, conforming to IS 1161 – 1979

Grade	Condition of tube	Bend test	
		Angle of bend	Radius of former[2]
HFW steel tubes			
YSt 210, YSt 240, YSt 310	Ungalvanized	180°	6*D*
	Galvanized	90°	8*D*
ERW steel tubes			
YSt 210, YSt 240, YSt 310	Ungalvanized	180°	6*D*
	Galvanized	90°	8*D*

(Contd.)

TABLE 3.45 Bend test (whole tube)[1] requirements for unalloyed steel tubes for structural purposes, conforming to IS 1161 – 1979 (continued)

[1] Applicable to tubes with inside diameter < 50 mm.
[2] D = outside diameter of the tube.

TABLE 3.46 Bend test (strip)[1] requirements for unalloyed steel tubes for structural purposes, conforming to IS 1161 – 1979

Grade	Bend mandrel diameter (180°) bend[2]
HFW steel tubes	
YSt 210, YSt 240, YSt 310	$3t$[3]/$8t$[4]
HFS steel tubes	
YSt 210, YSt 240	$3t$
YSt 310	$5t$

[1] For tubes with inside diameter > 100 mm.
[2] t = thickness of the tube.
[3] For test pieces not including the weld.
[4] For test pieces having the weld in the middle.

2.2 Steel tubes for mechanical and general engineering purposes

2.2.1 Chemical composition

TABLE 3.47 Chemical composition (cast analysis) of unalloyed steel tubes for mechanical and general engineering purposes, conforming to IS 3601 – 1984

Chemical composition [% (m/m)]				
C	Si	Mn	P	S
–	–	–	≤ 0.060	≤ 0.060

2.2.2 Mechanical properties

TABLE 3.48 Tensile properties of unalloyed steel tubes for mechanical and general engineering purposes, conforming to IS 3601 – 1984

Grade	R_m min. MPa	R_e min. MPa	A min. MPa
Welded steel tubes (WT)			
WT 160	310	160	$9500/R_m$
WT 200	330	200	$9500/R_m$
WT 240	410	240	$9500/R_m$
WT 310	540	310	$9500/R_m$
HFS steel tubes			
HFS 160	310	160	$9500/R_m$
HFS 200	330	200	$9500/R_m$
HFS 240	410	240	$9500/R_m$
HFS 310	540	310	$9500/R_m$
Cold drawn seamless (CDS) steel tubes			
Annealed or normalized			
CDS 160	310	160	$9500/R_m$
CDS 200	330	200	$9500/R_m$
CDS 240	410	240	$9500/R_m$
CDS 310	540	310	$9500/R_m$
As-drawn or as-drawn and tempered			
CDS 370	410	370	–
CDS 430	540	430	–
CDS 540	650	540	–

(Contd.)

TABLE 3.48 Tensile properties of unalloyed steel tubes for mechanical and general engineering purposes, conforming to IS 3601 – 1984 (continued)

Grade	R_m min. MPa	R_e min. MPa	A min. MPa
Cold drawn electric resistance welded (CEW) steel tubes			
Annealed or normalized			
CEW 160	310	160	$9500/R_m$
CEW 200	330	200	$9500/R_m$
CEW 240	410	240	$9500/R_m$
As-drawn or as-drawn and tempered			
CEW 370	410	370	–
CEW 430	540	430	–

TABLE 3.49 Flattening test[1] requirements for unalloyed steel tubes for mechanical and general engineering purposes, conforming to IS 3601 – 1984

Grade	Distance between platens or outside surfaces[1]
Welded steel tubes (WT)[2), 3)]	
WT 160	$3t$ or $2/3 \times$ inside diameter, whichever is smaller
WT 200	$4t$ or $2/3 \times$ inside diameter, whichever is smaller
WT 240	$6t$ or $3/4 \times$ inside diameter, whichever is smaller
WT 310	$8t$ or $7/8 \times$ inside diameter, whichever is smaller
HFS[4] steel tubes	
HFS 160	$3t$ or $1/2 \times$ inside diameter, whichever is smaller
HFS 200	$4t$ or $2/3 \times$ inside diameter, whichever is smaller
HFS 240	$6t$ or $3/4 \times$ inside diameter, whichever is smaller
HFS 310	$8t$ or $7/8 \times$ inside diameter, whichever is smaller

(Contd.)

TABLE 3.49 Flattening test[1] requirements for unalloyed steel tubes for mechanical and general engineering purposes, conforming to IS 3601 – 1984 (continued)

Grade	Distance between platens or outside surfaces[1]
CDS[5] steel tubes	
CDS 160	3*t* or 1/2 × inside diameter, whichever is smaller
CDS 200	4*t* or 2/3 × inside diameter, whichever is smaller
CDS 240	6*t* or 3/4 × inside diameter, whichever is smaller
CDS 310	8*t* or 7/8 × inside diameter, whichever is smaller
CDS 370	4*t* or 2/3 × inside diameter, whichever is smaller
CDS 430	6*t* or 3/4 × inside diameter, whichever is smaller
CDS 540	8*t* or 7/8 × inside diameter, whichever is smaller

[1] *t* = thickness of the tube.

[2] In case of oxy-acetylene welded (OAW) steel tubes, the distance between platens or outside surfaces for grade WT 160 should be 5*t*.

[3] In case of HFW steel tubes, the flattening test should be carried out on tubes with outside diameter ≥ 60.3 mm.

[4] Applicable to HFS steel tubes with outside diameter ≤ 114.3 mm.

[5] Applicable to CDS steel tubes with outside diameter ≤ 114.3 mm.

TABLE 3.50 Bend test (whole tube)[1] requirements for unalloyed welded steel tubes (WT) conforming to IS 3601 – 1984

Grade	Condition of tube	Bend test	
		Angle of bend	Radius of former[2]
WT 160, WT 200, WT 240, WT 310	Ungalvanized	180°	6*D*
	Galvanized	90°	8*D*

[1] For tubes with outside diameter ≤ 60.3 mm and manufactured from HFW tubes only.

[2] *D* = outside diameter of the tube.

TABLE 3.51 Bend test (strip)[1] requirements for unalloyed steel tubes for mechanical and general engineering purposes, conforming to IS 3601 – 1984

Designation	Bend mandrel diameter (180° bend) for tube thickness, in mm, of[2]	
	≤9.5	>9.5
	HFS steel tubes	
HFS 160	3*t*	4*t* or 2/3 × inside diameter, whichever is smaller
HFS 200	3*t*	4t or 2/3 × inside diameter, whichever is smaller
HFS 240	4*t*	5t or 7/10 × inside diameter, whichever is smaller
HFS 310	5*t*	6t or 3/4 × inside diameter, whichever is smaller
	CDS steel tubes	
CDS 160	3*t*	4t or 2/3 × inside diameter, whichever is smaller
CDS 200	3*t*	4t or 2/3 × inside diameter, whichever is smaller
CDS 240	4*t*	5t or 7/10 × inside diameter, whichever is smaller
CDS 310	5*t*	6t or 3/4 × inside diameter, whichever is smaller

1) Applicable to tubes with outside diameter > 114.3 mm.

2) *t* = thickness of the tube.

TABLE 3.52 Crushing test requirements for unalloyed cold drawn seamless (CDS) steel tubes, conforming to IS 3601 – 1984

Grade	Minimum reduction of original length after crushing %
CDS 160	50
CDS 200	50
CDS 240	25
CDS 310	25
CDS 370	25
CDS 430	25

TABLE 3.53 Crushing test requirements for unalloyed cold drawn electric resistance welded (CEW) steel tubes, conforming to IS 3601 – 1984

Grade	Minimum reduction of original length after crushing for the treatment condition	
	Annealed or normalized %	As-drawn or as-drawn and tempered %
CEW 160	50	25
CEW 200	50	25
CEW 240	50	25
CEW 370	25	25
CEW 430	25	25

3 REFERENCES

Book

1. *ASM Handbook*, vol. 1, ASM International, Materials Park, Ohio, U.S.A., 1990.

Standards

2. IS 1161 – 1979: *Specification for Steel Tubes for Structural Purposes.*
3. IS 3601 – 1984: *Specification for Steel Tubes for Mechanical and General Engineering Purposes.*

3.10

FREE-CUTTING STEELS

1 INTRODUCTION

Free-cutting steels are steels characterized by good machinability, essentially obtained by higher than normal contents of sulphur, sulphur and phosphorus, or sulphur and lead.

Sulphur is present in free-cutting steels in the form of manganese sulphide inclusions, in amounts of up to 0.33 %. These sulphide inclusions improve machinability by causing the formation of a broken chip and by providing a built-in lubricant that prevents the chips from sticking to the edge and undermining the cutting edge.

Phosphorus is often added to low-carbon resulphurized steels in amounts of 0.04-0.09 % or 0.07-0.12 % to improve their machining characteristics. Phosphorus increases the strength and hardness of the ferrite, an effect that promotes chip breaking in cutting operations. Excessive phosphorus contents, however impairs machining characteristics, and reduces ductility and toughness.

Most of the free-cutting steels can be produced with an addition of 0.15-0.35 % lead. Lead augments the effect of sulphur, permitting a further increase in machining speeds and better surface finish.

Free-cutting steels are more expensive than the corresponding unalloyed steels. Therefore, they should be selected only when the higher cost of the steel can be offset by the lower machining cost.

2 SPECIFICATION

2.1 Chemical composition

TABLE 3.54 Chemical composition (cast analysis) of free-cutting steels, conforming to IS 4431 – 1978

Steel designation	Chemical composition[1)] [% (*m*/*m*)]				
	C	Si	Mn	P	S
10C8S10	≤0.15	0.05–0.30	0.60–0.90	≤0.060	0.08–0.13
14C14S14	0.10–0.18	0.05–0.30	1.20–1.50	≤0.060	0.10–0.18
25C12S14	0.20–0.30	≤0.25	1.00–1.50	≤0.060	0.10–0.18
40C10S18	0.35–0.45	≤0.25	0.80–1.20	≤0.060	0.14–0.22
11C10S25	0.08–0.15	≤0.10	0.80–1.20	≤0.060	0.20–0.30
40C15S12	0.35–0.45	≤0.25	1.30–1.70	≤0.060	0.08–0.15

1) Residual elements up to the following limits are acceptable:
Cr ≤ 0.25 %; Mo < 0.05 %; Ni ≤ 0.25 %; Cu ≤ 0.35 %; V ≤ 0.05 %.
Cr + Mo + Ni + Cu + V ≤ 0.80%

Notes:
1. The steels may be supplied in the semi-killed or fully killed condition. Fully killed steels should contain ≥ 0.10 % Si.
2. If agreed in the order, the manufacturer may add Pb and other additives for the purpose of improving the machinability.

2.2 Mechanical properties

TABLE 3.55 Mechanical properties[1)] of free-cutting steels, conforming to IS 4431 – 1978, in the hot-rolled or normalized condition

Steel designation	R_m MPa	A min. %
10C8S10	365–480	24
14C14S14	430–530	22
25C12S14	490–590	20
40C10S18	540–795	17
11C10S25	365–480	22
40C15S12	640–830	15

1) For diameter or width across flats ≤ 150 mm.

Note: R_e min. = 0.55 x R_m.

TABLE 3.56 Mechanical properties of free-cutting steels, conforming to IS 4431 – 1978, in the cold-drawn condition

Steel designation	Mechanical properties for a diameter or width across flats, in mm, of							
	≤ 20		> 20 ≤ 40		> 40 ≤ 63		> 63	
	R_m min. MPa	A min. %	R_m min. MPa	A min. %	R_m min. MPa	A min. %	R_m min. MPa	A min. %
10C8S10	490	10	450	10	410	13	360	17
14C14S14	540	10	510	11	470	12	430	15
25C12S14	610	8	550	10	510	11	490	13
40C10S18	630	8	590	10	550	11	540	11
11C10S25	490	8	430	11	390	13	360	13
40C15S12	670	7	630	8	610	10	590	11

TABLE 3.57 Mechanical properties of free-cutting steels, conforming to IS 4431 – 1978, in the hardened and tempered condition

Steel designation	Limiting ruling section mm	R_m MPa	R_e min. MPa	A min. %	Minimum Izod impact strength J
40C10S18	30	700–850	480	17	35
	60	600–750	380	18	41
40C15S12	30	800–950	560	16	41
	60	700–850	500	18	48
	100	600–750	420	18	48

3 RECOMMENDED HOT WORKING AND HEAT TREATING TEMPERATURES

TABLE 3.58 Recommended hot working and heat treating temperatures for free-cutting steels, conforming to IS 4431 – 1978

Steel designation	Hot-working temperature °C	Annealing temperature °C	Softening temperature °C	Normalizing temperature °C	Hardening temperature °C	Quenching agent	Tempering temperature °C
10C8S10	1100–850	900–930	650–700	880–910	–	–	–
14C14S14	1100–850	900–930	650–700	880–910	–	–	–
25C12S14	1200–850	860–890	650–700	860–890	860–890	Oil or water	550–660
40C10S18	1200–850	830–860	630–700	830–860	830–860	Oil	550–660
11C10S25	1100–850	900–930	650–700	880–910	–	–	–
40C15S12	1200–850	840–870	600–660	840–870	840–870	Oil	550–660

4 EQUIVALENT GRADES

TABLE 3.59 Cmparison of grades of free-cutting steels specified in IS 4431 – 1978 with nearest equivalent grades specified in other national and international standards

IS 4431–78	BS 970 (1)–91 BS 970 (3)–91	DIN 1651–88	GOST 1414–75	ISO 683 (9)–88	JIS G 4804–83	SAE J403–94
10C8S10	–	15 S 10	–	–	SUM 12	–
14C14S14	214A15 214M15	–	–	17 SMn 20	–	1118
25C12S14	–	–	–	–	–	–
40C10S18	212A42	35 S 20	–	35 SMn 20	–	–
11C10S25	–	10 S 20	–	10 S 20	–	–
40C15S12	–	–	–	–	SUM 42	1141

5 REFERENCES

Book

1. *ASM Handbook*, vol. 1, ASM International, Materials Park, Ohio, U.S.A., 1990.

Standards

2. IS 4431 – 1978: *Specification for Carbon and Carbon-Manganese Free-Cutting Steels.*
3. ISO 683 (9) – 1988: *Heat-treatable Steels, Alloy Steels and Free-Cutting Steels—Part 9: Wrought Free-Cutting Steels.*

3.11

BRIGHT BARS

1 INTRODUCTION

Bright bars are unalloyed and alloy steel bars which are cold finished by cold drawing, turning, grinding or polishing, or by a combination of these processes. These bars are characterized by one or more of the following attributes:

a) Bright, smooth, scale free finish;
b) Closer size tolerance, concentricity and straightness;
c) Improved mechanical properties;
d) Improved machinability; and
e) Freedom from surface imperfections and decarburization.

Cold drawn bars: Bars of various cross-sectional shapes obtained, after descaling, by drawing of hot-rolled bars or rods through a die. Cold drawing of hot-rolled bars increases the strength and hardness. At the same time it also improves surface finish, dimensional accuracy and machinability.

Turned (or peeled) bars: Bars of circular cross-section produced by turning on a lathe (or peeling) followed by straightening and polishing. The surface finish of these bars is smoother and brighter than that obtained by cold drawing, but the size tolerances are the same as those for cold-drawn bars. Turned bars are generally free from rolling defects and decarburization. The mechanical properties of these turned bars are essentially the same as those of the bar steel used.

Ground bars: Drawn or turned bars of circular cross-section, whose surface quality and dimensional accuracy are improved by grinding or by grinding and polishing.

2 SPECIFICATIONS

2.1 Chemical composition

TABLE 3.60 Chemical composition of bright bars specified in Indian Standards

Steel designation	Chemical composition [% (*m*/*m*)]				
	C	Si	Mn	P	S
Unalloyed steels (IS 1570 (II) – 1979)[1), 2)]					
10C4	≤0.15	–	0.30–0.60	–	–
15C8	0.10–0.20	–	0.60–0.90	–	–
20C8	0.15–0.25	–	0.60–0.90	–	–
30C8	0.25–0.35	–	0.60–0.90	–	–
40C8	0.35–0.45	–	0.60–0.90	–	–
50C4	0.45–0.55	–	0.30–0.60	–	–
55C8	0.50–0.60	–	0.60–0.90	–	–
Free-cutting steels (IS 4431 – 1978)[3), 4), 5)]					
10C8S10	≤0.15	0.05–0.30	0.60–0.90	0.060	0.08–0.13
14C14S14	0.10–0.18	0.05–0.30	1.20–1.50	0.060	0.10–0.18
25C12S14	0.20–0.30	≤0.25	1.00–1.50	0.060	0.10–0.18
40C10S18	0.35–0.45	≤0.25	0.80–1.20	0.060	0.14–0.22
11C10S25	0.08–0.15	≤0.10	0.80–1.20	0.060	0.20–0.30
40C15S12	0.35–0.45	≤0.25	1.30–1.70	0.060	0.08–0.15

1) The steels may be supplied in the semi-killed or fully killed condition.
2) The limits for P and S may be agreed between the purchaser and the manufacturer.
3) Residual elements up to the following limits are acceptable:
Cr ≤ 0.25 %; Mo ≤ 0.05 %; Ni ≤ 0.25 %; Cu ≤ 0.35 %; V ≤ 0.05 %.
Cr + Mo + Ni + Cu + V ≤ 0.80 %.
4) The steels may be supplied in the semi-killed or fully killed condition. Fully killed steels should contain ≥ 0.10 % Si.
5) If agreed in the order, the manufacturer may add Pb and other additives for the purpose of improving the machinability.

2.2 Mechanical properties

TABLE 3.61 Mechanical properties of bright bars specified in Indian Standards

Steel designation	Normalized		Cold-drawn							
	Mechanical properties for a diameter or width across flats, in mm, of									
	≤ 150		≤ 20		> 20 ≤ 40		> 40 ≤ 63		> 63	
	R_m MPa	A min. %	R_m min. MPa	A min. %	R_m min. MPa	A min. %	R_m min. MPa	A min. %	R_m min. MPa	A min. %
Unalloyed steels (IS 1570 (II) – 1979)										
10C4	340–420	26	490	11	450	13	410	15	360	18
15C8	420–500	25	540	11	510	13	470	15	430	18
20C8	440–520	24	540	10	510	12	470	15	430	18
30C8	500–600	21	610	9	570	10	530	12	490	15
40C8	580–680	18	640	8	610	9	570	10	540	12
50C4	660–780	13	670	7	630	8	610	9	590	10
55C8	≥ 720	13	730	7	690	8	670	9	630	10
Free-cutting steels (IS 4431 – 1978)										
10C8S10	365–480	24	490	10	450	10	410	13	360	17
14C14S14	430–530	22	540	10	510	11	470	12	430	15
25C12S14	490–590	20	610	8	550	10	510	11	490	13

(Contd.)

TABLE 3.61 Mechanical properties of bright bars specified in Indian Standards (continued)

Steel designation	Normalized		Cold-drawn							
	Mechanical properties for a diameter or width across flats, in mm, of									
	≤150		≤20		>20≤40		>40≤63		>63	
	R_m MPa	A min. %	R_m min. MPa	A min. %	R_m min. MPa	A min. %	R_m min. MPa	A min. %	R_m min. MPa	A min. %
Free-cutting steels (IS 4431 – 1978)										
40C10S18	540–795	17	630	8	590	10	550	11	540	11
11C10S25	365–480	22	490	8	430	11	390	13	360	13
40C15S12	640–830	15	670	7	630	8	610	10	590	11

Notes:
1) R_e min. for normalized bars ≈ 0.55 × R_m min.
2) R_e min. for cold drawn bars ≈ 0.65–0.80 × R_m min.

3 EQUIVALENT GRADES

TABLE 3.62 Comparison of grades of unalloyed steel bright bars specified in IS 1570 (II) – 1979 with nearest equivalent grades specified in other national and international standards

IS 1570 (II)–79	ASTM A 29/A 29M–93 A 108–93 A 311/A 311M–90b	BS 970 (3)–91	DIN 1652 (3)–90 DIN 1652 (4)–90	GOST 1050–88 GOST 4543–71	ISO 683 (XVIII)–76	JIS G 4051–79	SAE J403–94
10C4	1010 1012	–	C 10 Ck 10	10ЧС 10	C10e C10ea C10eb C10	S 10 C S 12 C S 09 CK	1010 1012
15C8	1013 1016 1018	080A15 080M15	–	15Г	–	–	1016 1018
20C8	1018 1019 1021 1022	070M20	–	20Г	–	–	1018 1021 1022
30C8	1029 1030	080A30 080M30	–	30 30Г	C30e C30ea C30eb C30	S 28 C S 30 C S 33 C	1029 1030

(Contd.)

TABLE 3.62 Comparison of grades of unalloyed steel bright bars specified in IS 1570 (II) – 1979 with nearest equivalent grades specified in other national and international standards (continued)

IS 1570 (II)–79	ASTM A 29/A 29M–93 A 108–93 A 311/A 311M–90b	BS 970 (3)–91	DIN 1652 (3)–90 DIN 1652 (4)–90	GOST 1050–88 GOST 4543–71	ISO 683 (XVIII)–76	JIS G 4051–79	SAE J403–94
40C8	1038 1039 1040	080M40 080A42	–	40 40Г	C40e C40ea C40eb C40	S 38 C S 40 C S 43 C	1038 1039 1040
50C4	–	–	–	–	–	–	–
55C8	1055	070M55	–	55	C55e C55ea C55eb C55	S 53 C S 55 C S 58 C	1055

TABLE 3.63 Comparison of grades of free-cutting steel bright bars specified in IS 4431 – 1978 with nearest equivalent grades specified in other national and international standards

IS 4431–78	ASTM A 29/A 29M–93a A 108–93 A 311/A 311M–90b	BS 970 (3)–91	DIN 1651–88	GOST 1414–75	ISO 683 (9)–88	JIS G 4804–83	SAE J403–94
10C8S10	1108 1109	–	15 S 10	–	–	SUM 12	–
14C14S14	1118	–	–	–	17 SMn 20	–	1118
25C12S14	–	–	–	–	–	–	–
40C10S18	–	212A42	35 S 20	–	35 SMn 20	–	–
11C10S25	–	–	10 S 20	–	10 S 20	–	–
40C15S12	1141	–	–	–	–	SUM 42	1141

4 REFERENCES

Book/Article

1. *ASM Handbook*, vol. 1, ASM International, Materials Park, Ohio, U.S.A., 1990.
2. B.A. MacKenzie, "Cutting Costs With Prefinished Steel Bar", *Machine Design*, 53(7), 83–88, 1981.

Standards

3. IS 1570 (II) – 1979: *Schedules for Wrought Steels—Part II: Carbon Steel (Unalloyed Steels).*
4. IS 1570 (III) – 1979: *Schedules for Wrought Steels—Part III: Carbon and Carbon-Manganese Free-Cutting Steels.*
5. ISO 6929 – 1987: *Steel products—Definitions and Classification.*

3.12

STEELS FOR COLD HEADING AND COLD EXTRUDING

1 INTRODUCTION

Steels for cold heading and cold extruding are unalloyed, low-alloy and stainless steels which are used in the production of solid or hollow shapes by means of cold plastic deformation by heading or cold extrusion. Steels of this quality are produced by closely controlled steelmaking practices to ensure internal soundness, freedom from injurious segregation, freedom from detrimental surface imperfections and low non-metallic inclusion content.

Most cold heading and cold extruding steels are low-carbon and medium-carbon grades. For grades with a carbon content of 0.30 % or more, an anneal or spheroidize anneal may be required to obtain the proper hardness and microstructure for cold working.

A steel selected for cold heading or cold extrusion should possess low strength and hardness for ease of forming, high ductility to avoid fracture, and high machinability rating for finish machining. If a part is to be heat treated after cold forming, these properties must be balanced against the response needed to develop final desired level of strength, hardness, toughness and wear resistance.

2 SPECIFICATIONS

2.1 Unalloyed and alloy steels for cold heading and cold extruding

2.1.1 Chemical composition

TABLE 3.64 Chemical composition (cast analysis) of unalloyed and alloy steels for cold heading and cold extruding, conforming to IS 11169 (1) – 1984

Steel designation	Chemical composition[1)] [% (m/m)]								
	C	Si	Mn	P max.	S max.	Cr	Mo	Ni	B
Steels not intended for heat treatment[2)]									
4C2	≤0.08	–	≤0.40	0.035	0.035	–	–	–	–
5C4	≤0.10	–	≤0.50	0.035	0.035	–	–	–	–
7C4	≤0.12	–	≤0.50	0.035	0.035	–	–	–	–
10C4	≤0.15	–	0.30–0.60	0.035	0.035	–	–	–	–
14C6	0.10–0.18	–	0.40–0.70	0.035	0.035	–	–	–	–
15C4	≤0.20	–	0.30–0.60	0.035	0.035	–	–	–	–
15C8	0.10–0.20	–	0.60–0.90	0.035	0.035	–	–	–	–
20C8	0.15–0.25	–	0.60–0.90	0.035	0.035	–	–	–	–
25C4	0.20–0.30	–	0.30–0.60	0.035	0.035	–	–	–	–
25C8	0.20–0.30	–	0.60–0.90	0.035	0.035	–	–	–	–
Steels for hardening and tempering (including boron-treated steels)[3)]									
20C8	0.15–0.25	–	0.60–0.90	0.035	0.035	–	–	–	–
25C8	0.20–0.30	–	0.60–0.90	0.035	0.035	–	–	–	–
30C8	0.25–0.35	0.10–0.35	0.60–0.90	0.035	0.035	–	–	–	–
35C8	0.30–0.40	0.10–0.35	0.60–0.90	0.035	0.035	–	–	–	–
40C8	0.35–0.45	0.10–0.35	0.60–0.90	0.035	0.035	–	–	–	–

(Contd.)

TABLE 3.64 Chemical composition (cast analysis) of unalloyed and alloy steels for cold heading and cold extruding, conforming to IS 11169 (1) – 1984 (continued)

Steel designation	Chemical composition[1)] [% (m/m)]								
	C	Si	Mn	P max.	S max.	Cr	Mo	Ni	B
Steels for hardening and tempering (including boron-treated steels)[3)]									
45C8	0.40–0.50	0.10–0.35	0.60–0.90	0.035	0.035	–	–	–	–
20C15	0.16–0.24	0.10–0.35	1.30–1.70	0.035	0.035	–	–	–	–
27C15	0.22–0.32	0.10–0.35	1.30–1.70	0.035	0.035	–	–	–	–
37C15	0.32–0.42	0.10–0.35	1.30–1.70	0.035	0.035	–	–	–	–
35Mn6Mo3	0.30–0.40	0.10–0.35	1.30–1.80	0.035	0.035	–	0.20–0.35	–	–
35Mn6Mo4	0.30–0.40	0.10–0.35	1.30–1.80	0.035	0.035	–	0.35–0.55	–	–
40Cr4	0.35–0.45	0.10–0.35	0.60–0.90	0.035	0.035	0.90–1.20	–	–	–
40Cr4Mo3	0.35–0.45	0.10–0.35	0.50–0.80	0.035	0.035	0.90–1.20	0.20–0.35	–	–
25Cr13Mo6	0.20–0.30	0.10–0.35	0.40–0.70	0.035	0.035	2.90–3.40	0.45–0.65	–	–
40Ni14	0.35–0.45	0.10–0.35	0.50–0.80	0.035	0.035	≤0.30	–	3.20–3.60	–
35Ni5Cr2	0.30–0.40	0.10–0.35	0.60–0.90	0.035	0.035	0.45–0.75	–	1.00–1.50	–
30Ni16Cr5	0.26–0.34	0.10–0.35	0.40–0.70	0.035	0.035	1.10–1.40	–	3.90–4.30	–
21C10BT	0.18–.23	0.15–0.30	0.80–1.10	0.035	0.035	–	–	–	0.0005–0.003
26C10BT	0.23–0.29	0.15–0.30	0.90–1.20	0.035	0.035	–	–	–	0.0005–0.003
34C14BT	0.32–0.37	0.15–0.30	1.20–1.50	0.035	0.035	–	–	–	0.0005–0.003
38Cr4Mn2BT	0.35–0.40	0.15–0.30	0.30–0.50	0.035	0.035	0.95–1.15	–	–	0.0005–0.003
Case hardening steels[3)]									
10C4	≤0.15	0.05–0.35	0.30–0.60	0.035	0.035	–	–	–	–
14C6	0.10–0.18	0.05–0.35	0.40–0.70	0.035	0.035	–	–	–	–

(Contd.)

TABLE 3.64 Chemical composition (cast analysis) of unalloyed and alloy steels for cold heading and cold extruding, conforming to IS 11169 (1) – 1984 (continued)

Steel designation	Chemical composition[1] [% (*m*/*m*)]								
	C	Si	Mn	P max.	S max.	Cr	Mo	Ni	B
Case hardening steels[3]									
15C8	0.10–0.20	0.10–0.35	0.60–0.90	0.035	0.035	–	–	–	–
11C15	≤0.16	0.10–0.35	1.30–1.70	0.035	0.035	–	–	–	–
15Cr3	0.12–0.18	0.10–0.35	0.40–0.60	0.035	0.035	0.50–0.80	–	–	–
16Mn5Cr4	0.14–0.19	0.10–0.35	1.00–1.30	0.035	0.035	0.80–1.10	–	–	–
20Mn5Cr5	0.17–0.22	0.10–0.35	1.00–1.40	0.035	0.035	1.00–1.30	–	–	–
16Ni3Cr2	0.12–0.20	0.10–0.35	0.60–1.00	0.035	0.035	0.40–0.80	–	0.60–1.00	–
16Ni4Cr3	0.12–0.20	0.10–0.35	0.60–1.00	0.035	0.035	0.60–1.00	–	0.80–1.20	–
13Ni13Cr3	0.10–0.15	0.10–0.35	0.40–0.70	0.035	0.035	0.60–1.00	–	3.00–3.50	–
15Ni16Cr5	0.12–0.18	0.10–0.35	0.40–0.70	0.035	0.035	1.00–1.40	–	3.80–4.30	–
20Ni7Mo2	0.17–0.22	0.10–0.35	0.45–0.70	0.035	0.035	–	0.20–0.30	1.65–2.00	–
20Ni2Cr2Mo2	0.18–0.23	0.10–0.35	0.70–0.90	0.035	0.035	0.40–0.60	0.15–0.25	0.40–0.70	–
15Ni5Cr4Mo1	0.12–0.18	0.10–0.35	0.60–1.00	0.035	0.035	0.75–1.25	0.08–0.15	1.00–1.50	–
15Ni7Cr4Mo2	0.12–0.18	0.10–0.35	0.60–1.00	0.035	0.035	0.75–1.25	0.10–0.20	1.50–2.00	–
16Ni8Cr6Mo2	0.12–0.20	0.10–0.35	0.40–0.70	0.035	0.035	1.40–1.70	0.15–0.25	1.80–2.20	–

1) Residual elements up to the following limits are acceptable:
Cr ≤ 0.15 %; Mo ≤ 0.05 %; Ni ≤ 0.15 %; Cu ≤ 0.15 %; Sn ≤ 0.02 %; V ≤ 0.05 %.
Cr + Mo + Ni+ Cu + Sn + V ≤ 0.40 %.
2) The steels may be supplied in the fully killed, semi-killed or rimmed condition.
3) All steels should be fully killed.

2.1.2 Hardness

TABLE 3.65 Hardness requirements for cold heading and cold extruding steels, conforming to IS 11169 (1) – 1984, in the annealed condition

Steel designation	Maximum Brinell hardness HB		Steel designation	Maximum Brinell hardness HB
Steels not intended for heat treatment				
4C2	120		**15C4**	140
5C4	120		**15C8**	140
7C4	120		**20C8**	150
10C4	130		**25C4**	150
14C6	140		**25C8**	150
Steels for hardening and tempering (including boron- treated steels)				
20C8	150		**35Mn6Mo4**	190
25C8	150		**40Cr4**	200
30C8	160		**40Cr4Mo3**	200
35C8	160		**25Cr13Mo6**	190
40C8	170		**40Ni 14**	190
45C8	170		**35Ni5Cr2**	190
20C15	170		**30Ni16Cr5**	200
27C15	190		**21C10BT**	160
37C15	190		**26C10BT**	160
35Mn6Mo3	190		**34C14BT**	190
			38Cr4Mn2BT	200

(Contd.)

TABLE 3.65 Hardness requirements for cold heading and cold extruding steels, conforming to IS 11169 (1) – 1984, in the annealed condition (continued)

Steel designation	Maximum Brinell hardness HB		Steel designation	Maximum Brinell hardness HB
Case hardening steels				
10C4	130		**16Ni4Cr3**	180
14C6	140		**13Ni13Cr3**	190
15C8	140		**15Ni16Cr5**	200
11C15	170		**20Ni7Mo2**	180
15Cr3	140		**20Ni2Cr2Mo2**	180
16Mn5Cr4	160		**15Ni5Cr4Mo1**	180
20Mn5Cr5	170		**15Ni7Cr4Mo2**	180
16Ni3Cr2	160		**16Ni8Cr6Mo2**	190

2.1.3 Mechanical properties

I. Steels for hardening and tempering

TABLE 3.66 Mechanical properties of cold heading and cold extruding steels for hardening and tempering, conforming to IS 11169 (1) – 84, in the hardened and tempered condition

Steel designation	Limiting ruling section mm	R_m MPa	$R_{p0.2}$ min. MPa	*A* min. %	*Z* min. %	Minimum Izod impact strength J
20C8	16	540–690	370	20	45	39
	40	490–640	295	22	50	39
	100	–	–	–	–	–

(Contd.)

TABLE 3.66 Mechanical properties of cold heading and cold extruding steels for hardening and tempering, conforming to IS 11169 (1) – 84, in the hardened and tempered condition (continued)

Steel designation	Limiting ruling section mm	R_m MPa	$R_{p0.2}$ min. MPa	*A* min. %	*Z* min. %	Minimum Izod impact strength J
25C8	16	580–730	390	18	45	–
	40	540–690	330	20	50	30
	100	–	–	–	–	–
35C8	16	620–770	420	17	40	29
	40	580–730	365	19	45	29
	100	540–690	325	20	50	29
40C8	16	660–810	450	16	40	–
	40	620–770	390	18	45	20
	100	580–730	340	19	50	20
45C8	16	700–850	460	14	35	20
	40	660–810	410	16	40	20
	100	620–770	375	17	45	20
40Cr4	16	1000–1200	800	11	40	25
	40	900–1180	680	12	45	29
	100	800–950	570	14	50	29
40Cr4Mo3	16	1100–1300	900	10	40	25
	40	1000–1200	750	11	45	29
	100	900–1100	700	12	50	34

II. Case hardening steels

TABLE 3.67 Mechanical properties of cold heading and cold extruding case hardening steels, conforming to IS 11169 (1) – 84, in the simulated case hardened condition

Steel designation	Limiting ruling section mm	R_m min. MPa	*A* min. %	Minimum Izod impact strength J
10C4	15	500	17	55
14C6	30	500	17	55
15C8	30	550	14	55
11C15	30	600	17	55
15Cr3	30	600	13	48
16Mn5Cr4	30	800	10	35
20Mn5Cr5	30	1000	8	38
16Ni3Cr2	30	700	15	41
16Ni4Cr3	30	850	12	41
	60	800	–	–
	90	750	–	–
13Ni13Cr3	60	850	12	48
	100	800	–	–
15Ni16Cr5	30	1350	9	35
	60	1200	–	–
	90	1150	–	–
20Ni7Mo2	30	850	12	62
	60	700	–	–

TABLE 3.67 Mechanical properties of cold heading and cold extruding case hardening steels, conforming to IS 11169 (1) – 84, in the simulated case hardened condition (continued)

Steel designation	Limiting ruling section mm	R_m min. MPa	*A* min. %	Minimum Izod impact strength J
20Ni2Cr2Mo2	30	900	11	41
	60	800	–	–
	90	750	–	–
15Ni5Cr4Mo1	30	1000	9	41
	90	950	–	–
15Ni7Cr4Mo2	30	1100	9	35
	60	1000	–	–
	90	950	–	–
16Ni8Cr6Mo2	30	1350	9	35
	60	1200	–	–
	90	1150	–	–

2.2 Cold heading and cold extruding stainless steels

2.2.1 Chemical composition

TABLE 3.68 Chemical composition (cast analysis) of cold heading and cold extruding stainless steels, conforming to IS 11169 (2) – 1989

Steel designation	Chemical composition [% (m/m)]								
	C	Si max.	Mn max.	P max.	S max.	Cr	Mo	Ni	Ti
Ferritic stainless steel									
X07Cr17	≤0.12	1.00	1.00	0.045	0.030	16.0–18.0	–	–	–
Martensitic stainless steel									
X12Cr12	0.08–0.15	1.00	1.00	0.045	0.030	11.5–13.5	–	–	–
Austenitic stainless steels									
X02Cr19Ni10	≤0.030	1.00	2.00	0.045	0.030	17.5–20.0	–	8.0–12.0	–
X04Cr19Ni10	≤0.08	1.00	2.00	0.045	0.030	17.0–20.0	–	10.5–12.0	–
X04Cr17Ni12Mo2	≤0.08	1.00	2.00	0.045	0.030	16.0–18.0	2.0–3.0	10.0–14.0	–
X04Cr18Ni10Ti	≤0.08	1.00	2.00	0.045	0.030	17.0–19.0	–	9.0–12.0	≥5× %C≤ 0.80
X04Cr17Ni12Mo2Ti	≤0.08	1.00	2.00	0.045	0.030	16.0–18.0	2.0–3.0	10.0–14.0	≥5× %C≤ 0.80
X02Cr17Ni12Mo2	≤0.030	1.00	2.00	0.045	0.030	16.0–18.0	2.0–3.0	10.0–14.0	–

2.2.2 *Mechanical properties*

TABLE 3.69 Mechanical properties of cold heading and cold extruding stainless steels, conforming to IS 11169 (2) – 1989, in the treatment condition in which they are usually supplied

Steel designation	Treatment condition[1] for ferritic and martensitic stainless steels					
	AC or AC + P		C + AC		C + AC + LC	
	Treatment condition[1] for austenitic stainless steels					
	Q or Q + P		C + Q		C + Q + LC	
	R_m max. MPa	Z min. %	R_m max. MPa	Z min. %	R_m max. MPa	Z min. %
Ferritic stainless steel						
X07Cr17	570	63	570	60	620	65
Martensitic stainless steel						
X12Cr12	600	60	600	62	640	62
Austenitic stainless steels						
X02Cr19Ni10	630	55	630	55	680	55
X04Cr19Ni10	680	55	680	55	730	55
X04Cr17Ni12Mo2	680	55	680	55	730	55
X04Cr18Ni10Ti	680	55	680	55	730	55
X04Cr17Ni12Mo2Ti	680	55	680	55	730	55
X02Cr17Ni12Mo2	680	55	680	55	730	55

[1] AC = annealed; P = peeled; C = cold-drawn; LC = lightly cold-reduced; Q = quenched.

TABLE 3.70 Mechanical properties of cold heading and cold extruding stainless steels conforming to IS 11169 (2) – 1989, in the "annealed" (AC), "quenched and tempered" (Q + T) or "quenched" (Q) condition

Steel designation	Heat treatment condition[1]	R_m MPa	$R_{p0.2}$ min. MPa	A min. %	KU min. J
Ferritic stainless steel					
X07Cr17	AC	450–600	270	20	–
Martensitic stainless steel					
X12Cr12	Q + T	600–750	450	18	50
Austenitic stainless steels					
X02Cr19Ni10	Q	450–700	175	50	60
X04Cr19Ni10	Q	500–700	185	50	60
X04Cr17Ni12Mo2	Q	500–700	205	45	60
X04Cr18Ni10Ti	Q	500–750	205	40	60
X04Cr17Ni12Mo2Ti	Q	500–750	225	40	60
X02Cr17Ni12Mo2	Q	500–750	225	40	60

[1] AC = annealed; Q = quenched; T = tempered.

3 EQUIVALENT GRADES

TABLE 3.71 Comparison of grades of cold heading and cold extruding steels not intended for heat treatment specified in IS 11169 (1) – 1984 with nearest equivalent grades specified in other national and international standards

IS 11169 (1)–84	ASTM	BS 3111 (1)–87	DIN 1654 (2)–89	ISO 4954–93	JIS G 3507–91	SAE J403–94
4C2	–	–	QSt 32–3	CC 4 X CC 4 A	SWRCH6R SWRCH6A	1006
5C4	–	–	QSt 34–3	CC 8 X CC 8 A	SWRCH8R SWRCH8A	1008

(Contd.)

TABLE 3.71 Comparison of grades of cold heading and cold extruding steels not intended for heat treatment specified in IS 11169 (1) – 1984 with nearest equivalent grades specified in other national and international standards (continued)

IS 11169 (1)–84	ASTM	BS 3111 (1)–87	DIN 1654 (2)–89	ISO 4954–93	JIS G 3507–91	SAE J403–94
7C4	–	–	QSt 36–3	CC 11 X CC 11 A	SWRCH10R SWRCH10A SWRCH10K	1010
10C4	–	0	QSt 36–3	CC 11 X CC 11 A	SWRCH10R SWRCH10A SWRCH10K SWRCH12R SWRCH12A SWRCH12K	1010 1012
14C6	–	0	–	–	SWRCH15R SWRCH15A SWRCH15K	1015
15C4	–	0	QSt 38–3	CC 15 X CC 15 K CC 15 A	SWRCH12R SWRCH12A SWRCH12K SWRCH15R SWRCH15A SWRCH15K SWRCH17R SWRCH17K	1012 1015 1017
15C8	–	0	–	–	SWRCH16A SWRCH16K SWRCH18A SWRCH18K SWRCH19A	1016 1018
20C8	–	0	–	–	SWRCH18A SWRCH18K SWRCH19A SWRCH22A SWRCH22K	1018 1021 1022
25C4	–	0	–	–	SWRCH25K	1023 1025
25C8	–	0	–	–	–	1026 1029

TABLE 3.72 Comparison of grades of cold heading and cold extruding steels for hardening and tempering specified in IS 11169 (1) – 1984 with nearest equivalent grades specified in other national and international standards

IS 11169 (1)–84	ASTM	BS 3111 (1)–87	DIN 1654 (4)–89	ISO 4954–93	JIS G 3507–91 JIS G 3508–91 JIS G 3545–91	SAE J403–94 SAE J404–94 SAE J1268–93
20C8	–	0/4	–	–	SWRCH18K SWRCH22K	1018 1021 1022
25C8	–	–	–	–	–	1026 1029
30C8	–	1/1	–	CE 28 E4	SWRCH30K SWRCH33K	1029 1030
35C8	–	1/1 1/2	Cq 35	CE 35 E4	SWRCH33K SWRCH35K SWRCH38K	1035
40C8	–	1/2 1/3	–	CE 40 E4	SWRCH38K SWRCH40K SWRCH43K	1038 1038H 1039 1040
45C8	–	1/3	Cq 45	CE 45 E4	SWRCH43K SWRCH45K SWRCH48K	1042 1043 1045 1045H 1046
20C15	–	–	–	–	SWRCH24K	1524 1524H
27C15	–	–	–	–	SWRCH27K	1527
37C15	–	–	–	–	SWRCH41K	1541 1541H
35Mn6Mo3	–	–	–	–	–	–
35Mn6Mo4	–	–	–	–	–	–

(Contd.)

TABLE 3.72 Comparison of grades of cold heading and cold extruding steels for hardening and tempering specified in IS 11169 (1) – 1984 with nearest equivalent grades specified in other national and international standards (continued)

IS 11169 (1)–84	ASTM	BS 3111 (1)–87	DIN 1654 (4)–89	ISO 4954–93	JIS G 3507–91 JIS G 3508–91 JIS G 3545–91	SAE J403–94 SAE J404–94 SAE J1268–93
40Cr4	–	3/1 3/2	37 Cr 4 41 Cr 4	37 Cr 4 E 41 Cr 4 E	–	5140 5140H
40Cr4Mo3	–	5/1 5/2	42 CrMo 4	42 CrMo 4 E	–	4140 4140H 4142 4142H
25Cr13Mo6	–	–	–	–	–	–
40Ni14	–	–	–	–	–	–
35Ni5Cr2	–	–	–	–	–	–
30Ni16Cr5	–	–	–	–	–	–
21C10BT	–	9/1	19 MnB 4	CE 20 B G2	SWRCHB 420	15B21H
26C10BT	–	–	–	–	SWRCHB 526	–
34C14BT	–	–	–	35 MnB 5 E	SWRCHB 734	15B34
38Cr4Mn2BT	–	–	–	–	–	–

TABLE 3.73 Comparison of grades of cold heading and cold extruding case hardening steels specified in IS 11169 (1) – 1984 with nearest equivalent grades specified in other national and international standards

IS 11169 (1)–84	ASTM	BS 3111 (1)–87	DIN 1654 (3)–89	ISO 4954–93	JIS G 3507–91 JIS G 3539–91	SAE J403–94 SAE J404–94 SEA J1268–93
10C4	–	0/2	–	CE 10	SWRCH10K SWCH10K SWRCH12K SWCH12K	1010 1012

(Contd.)

TABLE 3.73 Comparison of grades of cold heading and cold extruding case hardening steels specified in IS 11169 (1) – 1984 with nearest equivalent grades specified in other national and international standards (continued)

IS 11169 (1)–84	ASTM	BS 3111 (1)–87	DIN 1654 (3)–89	ISO 4954–93	JIS G 3507–91 JIS G 3539–91	SAE J403–94 SAE J404–94 SEA J1268–93
14C6	–	–	Cq 15	CE 15 E4	SWRCH15K SWCH15K	1015
15C8	–	0/3	–	CE 16 E4	SWRCH16K SWCH16K SWRCH18K SWCH18K	1016 1018
11C15	–	–	–	–	–	–
15Cr3	–	–	17 Cr 3	–	–	–
16Mn5Cr4	–	–	16 MnCr 5	16 MnCr 5 E	–	–
20Mn5Cr5	–	–	–	–	–	–
16Ni3Cr2	–	–	–	–	–	–
16Ni4Cr3	–	–	–	–	–	–
13Ni13Cr3	–	–	–	–	–	–
15Ni16Cr5	–	–	–	–	–	–
20Ni7Mo2	–	–	–	–	–	4620 4620 H
20Ni2Cr2Mo2	–	–	20 NiCrMo 2	20 NiCrMo 2 E	–	8620 8620 H
15Ni5Cr4Mo1	–	–	–	–	–	–
15Ni7Cr4Mo2	–	–	–	–	–	–
16Ni8Cr6Mo2	–	–	–	–	–	–

TABLE 3.74 Comparison of grades of cold heading and cold extruding stainless steels specified in IS 11169 (2) – 1989 with nearest equivalent grades specified in other national and international standards

IS 11169(2)–89	ASTM A 493–93	BS 3111 (2)–79	DIN 1654 (5)–89	ISO 4954–93	JIS G 4315–94	SAE J405–89
X07Cr17	S43000	430S15	X 6 Cr 17	X 6 Cr 17 E	SUS 430	51430
X12Cr12	S41000	410S22	X 10 Cr 13	X 12 Cr 13 E	SUS 410	51410
X02Cr19Ni10	–	–	X2 CrNi 19 11	X 2 CrNi 18 10 E	–	30304L
X04Cr19Ni10	S30500	305S17	X 5 CrNi 18 12	X 5 CrNi 18 12 E	SUS 305J1	30305
X04Cr17Ni12Mo2	S31600	316S17	X 5 CrNiMo 17 12 2	X 5 CrNiMo 17 12 2 E	SUS 316	30316
X04Cr18Ni10Ti	–	–	X 6 CrNiTi 18 10	X 6 CrNiTi 18 10 E	–	30321
X04Cr17Ni12Mo2Ti	–	–	X 6 CrNiMoTi 17 12 2	X 6 CrNiMoTi 17 12 2 E	–	–
X02Cr17Ni12Mo2	–	–	–	X 2 CrNiMo 17 13 3 E	–	30316L

4 REFERENCES

Books

1. *ASM Handbook*, vol. 1, ASM International, Materials Park, Ohio, U.S.A., 1990.
2. *Source Book on Cold Forming*, American Society for Metals, Metals Park, Ohio, U.S.A., 1975.

Standards

3. IS 11169 (1) – 1984: *Specification for Steels for Cold Heading/Cold Extrusion Applications—Part 1: Wrought Carbon and Low Alloy Steels.*
4. IS 11169 (2) – 1989: *Steels for Cold Heading/Cold Extrusion Applications – Specification—Part 2: Stainless Steels.*

3.13

Microalloyed Ferritic-Pearlitic Forging Steels

1 INTRODUCTION

Microalloyed ferritic-pearlitic forging steels are alloy steels which contain small amounts (most commonly up to 0.25 %) of niobium, vanadium, titanium, or combinations of these elements. These steels transform to a ferrite–pearlite microstructure during controlled cooling from the hot working temperature and develop, essentially by precipitation strengthening, strengths comparable to those of hardened and tempered unalloyed and low-alloy steels, but with decidedly inferior ductility and impact strength. Hence, for applications where these deficiencies can be tolerated these steels can replace hardened and tempered steels with the following benefits:

a) Lower raw material cost. This is, however, true only when alloy steels are replaced;
b) Elimination of hardening and tempering;
c) Elimination of distortion arising from heat treatment; and
d) Greater degree of microstructural uniformity in the as-forged condition over a wide range of sizes and shapes.

2 SPECIFICATIONS

2.1 Chemical composition

TABLE 3.75 Chemical composition (cast analysis) of microalloyed ferritic-pearlitic forging steels

Steel designation	Chemical composition [% (m/m)]						
	C	Si	Mn	P max.	S	Al	V
United Kingdom (BS 970 (1)–1991)							
280M01	0.30–0.55	0.15–0.60	0.60–1.50	0.035	0.045–0.065[1)]	≤ 0.035	0.08–0.20[2)]
Germany (SEW 101)							
49 MnVS 3	0.44–0.50	≤ 0.50	0.40–1.00	0.035	0.030–0.065	–	0.08–0.13
38 MnSiVS 5	0.35–0.40	0.50–0.80	1.20–1.50	0.035	0.030–0.065	–	0.08–0.13
27 MnSiVS 6	0.25–0.30	0.50–0.80	1.30–1.60	0.035	0.030–0.050	–	0.08–0.13
44 MnSiVS 6	0.42–0.47	0.50–0.80	1.30–1.60	0.035	0.020–0.035	–	0.10–0.15

1) The steel may be supplied with a S content of ≤ 0.050 % or, to obtain improved machinability with 0.065–0.10 % S.

2) Other microalloying additions (such as Nb, Ti) may be made either singly or in combination, in which case the total, as determined by product analysis, should be in the range of 0.08–0.20 %.

2.2 Mechanical properties

TABLE 3.76 Mechanical properties of microalloyed ferritic-pearlitic forging steels in the precipitation strengthened condition

Steel designation	Limiting ruling section mm	R_m MPa	R_e min. MPa	A min. %	Z min. %	KV min. J	Brinell hardness HB
BS 970 (1) – 1991							
280M01	100	775–925	530	14	–	10	223–277
	100	850–1000	560	12	–	8	248–302
	100	925–1075	600	10	–	8	269–331

3 REFERENCES

Books/Articles

1. *ASM Handbook*, vol. 1, ASM International, Materials Park, Ohio, U.S.A., 1990.
2. *Vanadium "As Forged" Steels for Automobile Components*, VANITEC Monograph No. 3, VANITEC, Westerham, Kent, U.K.
3. B.A. MacKenzie, "Stronger Steels by Microalloying", *Machine Design*, 55(18), 71–73, 1983.
4. C.W. Wegst, *Stahlschlüssel*, 16th ed., Verlag Stahlschlüssel Wegst, Marbach, Germany, 1992.

Standards

5. BS 970 (1) – 1991: *Specification for Wrought Steels for Mechanical and Allied Engineering Purposes—Part 1. General Inspection and Testing Procedures and Specific Requirements for Carbon, Carbon-Manganese, Alloy and Stainless Steels.*
6. ISO 11692 – 1994: *Ferritic-Pearlitic Engineering Steels for Precipitation Hardening from Hot-Working Temperatures.*
7. ASTM A 920/A 920M–1993: *Specification for Steel Bars, Microalloy, Hot-Wrought, Special Quality, Mechanical Properties.*
8. ASTM A 921/A 921M–1993: *Specification for Steel Bars, Microalloy, Hot-Wrought, Special Quality, for Subsequent Hot Forging.*

3.14

STEELS FOR HARDENING AND TEMPERING

1 INTRODUCTION

Steels for hardening and tempering are unalloyed and low-alloy steels which in the quenched and tempered condition possess higher strength and toughness than can be achieved by normalizing.

The steel selected for a part to be hardened and tempered should possess sufficient carbon and hardenability so that the desired mechanical properties and microstructure can be obtained through a quench that is appropriate to the size and configuration of the part.

Steels for hardening and tempering usually contain between 0.25 % and 0.60 % carbon. The unalloyed steels, because of their lower hardenability, are used for parts of small cross-section requiring through-hardening, or for thicker sections where a lower ratio of yield strength to tensile strength is acceptable. Alloy steels are used if a higher strength is required than can be achieved with unalloyed steels in the section sizes involved, or if restrictions on distortion, or the hazard of cracking precludes the use of unalloyed steels quenched in water.

2 SPECIFICATION

2.1 Chemical composition

TABLE 3.77 Chemical composition (cast analysis) of steels for hardening and tempering, conforming to IS 5517 – 1993

Steel designation	Chemical composition[1)] [% (m/m)]									
	C[2)]	Si	Mn	P max.	S[3)]	Cr	Mo	Ni	Al	V
30C8	0.25–0.35	0.10–0.35	0.60–0.90	0.035	≤ 0.035	–	–	–	–	–
35C8	0.30–0.40	0.10–0.35	0.60–0.90	0.035	≤ 0.035	–	–	–	–	–
40C8	0.35–0.45	0.10–0.35	0.60–0.90	0.035	≤ 0.035	–	–	–	–	–
45C8	0.40–0.50	0.10–0.35	0.60–0.90	0.035	≤ 0.035	–	–	–	–	–
50C8	0.45–0.55	0.10–0.35	0.60–0.90	0.035	≤ 0.035	–	–	–	–	–
55C8	0.50–0.60	0.10–0.35	0.60–0.90	0.035	≤ 0.035	–	–	–	–	–
20C15	0.16–0.24	0.10–0.35	1.30–1.70	0.035	≤ 0.035	–	–	–	–	–
27C15	0.22–0.32	0.10–0.35	1.30–1.70	0.035	≤ 0.035	–	–	–	–	–
37C15	0.32–0.42	0.10–0.35	1.30–1.70	0.035	≤ 0.035	–	–	–	–	–
40C10S18	0.35–0.45	≤ 0.25	0.80–1.20	0.060	0.14–0.22	–	–	–	–	–
40C15S12	0.35–0.45	≤ 0.25	1.30–1.70	0.035	0.08–0.15	–	–	–	–	–

(Contd.)

TABLE 3.77 Chemical composition (cast analysis) of steels for hardening and tempering, conforming to IS 5517 – 1993 (continued)

Steel designation	Chemical composition[1] [% (*m*/*m*)]									
	C[2]	Si	Mn	P max.	S[3]	Cr	Mo	Ni	Al	V
55Si6Cr3	0.50–0.60	1.20–1.60	0.50–0.80	0.035	≤ 0.035	0.50–0.80	–	–	–	–
37Mn5Si5	0.33–0.41	1.10–1.40	1.10–1.40	0.035	≤ 0.035	–	–	–	–	–
35Mn6Mo3	0.30–0.40	0.10–0.35	1.30–1.80	0.035	≤ 0.035	–	0.20–0.35	–	–	–
35Mn6Mo4	0.30–0.40	0.10–0.35	1.30–1.80	0.035	≤ 0.035	–	0.35–0.55	–	–	–
55Cr3	0.50–0.60	0.10–0.35	0.60–0.80	0.035	≤ 0.035	0.60–0.80	–	–	–	–
40Cr4	0.35–0.45	0.10–0.35	0.60–0.90	0.035	≤ 0.035	0.90–1.20	–	–	–	–
X45Cr9Si3	0.40–0.50	2.75–3.25	0.30–0.60	0.035	≤ 0.035	8.50–9.50	–	–	–	–
42Cr4Mo2	0.38–0.45	0.10–0.35	0.60–0.90	0.035	≤ 0.035	0.90–1.20	0.15–0.30	–	–	–
15Cr13Mo6	0.10–0.20	0.10–0.35	0.40–0.70	0.035	≤ 0.035	2.90–3.40	0.45–0.65	–	–	–
25Cr13Mo6	0.20–0.30	0.10–0.35	0.40–0.70	0.035	≤ 0.035	2.90–3.40	0.45–0.65	–	–	–
40Cr7Al10Mo2	0.35–0.45	0.10–0.35	0.40–0.70	0.035	≤ 0.035	1.50–1.80	0.10–0.25	–	0.90–1.30	–
40Cr13Mo10V2	0.35–0.45	0.10–0.35	0.40–0.70	0.035	≤ 0.035	3.00–3.50	0.90–1.10	–	–	0.15–0.25
58Cr4V1	0.53–0.63	0.15–0.50	0.80–1.10	0.035	≤ 0.035	0.90–1.20	–	–	–	0.07–0.12
50Cr4V2	0.45–0.55	0.15–0.40	0.70–1.10	0.035	≤ 0.035	0.90–1.20	–	–	–	0.10–0.20
42Cr6V1	0.37–0.47	0.15–0.35	0.50–0.80	0.035	≤ 0.035	1.40–1.70	–	–	–	0.07–0.12

(Contd.)

Table 3.77 Chemical composition (cast analysis) of steels for hardening and tempering, conforming to IS 5517 – 1993 (continued)

Steel designation	Chemical composition[1)] [% (m/m)]									
	C[2)]	Si	Mn	P max.	S[3)]	Cr	Mo	Ni	Al	V
40Ni14	0.35–0.45	0.10–0.35	0.50–0.80	0.035	≤ 0.035	≤ 0.30	–	3.20–3.60	–	–
35Ni5Cr2	0.30–0.40	0.10–0.35	0.60–0.90	0.035	≤ 0.035	0.45–0.75	–	1.00–1.50	–	–
30Ni16Cr5	0.26–0.34	0.10–0.35	0.40–0.70	0.035	≤ 0.035	1.10–1.40	–	3.90–4.30	–	–
40Ni6Cr4Mo2	0.35–0.45	0.10–0.35	0.40–0.70	0.035	≤ 0.035	0.90–1.30	0.10–0.20	1.20–1.60	–	–
40Ni6Cr4Mo3	0.35–0.45	0.10–0.35	0.40–0.70	0.035	≤ 0.035	0.90–1.30	0.20–0.35	1.25–1.75	–	–
31Ni10Cr3Mo6	0.27–0.35	0.10–0.35	0.40–0.70	0.035	≤ 0.035	0.50–0.80	0.40–0.70	2.25–2.75	–	–
40Ni10Cr3Mo6	0.36–0.44	0.10–0.35	0.40–0.70	0.035	≤ 0.035	0.50–0.80	0.40–0.70	2.25–2.75	–	–

1) Residual elements up to the following limits are acceptable:
$Cr \leq 0.25$ %; $Mo \leq 0.05$ %; $Ni \leq 0.25$ %; $B \leq 0.0003$ %; $Cu \leq 0.35$ %; $Sn \leq 0.05$ %; $V \leq 0.05$ %.
$Cr + Mo + Ni + B + Cu + Sn + V \leq 0.80$ %.
% Cu + 10 × % Sn ≤ 0.60 %.

2) The steels may be ordered to a restricted C range if agreed between the purchaser and the manufacturer.

3) The steels, other than free-cutting steels, may also be ordered to a restricted S range of 0.020–0.035 % or to any other range as agreed between the purchaser and the manufacturer.

Note: All steels should be fully killed.

2.2 Hardness

TABLE 3.78 Hardness requirements for steels for hardening and tempering, conforming to IS 5517 – 1993, in the "annealed to maximum hardness" (A) or "treated to improve machinability" (M) condition

Steel designation	Maximum Brinell hardness, in HB, in the treatment condition		Steel designation	Maximum Brinell hardness, in HB, in the treatment condition	
	A	M		A	M
30C8	187	170	**40Cr4**	241	241
35C8	197	183	**X45Cr9Si3**	240	–
40C8	207	197	**42Cr4Mo2**	241	241
45C8	207	207	**15Cr13Mo6**	200	–
50C8	210	217	**25Cr13Mo6**	230	–
55C8	220	229	**40Cr7Al10Mo2**	230	–
20C15	200	220	**40Cr13Mo10V2**	230	–
27C15	200	223	**58Cr4V1**	250	–
37C15	220	229	**50Cr4V2**	250	–
40C10S18	200	–	**42Cr6V1**	220	–
40C15S12	200	–	**40Ni14**	229	–
55Si6Cr3	220	220	**35Ni5Cr2**	229	–
37Mn5Si5	220	220	**30Ni16Cr5**	250	–
35Mn6Mo3	220	–	**40Ni6Cr4Mo2**	241	–
35Mn6Mo4	220	–	**40Ni6Cr4Mo3**	241	–
55Cr3	220	–	**31Ni10Cr3Mo6**	269	–
			40Ni10Cr3Mo6	269	–

2.3 Mechanical properties

TABLE 3.79 Mechanical properties of steels for hardening and tempering, conforming to IS 5517 – 1993, in the hardened and tempered condition

Steel designation	Limiting ruling section mm	R_m MPa	$R_{p0.2}$ min. MPa	A min. %	Minimum Izod impact strength[1)] J
30C8	30	600–750	400	18	55
35C8	63	600–750	400	18	55
40C8	30	700–850	480	17	35
	63	600–750	380	18	41
45C8	30	700–850	480	15	35
	100	600–750	380	17	41
50C8	30	800–950	540	13	–
	63	700–850	460	15	–
55C8	30	800–950	540	13	–
	63	700–850	460	15	–
20C15	15	700–850	500	16	48
	30	600–750	440	18	48
27C15	30	700–850	500	16	48
	63	600–750	440	18	48
37C15	15	900–1050	700	15	41
	30	800–950	600	16	48
	63	700–850	540	18	48
	100	600–750	440	18	48

(Contd.)

TABLE 3.79 Mechanical properties of steels for hardening and tempering, conforming to IS 5517 – 1993, in the hardened and tempered condition (continued)

Steel designation	Limiting ruling section mm	R_m MPa	$R_{p0.2}$ min. MPa	A min. %	Minimum Izod impact strength[1)] J
40C10S18	30	700–850	480	17	35
	60	600–750	380	18	41
40C15S12	60	700–850	500	18	48
55Si6Cr 3	16	1400–1600	1200	6	–
37Mn5Si5	16	1000–1200	800	11	21
	40	900–1050	650	12	27
	100	800–950	550	14	34
35Mn6Mo3	30	1000–1150	800	13	48
	63	900–1050	700	15	55
	100	800–950	600	16	55
	150	700–850	540	18	55
35Mn6Mo4	63	1000–1150	800	15	48
	100	900–1050	700	16	55
	150	800–950	600	16	55
55Cr3	63	1000–1150	740	10	17
	63	900–1050	660	12	35
40Cr4	30	900–1050	700	15	55
	63	800–950	600	16	55
	100	700–850	540	18	55
X45Cr 9Si3	–	880–1030	685	10	–

(Contd.)

TABLE 3.79 Mechanical properties of steels for hardening and tempering, conforming to IS 5517 – 1993, in the hardened and tempered condition (continued)

Steel designation	Limiting ruling section mm	R_m MPa	$R_{p0.2}$ min. MPa	A min. %	Minimum Izod impact strength[1)] J
42Cr4Mo2	30	1000–1150	750	10	48
	63	900–1050	650	11	50
	100	800–950	550	12	50
	150	700–850	490	13	55
15Cr13Mo6	63	≥1550	1300	8	14
	100	1100–1250	880	12	41
	150	1000–1150	800	13	48
	150	900–1050	700	15	55
	150	800–950	600	16	55
	150	700–850	540	18	55
25Cr13Mo6	63	≥1550	1300	8	14
	100	1100–1250	880	12	41
	150	1000–1150	800	13	48
	150	900–1050	700	15	55
	150	800–950	600	16	55
	150	700–850	540	18	55
40Cr7Al10Mo2	63	900–1050	700	15	48
	100	800–950	600	16	55
	150	700–850	540	18	55

(Contd.)

TABLE 3.79 Mechanical properties of steels for hardening and tempering, conforming to IS 5517 – 1993, in the hardened and tempered condition (continued)

Steel designation	Limiting ruling section	R_m	$R_{p0.2}$ min.	A min.	Minimum Izod impact strength[1)]
	mm	MPa	MPa	%	J
40Cr13Mo10V2	30	≥ 1550	1300	8	14
	63	≥ 1350	1120	8	21
58Cr4V1	16	1320–1570	1080	7	21
	40	1180–1370	980	8	27
	100	1080–1270	885	10	34
	250	980–1180	735	12	41
50Cr4V2	16	1080–1270	880	9	34
	40	980–1180	780	10	34
	100	900–1080	690	12	34
	250	780–980	590	14	34
42Cr6V1	16	1080–1270	885	10	34
	40	980–1180	785	11	41
	100	880–1030	685	12	48
	160	740–880	540	14	55
40Ni14	63	900–1050	700	15	55
	100	800–950	600	16	55
35Ni5Cr2	100	800–950	600	16	55
	150	700–850	540	18	55
30Ni16Cr5	63[2)]	≥ 1550	1300	8	14
	150[3)]	≥ 1550	1300	8	14

(Contd.)

TABLE 3.79 Mechanical properties of steels for hardening and tempering, conforming to IS 5517 – 1993, in the hardened and tempered condition (continued)

Steel designation	Limiting ruling section mm	R_m MPa	$R_{p0.2}$ min. MPa	A min. %	Minimum Izod impact strength[1)] J
40Ni6Cr4Mo2	30	1100–1250	880	11	41
	63	1000–1150	800	13	48
	100	900–1050	700	15	55
	150	800–950	600	16	55
40Ni6Cr4Mo3	30	≥ 1550	1300	6	11
	30	1200–1350	1000	10	30
	63	1100–1250	880	11	41
	100	1000–1150	800	13	48
	150	900–1050	700	15	55
	150	800–950	600	16	55
31Ni10Cr3Mo6	63	≥ 1550	1300	8	14
	63	1200–1350	1000	10	35
	100	1100–1250	880	11	41
	150	1000–1150	800	12	48
	150	900–1050	700	15	55
40Ni10Cr3Mo6	100	≥ 1550	1300	8	14
	150	1200–1350	1000	10	35
	150	1100–1250	880	11	41
	150	1000–1150	800	12	48

(Contd.)

TABLE 3.79 Mechanical properties of steels for hardening and tempering, conforming to IS 5517 – 1993, in the hardened and tempered condition (continued)

1) For fine grain steels.
2) Air-hardened.
3) Oil-hardened.

3 RECOMMENDED HOT WORKING AND HEAT TREATING TEMPERATURES

TABLE 3.80 Recommended hot working and heat treating temperatures for steels for hardening and tempering, conforming to IS 5517 – 1993

Steel designation	Hot working temperature °C	Softening temperature °C	Normalizing temperature °C	Hardening temperature °C	Quenching agent	Tempering temperature °C
30C8	1200–850	650–700	860–890	860–890	Water or oil	550–660
35C8	1200–850	650–700	850–880	850–880	Water or oil	550–660
40C8	1200–850	650–700	830–860	830–860	Water or oil	550–660
45C8	1200–850	650–700	830–860	830–860	Water or oil	550–660
50C8	1150–850	650–700	810–840	810–840	Oil	550–660
55C8	1150–850	650–700	810–840	810–840	Oil	550–660
20C15	1200–850	600–660	860–900	860–900	Water or oil	550–660
27C15	1200–850	600–660	840–880	840–880	Water or oil	550–660
37C15	1200–850	600–660	840–870	840–870	Water or oil	550–660
40C10S18	1200–850	630–700	830–860	830–860	Oil	550–660
40C15S12	1200–850	600–660	840–870	840–870	Oil	550–660
55Si6Cr3	1050–850	640–680	850–880	830–860	Oil	430–500
37Mn5Si5	1100–850	680–720	860–890	830–860	Water or Oil	480–650
35Mn6Mo3	1200–900	640–660	830–860	830–860	Oil	550–660

(Contd.)

TABLE 3.80 Recommended hot working and heat treating temperatures for steels for hardening and tempering, conforming to IS 5517 – 1993 (continued)

Steel designation	Hot working temperature °C	Softening temperature °C	Normalizing temperature °C	Hardening temperature °C	Quenching agent	Tempering temperature °C
35Mn6Mo4	1200–900	640–660	830–860	830–860	Oil	550–660
55Cr3	1100–800	650–700	800–850	800–850	Oil	500–700
40Cr4	1200–850	650–700	850–880	850–880	Oil	550–700
X45Cr9Si3	1100–900	780–820	–	1020–1070	Oil	720–820
42Cr4Mo2	1200–850	650–700	850–880	850–880	Oil	550–720
15Cr13Mo6	1200–850	700–750	–	880–910	Oil	550–700
25Cr13Mo6	1200–850	700–750	–	880–910	Oil	550–700
40Cr7Al10Mo2	1200–850	700–750	–	880–910	Oil	550–700
40Cr13Mo10V2	1200–850	700–750	–	900–940	Oil	570–650
58Cr4V1	1050–850	680–720	850–880	820–850	Oil	480–650
50Cr4V2	1050–850	680–720	840–880	830–860	Oil	540–680
42Cr6V1	1100–850	680–720	850–880	830–860	Oil	480–650
40Ni14	1150–900	630–650	830–860	830–860	Oil	550–650
35Ni5Cr2	1200–900	630–670	–	820–850	Oil or water	550–660
30Ni16Cr5	1200–850	610–630	–	810–830	Air or oil[1)]	≤250
40Ni6Cr4Mo2	1200–900	630–670	–	820–850	Oil	550–660
40Ni6Cr4Mo3	1200–900	630–660	–	820–850	Oil	550–660[2)]
31Ni10Cr3Mo6	1200–850	630–650	–	820–850	Oil	≤660
40Ni10Cr3Mo6	1200–850	630–650	–	820–850	Oil	≤660

1) For full hardening of ruling sections > 63 mm oil hardening is recommended.
2) Or 150–200 °C, depending on the tensile strength or hardness required.

4 EQUIVALENT GRADES

TABLE 3.81 Comparison of grades of steels for hardening and tempering specified in IS 5517 – 1993 with nearest equivalent grades specified in other national and international standards

IS 5517–93	BS EN 10083 (1)–91 BS EN 10083 (2)–91 BS 970 (1)–91 BS 970 (2)–88 BS 970 (4)–70	DIN EN 10083 (1)–91 DIN EN 10083 (2)–91 DIN 1651–88 DIN 17211–87 DIN 17221–88 DIN 17480–92	GOST 1050–88 GOST 1414–75 GOST 4543–71 GOST 5632–72 GOST 14959–79	ISO 683 (1)–87 ISO 683 (9)–88 ISO 683 (10)–87 ISO 683 (14)–92 ISO 683 (15)–92	JIS G 4051–79 JIS G 4052–79 JIS G 4102–79 JIS G 4103–79 JIS G 4104–79 JIS G 4105–79 JIS G 4106–79 JIS G 4202–79 JIS G 4311–91 JIS G 4801–84 JIS G 4804–83	SAE J403–94 SAE J404–94 SAE J775–93 SAE J1268–93
30C8	2 C 30 3 C 30	2 C 30 3 C 30	30 30Г	C 30 E4 C 30 M2	S 28 C S 30 C S 33 C	1029 1030
35C8	2 C 35 3 C 35	2 C 35 3 C 35	35 35Г	C 35 E4 C 35 M2	S 33 C S 35 C S 38 C	1035
40C8	2 C 40 3 C 40	2 C 40 3 C 40	40 40Г	C 40 E4 C 40 M2	S 38 C S 40 C S 43 C	1038 1038H 1039 1040
45C8	2 C 45 3 C 45 080M42 080H42	2 C 45 3 C 45	45 45Г	C 45 E4 C 45 M2	S 43 C S 45 C S 48 C	1042 1043 1045 1045H 1046
50C8	2 C 50 3 C 50	2 C 50 3 C 50	50 50Г	C 50 E4 C 50 M2	S 48 C S 50 C S 53 C	1049 1050 1053

(Contd.)

TABLE 3.81 Comparison of grades of steels for hardening and tempering specified in IS 5517 – 1993 with nearest equivalent grades specified in other national and international standards (continued)

IS 5517–93	BS EN 10083 (1)–91 BS EN 10083 (2)–91 BS 970 (1)–91 BS 970 (2)–88 BS 970 (4)–70	DIN EN 10083 (1)–91 DIN EN 10083 (2)–91 DIN 1651–88 DIN 17211–87 DIN 17221–88 DIN 17480–92	GOST 1050–88 GOST 1414–75 GOST 4543–71 GOST 5632–72 GOST 14959–79	ISO 683 (1)–87 ISO 683 (9)–88 ISO 683 (10)–87 ISO 683 (14)–92 ISO 683 (15)–92	JIS G 4051–79 JIS G 4052–79 JIS G 4102–79 JIS G 4103–79 JIS G 4104–79 JIS G 4105–79 JIS G 4106–79 JIS G 4202–79 JIS G 4311–91 JIS G 4801–84 JIS G 4804–83	SAE J403–94 SAE J404–94 SAE J775–93 SAE J1268–93
55C8	2 C 55 3 C 55	2 C 55 3 C 55	55	C 55 E4 C 55 M2	S 53 C S 55 C S 58 C	1055
20C15	150M19 150H19	–	–	22 Mn 6	SMn 420 SMn 420 H	1524 1524H
27C15	28 Mn 6	28 Mn 6	30Г 2	28 Mn 6	–	1527
37C15	150M36 150H36	–	35Г 2	36 Mn 6	SMn 438 SMn 438 H	1541 1541H
40C10S18	–	35 S 20	–	35 SMn 20	–	–
40C15S12	–	–	–	–	SUM 42	1141
55Si6Cr3	685A57 685H57	54 SiCr 6	–	55 SiCr 6 3	SUP 12	–
37Mn5Si5	–	–	–	–	–	–
35Mn6Mo3	605M36 605H36	–	–	–	–	–

(Contd.)

TABLE 3.81 Comparison of grades of steels for hardening and tempering specified in IS 5517 – 1993 with nearest equivalent grades specified in other national and international standards (continued)

IS 5517–93	BS EN 10083 (1)–91 BS EN 10083 (2)–91 BS 970 (1)–91 BS 970 (2)–88 BS 970 (4)–70	DIN EN 10083 (1)–91 DIN EN 10083 (2)–91 DIN 1651–88 DIN 17211–87 DIN 17221–88 DIN 17480–92	GOST 1050–88 GOST 1414–75 GOST 4543–71 GOST 5632–72 GOST 14959–79	ISO 683 (1)–87 ISO 683 (9)–88 ISO 683 (10)–87 ISO 683 (14)–92 ISO 683 (15)–92	JIS G 4051–79 JIS G 4052–79 JIS G 4102–79 JIS G 4103–79 JIS G 4104–79 JIS G 4105–79 JIS G 4106–79 JIS G 4202–79 JIS G 4311–91 JIS G 4801–84 JIS G 4804–83	SAE J403–94 SAE J404–94 SAE J775–93 SAE J1268–93
35Mn6Mo4	–	–	–	–	–	–
55Cr3	–	55 Cr 3	–	55 Cr 3	SUP 9	5155H
40Cr4	37 Cr 4 37 CrS 4 41 Cr 4 41 CS 4	37 Cr 4 37 CrS 4 41 Cr 4 41 CrS 4	38XA 40X	37 Cr 4 37 CrS 4 41 Cr 4 41 CrS 4	SCr 440 SCr 440 H	5140 5140H
X45Cr9Si3	401S45	X 45 CrSi 9 3	40X9C2	X 45 CrSi 9 3	SUH 1	S65007
42Cr4Mo2	42 CrMo 4 42 CrMoS 4 709M40 709H40	42 CrMo 4 42 CrMoS 4	38XM	42 CrMo 4 42 CrMoS 4	SCM 440 SCM 440 H	4140 4140H 4142 4142H
15Cr13Mo6	–	–	–	–	–	–
25Cr13Mo6	722M24 722H24	31 CrMo 12	–	31 CrMo 12	–	–
40Cr7Al10Mo2	897M39	–	**38X2MЮA**	41 CrAlMo 7 4	SACM 645	

(Contd.)

TABLE 3.81 Comparison of grades of steels for hardening and tempering specified in IS 5517 – 1993 with nearest equivalent grades specified in other national and international standards (continued)

IS 5517–93	BS EN 10083 (1)–91 BS EN 10083 (2)–91 BS 970 (1)–91 BS 970 (2)–88 BS 970 (4)–70	DIN EN 10083 (1)–91 DIN EN 10083 (2)–91 DIN 1651–88 DIN 17211–87 DIN 17221–88 DIN 17480–92	GOST 1050–88 GOST 1414–75 GOST 4543–71 GOST 5632–72 GOST 14959–79	ISO 683 (1)–87 ISO 683 (9)–88 ISO 683 (10)–87 ISO 683 (14)–92 ISO 683 (15)–92	JIS G 4051–79 JIS G 4052–79 JIS G 4102–79 JIS G 4103–79 JIS G 4104–79 JIS G 4105–79 JIS G 4106–79 JIS G 4202–79 JIS G 4311–91 JIS G 4801–84 JIS G 4804–83	SAE J403–94 SAE J404–94 SAE J775–93 SAE J1268–93
40Cr13Mo10V2	905M39	–	–	–	–	–
58Cr4V1	–	–	–	–	–	–
50Cr4V2	51 CrV 4	51 CrV 4	50ХФА 50ХГФА	51CrV 4	SUP 10	6150 6150H
42Cr6V1	–	–	–	–	–	–
40Ni14	–	–	–	–	–	–
35Ni5Cr2	–	–	40ХН	–	SNC 236	–
30Ni16Cr5	–	–	–	–	–	–
40Ni6Cr4Mo2	817M40 817H40	–	–	–	–	–
40Ni6Cr4Mo3	–	–	–	–	–	–
31Ni10Cr3Mo6	–	–	–	–	–	–
40Ni10Cr3Mo6	826M40 826H40	–	–	–	–	–

5 REFERENCES

Books

1. *ASM Handbook*, vol. 1, ASM International, Materials Park, Ohio, U.S.A., 1990.
2. *ASM Handbook*, vol. 4, ASM International, Materials Park, Ohio, U.S.A., 1991.
3. *Source Book on Heat Treating—Volume I: Materials and Processes*, American Society for Metals, Metals Park, Ohio, U.S.A., 1975.

Standards

4. IS 5517 – 1993: *Steels for Hardening and Tempering - Specification.*
5. ISO 683 (1) – 1987: *Heat-Treatable Steels, Alloy Steels and Free-Cutting Steels—Part 1: Direct-Hardening Unalloyed and Low-Alloyed Wrought Steel in Form of Different Black Products.*

3.15

CASE HARDENING STEELS

1 INTRODUCTION

Case hardening steels are unalloyed and low-alloy steels which are used in the carburized or carbonitrided, and subsequently hardened and tempered condition. These steels in the case hardened condition are characterized by a high-carbon martensitic case with good wear resistance and fatigue strength, superimposed on a tough low-carbon core.

Case hardening steels usually contain between 0.08 % and 0.20 % carbon, with or without alloying elements. The steel selected for a part to be case hardened should possess sufficient case and core hardenability to achieve the desired case and core microstructure through a quench which is appropriate to the size and configuration of the part.

Unalloyed case hardening steels are usually water quenched and are used for smaller and less heavily loaded parts. The higher hardenability alloy steels are used for parts which require higher core strength, and for parts requiring a slower quench to minimize distortion and eliminate cracking.

2 SPECIFICATION

2.1 Chemical composition

TABLE 3.82 Chemical composition (cast analysis) of case hardening steels, conforming to IS 4432 – 1988

Steel designation	Chemical composition[1] [% (*m*/*m*)]							
	C	Si	Mn	P[2] max.	S[2]	Cr	Mo	Ni
10C4	≤0.15	0.15–0.35	0.30–0.60	0.045	≤0.045	–	–	–
15C8	0.10–0.20	0.15–0.35	0.60–0.90	0.035	≤0.035	–	–	–
10C8S10	≤0.15	0.15–0.35	0.60–0.90	0.035	0.08–0.13	–	–	–
11C10S25	0.08–0.15	0.10–0.35	0.80–1.20	0.045	0.20–0.30	–	–	–
14C14S14	0.10–0.18	0.10–0.35	1.20–1.50	0.045	0.10–0.18	–	–	–
15Cr3	0.12–0.18	0.15–0.35	0.40–0.60	0.035	≤0.035	0.50–0.80	–	–
16Mn5Cr4	0.14–0.19	0.15–0.35	1.00–1.30	0.035	≤0.035	0.80–1.10	–	–
20Mn5Cr5	0.17–0.22	0.15–0.35	1.00–1.40	0.035	≤0.035	1.00–1.30	–	–
21Cr4Mo2	≤0.26	0.10–0.35	0.60–0.90	0.035	≤0.035	0.90–1.20	0.15–0.30	–
16Ni3Cr2	0.12–0.20	0.15–0.35	0.60–1.00	0.035	≤0.035	0.40–0.80	–	0.60–1.00
14Cr6Ni6	0.12–0.17	0.15–0.40	0.40–0.60	0.035	≤0.035	1.40–1.70	–	1.40–1.70
13Ni13Cr3	0.10–0.15	0.15–0.35	0.40–0.70	0.035	≤0.035	0.60–1.00	–	3.00–3.50
20Ni7Mo2	0.17–0.22	0.15–0.35	0.40–0.65	0.035	≤0.035	–	0.20–0.30	1.65–2.00
15Ni5Cr4Mo1	0.12–0.18	0.15–0.35	0.60–1.00	0.035	≤0.035	0.75–1.25	0.08–0.15	1.00–1.50
15Ni7Cr4Mo2	0.12–0.18	0.15–0.35	0.60–1.00	0.035	≤0.035	0.75–1.25	0.10–0.20	1.50–2.00
20Ni2Cr2Mo2	0.18–0.23	0.15–0.35	0.70–0.90	0.035	≤0.035	0.40–0.60	0.15–0.25	0.40–0.70
20Ni7Cr2Mo2	0.17–0.22	0.15–0.35	0.45–0.65	0.035	≤0.035	0.40–0.60	0.20–0.30	1.65–2.00

[1] Residual elements up to the following limits are acceptable:
Cr ≤ 0.30 %; Mo ≤ 0.05 %; Ni ≤ 0.30 %; Cu ≤ 0.25 %; V ≤ 0.05 %.
Cr + Mo + Ni ≤ 0.50 %.
% Cu + 10 × % Sn ≤ 0.50 %.

TABLE 3.82 Chemical composition (cast analysis) of case hardening steels, conforming to IS 4432 – 1988 (continued)

2) Steels, other than free-cutting steels, can also be ordered to the following limits of phosphorus and sulphur:

	% P	% S
Limit 1	≤ 0.045	≤ 0.045
Limit 2	≤ 0.035	0.020–0.035

Notes:
1) For steels ordered to a restricted sulphur range (according to Limit 2) the steel designation should be appended by the letter S, for example 20Mn5Cr5-S.
2) All steels should be fully killed.

2.2 Hardness

TABLE 3.83 Hardness requirements for case hardening steels, conforming to IS 4432 – 1988, in the "annealed to maximum hardness" (A) or "treated to improve machinability" (M) condition

Steel designation	Brinell hardness, in HB, in the treatment condition		Steel designation	Brinell hardness, in HB, in the treatment condition	
	A	M		A	M
10C4	≤ 131	90–145	**16Ni3Cr2**	≤ 207	150–202
15C8	≤ 150	105–170	**14Cr6Ni6**	≤ 217	160–210
10C8S10	≤ 150	105–170	**13Ni13Cr3**	≤ 217	160–210
11C10S25	≤ 150	105–170	**20Ni7Mo2**	≤ 217	160–210
14C14S14	≤ 150	105–170	**15Ni5Cr4Mo1**	≤ 229	170–220
15Cr3	≤ 174	120–180	**15Ni7Cr4Mo2**	≤ 229	170–220
16Mn5Cr4	≤ 207	150–202	**20Ni2Cr2Mo2**	≤ 217	170–220
20Mn5Cr5	≤ 217	160–210	**20Ni7Cr2Mo2**	≤ 229	170–220
21Cr4Mo2	≤ 217	170–210			

2.3 Mechanical properties

TABLE 3.84 Mechanical properties of case hardening steels, conforming to IS 4432 – 1988, in the simulated case hardened condition (for guidance only)

Steel designation	Mechanical properties of reference test bars of diameter											
	16 mm				30 mm				63 mm			
	R_m MPa	R_e min. MPa	A min. %	Z min. %	R_m MPa	R_e min. MPa	A min. %	Z min. %	R_m MPa	R_e min. MPa	A min. %	Z min. %
10C4	550–800	330	13	40	500–650	300	16	45	–	–	–	–
15C8	600–850	400	12	35	550–800	330	14	40	–	–	–	–
10C8S10	600–850	400	12	35	550–800	330	14	40	–	–	–	–
11C10S25	600–850	400	12	35	550–800	330	14	40	–	–	–	–
14C14S14	650–900	430	12	35	600–900	360	14	40	–	–	–	–
15Cr3	650–900	430	12	35	600–900	360	14	40	–	–	–	–
16Mn5Cr4	850–1100	620	9	30	800–1050	600	10	40	650–950	450	11	40
20Mn5Cr5	1000–1300	750	7	25	1000–1300	700	8	30	800–1100	550	10	35
21Cr4Mo2	1100–1400	750	8	35	1000–1350	700	9	40	950–1250	650	10	40
16Ni3Cr2	1100–1400	750	8	35	1050–1400	720	9	40	900–1200	600	11	45
14Cr6Ni6	1050–1350	720	8	35	970–1300	700	9	40	800–1100	550	11	40

(Contd.)

TABLE 3.84 Mechanical properties of case hardening steels, conforming to IS 4432 – 1988, in the simulated case hardened condition (for guidance only) (continued)

Steel designation	Mechanical properties of reference test bars of diameter											
	16 mm				30 mm				63 mm			
	R_m MPa	R_e min. MPa	*A* min. %	*Z* min. %	R_m MPa	R_e min. MPa	*A* min. %	*Z* min. %	R_m MPa	R_e min. MPa	*A* min. %	*Z* min. %
13Ni13Cr3	1000–1300	720	8	35	900–1250	650	9	35	850–1200	600	10	40
20Ni7Mo2	800–1150	550	8	35	700–1000	500	9	35	650–950	450	10	40
15Ni5Cr4Mo1	1050–1350	720	8	35	1000–1350	700	9	35	900–1200	600	11	40
15Ni7Cr4Mo2	1100–1400	750	9	40	1050–1400	730	10	40	800–1100	550	12	45
20Ni2Cr2Mo2	850–1200	600	8	35	800–1150	550	9	35	750–1050	520	10	40
20Ni7Cr2Mo2	950–1350	650	8	35	850–1200	600	9	35	800–1150	550	10	40

3 RECOMMENDED HOT WORKING AND HEAT TREATING TEMPERATURES

TABLE 3.85 Recommended hot working and heat treating temperatures for case hardening steels, conforming to IS 4432 – 1988

Steel designation	Hot working temperature	Annealing temperature	Softening temperature	Normalizing temperature	Carburizing temperature	Direct and single hardening temperature	Double hardening		Quenching agent[1)]	Tempering temperature
							Core hardening temperature	Case hardening temperature		
	°C	°C	°C	°C	°C	°C	°C	°C		°C
10C4	1100–850	900–930	650–700	880–910	900–940	830–860	880–900	780–820	Water	150–180
15C8	1100–850	900–930	650–700	880–910	900–940	830–860	870–890	780–820	Water	150–180
10C8S10	1100–850	900–930	650–700	880–910	900–940	830–860	870–890	780–820	Water	150–180
11C10S25	1100–850	900–930	650–700	880–910	900–940	830–860	870–890	780–820	Water	150–180
14C14S14	1100–850	900–930	650–700	880–910	890–940	820–850	870–890	780–820	Oil or water	170–200
15Cr3	1150–850	880–910	650–700	880–910	890–940	820–850	870–890	780–820	Oil or water	170–200
16Mn5Cr4	1150–850	860–900	650–700	840–870	890–940	810–840	860–880	790–830	Oil	170–200
20Mn5Cr5	1150–850	860–900	650–700	840–870	890–940	810–840	860–880	790–830	Oil	170–200
21Cr4Mo2	1150–850	860–880	650–700	840–870	890–940	810–840	840–880	780–820	Oil	180–220
16Ni3Cr2	1150–850	860–880	650–700	840–870	890–940	810–840	840–880	780–820	Oil	180–220

(Contd.)

TABLE 3.85 Recommended hot working and heat treating temperatures for case hardening steels, conforming to IS 4432 – 1988 (continued)

Steel designation	Hot working temperature °C	Annealing temperature °C	Softening temperature °C	Normalizing temperature °C	Carburizing temperature °C	Direct and single hardening temperature °C	Double hardening		Quenching agent[1)]	Tempering temperature °C
							Core hardening temperature °C	Case hardening temperature °C		
14Cr6Ni6	1150–850	860–900	650–700	840–870	890–940	810–840	860–880	790–830	Oil	170–200
13Ni13Cr3	1150–850	860–880	650–700	840–870	890–940	810–840	860–880	790–830	Oil	180–220
20Ni7Mo2	1150–850	860–880	650–700	840–870	890–940	810–840	830–870	780–820	Oil	180–220
15Ni5Cr4Mo1	1150–850	860–880	650–700	840–870	890–940	810–840	840–880	780–820	Oil	180–220
15Ni7Cr4Mo2	1150–850	860–880	650–700	840–870	890–940	810–840	840–880	780–820	Oil	180–220
20Ni2Cr2Mo2	1150–850	860–880	650–700	840–870	890–940	810–840	830–870	780–820	Oil	180–220
20Ni7Cr2Mo2	1150–850	860–880	650–700	840–870	890–940	810–840	830–870	780–820	Oil	180–220

1) Other quenching agents such as polymer quenchants may also be used.

4 EQUIVALENT GRADES

TABLE 3.86 Comparison of grades of case hardening steel specified in IS 4432 – 1988 with nearest equivalent grades specified in other national and international standards

IS 4432–88	BS 970 (1)–91	DIN 1651–88 DIN 17210–86	GOST 1050–88 GOST 1414–75 GOST 4543–71	ISO 683 (9)–88 ISO 683 (11)–87	JIS G 4051–79 JIS G 4052–79 JIS G 4102–79 JIS G 4103–79 JIS G 4105–79 JIS G 4804–83	SAE J403–94 J404–94 J1268–93
10C4	045A10 045M10	C 10 Ck 10	10	C 10	S 10 C S 12 C S 09 CK	1010 1012
15C8	080A15 080M15	–	15Г	C 16 E4 C 16 M2	–	1016 1018
10C8S10	–	15 S 10	–	–	SUM 12	–
11C10S25	–	10 S 20	–	10 S 20	–	–
14C14S14	214A15 214M15	–	–	17 SMn 20	–	1118
15Cr3	523H15 523M15	17 Cr 3	15X	–	–	–
16Mn5Cr4	590H17 590M17	16 MnCr 5 16 MnCrS 5	18XГ	16 Mn Cr 5 16 MnCrS 5	–	–
20Mn5Cr5	–	20 MnCr 5 20 MnCrS 5	–	20 MnCr 5 20 MnCrS 5	–	–
21Cr4Mo2	708H20 708M20	–	20XM	18 CrMo 4 18 CrMoS 4	SCM 420 SCM 420 H SCM 421	–
16Ni3Cr2	635A14 635H15 635M15	–	12XH	–	–	–
14Cr6Ni6	–	15 CrNi 6	–	–	–	–

(Contd.)

TABLE 3.86 Comparison of grades of case hardening steel specified in IS 4432 – 1988 with nearest equivalent grades specified in other national and international standards (continued)

IS 4432–88	BS 970 (1)–91	DIN 1651–88 DIN 17210–86	GOST 1050–88 GOST 1414–75 GOST 4543–71	ISO 683 (9)–88 ISO 683 (11)–87	JIS G 4051–79 JIS G 4052–79 JIS G 4102–79 JIS G 4103–79 JIS G 4105–79 JIS G 4804–83	SAE J403–94 J404–94 J1268–93
13Ni13Cr3	655H13 655M13	–	12XH3A	15 NiCr 13	SNC 815 SNC 815 H	–
20Ni7Mo2	665H20 665M20	–	20H2M	–	–	4620 4620H
15Ni5Cr4Mo1	815H17 815M17	–	–	17 NiCrMo 6	–	–
15Ni7Cr4Mo2	820H17 820M17	–	–	–	–	–
20Ni2Cr2Mo2	805A20 805H20 805M20	21 NiCrMo 2 21 NiCrMoS 2	20XГHM	20 NiCrMo 2 20 NiCrMoS 2	SNCM 220 SNCM 220 H	8620 8620H
20Ni7Cr2Mo2	–	–	20XH2M	–	SNCM 420 SNCM 420 H	4320 4320H

5 REFERENCES

Books

1. *ASM Handbook*, vol. 1, ASM International, Materials Park, Ohio, U.S.A., 1990.
2. *ASM Handbook*, vol. 4, ASM International, Materials Park, Ohio, U.S.A., 1991.
3. *Source Book on Heat Treating—Volume I: Materials and Processes*, American Society for Metals, Metals Park, Ohio, U.S.A., 1975.

Standards

4. IS 4432 – 1988: *Specification for Case Hardening Steels.*
5. ISO 683 (11) – 1987: *Heat-Treatable Steels, Alloy Steels and Free-Cutting Steels—Part 11: Wrought Case Hardening Steels.*

3.16

NITRIDING STEELS

1 INTRODUCTION

Nitriding steels are heat-treatable steels which contain controlled amounts of nitride forming elements such as aluminium, chromium, molybdenum and vanadium. These steels, in the nitrided condition, are used in many applications where a hard, wear resistant case, increased fatigue strength, decreased notch sensitivity, and improved corrosion resistance are important.

The choice of a satisfactory grade of nitriding steel depends primarily on the desired case hardness, the required hardenability and the core strength needed to support the nitrided case in service. Aluminium-containing nitriding steels produce a nitrided case of very high hardness (≈ 950 HV), excellent wear resistance, but low ductility. In contrast, the chromium-containing nitriding steels provide a nitrided case with lower hardness (≈ 800 HV) but considerably higher ductility. Addition of molybdenum to these steels reduces the likelihood of temper embrittlement which may occur as a result of the nitriding process.

2 SPECIFICATION

2.1 *Chemical composition*

TABLE 3.87 Chemical composition (cast analysis) of nitriding steels, conforming to IS 5517 – 1993

Steel designation	Chemical composition[1] [% (*m*/*m*)]								
	C[2]	Si	Mn	P max.	S[3] max.	Cr	Mo	Al	V
15Cr13Mo6	0.10–0.20	0.10–0.35	0.40–0.70	0.035	0.035	2.90–3.40	0.45–0.65	–	–
25Cr13Mo6	0.20–0.30	0.10–0.35	0.40–0.70	0.035	0.035	2.90–3.40	0.45–0.65	–	–
40Cr7Al10Mo2	0.35–0.45	0.10–0.35	0.40–0.70	0.035	0.035	1.50–1.80	0.10–0.25	0.90–1.30	–
40Cr13Mo10V2	0.35–0.45	0.10–0.35	0.40–0.70	0.035	0.035	3.00–3.50	0.90–1.10	–	0.15–0.25

[1] Residual elements up to the following limits are acceptable:
$Ni \leq 0.25$ %; $B \leq 0.0003$ %; $Cu \leq 0.35$ %; $Sn \leq 0.05$ %; $V \leq 0.05$ %.
% Cu + 10 × % Sn ≤ 0.60 %.
[2] The steels may be ordered to a restricted C range if agreed between the purchaser and the manufacturer.
[3] The steels may be ordered to a restricted S range of 0.20–0.035 %, or any other range as agreed between the purchaser and the manufacturer.

Note: All steels should be fully killed.

2.2 *Hardness*

TABLE 3.88 Hardness requirements for nitriding steels, conforming to IS 5517 – 1993, in the "annealed to maximum hardness" condition

Steel designation	Maximum Brinell hardness HB	Steel designation	Maximum Brinell hardness HB
15Cr13Mo6	200	**40Cr 7Al10Mo2**	230
25Cr13Mo6	230	**40Cr13Mo10V2**	230

2.3 Mechanical properties

TABLE 3.89 Mechanical properties of nitriding steels, conforming to IS 5517 – 1993, in the hardened and tempered condition

Steel designation	Limiting ruling section mm	R_m MPa	$R_{p0.2}$ min. MPa	*A* min. %	Minimum Izod impact strength[1)] J
15Cr13Mo6	63	≥ 1550	1300	8	14
	100	1100–1250	880	12	41
	150	1000–1150	800	13	48
	150	900–1050	700	15	55
	150	800–950	600	16	55
	150	700–850	540	18	55
25Cr13Mo6	63	≥ 1550	1300	8	14
	100	1100–1250	880	12	41
	150	1000–1150	800	13	48
	150	900–1050	700	15	55
	150	800–950	600	16	55
	150	700–850	540	18	55
40Cr7Al10Mo2	63	900–1050	700	15	48
	100	800–950	600	16	55
	150	700–850	540	18	55
40Cr13Mo10V2	30	≥ 1550	1300	8	14
	63	≥ 1350	1120	8	21

1) For fine grain steels.

3 RECOMMENDED HOT WORKING AND HEAT -TREATING TEMPERATURES

TABLE 3.90 Recommended hot working and heat treating temperatures for nitriding steels, conforming to IS 5517 – 1993

Steel designation	Hot-working temperature °C	Softening temperature °C	Hardening temperature °C	Quenching agent	Tempering temperature °C	Stabilizing temperature °C	Nitriding temperature °C
15Cr13Mo6	1200–850	700–750	880–910	Oil	550–700	520–550	490–510
25Cr13Mo6	1200–850	700–750	880–910	Oil	550–700	520–550	490–510
40Cr7Al10Mo2	1200–850	700–750	880–910	Oil	550–700	530–550	500–520
40Cr13Mo10V2	1200–850	700–750	900–940	Oil	570–650	520–550	490–510

Source: References 3 and 5

4 EQUIVALENT GRADES

TABLE 3.91 Comparison of grades of nitriding steels specified in IS 5517 – 1993 with nearest equivalent grades specified in other national and international standards

IS 5517–93	ASTM A355–89	BS EN 10083 (1)–91 970 (1)–91	DIN 17211–87	GOST 4543–71	ISO 683 (10)–87	JIS G 4202–79
15Cr13Mo6	–	–	–	–	–	–
25Cr13Mo6	–	722M24 722H24	31 CrMo 12	–	31 CrMo 12	–
40Cr7Al10Mo2	Class A	897M39	–	38X2MЮA	41 CrAlMo 7 4	SACM 645
40Cr13Mo10V2	–	905M39	–	–	–	–

5 REFERENCES

Books

1. *ASM Handbook*, vol. 1, ASM International, Materials Park, Ohio, U.S.A., 1990.
2. *ASM Handbook*, vol. 4, ASM International, Materials Park, Ohio, U.S.A., 1991.
3. J.Woolman and R.A.Mottram, *The Mechanical and Physical Properties of the British Standard En Steels*, 1st ed., vol. 3, The British Iron and Steel Research Association, Pergamon Press, Oxford, U.K., 1968.
4. *Three Keys to Satisfaction*, Climax Molybdenum, New York, U.S.A., 1961.

Standards

5. IS 5517 – 1993: *Steels for Hardening and Tempering—Specification.*
6. ISO 683 (10) – 1987: *Heat-Treatable Steels, Alloy Steels and Free-Cutting Steels—Part 10: Wrought Nitriding Steels.*

3.17

STEELS FOR FLAME AND INDUCTION HARDENING

1 INTRODUCTION

Steels for flame and induction hardening are unalloyed and alloy steels which are suitable for surface hardening. These steels in the surface hardened condition develop a hard, wear-resistant surface and improved fatigue strength, without affecting the properties of the core. Since flame or induction hardening does not change the composition of the base metal, these steels must contain sufficient carbon to produce a martensitic case of the desired hardness. At the same time, the carbon content should not be so high that quench cracking during surface hardening may result. Because hardness levels below 50 HRC are generally inadequate for applications require surface hardening, steels containing at least 0.30 % carbon are usually specified for these applications.

Unalloyed steels are generally preferred for flame or induction hardening as they go into solution much faster and at a lower temperature than alloy steels. Moreover, a water quench, which is more economical than an oil quench, can often be used.

Alloy steels are used if a higher core strength is required than can be achieved with unalloyed steels in the section sizes involved, or if restrictions on distortion, or the hazard of cracking precludes the use of unalloyed steels quenched in water.

For best results the steel to be flame or induction hardened should be in the normalized or hardened and tempered condition so as to permit rapid and complete austenitization and full hardening. Necessary steps should also be taken to ensure that areas to be hardened are free from decarburization.

2 SPECIFICATION

2.1 Chemical composition

TABLE 3.92 Chemical composition (cast analysis) of steels for flame and induction hardening, conforming to IS 3930 – 1994

Steel designation	Chemical composition[1)] [% (*m*/*m*)]								
	C[2)]	Si	Mn	P max.	S	Cr	Mo	Ni	V
30C8	0.27–0.35	0.10–0.35	0.60–0.90	0.035	≤0.035	–	–	–	–
35C8	0.32–0.40	0.10–0.35	0.60–0.90	0.035	≤0.035	–	–	–	–
45C8	0.42–0.50	0.10–0.35	0.60–0.90	0.035	≤0.035	–	–	–	–
55C6	0.50–0.57	0.10–0.35	0.40–0.70	0.035	≤0.035	–	–	–	–
37C15	0.32–0.40	0.10–0.35	1.30–1.70	0.035	≤0.035	–	–	–	–
47C15	0.42–0.50	0.10–0.35	1.30–1.70	0.035	≤0.035	–	–	–	–
40C15S12	0.35–0.45	≤0.25	1.30–1.70	0.035	0.08–0.15	–	–	–	–
35Mn6Mo3	0.32–0.40	0.10–0.35	1.30–1.80	0.035	≤0.035	–	0.20–0.35	–	–
35Mn6Mo4	0.32–0.40	0.10–0.35	1.30–1.80	0.035	≤0.035	–	0.35–0.55	–	–
40Cr4	0.38–0.44	0.10–0.35	0.60–0.90	0.035	≤0.035	0.90–1.20	–	–	–
50Cr4	0.45–0.55	0.10–0.35	0.60–0.90	0.035	≤0.035	0.90–1.20	–	–	–
42Cr4Mo2	0.38–0.45	0.10–0.35	0.50–0.80	0.035	≤0.035	0.90–1.20	0.15–0.30	–	–
50Cr4V2	0.46–0.54	0.10–0.35	0.50–0.80	0.035	≤0.035	0.90–1.20	–	–	0.15–0.30
40Ni14	0.35–0.45	0.10–0.35	0.50–0.80	0.035	≤0.035	≤0.30	–	3.20–3.60	–
35Ni5Cr2	0.32–0.40	0.10–0.35	0.60–0.90	0.035	≤0.035	0.45–0.75	–	1.00–1.50	–
40Ni6Cr4Mo2	0.35–0.45	0.10–0.35	0.40–0.70	0.035	≤0.035	0.90–1.30	0.10–0.20	1.20–1.60	–
40Ni6Cr4Mo3	0.35–0.45	0.10–0.35	0.40–0.70	0.035	≤0.035	0.90–1.30	0.20–0.35	1.25–1.75	–
31Ni10Cr3Mo6	0.27–0.33	0.10–0.35	0.40–0.70	0.035	≤0.035	0.50–0.80	0.40–0.70	2.25–2.75	–

(Contd.)

TABLE 3.92 Chemical composition (cast analysis) of steels for flame and induction hardening, conforming to IS 3930 – 1994 (continued)

[1] Residual elements up to the following limits are acceptable:
$Cr \leq 0.25$ %; $Mo \leq 0.05$ %; $Ni \leq 0.25$ %; $Cu \leq 0.35$ %; $Sn \leq 0.05$ %; $V \leq 0.05$ %.
$Cr + Mo + Ni + Cu + Sn + V \leq 0.80$ %;
% Cu + 10 × % Sn ≤ 0.60 %.
[2] The steels may be ordered to a restricted C range to facilitate flame or induction hardening.

Note: All steels should be fully killed.

2.2 Hardness

TABLE 3.93 Hardness requirements for steels for flame and induction hardening, conforming to IS 3930 – 1994, in the "treated to improve machinability" condition

Steel designation	Maximum Brinell hardness HB	Steel designation	Maximum Brinell hardness HB
30C8	187	**40Cr4**	241
35C8	197	**50Cr4**	250
45C8	207	**42Cr4Mo2**	241
55C6	229	**50Cr4V2**	250
37C15	229	**40Ni14**	229
47C15	229	**35Ni5Cr2**	229
40C15S12	229	**40Ni6Cr4Mo2**	241
35Mn6Mo3	229	**40Ni6Cr4Mo3**	241
35Mn6Mo4	229	**31Ni10Cr3Mo6**	269

2.3 Mechanical properties

TABLE 3.94 Mechanical properties of steels for flame and induction hardening, conforming to IS 3930 – 1994, in the hardened and tempered condition

Steel designation	Limiting ruling section mm	R_m MPa	$R_{p0.2}$ min. MPa	A min. %	Minimum Izod impact strength J
30C8	30	600–750	400	18	55
35C8	63	600–750	400	18	55
45C8	30	700–850	480	15	35
	63	650–800	410	16	35
	100	600–750	370	17	41
55C6	30	700–850	430	14	28
	63	650–800	410	15	28
37C15	30	800–950	540	14	41
	63	700–850	490	16	41
	100	600–750	440	18	48
47C15	30	800–950	540	14	35
	63	750–900	470	15	41
	100	700–850	430	16	41
40C15S12	30	800–950	560	16	28
	60	700–850	500	18	35
35Mn6Mo3	30	925–1075	755	12	42
	63	850–1000	680	13	50
	100	800–950	585	15	50
	150	700–850	525	17	50

(Contd.)

TABLE 3.94 Mechanical properties of steels for flame and induction hardening, conforming to IS 3930 – 1994, in the hardened and tempered condition (continued)

Steel designation	Limiting ruling section mm	R_m MPa	$R_{p0.2}$ min. MPa	A min. %	Minimum Izod impact strength J
35Mn6Mo4	63	1000–1150	755	12	42
	100	900–1050	680	13	50
	150	800–950	585	15	50
40Cr4	30	900–1050	660	13	50
	63	800–950	560	14	50
	100	700–850	540	15	50
50Cr4	150	800–950	480	13	28
42Cr4Mo2	30	925–1075	750	10	42
	63	850–1000	650	11	42
	100	800–950	550	12	50
	150	700–850	490	13	50
50Cr4V2	40	980–1180	780	10	21
	100	880–1080	685	12	21
	160	830–980	640	13	21
	250	780–930	480	13	21
40Ni14	63	900–1050	660	15	55
	100	800–950	560	16	55
35Ni5Cr2	100	800–950	560	16	55
	150	700–850	500	18	55

(Contd.)

TABLE 3.94 Mechanical properties of steels for flame and induction hardening, conforming to IS 3930 – 1994, in the hardened and tempered condition (continued)

Steel designation	Limiting ruling section mm	R_m MPa	$R_{p0.2}$ min. MPa	A min. %	Minimum Izod impact strength J
40Ni6Cr4Mo2	30	1000–1150	830	12	48
	63	925–1075	740	12	48
	100	850–1000	680	13	55
	150	800–950	560	15	55
40Ni6Cr4Mo3	30	1080–1240	925	11	55
	63	1000–1150	830	12	42
	100	925–1075	740	12	42
	150	850–1000	670	13	50
31Ni10Cr3Mo6	63	≥ 1550	1125	5	9
	63	1150–1300	1000	10	28
	100	1080–1240	880	11	35
	150	1000–1150	830	12	42
	150	925–1075	740	12	42
	150	850–1000	670	13	50

3 SURFACE HARDNESS OF FLAME AND INDUCTION HARDENED ZONES

TABLE 3.95 Surface hardness of steels for flame and induction hardening, conforming to IS 3930 – 1979, after induction hardening and stress relieving[1)]

Steel designation	Minimum surface hardness[2)] HRC	Steel designation	Minimum surface hardness[2)] HRC
30C8	45	**40Cr4**	54
35C8	51	**50Cr4**	57
45C8	55	**42Cr4Mo2**	54
55C6	57	**50Cr4V2**	57
37C15	53	**40Ni14**	54
47C15	54	**35Ni5Cr2**	54
40C15S12	53	**40Ni6Cr4Mo2**	54
35Mn6Mo3	53	**40Ni6Cr4Mo3**	54
35 Mn6Mo4	53	**31Ni10Cr3Mo6**	49

1) At 150–180 °C for about 1 h.
2) Typical values.

4 RECOMMENDED HOT WORKING AND HEAT TREATING TEMPERATURES

TABLE 3.96 Recommended hot working and heat treating temperatures for steels for flame and induction hardening, conforming to IS 3930 – 1994

Steel designation	Hot working temperature °C	Softening temperature °C	Normalizing temperature °C	Hardening temperature °C	Quenching agent	Tempering temperature °C
30C8	1200–850	650–700	860–890	860–890	Water or oil	550–660
35C8	1200–850	650–700	850–880	850–880	Water or oil	550–660
45C8	1200–850	650–700	830–860	830–860	Water or oil	550–660
55C6	1150–850	650–700	810–840	810–840	Oil	550–660
37C15	1200–850	600–660	840–870	840–870	Water or oil	550–660
47C15	1200–850	600–660	830–860	830–860	Oil	550–660
40C15S12	1200–850	600–660	840–870	840–870	Oil	550–660
35Mn6Mo3	1200–900	640–660	830–860	830–860	Oil	550–660
35Mn6Mo4	1200–900	640–660	830–860	830–860	Oil	550–660
40Cr4	1200–850	650–700	850–880	850–880	Oil	550–700
50Cr4	1150–850	670–700	830–880	830–880	Oil	540–680
42Cr4Mo2	1200–850	650–700	850–880	850–880	Oil	550–720
50Cr4V2	1050–850	680–720	840–880	830–860	Oil	540–680
40Ni14	1150–900	630–650	830–860	830–860	Oil	550–650
35Ni5Cr2	1200–900	630–670	–	820–850	Oil or water	550–660
40Ni6Cr4Mo2	1200–900	630–670	–	820–850	Oil	550–660
40Ni6Cr4Mo3	1200–900	630–660	–	820–850	Oil	550–660

(Contd.)

TABLE 3.96 Recommended hot working and heat treating temperatures for steels for flame and induction hardening, conforming to IS 3930 – 1994 (continued)

Steel designation	Hot working temperature °C	Softening temperature °C	Normalizing temperature °C	Hardening temperature °C	Quenching agent	Tempering temperature °C
31Ni10Cr3Mo6	1200–850	630–650	–	820–850	Oil	550–660

Note: For flame or induction hardening, the hardening temperature is about 50 °C higher than the limits indicated above for hardening. Quenching is usually effected in water; in special cases, however, a less drastic quenching agent may be used. Stress relieving after induction hardening should be performed at 150–180 °C.

5 EQUIVALENT GRADES

TABLE 3.97 Comparison of grades of steels for flame and induction hardening specified in IS 3930 – 1994 with nearest equivalent grades specified in other national and international standards

IS 3930–94	BS EN 10083 (1)–91 BS EN 10083 (2)–91 BS 970 (1)–91	DIN 17212–72	GOST 1050–88 GOST 1414–75 GOST 4543–71 GOST 14959–79	ISO 683 (1)–87 ISO 683 (9)–88	JIS G 4051–79 JIS G 4052–79 JIS G 4102–79 JIS G 4103–79 JIS G 4104–79 JIS G 4105–79 JIS G 4106–79 JIS G 4801–84 JIS G 4804–84	SAE J403–94 SAE J404–94 SAE J1268–93
30C8	2 C 30	–	30 30Г	C 30 E4	S 30 C S 33 C	1030
35C8	2 C 35	Cf 35	35 35Г	C 35 E4	S 35 C S 38 C	1035
45C8	2 C 45 3 C 45 080M42 080H42	Cf 45	45 45Г	C 45 E 4	S 45 C S 48 C	1045 1045H
55C6	2 C 50	Cf 53	50 50Г	C 50 E4	S 53 C	1050

(Contd.)

TABLE 3.97 Comparison of grades of steels for flame and induction hardening specified in IS 3930 – 1994 with nearest equivalent grades specified in other national and international standards (continued)

IS 3930–94	BS EN 10083 (1)–91 BS EN 10083 (2)–91 BS 970 (1)–91	DIN 17212–72	GOST 1050–88 GOST 1414–75 GOST 4543–71 GOST 14959–79	ISO 683 (1)–87 ISO 683 (9)–88	JIS G 4051–79 JIS G 4052–79 JIS G 4102–79 JIS G 4103–79 JIS G 4104–79 JIS G 4105–79 JIS G 4106–79 JIS G 4801–84 JIS G 4804–84	SAE J403–94 SAE J404–94 SAE J1268–93
37C15	150M36 150H36	–	35Г2	36 Mn 6	SMn 438 SMn 438 H	1541 1541H
47C15	–	–	45Г2	–	–	1548
40C15S12	–	–	–	–	SUM 42	1141
35Mn6Mo3	605M36 605H36	–	–	–	–	–
35Mn6Mo4	–	–	–	–	–	–
40Cr4	41 Cr 4	42 Cr 4	40X	41 Cr 4	SCr 440 SCr 440 H	5140 5140H
50Cr4	–	–	50X	–	–	5150 5150H
42Cr4Mo2	42 CrMo 4 709M40 709H40	42 CrMo 4	–	42 CrMo 4	SCM 440 SCM 440 H	4140 4140H 4142 4142H

(Contd.)

TABLE 3.97 Comparison of grades of steels for flame and induction hardening specified in IS 3930 – 1994 with nearest equivalent grades specified in other national and international standards (continued)

IS 3930–94	BS EN 10083 (1)–91 BS EN 10083 (2)–91 BS 970 (1)–91	DIN 17212–72	GOST 1050–88 GOST 1414–75 GOST 4543–71 GOST 14959–79	ISO 683 (1)–87 ISO 683 (9)–88	JIS G 4051–79 JIS G 4052–79 JIS G 4102–79 JIS G 4103–79 JIS G 4104–79 JIS G 4105–79 JIS G 4106–79 JIS G 4801–84 JIS G 4804–84	SAE J403–94 SAE J404–94 SAE J1268–93
50Cr4V2	51 CrV 4	–	50XΦA 50XΓΦ A	51 CrV 4	SUP 10	6150 6150H
40Ni14	–	–	–	–	–	–
35Ni5Cr2	–	–	40XH	–	SNC 236	–
40Ni6Cr4Mo2	–	–	–	–	–	–
40Ni6Cr4Mo3	817M40 817H40	–	–	–	–	–
31Ni10Cr3Mo6	–	–	–	–	–	–

6 REFERENCES

Books

1. *ASM Handbook*, vol. 1, ASM International, Materials Park, Ohio, U.S.A., 1990.
2. *ASM Handbook*, vol. 4, ASM International, Materials Park, Ohio, U.S.A., 1991.
3. *Source Book on Heat Treating—Volume I: Materials and Processes*, American Society for Metals, Metals Park, Ohio, U.S.A., 1975.

Standard

4. IS 3930 – 1994: *Flame and Induction Hardening Steels—Specification.*

3.18

SPRING STEELS

1 INTRODUCTION

Spring steels are unalloyed, low-alloy and stainless steels which are characterized by high elastic limit, high tensile and fatigue strength, good toughness and good formability. These steels are produced using special steelmaking and conditioning practices to ensure freedom from harmful surface imperfections and decarburization, and low non-metallic inclusion content. Steels selected for hardened and tempered springs should possess adequate hardenability to guarantee the desired microstructure.

Unalloyed spring steels generally contain carbon in the range 0.45 to 1.00 %. They are used in the work hardened, or oil hardened and tempered condition.

The low-alloy spring steels are essentially silicon-manganese, chromium-vanadium and silicon-chromium steels which are used in the oil hardened and tempered condition. These steels are superior to the unalloyed steels with respect to resistance to stress relaxation and shock loading.

Stainless spring steels are the most expensive of the spring steels . Nevertheless, they are increasingly in demand because of their corrosion resistance coupled with their ability to withstand elevated temperatures.

2 SPECIFICATIONS

2.1 Spring steel wire for cold-formed springs

2.1.1 Patented and cold-drawn unalloyed steel spring wire

I. Chemical composition

TABLE 3.98 Chemical composition (cast analysis) of patented and cold-drawn unalloyed steel spring wire, conforming to IS 4454 (I) – 1981

Grade	Chemical composition [% (m/m)]						
	C	Si	Mn max.	P max.	S max.	P+S max.	Cu max.
1	0.50–0.75	0.15–0.35	1.00	0.040	0.050	–	0.20
2	0.60–0.85	0.15–0.35	0.80	0.040	0.040	–	0.15
3	0.75–1.00	0.15–0.35	0.80	0.030	0.030	0.050	0.12
4	0.75–1.00	0.15–0.35	0.80	0.025	0.025	0.040	0.12
Note: All steels should be fully killed.							

II. Mechanical properties

TABLE 3.99 Tensile strength, percentage reduction of area and torsion test requirements of patented and cold-drawn unalloyed steel spring wire, conforming to IS 4454 (I) – 1981

Nominal diameter	R_m min. for grades				Z min. for wire grades 1–4	Minimum number of turns in torsion for wire grades[1)] 1–4
	1	**2**	**3**	**4**		
mm	MPa	MPa	MPa	MPa	%	
0.07	–	–	2550	–	–	–
0.08	–	–	2540	–	–	–
0.09	–	–	2530	–	–	–

(Contd.)

TABLE 3.99 Tensile strength, percentage reduction of area and torsion test requirements of patented and cold-drawn unalloyed steel spring wire, conforming to IS 4454 (I) – 1981 (continued)

Nominal diameter	R_m min. for grades				Z min. for wire grades 1–4	Minimum number of turns in torsion for wire grades[1)] 1–4
	1	2	3	4		
mm	MPa	MPa	MPa	MPa	%	
0.10	–	–	2530	–	–	–
0.11	–	–	2520	–	–	–
0.12	–	–	2520	–	–	–
0.14	–	–	2510	–	–	–
0.16	–	–	2500	–	–	–
0.18	–	–	2500	–	–	–
0.20	–	–	2490	2700	–	–
0.22	–	–	2480	2680	–	–
0.25	–	–	2470	2670	–	–
0.28	–	–	2460	2660	–	–
0.30	1720	2060	2460	2660	–	–
0.32	1710	2050	2450	2650	–	–
0.34	1710	2050	2450	2640	–	–
0.36	1700	2040	2440	2630	–	–
0.38	1700	2040	2430	2620	–	–
0.40	1700	2040	2430	2620	–	–
0.43	1690	2030	2420	2610	–	–
0.45	1680	2020	2410	2600	–	–
0.48	1680	2020	2400	2590	–	–
0.50	1670	2010	2390	2580	–	25

(Contd.)

TABLE 3.99 Tensile strength, percentage reduction of area and torsion test requirements of patented and cold-drawn unalloyed steel spring wire, conforming to IS 4454 (I) – 1981 (continued)

Nominal diameter	R_m min. for grades				Z min. for wire grades 1–4	Minimum number of turns in torsion for wire grades[1)] 1–4
	1	2	3	4		
mm	MPa	MPa	MPa	MPa	%	
0.53	1660	2000	2380	2570	–	25
0.56	1660	2000	2370	2560	–	25
0.60	1650	1990	2360	2550	–	25
0.63	1640	1980	2340	2540	–	25
0.65	1640	1980	2330	2540	–	25
0.70	1630	1970	2320	2530	–	25
0.75	1620	1960	2300	2500	–	25
0.80	1610	1950	2280	2480	–	25
0.85	1600	1930	2260	2460	–	25
0.90	1590	1920	2250	2440	–	25
0.95	1580	1910	2250	2420	–	25
1.00	1570	1900	2240	2400	–	25
1.05	1560	1890	2210	2380	40	22
1.10	1550	1880	2190	2370	40	22
1.20	1540	1860	2170	2340	40	22
1.25	1530	1850	2140	2320	40	22
1.30	1520	1840	2130	2300	40	22
1.40	1500	1820	2110	2290	40	22
1.50	1490	1800	2100	2260	40	22
1.60	1470	1780	2080	2250	40	22

(Contd.)

TABLE 3.99 Tensile strength, percentage reduction of area and torsion test requirements of patented and cold-drawn unalloyed steel spring wire, conforming to IS 4454 (I) – 1981 (continued)

Nominal diameter	R_m min. for grades				Z min. for wire grades 1–4	Minimum number of turns in torsion for wire grades[1)] 1–4
	1	2	3	4		
mm	MPa	MPa	MPa	MPa	%	
1.70	1460	1760	2050	2220	40	22
1.80	1440	1750	2030	2190	40	22
1.90	1430	1730	2010	2180	40	22
2.00	1420	1720	1990	2160	40	22
2.10	1410	1700	1960	2130	40	22
2.25	1400	1680	1940	2100	40	22
2.40	1380	1660	1910	2070	40	22
2.50	1370	1640	1890	2050	40	22
2.60	1360	1620	1860	2030	40	22
2.80	1340	1600	1840	2000	40	22
3.00	1320	1570	1830	1980	40	22
3.20	1310	1550	1790	1960	40	20
3.40	1290	1530	1760	1920	40	20
3.60	1270	1510	1750	1890	40	20
3.80	1260	1490	1720	1860	40	20
4.00	1250	1480	1700	1840	35	20
4.25	1250	1460	1680	1820	35	18
4.50	1230	1440	1660	1800	35	18
4.75	1210	1420	1620	1770	35	18
5.00	1190	1390	1600	1750	35	18

(Contd.)

TABLE 3.99 Tensile strength, percentage reduction of area and torsion test requirements of patented and cold-drawn unalloyed steel spring wire, conforming to IS 4454 (I) – 1981 (continued)

Nominal diameter	R_m min. for grades				Z min. for wire grades 1–4	Minimum number of turns in torsion for wire grades[1)] 1–4
	1	2	3	4		
mm	MPa	MPa	MPa	MPa	%	
5.30	1170	1370	1570	1720	35	13
5.60	1150	1350	1550	1690	35	13
6.00	1130	1320	1530	1670	35	13
6.30	1120	1310	1500	1640	35	13
6.50	1110	1290	1480	1620	35	13
7.00	1090	1260	1460	1610	35	9
7.50	1070	1250	1430	1570	30	9
8.00	1050	1220	1400	1540	30	9
8.50	1020	1200	1370	1500	30	6
9.00	1000	1180	1350	1480	30	6
9.50	990	1150	1310	–	–	6
10.0	980	1130	1290	–	–	6
10.5	–	1100	–	–	–	–
11.0	–	1080	–	–	–	–
12.0	–	1040	–	–	–	–
12.5	–	1030	–	–	–	–
13.0	–	1020	–	–	–	–
14.0	–	990	–	–	–	–
15.0	–	970	–	–	–	–
16.0	–	960	–	–	–	–

(Contd.)

TABLE 3.99 Tensile strength, percentage reduction of area and torsion test requirements of patented and cold-drawn unalloyed steel spring wire, conforming to IS 4454 (I) – 1981 (continued)

Nominal diameter	R_m min. for grades				Z min. for wire grades	Minimum number of turns in torsion for wire grades[1)]
	1	**2**	**3**	**4**	**1–4**	**1–4**
mm	MPa	MPa	MPa	MPa	%	
17.0	–	950	–	–	–	–

[1)] $L = 100d$, but < 500 mm.

2.1.2 *Oil-hardened and tempered unalloyed steel spring wire and valve spring wire*

I. Chemical composition

TABLE 3.100 Chemical composition (cast analysis) of oil-hardened and tempered unalloyed steel spring wire and value spring wire, conforming to IS 4454 (II) – 1975

·Grade	Chemical composition [% (*m*/*m*)]					
	C	Si	Mn	P	S	Cu
				max.	max.	max.
SW[1)]	0.55–0.75	0.10–0.35	0.60–0.90	0.040	0.040	0.15
VW[2)]	0.60–0.70	0.10–0.25	0.60–0.90	0.025	0.020	0.06

[1)] Oil-hardened and tempered unalloyed steel spring wire.
[2)] Oil-hardened and tempered unalloyed steel valve spring wire.

Note: All steels should be fully killed.

II. *Mechanical properties*

1 *Tensile properties*

TABLE 3.101 Tensile strength and percentage reduction of area of oil-hardened and tempered unalloyed steel spring wire and valve spring wire, conforming to IS 4454 (II) – 1975

Nominal diameter	R_m min. for grades		Z min. for grades	
	SW	**VW**	**SW**	**VW**
mm	MPa	MPa	%	%
1.00	1760–1960	1670–1810	45	45
1.05	1760–1960	1670–1810	45	45
1.10	1760–1960	1670–1810	45	45
1.20	1720–1910	1620–1760	45	45
1.25	1720–1910	1620–1760	45	45
1.30	1720–1910	1620–1760	45	45
1.40	1720–1910	1620–1760	45	45
1.50	1670–1860	1570–1690	45	45
1.60	1670–1860	1570–1690	45	45
1.70	1670–1860	1570–1690	45	45
1.80	1670–1860	1570–1690	45	45
1.90	1670–1860	1570–1690	45	45
2.00	1620–1760	1520–1620	45	45
2.10	1620–1760	1520–1620	45	45
2.25	1620–1760	1520–1620	45	45
2.40	1620–1760	1520–1620	45	45
2.50	1570–1720	1470–1620	45	45
2.60	1570–1720	1470–1620	45	45

(Contd.)

TABLE 3.101 Tensile strength and percentage reduction of area of oil-hardened and tempered unalloyed steel spring wire and valve spring wire, conforming to IS 4454 (II) – 1975 (continued)

Nominal diameter	R_m min. for grades		Z min. for grades	
	SW	**VW**	**SW**	**VW**
mm	MPa	MPa	%	%
2.80	1570–1720	1470–1620	45	45
3.00	1520–1670	1430–1530	45	45
3.20	1520–1670	1430–1530	45	45
3.40	1520–1670	1430–1530	45	45
3.60	1480–1630	1400–1500	45	45
3.80	1480–1630	1400–1500	45	45
4.00	1480–1630	1400–1500	42	42
4.25	1440–1590	1370–1470	42	42
4.50	1440–1590	1370–1470	42	42
4.75	1440–1590	1370–1470	40	40
5.00	1440–1590	1370–1470	40	40
5.30	1400–1550	1340–1440	40	40
5.60	1400–1550	1340–1440	40	40
6.00	1400–1550	1340–1440	40	40
6.30	1360–1510	1300–1400	40	40
6.50	1360–1510	1300–1400	38	38
7.00	1360–1510	1300–1400	38	38
7.50	1360–1510	1300–1400	38	38
8.00	1290–1440	–	38	38

(Contd.)

TABLE 3.101 Tensile strength and percentage reduction of area of oil-hardened and tempered unalloyed steel spring wire and valve spring wire, conforming to IS 4454 (II) – 1975 (continued)

Nominal diameter	R_m min. for grades		Z min. for grades	
	SW	**VW**	**SW**	**VW**
mm	MPa	MPa	%	%
8.50	1290–1440	–	38	–
9.00	1290–1440	–	35	–
9.50	1290–1440	–	35	–
10.0	1250–1400	–	35	–
10.5	1250–1400	–	35	–
11.0	1250–1400	–	35	–
12.0	1250–1400	–	35	–
12.5	1250–1400	–	35	–
13.0	1250–1400	–	35	–
14.0	1250–1400	–	35	–

2 *Reverse torsion test requirements*

TABLE 3.102 Reverse torsion test requirements of oil-hardened and tempered unalloyed steel spring wire and valve spring wire, conforming to, IS 4454 (II) – 1975

Nominal diameter	Test length[1)]	Minimum number of turns	
		Forward	Reverse
mm	mm		
Grade SW			
All diameters	100*d*	7	7

(Contd.)

TABLE 3.102 Reverse torsion test requirements of oil-hardened and tempered unalloyed steel spring wire and valve spring wire, conforming to IS 4454 (II) – 1975 (continued)

Nominal diameter mm	Test length[1] mm	Minimum number of turns	
		Forward	Reverse
Grade VW			
> 1.0 ≤ 1.2	100	6	10
≥ 1.2 ≤ 1.5	100	6	9
≥ 1.5 ≤ 2.0	100	6	8
≥ 2.0 ≤ 3.4	50*d*	6	8
≥ 3.4 ≤ 7.5	50*d*	6	10

[1] *d* = nominal wire diameter.

2.2.1 Oil-hardened and tempered low-alloy steel spring wire and valve spring wire

I. Chemical composition

TABLE 3.103 Chemical composition (cast analysis) of oil-hardened and tempered low-alloy steel spring wire and valve spring wire, conforming to IS 4454 (III) – 1975

Grade	Chemical composition [% (*m*/*m*)]						
	C	Si	Mn	P max.	S max.	Cr	V
1 S[1]	0.45–0.55	0.15–0.35	0.60–0.90	0.040	0.040	0.90–1.20	0.15–0.30
1 D[2]	0.45–0.55	0.15–0.35	0.60–0.90	0.020	0.020	0.90–1.20	0.15–0.30
2 S[1]	0.50–0.60	1.20–1.60	0.50–0.80	0.040	0.040	0.50–0.80	–
2 D[2]	0.50–0.60	1.20–1.60	0.50–0.80	0.025	0.025	0.50–0.80	–

[1] Oil-hardened and tempered low-alloy steel spring wire.
[2] Oil-hardened and tempered low-alloy steel valve spring wire.

Note: All steels should be fully killed.

II. Mechanical properties

TABLE 3.104 Tensile strength and percentage reduction of area of oil-hardened and tempered low-alloy steel spring wire and valve spring wire, conforming to IS 4454 (III) – 1975

Nominal diameter	R_m min. for grades				Z min.
	1 S	1 D	2 S	2 D	
mm	MPa	MPa	MPa	MPa	%
1.00	–	–	2060	2060	50
1.05	1860	1760	1960	1960	50
1.10	1860	1760	1960	1960	50
1.20	1860	1760	1960	1960	50
1.25	1860	1760	1960	1960	50
1.30	1860	1760	1960	1960	50
1.40	1860	1760	1960	1960	50
1.50	1860	1760	1960	1960	50
1.60	1860	1760	1960	1960	50
1.70	1860	1760	1960	1960	50
1.80	1860	1760	1960	1960	50
1.90	1860	1760	1960	1960	50
2.00	1860	1760	1960	1960	50
2.10	1760	1670	1910	1910	50
2.25	1760	1670	1910	1910	50
2.40	1760	1670	1910	1910	50
2.50	1760	1670	1910	1910	50
2.60	1760	1670	1910	1910	50

(Contd.)

TABLE 3.104 Tensile strength and percentage reduction of area of oil-hardened and tempered low-alloy steel spring wire and valve spring wire, conforming to IS 4454 (III) – 1975 (continued)

Nominal diameter	R_m min. for grades				Z min.
	1 S	1 D	2 S	2 D	
mm	MPa	MPa	MPa	MPa	%
2.80	1760	1670	1910	1910	50
3.00	1760	1670	1910	1910	50
3.20	1760	1670	1910	1910	50
3.40	1670	1570	1860	1860	45
3.60	1670	1570	1860	1860	45
3.80	1670	1570	1860	1860	45
4.00	1670	1570	1860	1860	45
4.25	1590	1520	1810	1810	45
4.50	1590	1520	1810	1810	45
4.75	1590	1520	1810	1810	45
5.00	1590	1520	1810	1810	45
5.30	1520	1470	1760	1760	45
5.60	1520	1470	1760	1760	45
6.00	1520	1470	1760	1760	45
6.30	1470	1420	1720	1720	42
6.50	1470	1420	1720	1720	42
7.00	1470	1420	1720	1720	42
7.50	–	–	1670	1670	42
8.00	–	–	1670	1670	42

2.2.2 Stainless steel spring wire

I. Chemical composition

TABLE 3.105 Chemical composition (cast analysis) of stainless steel spring wire, conforming to IS 4454 (IV) – 1975

Grade	Chemical composition [% (m/m)]							
	C max.	Si max.	Mn max.	P max.	S max.	Cr	Ni	Al
1	0.15	1.00	2.00	0.045	0.030	17.0–19.0	8.00–10.0	–
2	0.09	1.00	1.00	0.040	0.030	16.0–18.0	6.75–7.75	0.75–1.25

II. Mechanical properties

TABLE 3.106 Tensile strength and percentage reduction of area of hard-drawn stainless steel spring wire, conforming to IS 4454 (IV)–1975

Nominal diameter mm.	R_m min. MPa	Z min. %	Nominal diameter mm.	R_m min. MPa	Z min. %
0.10	2060	–	0.28	1960	–
0.11	2010	–	0.30	1960	–
0.125	2010	–	0.32	1960	–
0.14	2010	–	0.34	1910	–
0.16	2010	–	0.36	1910	–
0.18	2010	–	0.38	1910	–
0.20	1960	–	0.40	1910	–
0.22	1960	–	0.43	1910	–
0.25	1960	–	0.45	1910	–

(Contd.)

TABLE 3.106 Tensile strength and percentage reduction of area of hard-drawn stainless steel spring wire, conforming to IS 4454 (IV) – 1975 (continued)

Nominal diameter mm.	R_m min. MPa	Z min. %	Nominal diameter mm.	R_m min. MPa	Z min. %
0.48	1910	–	1.40	1670	40
0.50	1910	–	1.50	1670	40
0.53	1860	–	1.60	1670	40
0.56	1860	–	1.70	1670	40
0.60	1860	–	1.80	1670	40
0.63	1860	–	1.90	1670	40
0.65	1860	–	2.00	1670	40
0.70	1860	–	2.10	1570	40
0.75	1860	–	2.25	1570	40
0.80	1860	–	2.40	1570	40
0.85	1760	–	2.50	1570	40
0.90	1760	–	2.60	1570	40
0.95	1760	–	2.80	1570	40
1.00	1760	–	3.00	1570	40
1.05	1760	40	3.15	1570	40
1.10	1760	40	3.20	1470	40
1.15	1760	40	3.40	1470	40
1.20	1760	40	3.60	1470	40
1.25	1760	40	3.80	1470	40
1.30	1670	40	4.00	1470	40

(Contd.)

TABLE 3.106 Tensile strength and percentage reduction of area of hard-drawn stainless steel spring wire, conforming to IS 4454 (IV)–1975 (continued)

Nominal diameter mm.	R_m min. MPa	Z min. %		Nominal diameter mm.	R_m min. MPa	Z min. %
4.25	1470	40		6.50	1320	40
4.50	1470	40		7.00	1320	40
4.75	1470	40		7.50	1320	40
5.00	1470	40		8.00	1320	40
5.30	1320	40		8.50	1270	40
5.60	1320	40		9.00	1270	40
6.00	1320	40		9.50	1270	40
6.30	1320	40		10.0	1270	40

2.3 Cold-rolled steel strips for springs

I. Chemical composition

TABLE 3.107 Chemical composition (cast analysis) of cold-rolled unalloyed and low-alloy steel strips for springs, conforming to IS 2507 – 1975

Steel designation	Chemical composition [% (*m*/*m*)]						
	C	Si	Mn	P max.	S max.	Cr	V
45C8	0.40–0.50	0.10–0.35	0.60–0.90	0.050	0.050	–	–
55C6	0.50–0.60	0.10–0.35	0.50–0.65	0.050	0.050	–	–
65C6	0.60–0.70	0.10–0.35	0.50–0.80	0.050	0.050	–	–
70C6	0.65–0.75	0.10–0.35	0.50–0.80	0.050	0.050	–	–

(Contd.)

TABLE 3.107 Chemical composition (cast analysis) of cold-rolled unalloyed and low-alloy steel strips for springs, conforming to IS 2507 – 1975 (continued)

Steel designation	Chemical composition [% (*m*/*m*)]						
	C	Si	Mn	P max.	S max.	Cr	V
75C6	0.70–0.80	0.10–0.35	0.50–0.80	0.050	0.050	–	
80C6	0.75–0.85	0.10–0.35	0.50–0.80	0.050	0.050	–	–
85C6	0.80–0.90	0.10–0.35	0.50–0.80	0.050	0.050	–	–
98C6	0.90–1.05	0.10–0.35	0.50–0.80	0.050	0.050	–	–
55Si7	0.50–0.60	1.50–2.00	0.80–1.00	0.050	0.050	–	–
50Cr4	0.45–0.55	0.10–0.35	0.60–0.90	0.050	0.050	0.90–1.20	–
50Cr4V2	0.45–0.55	0.10–0.35	0.50–0.80	0.050	0.050	0.90–1.20	0.15–0.30

Note: All steels should be fully killed.

II. Mechanical properties

TABLE 3.108 Mechanical properties of cold-rolled unalloyed and low-alloy steel strips for springs, conforming to IS 2507 – 1975, in the heat treatment conditions “annealed” or “hardened and tempered” (for information only)

Steel designation	Mechanical properties in the heat treatment condition							
	Annealed				Hardened and tempered			
	R_m max. MPa	R_e max. MPa	*A* min. %	Maximum Vickers hardness HV	R_m MPa	R_e min. MPa	*A* min. %	Vickers hardness HV
45C8	590	270	30	185	1180–1420	1030	6	350–425
55C6	590	270	30	195	1180–1420	1030	6	350–425

(Contd.)

TABLE 3.108 Mechanical properties of cold-rolled unalloyed and low-alloy steel strips for springs, conforming to IS 2507 – 1975, in the heat treatment conditions "annealed" or "hardened and tempered" (for information only) (continued)

Steel designation	Mechanical properties in the heat treatment condition							
	Annealed				Hardened and tempered			
	R_m max. MPa	R_e max. MPa	A min. %	Maximum Vickers hardness HV	R_m MPa	R_e min. MPa	A min. %	Vickers hardness HV
65C6	590	270	30	205	1180–1420	1030	6	350–425
70C6	590	270	30	215	1180–1420	1030	6	350–425
75C6	640	290	25	220	1180–1570	1080	6	350–475
80C6	640	290	25	220	1180–1570	1080	6	350–475
85C6	640	290	25	220	1180–1570	1080	6	350–475
98C6	690	310	25	225	1570–1760	1470	6	475–540
55Si7	780	340	20	240	1570–1960	1470	6	475–570
50Cr4	780	340	20	240	1670–2260	1570	5	505–650
50Cr4V2	780	340	20	240	1880–2360	1760	4	555–670

2.3.1 Unalloyed and alloy steels for hot-formed springs

I. Chemical composition

TABLE 3.109 Chemical composition (cast analysis) of unalloyed and alloy steels for hot-formed springs specified in Indian Standards

Steel designation	Chemical composition [% (*m/m*)]							
	C	Si	Mn	P max.	S max.	Cr	Mo	V
55C6	0.50–0.60	0.15–0.35	0.50–0.65	0.050	0.050	–	–	–
75C6	0.70–0.80	0.15–0.35	0.50–0.80	0.050	0.050	–	–	–
98C6	0.90–1.05	0.15–0.35	0.50–0.80	0.050	0.050	–	–	–
113C6	1.05–1.20	0.15–0.35	0.50–0.80	0.050	0.050	–	–	–
40Si7	0.35–0.45	1.50–2.00	0.80–1.00	1a)	1a)	–	–	–
55Si7	0.50–0.60	1.50–2.00	0.80–1.00	1b)	1b)	–	–	–
60Si7	0.55–0.65	1.50–2.00	0.80–1.00	1b)	1b)	–	–	–
65Si7	0.60–0.70	1.50–2.00	0.80–1.00	1b)	1b)	–	–	–
50Cr4V2	0.45–0.55	0.15–0.35	0.50–0.80	1b)	1b)	0.90–1.20	–	0.15–0.30
60Cr4V2	0.55–0.65	0.15–0.35	0.80–1.00	1b)	1b)	0.90–1.20	–	≥0.15
51Cr4Mo2V1	0.48–0.56	0.15–0.40	0.70–1.10	0.035	0.035	0.90–1.20	0.15–0.25	0.07–0.12

[1] The steels should be ordered to the following limits of phosphorus and sulphur:

	a) % P max.	a) % S max.	b) % P max.	b) % S max.
IS 3885 (I)–1992	0.050	0.050	0.050	0.050
IS 3885 (II)–1992	0.045	0.045	0.045	0.045
IS 3431–1982	–	–	0.045	0.045
IS 3195–1992	–	–	0.040	0.040

Source: References 4–7

II. Mechanical properties

TABLE 3.110 Mechanical properties of unalloyed and alloy steels for hot-formed springs specified in the Indian Standards, in the treatment conditions "untreated", "soft annealed" or "hardened and tempered"

Steel designation	Mechanical properties in the treatment condition				
	Untreated[1)]	Soft annealed	Hardened and Tempered[2)]		
	Brinell hardness HB	Brinell hardness HB	R_m MPa	R_e min. MPa	A min. %
55C6	–	≤ 240	–	–	–
75C6	–	≤ 240	–	–	–
98C6	–	≤ 240	–	–	–
113C6	–	≤ 240	–	–	–
40Si7	≈ 270	≤ 245	–	–	–
55Si7	≈ 270	≤ 245	1320–1570	1130	6
60Si7	≈ 310	≤ 255	1320–1570	1130	6
65Si7	≈ 310	≤ 255	1370–1620	1180	6
50Cr4V2	≈ 310	≤ 245	1370–1620	1180	6
60Cr4V2	≈ 310	≤ 255	1370–1620	1180	6
51Cr4Mo2V1	≈ 310	â 255	1370–1670	1180	6

1) For information only.
2) For ruling sections ≤ 10 mm.

Source: References 2 and 5

III. Hardenability

TABLE 3.111 Limiting dimensions for through hardenability of steels for hot-formed springs

Steel designation	Minimum core hardness after quenching	Maximum dimension for		Heat treatment for verifying the maximum dimensions	
		Flats (thickness)	Rounds (diameter)	Hardening temperature	Quenching agent
	HRC	mm	mm	°C	
55Si7	54	16	20	830–860	Oil
60Si7	54	16	20	830–860	Oil
65Si7	54	16	25	830–860	Oil
50Cr4V2	54	16	25	830–860	Oil
60Cr4V2	54	25	40	830–860	Oil
51Cr4Mo2V1	54	40	60	830–860	Oil

Source: References 2, 5 and 9.

3 RECOMMENDED HOT WORKING AND HEAT TREATING TEMPERATURES

TABLE 3.112 Recommended hot working and heat treating temperatures for spring steels specified in Indian Standards

Steel designation	Hot working temperature °C	Softening temperature °C	Normalizing temperature °C	Hardening temperature °C	Quenching agent	Tempering temperature °C
45C8	1100–850	650–690	830–860	830–860	Oil or water	300–500
55C6	1050–850	650–690	810–840	810–840	Oil	300–500
65C6	1050–850	650–690	810–840	810–840	Oil	300–500
70C6	1050–850	650–690	810–840	810–840	Oil	300–500

(Contd.)

TABLE 3.112 Recommended hot working and heat-treating temperatures for spring steels specified in Indian Standards (continued)

Steel designation	Hot working temperature °C	Softening temperature °C	Normalizing temperature °C	Hardening temperature °C	Quenching agent	Tempering temperature °C
80C6	1050–850	650–690	810–840	810–840	Oil	300–500
85C6	1050–850	650–690	800–830	800–830	Oil	300–500
98C6	1050–850	640–680	780–810	780–810	Oil	430–500
113C6	1050–850	640–680	780–810	780–810	Oil	430–500
40Si7	1050–850	640–680	850–880	830–860	Water	350–550
55Si7	1050–850	640–680	850–880	830–860	Oil	430–500
60Si7	1050–850	640–680	850–880	830–860	Oil	430–500
65Si7	1050–850	640–680	850–880	830–860	Oil	430–500
50Cr4	1150–850	670–700	830–880	830–880	Oil	400–550
50Cr4V2	1050–850	640–680	850–880	830–860	Oil	350–550
60Cr4V2	1050–850	640–680	850–880	820–850	Oil	350–550
50Cr4Mo2V1	1050–850	640–680	850–880	830–860	Oil	350–550

4 EQUIVALENT STANDARDS/GRADES

TABLE 3.113 Comparison of Indian Standards on spring steel wire for cold-formed springs with nearest equivalent national and international standards

IS	ASTM	BS	DIN	ISO	JIS	SAE
Patented and cold-drawn unalloyed steel spring wire						
4454 (I)–81	A 227/A 227M–93 A 228/A 228M–93 A 679/A 679M–93	5216–91	17223 (1)–84	8458 (2)–89	G 3521–91 G 3522–91	J113–94 J178–94 J271–94

(Contd.)

Table 3.113 Comparison of Indian Standards on spring steel wire for cold-formed springs with nearest equivalent national and international standards (continued)

IS	ASTM	BS	DIN	ISO	JIS	SAE
Oil-hardened and tempered unalloyed steel spring wire and valve spring wire						
4454 (II)–75	A 229/A 229M–93 A 230/A 230M–93	2803–80	17223 (2)–90	8458 (3)–92	G 3560–94 G 3561–94	J316–94 J351–94
Oil-hardened and tempered alloy steel spring wire and valve spring wire						
4454 (III)–75	A 231/A 231M–93 A 232/A 232M–93 A 401/A 401M–95a A 877/A 877M–93	2803–80	17223 (2)–90	8458 (3)–92	G 3565–88 G 3566–88	J132–94 J157–94
Spring steel spring wire						
4454 (IV)–75	A 313/A 313M–95	2056–91	17224–82	6931 (1)–94	G 4314–94	J230–94 J217–94

TABLE 3.114 Comparison of grades of cold-rolled steel strips for springs specified in IS 2507 – 1975 with nearest equivalent grades specified in other national and international standards

IS 2507–75	ASTM A 682M–91	BS 5770 (2)–81 BS 5770 (3)–81	DIN 17222–79	ISO 4960–86	JIS G 4802–83	SAE J403–94 SAE J404–94
45C8	1045	–	–	CS 45	–	1045 1046
55C6	1055	–	C 55 Ck 55	CS 55	S 55 C - CSP	1055
65C6	1064	–	–	CS 65	S 65 C - CSP	1065
70C6	1070	70	C 67 Ck 67	CS 70	S 70 C - CSP	1070
75C6	1074	–	C 75 Ck 75	CS 75	–	1074 1075
80C6	1080	80	–	–	–	1080

(Contd.)

TABLE 3.114 Comparison of grades of cold-rolled steel strips for springs specified in IS 2507–1975 with nearest equivalent grades specified in other national and international standards (continued)

IS 2507–75	ASTM A 682M–91	BS 5770 (2)–81 BS 5770 (3)–81	DIN 17222–79	ISO 4960–86	JIS G 4802–83	SAE J403–94 SAE J404–94
85C6	1085	–	Ck 85	CS 85	SK 5 - CSP	1084 1085
98C6	1095	95	Ck 101	CS 95	SK 4 - CSP	1095
55Si7	–	–	55 Si 7	–	–	–
50Cr4	–	–	–	–	–	–
50Cr4V2	–	735M50	50 CrV 4	–	SUP 10 - CSP	–

TABLE 3.115 Comparison of grades of unalloyed and alloy steels for hot-formed springs specified in Indian Standards with nearest equivalent grades specified in other national and international standards

IS 3195–92 IS 3431–82 IS 3885 (I)–92 IS 3885 (II)–92	ASTM A 322–91 ASTM A 576–90b	BS 970 (2)–88	DIN 17221–88	GOST 14959–79	ISO 683 (14)–92	JIS G 4801–84	SAE J403–94 J404–94 J1268–93
55C6	1055	–		–	–	–	1055
75C6	1074	–		75	–	–	1074
98C6	1095	–	–	–	–	–	1095
113C6	–	–	–	–	–	–	–
40Si7	–	–	38 Si 7	–	–	–	
55Si7	–	–	–	55C2 55C2A	–	–	–
60Si7	9260	251H60	–	60C2 60C2A 60C2Г	59 Si 7	SUP6 SUP7	9260 9260H
65Si7	–	–	–	–	–	–	–
50Cr4V2	6150	735A51 735H51	50 CrV 4	50XOA	51 CrV 4	SUP10	6150 6150H
60Cr4V2	–	–	–	–	–	–	–
51Cr4Mo2V1	–	–	51 CrMoV 4	–	52 CrMoV 4	–	–

5 REFERENCES

Books

1. *ASM Handbook*, vol. 1, ASM International, Materials Park, Ohio, U.S.A., 1990.
2. C.W. Wegst, *Stahlschlüssel*, 16th ed., Verlag Stahlschlüssel Wegst, Marbach, Germany, 1992.

Standards

3. IS 2507 – 1975: *Specification for Cold-Rolled Steel Strips for Springs.*
4. IS 3195 – 1992: *Steel for the Manufacture of Volute and Helical Springs (for Railway Rolling Stock)—Specification.*
5. IS 3431 – 1982: *Specification for Steel for the Manufacture of Volute, Helical and Laminated Springs for Automotive Suspension.*
6. IS 3885 (1) – 1992: *Steel for the Manufacture of Laminated Springs (Railway Rolling Stock)—Part 1: Flat Sections—Specification.*
7. IS 3885 (2) – 1992: *Steel for the Manufacture of Laminated Springs (Railway Rolling Stock)—Part 2: Rib and Groove Sections.*
8. IS 4454 (I) – 1981: *Specification for Steel Wires for Cold Formed Springs—Part I: Patented and Cold Drawn Steel Wires—Unalloyed.*
9. IS 4454 (II) – 1975: *Specification for Steel Wires for Cold Formed Springs—Part II: Oil Hardened and Tempered Spring Steel Wire and Valve Spring Wire—Unalloyed.*
10. IS 4454 (III) – 1975: *Specification for Steel Wires for Cold Formed Springs—Part III: Oil Hardened and Tempered Steel Wires—Alloyed.*
11. IS 4454 (IV) – 1975: *Specification for Steel Wires for Cold Formed Springs—Part IV: Stainless Spring Steel Wire for Normal Corrosion Resistance.*
12. DIN 17221 – 1988: *Hot rolled Steels for Springs Suitable for Quenching and Tempering—Technical delivery conditions.*
13. ISO 683 (14) – 1992: *Heat-treatable Steels, Alloy Steels and Free-cutting Steels—Part 14: Hot-Rolled Steels for Quenched and Tempered Springs.*

3.19

WROUGHT STAINLESS STEELS

1 INTRODUCTION

Stainless steels are alloy steels which contain up to 1.2 % carbon and at least 10.5 % chromium, with or without other elements. These steels exhibit a wide range of strength levels, excellent corrosion resistance, good forming characteristics, and an aesthetically pleasing appearance.

Stainless steels are commonly divided into the five groups discussed below:

Austenitic stainless steels

Austenitic stainless steels are stainless steels which have been sufficiently alloyed with nickel, or by a combination of nickel, manganese and nitrogen, to produce an austenitic structure at ambient temperature. Austenitic steels cannot be hardened by heat treatment, but they can be significantly strengthened by cold working. They also exhibit good ductility and toughness even at high strength levels. Austenitic stainless steels have excellent corrosion resistance and good forming characteristics. In the annealed condition these steels are essentially nonmagnetic, although some of them may become slightly magnetic by cold working.

Ferritic stainless steels

Ferritic stainless steels are stainless steels which are used in a condition in which the microstructure consists essentially of ferrite and carbides. These steels cannot be strengthened by heat treatment. However, they can be strengthened moderately by cold working. They are magnetic, have good ductility, and possess good resistance to corrosion and oxidation.

Martensitic stainless steels

Martensitic stainless steels are stainless steels which are hardenable by heat-treatment, i.e., they develop a martensitic structure when rapidly cooled from the austenitizing temperature. These steels in the hardened and tempered condition possess high strength and hardness, fairly good ductility, adequate toughness and good wear resistance. They also resist corrosion in mild environments. They are magnetic.

Duplex stainless steels

Duplex stainless steels are stainless steels which have a mixed structure of ferritic and austenite. Compared to austenitic stainless steels they possess higher tensile and yield strengths, and improved resistance to stress-corrosion cracking. The toughness of these steels is between ferritic and austenitic stainless steels. They are magnetic.

Precipitation hardening stainless steels

Precipitation hardening stainless steels are stainless steels which are strengthened by solution treatment and ageing. They possess high strength at ambient and elevated temperatures, relatively good ductility and good corrosion resistance.

2 SPECIFICATIONS

2.1 Chemical Composition

TABLE 3.116 Chemical composition (cast analysis) of stainless steels specified in Indian Standards

Steel designation	Chemical composition [% (*m*/*m*)]								
	C	Si max.	Mn	P max.	S max.	Cr	Mo	Ni	Other elements
Ferritic stainless steels									
X04Cr12	≤ 0.08	1.00	≤ 1.00	0.040	0.030	11.5–13.5	–	–	Al = 0.10–0.30
X07Cr17	≤ 0.12	1.00	≤ 1.00	0.040	0.030	16.0–18.0	–	≤ 0.50	–
Martensitic stainless steels									
X12Cr12	0.08–0.15	1.00	≤ 1.00	0.040	0.030	11.5–13.5	–	≤ 1.00	–
X20Cr13	0.16–0.25	1.00	≤ 1.00	0.040	0.030	12.0–14.0	–	≤ 1.00	–
X30Cr13	0.26–0.35	1.00	≤ 1.00	0.040	0.030	12.0–14.0	–	≤ 1.00	–
X40Cr13	0.35–0.45	1.00	≤ 1.00	0.040	0.030	12.0–14.0	–	≤ 1.00	–
X15Cr16Ni2	0.10–0.20	1.00	≤ 1.00	0.045	0.030	15.0–17.0	–	1.25–2.50	–
X108Cr17Mo	0.95–1.20	1.00	≤ 1.00	0.045	0.030	16.0–18.0	≤ 0.75	≤ 0.50	–
Austenitic stainless steels									
X10Cr17Mn6Ni4N	≤ 0.20	1.00	4.0–8.0	0.045	0.030	16.0–18.0	–	3.5–5.5	N = 0.05–0.20
X07Cr17Mn12Ni4	≤ 0.12	1.00	10.0–14.0	0.045	0.030	16.0–18.0	–	3.5–5.5	–

(Contd.)

TABLE 3.116 Chemical composition (cast analysis) of stainless steels specified in Indian Standards (continued)

Steel designation	Chemical composition [% (m/m)]								
	C	Si max.	Mn	P max.	S max.	Cr	Mo	Ni	Other elements
Austenitic stainless steels									
X10Cr18Mn9Ni5	≤ 0.15	1.00	8.0–10.0	0.045	0.030	17.0–19.0	–	4.0–6.0	–
X10Cr17Ni7	≤ 0.15	1.00	≤ 2.00	0.045	0.030	16.0–18.0	–	6.0–8.0	–
X07Cr18Ni9	≤ 0.12	1.00	≤ 2.00	0.045	0.030	17.0–19.0	–	8.0–10.0	–
X04Cr19Ni9	≤ 0.08	1.00	≤ 2.00	0.045	0.030	17.5–20.0	–	8.0–10.0	–
X02Cr19Ni10	≤ 0.030	1.00	≤ 2.00	0.045	0.030	17.5–20.0	–	8.0–12.0	–
X15Cr24Ni13	≤ 0.20	1.50	≤ 2.00	0.045	0.030	22.0–25.0	–	11.0–15.0	–
X20Cr25Ni20	≤ 0.25	2.50	≤ 2.00	0.045	0.030	24.0–26.0	–	18.0–21.0	–
X04Cr17Ni12Mo2	≤ 0.08	1.00	≤ 2.00	0.045	0.030	16.0–18.0	2.0–3.0	10.0–14.0	–
X02Cr17Ni12Mo2	≤ 0.030	1.00	≤ 2.00	0.045	0.030	16.0–18.0	2.0–3.0	10.0–14.0	–
X04Cr17Ni12Mo2Ti	≤ 0.08	1.00	≤ 2.00	0.045	0.030	16.0–18.0	2.0–3.0	10.0–14.0	Ti ≥ 5 × (% C + % N) ≤ 0.80
X04Cr18Ni10Ti	≤ 0.08	1.00	≤ 2.00	0.045	0.030	17.0–19.0	–	9.0–12.0	Ti ≥ 5 × (% C + % N) ≤ 0.80
X04Cr18Ni10Nb	≤ 0.08	1.00	≤ 2.00	0.045	0.030	17.0–19.0	–	9.0–12.0	Nb ≥ 10 x % C ≤ 1.00

Source: References 4–8

2.2 Mechanical properties

2.2.1 Plate, sheet and strip

TABLE 3.117 Mechanical properties of stainless steel plate, sheet and strip specified in Indian Standards, in the annealed condition

Steel designation	R_m min. MPa	$R_{p0.2}$ min. MPa	$A_{50\,mm}$ min. %	Maximum hardness HB	Maximum hardness HRB	Bend mandrel diameter (180° bend)[1)]
Ferritic stainless steels						
X04Cr12	440	250	20	187	88	2*a*
X07Cr17	440	250	18	192	88	2*a*
Martensitic stainless steels						
X12Cr12	590	410	16	212	95	–
X20Cr13	690	490	14	229	–	–
X30Cr13	780	590	11	235	–	–
X40Cr13	–	–	–	255	–	–
X15Cr16Ni2	830	640	10	262	–	–
X108Cr17Mo	–	–	–	285	–	–
Austenitic stainless steels						
X10Cr17Mn6Ni4N	640	300	40	217	95	not required
X07Cr17Mn12Ni4	540	260	40	217	95	not required
X10Cr18Mn9Ni5	620	310	40	217	95	not required
X10Cr17Ni7	590	220	40	212	–	not required
X07Cr18Ni9	400	210	40	192	92	not required
X04Cr19Ni9	400	200	40	192	92	not required
X02Cr19Ni10	440	180	40	192	88	not required

(Contd.)

TABLE 3.117 Mechanical properties of stainless steel plate, sheet and strip specified in the Indian Standards, in the annealed condition (continued)

Steel designation	R_m min.	$R_{p0.2}$ min.	$A_{50\,mm}$ min.	Maximum hardness		Bend mandrel diameter (180° bend)[1]
	MPa	MPa	%	HB	HRB	
X15Cr24Ni13	490	210	40	217	95	not required
X20Cr25Ni20	490	210	40	217	95	not required
X04Cr17Ni12Mo2	490	210	40	192	95	not required
X02Cr17Ni12Mo2	440	200	40	192	95	not required
X04Cr17Ni12Mo2Ti	490	220	35	192	95	not required
X04Cr18Ni10Ti	490	210	35	192	95	not required
X04Cr18Ni10Nb	490	210	35	192	92	not required

1) a = thickness of the bend test piece.

Source: Reference 8

TABLE 3.118 Mechanical properties of martensitic stainless steel plate, sheet and strip specified in Indian Standards, in the hardened and tempered condition

Steel designation	R_m	$R_{p0.2}$ min.	A min.
	MPa	MPa	%
X12Cr13	590–780	410	16
X20Cr13	690–880	490	14
X30Cr13	780–980	590	11
X15Cr16Ni2	830–1030	640	10

Source: Reference 4

TABLE 3.119 Mechanical properties of austenitic stainless steel sheet and strip specified in Indian Standards, in the work-hardened condition

Steel designation	Limiting thickness mm	R_m min. MPa	$R_{p0.2}$ min. MPa	A min. %
X10Cr17Mn6Ni4N	2.3	1270	980	4
	2.4	1180	910	7
	2.9	1030	740	10
	3.3	830	490	20
X10Cr17Ni7	2.3	1270	960	4
	2.4	1180	910	5
	2.9	1030	740	10
	3.3	830	490	25
X07Cr18Ni9	2.4	1030	740	9
	2.8	830	490	12
X04Cr19Ni9	1.4	1270	960	3
	1.8	1180	910	7
	2.4	1030	740	8
	2.8	830	490	12

Source: Reference 8

2.2.2 Bars and flats

TABLE 3.120 Mechanical properties of ferritic stainless steel bars and flats specified in Indian Standards, in the annealed condition

Steel designation	R_m	$R_{p0.2}$ min.	*A* min. for a product thickness, in mm, of			Maximum Brinell hardness	Bend mandrel diameter (180° bend)[1),2)]
			Bars	Flats			
			≥5≤25	≥0.5≤3	>3 ≤10		
	MPa	MPa	%	%	%	HB	
X04Cr12	440–640	250	20	18	20	187	2*a*
X07Cr17	440–640	250	16	17	18	192	2*a*

1) *a* = thickness of the bend test piece.
2) Applicable to flat products with thickness ≥ 0.5 mm and ≤ 3 mm.

Source: Reference 5

TABLE 3.121 Mechanical properties of martensitic stainless steel bars and flats specified in Indian Standards, in the annealed and, hardened and tempered condition

Steel designation	Mechanical properties in the treatment condition					
	Annealed	Hardened and tempered				
	Maximum Brinell hardness	R_m	$R_{p0.2}$ min.	*A* min. for a product thickness, in mm, of		$KU^{1)}$ min.
				Bars	Flats	
				≥5≤100	≥3≤30	
	HB	MPa	MPa	%	%	J
X12Cr12	212	590–780	410	16	16	60
X20Cr13	229	690–880	490	14	14	40
X30Cr13	235	780–980	590	11	11	–
X40Cr13	255	–	–	–	–	–

(Contd.)

TABLE 3.121 Mechanical properties of martensitic stainless steel bars and flats specified in Indian Standards, in the annealed and, hardened and tempered condition (continued)

Steel designation	Mechanical properties in the treatment condition					
	Annealed	Hardened and tempered				
	Maximum Brinell hardness	R_m	$R_{p0.2}$ min.	A min. for a product thickness, in mm, of		KU[1] min.
				Bars	Flats	
				≥5≤100	≥3≤30	
	HB	MPa	MPa	%	%	J
X15Cr16Ni2	262	830–1030	640	10	10	30
X108Cr17Mo	285	–	–	–	–	–

1) Applicable to bars and flats with diameter or thickness ≥ 15 mm and ≤ 63 mm.

Source: Reference 5

TABLE 3.122 Mechanical properties of austenitic stainless steel bars and flats specified in Indian Standards, in the softened condition

Steel designation	R_m	$R_{p0.2}$ min.	A min. for a product thickness, in mm, of			Maximum Brinell hardness
			Bars	Flats		
			≥5≤100	≥0.5≤3	>3 ≤30	
	MPa	MPa	%	%	%	HB
X10Cr17Mn6Ni4N	640–830	300	40	38	40	217
X10Cr17Ni7	590–780	220	–	38	40	212
X07Cr18Ni9	490–690	210	40	38	40	192
X04Cr19Ni9	490–690	200	40	38	40	192
X02Cr19Ni10	440–650	180	40	38	40	192

(Contd.)

TABLE 3.122 Mechanical properties of austenitic stainless steel bars and flats specified in Indian Standards, in the softened condition (continued)

Steel designation	R_m	$R_{p0.2}$ min.	A min. for a product thickness, in mm, of			Maximum Brinell hardness
			Bars	Flats		
			≥5 ≤100	≥0.5 ≤3	>3 ≤30	
	MPa	MPa	%	%	%	HB
X04Cr17Ni12Mo2	490–690	210	40	38	40	192
X02Cr17Ni12Mo2	440–640	200	40	38	40	192
X04Cr17Ni12Mo2Ti	490–690	220	35	33	35	192
X04Cr18Ni10Ti	490–690	210	35	33	35	192
X04Cr18Ni10Nb	490–690	210	35	33	35	192

Source: Reference 5

TABLE 3.123 Mechanical properties of austenitic stainless steel bars specified in Indian Standards, in the cold-drawn condition

Steel designation	Limiting thickness/ diameter	R_m min.	$R_{p0.2}$ min.	A min.
	mm	MPa	MPa	%
X10Cr17Ni7	12	1270	960	12
	19	1180	910	12
	25	1030	740	15
	45	830	490	20
X07Cr18Ni9	25	1030	740	15
	45	830	490	20

(Contd.)

TABLE 3.123 Mechanical properties of austenitic stainless steel bars specified in Indian Standards, in the cold-drawn condition (continued)

Steel designation	Limiting thickness/ diameter mm	R_m min. MPa	$R_{p0.2}$ min. MPa	A min. %
X04Cr19Ni9	12	1270	960	12
	19	1180	910	12
	25	1030	740	15
	45	830	490	20
Source: Reference 5				

3 RECOMMENDED HEAT TREATING TEMPERATURES

TABLE 3.124 Recommended heat treating temperatures for stainless steels specified in Indian Standards

Steel designation	Annealing or softening			Hardening and tempering			
	Symbol[1)]	Temperature °C	Cooling[2)]	Symbol[1)]	Quenching temperature °C	Cooling[2)]	Tempering temperature °C
Ferritic stainless steels							
X04Cr12	A	750–800	f, a	–	–	–	–
X07Cr17	A	750–850	a, w	–	–	–	–
Martensitic stainless steels							
X12Cr12	A1	700–780	a	Q + T	950–1000	o, a	700–750
	A2	770–870	f				
X20Cr13	A	770–870	f	Q + T	980–1030	o, a	650–770

TABLE 3.124 Recommended heat treating temperatures for stainless steels specified in Indian Standards (continued)

Steel designation	Annealing or softening			Hardening and tempering			
	Symbol[1)]	Temperature °C	Cooling[2)]	Symbol[1)]	Quenching temperature °C	Cooling[2)]	Tempering temperature °C
Martensitic stainless steels							
X30Cr13	A	770–870	f	Q+T1	980–1030	o, a	630–700
				Q+T2	980–1030	o, a	100–250
X40Cr13	A	770–870	f	Q+T	1000–1050	o, a	100–250
X15Cr16Ni2	A	750–800	a	Q+T	980–1030	o, a	630–700
		620–670	a				
X108Cr17Mo	A	780–880	f	Q+T	1000–1030	o, a	100–250
Austenitic stainless steels							
X10Cr17Mn6Ni4N	S	1000–1120	w, o, a	–	–		
X07Cr17Mn12Ni4	S	1000–1120	w, o, a	–	–	–	–
X10Cr18Mn9Ni5	S	1000–1120	w, o, a	–	–	–	–
X10Cr17Ni7	S	1000–1120	w, o, a	–	–	–	–
X07Cr18Ni9	S	1000–1120	w, o, a	–	–	–	–
X04Cr19Ni9	S	1000–1120	w, o, a	–	–	–	–
X02Cr19Ni10	S	1000–1120	w, o, a	–	–	–	–
X15Cr24Ni13	S	1000–1120	w, o, a	–	–	–	–
X20Cr25Ni20	S	1000–1120	w, o, a	–	–	–	–
X04Cr17Ni12Mo2	S	1000–1120	w, o, a	–	–	–	–
X02Cr17Ni12Mo2	S	1000–1120	w, o, a	–	–	–	–
X04Cr17Ni12Mo2Ti	S	1000–1120	w, o, a	–	–	–	–

(Contd.)

TABLE 3.124 Recommended heat treating temperatures for stainless steels specified in Indian Standards (continued)

Steel designation	Annealing or softening			Hardening and tempering			
	Symbol[1)]	Temperature °C	Cooling[2)]	Symbol[1)]	Quenching temperature °C	Cooling[2)]	Tempering temperature °C
X04Cr18Ni10Ti	S	1000–1120	w, o, a	–	–	–	–
X04Cr18Ni10Nb	S	1000–1120	w, o, a	–	–	–	–

1) A = annealing; Q = quenching; T = tempering; S = softening.
2) f = furnace; a = air; o = oil; w = water.

Source: References 4 and 8

4 EQUIVALENT GRADES

TABLE 3.125 Comparison of grades of stainless steels specified in Indian Standards with nearest equivalent grades specified in other national and international standards

IS 1570 (5)–85 IS 6528–72 IS 6529–72 IS 6603–72 IS 6911–92	ASTM A 276–95 A 314–95 A 580/A 580M –95a A 666–94a	BS 970 (1)–91 BS 1449 (2)–83 BS 1554–90	DIN 17440–85	GOST 5632–72	ISO 683 (13)–86 4995–94	JIS G 4303–91 JIS G 4304–91 JIS G 4305–91 JIS G 4308–91 JIS G 4309–91 JIS G 4311–91 JIS G 4312–91 JIS G 4318–91 JIS G 4319–91	SAE J405–89
Ferritic stainless steels							
X04Cr12	405	405S17	X 6 CrAl 13	–	2	SUS 405	51405
X07Cr17	430	430S17, 430S18	X 6 Cr 17	12X17	8	SUS 430	51430
Martensitic stainless steels							
X12Cr12	410	410S21	X 10 Cr 13	12X13	3	SUS 410	51410
X20Cr13	420	420S37	X 20 Cr 13	20X13	4	SUS 420J1	51420
X30Cr13	–	420S45	X 30 Cr 13	30X13	5	SUS 420J2	–
X40Cr13	–	–	X 38 Cr 13	40X13	–	–	–

(Contd.)

TABLE 3.125 Comparison of grades of stainless steels specified in Indian Standards with nearest equivalent grades specified in other national and international standards (continued)

IS 1570 (5)–85 IS 6528–72 IS 6529–72 IS 6603–72 IS 6911–92	ASTM A 276–92 A 314–92 A 580/A 580M–93 A 666–93	BS 970 (1)–91 BS 1449 (2)–83 BS 1554–90	DIN 17440–85	GOST 5632–72	ISO 683 (13)–86 4995–94	JIS G 4303–91 JIS G 4304–91 JIS G 4305–91 JIS G 4308–91 JIS G 4309–91 JIS G 4311–91 JIS G 4312–91 JIS G 4318–91 JIS G 4319–91	SAE J405–89
Martensitic stainless steels							
X15Cr16Ni2	431	431S29	X 20 CrNi 172	14X17H2	9b	SUS 431	51431
X108Cr17Mo	440C	–	–	95X18	–	SUS 440C	51440C
Austenitic stainless steels							
X10Cr17Mn6Ni4N	201	–	–	–	A-2	SUS 201	30201
X07Cr17Mn12Ni4	–	–	–	–	–	–	–
X10Cr18Mn9Ni5	202	202SI6	–	12X17Г9AH4	A-3	SUS 202	30202
X10Cr17Ni7	301	301S21, 301S22	–	–	14	SUS 301	30301
X07Cr18Ni9	302	302S31	–	12X18H9	12	SUS 302	30302
X04Cr19Ni9	304	304S31	X 5 CrNi 18 10	08X18H10	11	SUS 304	30304
X02Cr19Ni10	304L	304S11	X 2 CrNi 19 11	03X18H11	10	SUS 304L	30304L

(Contd.)

TABLE 3.125 Comparison of grades of stainless steels specified in Indian Standards with nearest equivalent grades specified in other national and international standards (continued)

IS 1570 (5)–85 IS 6528–72 IS 6529–72 IS 6603–72 IS 6911–92	ASTM A 276–92 A 314–92 A 580/A 580M–93 A 666–89	BS 970 (1)–91 BS 1449 (2)–83 BS 1554–90	DIN 17440–85	GOST 5632–72	ISO 683 (13)–86 4995–94	JIS G 4303–91 JIS G 4304–91 JIS G 4305–91 JIS G 4308–91 JIS G 4309–91 JIS G 4311–91 JIS G 4312–91 JIS G 4318–91 JIS G 4319–91	SAE J405–89
Austenitic stainless steels							
X15Cr24Ni13	309	309S24	–	–	–	SUH 309	30309
X20Cr25Ni20	310	310S31	–	20X25H20C2	H 16	SUH 310	30310
X04Cr17Ni12Mo2	316	316S31, 316S33	X 5 CrNiMo 17 12 2 X 5 CrNiMo 17 13 3	–	20, 20a	SUS 316	30316
X02Cr17Ni12Mo2	316L	316S11, 316S13	X 2 CrNiMo 17 13 2 X 2 CrNiMo 18 14 3	03X17H14Mэ	19, 19a	SUS 316L	30316L
X04Cr17Ni12Mo2Ti	316Ti	320S31, 320S33	X 6 CrNiMo Ti 17 122	08X17H13M2T	21	–	–
X04Cr18Ni10Ti	321	321S31	X 6 CrNiTi 18 10	08X18H10T	15	SUS 321	30321
X04Cr18Ni10Nb	347	347S20, 347S31	X 6 CrNiNb 18 10	08X18H12Б	16	SUS 347	30347

5 REFERENCES

Books

1. *ASM Handbook*, vol. 1, ASM International, Materials Park, Ohio, U.S.A., 1990.
2. D. Peckner and I. M. Bernstein, *Handbook of Stainless Steels*, McGraw-Hill, New York, U.S.A., 1977.
3. *Source Book on Stainless Steels*, American Society for Metals, Metals Park, Ohio, U.S.A., 1976.

Standards

4. IS 1570 (5) – 1985: *Schedules for Wrought Steels—Part 5: Stainless and Heat-Resisting Steels.*
5. IS 6603 – 1972: *Specification for Stainless Steel Bars and Flats.*
6. IS 6528 – 1972: *Specification for Stainless Steel Wire.*
7. IS 6529 – 1972: *Specification for Stainless Steel Blooms, Billets and Slabs for Forgings.*
8. IS 6911 – 1992: *Stainless Steel Plate, Sheet and Strip—Specification.*

3.20

VALVE STEELS FOR INTERNAL COMBUSTION ENGINES

1 INTRODUCTION

Valve steels for internal combustion engines are martensitic and austenitic steels which exhibit, to varying degrees, resistance to heat, thermal cycling, corrosion, oxidation, fatigue loading, impact, and adhesive and abrasive wear. These steels are used for the manufacture of inlet and exhaust valves in reciprocating internal combustion engines.

Valve steels are divided into two categories based upon their microstructure:

Martensitic valve steels

Martensitic valve steels are unalloyed and alloy steels which are primarily used for inlet valves and the stem portion of exhaust valves. These steels are used in the hardened and tempered condition in the hardness range of 25-45 HRC so as to develop an optimum combination of strength, ductility, toughness, and wear resistance.

Austenitic valve steels

Austenitic valve steels are alloy steels which are primarily used for exhaust valves. These steels are used in the solution treated and precipitation hardened condition in the hardness range 20-40 HRC. In the heat treated condition these steels possess superior elevated temperature properties and corrosion resistance than martensitic valve steels.

2 SPECIFICATION

2.1 *Chemical composition*

TABLE 3.126 Chemical composition (cast analysis) of valve steels, conforming to IS 7494 – 1981

Valve steel no.	Steel designation	Chemical composition[1)] [% (m/m)]										
		C	Si	Mn	P max.	S max.	Cr	Mo	Ni	N	V	W
Martensitic valve steels												
V–1	**55C8**	0.50–0.60	0.10–0.35	0.60–0.90	0.040	0.040	–	–	–	–	–	–
V–2	**40Cr4**	0.35–0.45	0.10–0.35	0.60–0.90	0.040	0.040	0.90–1.20	–	–	–	–	–
V–3	**40Cr4Mo3**	0.35–0.45	0.10–0.35	0.50–0.80	0.040	0.040	0.90–1.20	0.20–0.35	–	–	–	–
V–4	**50Cr4V2**	0.45–0.55	0.10–0.35	0.50–0.80	0.040	0.040	0.90–1.20	–	–	–	0.15–0.30	–
V–5	**25Cr13Mo6**	0.20–0.30	0.10–0.35	0.40–0.70	0.040	0.040	2.90–3.40	0.45–0.65	≤ 0.30	–	–	–
V–6	**40Ni10Cr3Mo6**	0.35–0.45	0.10–0.35	0.40–0.70	0.040	0.040	0.50–0.80	0.40–0.70	2.25–2.75	–	–	–
V–7	**40Ni6Cr4Mo3**	0.35–0.45	0.10–0.35	0.40–0.70	0.040	0.040	0.90–1.30	0.20–0.35	1.25–1.75	–	–	–
V–8	**X45Cr9Si3**	0.40–0.50	2.75–3.75	≤ 0.80	0.040	0.035	7.50–9.50	–	≤ 0.50	–	–	–

(Contd.)

TABLE 3.126 Chemical composition (cast analysis) of valve steels, conforming to IS 7494 – 1981 (continued)

Valve steel no.	Steel designation	Chemical composition1) [% (m/m)]										
		C	Si	Mn	P max.	S max.	Cr	Mo	Ni	N	V	W
Martensitic valve steels												
V–9	**X80Cr20Si2Ni1**	0.75–0.85	1.75–2.50	≤ 0.80	0.040	0.035	19.0–21.0	–	1.00–1.70	–	–	–
V–10	**X85Cr18Mo2V**	0.80–0.90	≤ 1.00	≤ 1.50	0.040	0.035	16.5–18.5	≤ 2.50	–	–	≤0.60	–
Austenitic valve steels												
V–11	**X40Ni14Cr14W3Si2**	0.35–0.50	≤ 2.00	≤ 1.00	0.050	0.035	12.0–15.0	–	12.0–15.0	–	–	2.00–3.00
V–12	**X20Cr21Ni12N**	0.15–0.25	0.75–1.25	≤ 1.50	0.050	0.035	20.0–22.0	–	10.5–12.5	0.15–0.30	–	–
V–13	**X70Cr21Mn6Ni2N**	0.65–0.75	0.45–0.85	5.50–7.00	0.050	0.035	20.0–22.0	–	1.40–1.90	0.18–0.28	–	–
V–14	**X55Cr21Mn8Ni2N**	0.50–0.60	≤ 1.00	7.00–9.50	0.050	0.035	20.0–22.0	–	1.50–2.75	0.20–0.40	–	–
V–15	**X53Cr22Mn9Ni4N**	0.48–0.58	≤ 0.25	8.00–10.0	0.050	0.035	20.0–23.0	–	3.25–4.50	0.38–0.55	–	–

[1] Residual elements up to the following limits are acceptable: Cr ≤ 0.20 %; Mo ≤ 0.10 %; Ni ≤ 0.25 %; Cu ≤ 0.25 %; V ≤ 0.05 %.

Notes:
1) For improved machinability, valve steels V–13, V–14 and V–15 may be ordered with a S content of up to 0.090 %.
2) Valve steels V–1 to V–7 are generally used for inlet valves, and valve steels V–8 to V–15 are generally used for exhaust valves.

2.2 Hardness

TABLE 3.127 Typical hardness values[1] for valve steels, conforming to IS 7494 – 1981, at room temperature, delivered in the finally heat-treated condition

Steel designation	Heat treatment condition[2]	Brinell hardness HB	Steel designation	Heat treatment condition[2]	Brinell hardness HB
Martensitic valve steels			**Martensitic valve steels**		
55C8	Q + T	255–293	**X80Cr20Si2Ni1**	Q + T	255–306
40Cr4	Q + T	255–293	**X85Cr18Mo2V**	Q + T	255–306
40Cr4Mo3	Q + T	255–293	**Austenitic valve steels**		
50Cr4V2	Q + T	255–293	**X40Ni14Cr14W3Si2**	ST	≤ 302
25Cr13Mo6	Q + T	255–293	**X20Cr21Ni12N**	ST	≤ 302
40Ni10Cr3Mo6	Q + T	255–293	**X70Cr21Mn6Ni2N**	ST + P	≥ 321
40Ni6Cr4Mo3	Q + T	255–293	**X55Cr21Mn8Ni2N**	ST + P	≥ 321
X45Cr9Si3	Q + T	255–293	**X53Cr22Mn9Ni4N**	ST + P	≥ 321
				SR	≤ 390[3]

1) For diameters ≤ 40 mm.
2) Q = quenched; T = tempered; ST = solution treated; P = precipitation hardened; SR = stress relieved.
3) This hardness value does not take into account the effect of work hardening which can result in a higher hardness.

TABLE 3.128 Typical hardness values[1] for valve steels, conforming to IS 7494 – 1981, at room temperature, delivered in the hot-rolled condition and in the hot-rolled and annealed condition

Steel designation	Heat treatment condition[2]	Maximum Brinell hardness HB	Steel designation	Heat treatment condition[2]	Maximum Brinell hardness HB
Martensitic valve steels			**Martensitic valve steels**		
55C8	Annealed	220	**X80Cr20Si2Ni1**	Annealed	330
40Cr4	Annealed	220	**X85Cr18Mo2V**	Annealed	330
40Cr4Mo3	Annealed	220	**Austenitic valve steels**		
50Cr4V2	Annealed	220	**X40Ni14Cr14W3Si2**	Hot-rolled	302
25Cr13Mo6	Annealed	230	**X20Cr21Ni12N**	Hot-rolled	302
40Ni10Cr3Mo6	Annealed	250	**X70Cr21Mn6Ni2N**	Hot-rolled	390
40Ni6Cr4Mo3	Annealed	230	**X55Cr21Mn8Ni2N**	Hot-rolled	390
X45Cr9Si3	Annealed	320	**X53Cr22Mn9Ni4N**	Hot-rolled	390

[1] For diameters ≤ 40 mm.

2.3 Mechanical properties

TABLE 3.129 Typical mechanical properties[1] of valve steels, conforming to IS 7494 – 1981, at room temperature, in the finally heat-treated condition

Steel designation	Heat treatment condition[2]	R_m min. MPa	R_e min. MPa	*A* min. %	*Z* min. %
Martensitic valve steels					
55C8	Q + T	790	530	–	–
40Cr4	Q + T	790	590	–	–

(Contd.)

TABLE 3.129 Typical mechanical properties[1] of valve steels, conforming to IS 7494 – 1981, at room temperature, in the finally heat-treated condition (continued)

Steel designation	Heat treatment condition[2]	R_m min. MPa	R_e min. MPa	A min. %	Z min. %
Martensitic valve steels					
40Cr4Mo3	Q + T	890	690	–	–
50Cr4V2	Q + T	890	690	–	–
25Cr13Mo6	Q + T	890	690	–	–
40Ni10Cr3Mo6	Q + T	890	690	–	–
40Ni6Cr4Mo3	Q + T	890	690	–	–
X45Cr9Si3	Q + T	930	685	16	40
X80Cr20Si2Ni1	Q + T	930	735	10	15
X85Cr18Mo2V	Q + T	1080	835	12	15
Austenitic valve steels					
X40Ni14Cr14W3Si2	ST	785	345	35	40
X20Cr21Ni12N	ST	835	440	25	25
X70Cr21Mn6Ni2N	ST + P	1030	540	20	30
X55Cr21Mn8Ni2N	ST + P	1030	490	20	30
X53Cr22Mn9Ni4N	ST + P	1030	640	8	10
	SR	1080	830	20	10

1) For diameters ≤ 40 mm.
2) Q = quenched; T = tempered; ST = solution treated; P = precipitation hardened; SR = stress relieved.

TABLE 3.130 Typical tensile strength valves[1] for valve steels, conforming to IS 7494–1981, at elevated temperatures, in the finally heat-treated condition

Steel designation	Heat treatment condition[2]	R_m min., in MPa, at a temperature, in °C, of							
		500	550	600	650	700	750	800	850
Martensitic valve steels									
X45Cr9Si3	Q + T	490	365	245	155	110	70	–	–
X80Cr20Si2Ni1	Q + T	590	460	345	245	145	110	70	–
X85Cr18Mo2V	Q + T	540	410	295	235	175	135	100	–
Austenitic valve steels									
X40Ni14Cr14W3Si2	ST	540	510	490	390	315	225	195	–
X20Cr21Ni12N	ST + P	590	550	510	450	390	345	295	–
X70Cr21Mn6Ni2N	ST + P	640	590	540	490	440	365	295	–
X55Cr21Mn8Ni2N	ST + P	640	600	560	500	440	365	295	–
X53Cr22Mn9Ni4N	ST + P	640	600	560	500	440	365	325	245

1) For diameters ≤ 40 mm.
2) Q = quenched; T = tempered; ST = solution treated; P = precipitation hardened.

3 RECOMMENDED HOT WORKING AND HEAT TREATING TEMPERATURES

TABLE 3.131 Recommended hot working and heat treating temperatures for valve steels, conforming to IS 7494 – 1981

Steel designation	Hot working temperature °C	Soft annealing temperature °C	Hardening or solution treatment temperature °C	Quenching agent	Tempering or artificial ageing temperature °C
Martensitic valve steels					
55C8	1150–850	650–700	810–840	Oil	550–660
40Cr4	1200–850	650–700	850–880	Oil	550–700
40Cr4Mo3	1200–850	650–700	850–880	Oil	550–720
50Cr4V2	1050–850	680–720	830–860	Oil	540–680
25Cr13Mo6	1200–850	700–750	880–910	Oil	550–700
40Ni10Cr3Mo6	1200–850	630–650	820–850	Oil	500–660
40Ni6Cr4Mo3	1200–900	630–660	820–850	Oil	500–660
X45Cr9Si3	1100–900	780–820	1020–1070	Oil or air	720–820
X80Cr20Si2Ni1	1100–900	820–860	1050–1080	Oil or air	700–800
X85Cr18Mo2V	1100–900	820–860	1050–1080	Oil	700–750
Austenitic valve steels					
X40Ni14Cr14W3Si2	1150–900	–	980–1080	Water, oil or air	–
X20Cr21Ni12N	1150–950	–	1100–1200	Oil or water	700–800
X70Cr21Mn6Ni2N	1150–950	–	1100–1200	Water	730–780
X55Cr21Mn8Ni2N	1100–950	–	1100–1200	Water	730–780
X53Cr22Mn9Ni4N	1150–950	–	1100–1200	Water	730–780

4 EQUIVALENT GRADES

TABLE 3.132 Comparison of grades of valve steels specified in IS 7494 – 1981 with nearest equivalent grades specified in other national and international standards

IS 7494–81	BS 970 (4)–70	DIN 17480–92	GOST 5632–72	ISO 683 (15)–92	JIS G 4311–91	SAE J775–93	Trade designation
Martensitic valve steels							
55C8	–	–	–	–	–	–	–
40Cr4	–	–	–	–	–	–	–
40Cr4Mo3	–	–	–	–	–	H41400	–
50Cr4V2	–	–	–	–	–	–	–
25Cr13Mo6	–	–	–	–	–	–	–
40Ni10Cr3Mo6	–	–	–	–	–	–	–
40Ni6Cr4Mo3	–	–	–	–	–	–	–
X45Cr9Si3	401S45	X 45 CrSi 9 3	40X9C2	X 45 CrSi 9 3	SUH 1	S65007	Sil 1
X80Cr20Si2Ni1	443S65	–	–	–	SUH 4	S65006	Sil XB
X85Cr18Mo2V	–	X 85 CrMoV 18 2	–	X 85 CrMoV 18 2	–	–	–
Austenitic valve steels							
X40Ni14Cr14W3Si2	331S40	–	45X14H14B2M	–	SUH 31	–	TPA

(Contd.)

TABLE 3.132 Comparison of grades of valve steels specified in IS 7494 – 1981 with nearest equivalent grades specified in other national and international standards (continued)

IS 7494–81	BS 970 (4)–70	DIN 17480–92	GOST 5632–72	ISO 683 (15)–92	JIS G 4311–91	SAE J775–93	Trade designation
Austenitic valve steels							
X20Cr21Ni12N	381S34	–	–	–	SUH 37	S63017	21-12N
X70Cr21Mn6Ni2N	–	–	–	–	–	–	Sil 746
X55Cr21Mn8Ni2N	–	X 55 CrMnNiN 208	–	X 55 CrMnNiN 20 8	–	S63012	21-2N
X53Cr22Mn9Ni4N	349S52 349S54	X 53 CrMnNiN 219	55X20Г9АН4	X 53 CrMnNiN 21 9	SUH 35 SUH 36	S63008	21-4N

5 REFERENCES

Book

1. *Metals Handbook*, vol. 1, 8th ed., American Society for Metals, Metals Park, Ohio, U.S.A., 1961.

Standards

2. IS 7494 – 1981: *Specification for Steel for Valves for Internal Combustion Engines.*
3. ISO 683 (15) – 1992: *Heat-Treatable Steels, Alloy Steels and Free-Cutting Steels—Part 15: Valve Steels for Internal Combustion Engines.*
4. DIN 17480 – 1992: *Valve Materials; Technical Delivery Conditions.*
5. SAE J775 – 1993: *Engine Poppet Valve Information Report.*

3.21

BEARING STEELS

1 INTRODUCTION

Bearing steels are through-hardening steels, case hardening steels, stainless steels, and high-temperature steels which are used for ball and roller bearing components (balls, rollers, needles, races and discs). These steels are characterized by high yield and high creep strength, high fatigue strength, high impact strength, good wear resistance, good corrosion resistance, and high microstructural stability at operating temperatures. These steels are produced using special steelmaking and conditioning practices that are intended to ensure internal soundness, freedom from injurious segregation and harmful imperfections, and low non-metallic inclusion content.

Through-hardening bearing steels

Through-hardening bearing steels are alloy steels which contain about 1 % carbon and 1 to 2 % chromium. These steels are used in the through-hardened condition for normal service with operating temperatures less than 150 °C. Modification of the basic chemical composition, by an increase in manganese content or an addition of molybdenum, is used to increase the hardenability of heavier sections and ensure freedom from nonmartensitic transformations in hardening.

Case hardening bearing steels

Case hardening bearing steels are unalloyed and low-alloy steels which are used for the manufacture of bearing components of large dimensions, especially when conventional steels are difficult to through harden. These steels may also be used as an alternative to the through-hardening bearing steels. The use of case hardening bearing steels is restricted to normal service with operating temperatures less than 150 °C.

Stainless bearing steels

Stainless bearing steels are martensitic stainless steels which are used in corrosive environments.

High-temperature bearing steels

High-temperature bearing steels are essentially high-speed tool steels which are superior to other bearing materials at elevated temperatures, assuming adequate lubrication.

2 SPECIFICATION

2.1 Chemical composition

TABLE 3.133 Chemical composition (cast analysis) of bearing steels specified in Indian Standards

Steel designation	Chemical composition[1)] [% (m/m)]							
	C	Si	Mn	P max.	S max.	Cr	Mo	Ni
Through hardening bearing steels (IS 4398 – 1994)								
104Cr6	0.98–1.10	0.15–0.35	0.25–0.45	0.025	0.025	1.30–1.40	–	–
103Cr6	0.95–1.10	0.15–0.35	0.60–0.80	0.025	0.025	1.50–1.70	–	–
103Cr4Mn4	0.95–1.10	0.40–0.70	0.90–1.15	0.030	0.025	0.90–1.20	–	–
98Cr6Mn4	0.90–1.05	0.50–0.70	1.00–1.20	0.030	0.025	1.40–1.65		
Case hardening bearing steels (IS 5489 – 1975)								
14C6	0.10–0.18	0.15–0.35	0.40–0.70	0.025	0.025	–	–	–
28C6	0.25–0.32	0.15–0.35	0.40–0.70	0.025	0.025	–	–	–
17Mn5Cr4	0.14–0.19	0.15–0.35	1.00–1.30	0.025	0.025	0.80–1.10	–	–
20Ni7Mo2	0.17–0.22	0.15–0.35	0.40–0.70	0.025	0.025	–	0.20–0.30	1.65–2.00
20Ni2Cr2Mo2	0.18–0.23	0.20–0.35	0.70–0.90	0.025	0.025	0.40–0.60	0.15–0.25	0.40–0.70

1) Residual elements up to the following limits are acceptable:
(a) for through hardening bearing steels: Mo $\leq$ 0.10 %; Ni $\leq$ 0.25 %; Cu $\leq$ 0.30 %; Ti $\leq$ 0.00045 %; V $\leq$ 0.05 %;
(b) for case hardening bearing steels: Cr $\leq$ 0.20 %; Cu $\leq$ 0.25 %; V $\leq$ 0.05 %.

Note: All steels should be fully killed.

2.2 Hardness

TABLE 3.134 Hardness requirements for bearing steels specified in Indian Standards, in the "annealed" (A) or "annealed and cold-drawn" (A + C) condition

Steel designation	Maximum hardness in the treatment condition			
	A		A+C	
	HRB	HB	HRB	HB
Through hardening bearing steels (IS 4398 – 1994)				
104Cr6	92	207	–	250
103Cr6	92	207	–	250
103Cr4Mn4	92	207	–	250
98Cr6Mn4	92	207	–	250
Case hardening bearing steels (IS 5489 – 1975)				
14C6	–	200	–	–
28C6	–	200	–	–
17Mn5Cr4	–	200	–	–
20Ni7Mo2	–	200	–	–
20Ni2Cr2Mo2	–	200	–	–

3 RECOMMENDED HOT WORKING AND HEAT TREATING TEMPERATURES

TABLE 3.135 Recommended hot working and heat treating temperatures for through hardening bearing steels, conforming to IS 4398 – 1994

Steel designation	Hot working temperature °C	Softening temperature °C	Normalizing temperature °C	Hardening temperature °C	Quenching agent	Tempering temperature °C
104Cr6	1100–850	650–680	870–900	830–870	Oil	150–180
103Cr6	1100–850	650–680	870–900	830–870	Oil	150–180
103Cr4Mn4	1100–850	650–680	860–890	830–870	Oil	150–180
98Cr6Mn4	1100–850	650–680	860–890	830–870	Oil	150–180

TABLE 3.136 Recommended hot working and heat treating temperatures for case hardening bearing steels, conforming to IS 5489 – 1975

Steel designation	Hot working temper-ature °C	Annealing temper-ature °C	Softening temper-ature °C	Normalizing temper-ature °C	Carburizing temper-ature °C	Direct and single hardening temper-ature °C	Double hardening		Quenching agent	Tempering temper-ature °C
							Core hardening temper-ature °C	Case hardening temper-ature °C		
14C6	1100–850	900–930	650–700	880–910	900–940	830–860	870–890	780–820	Water	150–180
28C6	1100–850	890–920	650–700	880–910	900–940	830–860	870–890	780–820	Oil	150–180
17Mn5Cr4	1150–850	860–900	650–700	840–870	890–940	810–840	860–880	790–830	Oil	150–180
20Ni7Mo2	1150–850	860–880	650–700	840–870	890–940	810–840	830–870	780–820	Oil	150–180
20 Ni2Cr2Mo2	1150–850	860–880	650–700	840–870	890–940	810–840	830–870	780–820	Oil	150–180

4 EQUIVALENT GRADES

TABLE 3.137 Comparison of grades of through hardening bearing steels specified in the IS 4398 – 1994 with nearest equivalent grades specified in other national and international standards

IS 4398–94	ASTM A 295–94 ASTM A 485–89 ASTM A 535–85	BS	DIN 17230–80	GOST 801–78 GOST 21022–75	ISO 683 (XVII)–76	JIS G 4805–90	SAE J404–94
104Cr6	52100	–	100 Cr 6	ШХ15 ШХ15-ШД	1	SUJ 2	E52100
103Cr6	–	–	–	–	–	–	–
103Cr4Mn4	A 485 - Grade 1 A 535 - 52100 Mod. 1	–	–	–	2	SUJ 3	–
98Cr6Mn4	–	–	100 CrMn 6	–	3	–	–

TABLE 3.138 Comparison of grades of case hardening bearing steels specified in the IS 5489 – 1975 with nearest equivalent grades specified in other national and international standards

IS 5489–75	ASTM A 534–94 ASTM A 535–85	BS 970 (1)–91	DIN 17230–80	GOST 1050–88 GOST 4543–71	ISO 683 (XVII)–76	JIS G 4051–79 JIS G 4052–79 JIS G 4103–79	SAE J403–94 J404–94 J1268–93
14C6	–	–	–	15	–	S 15 CK	1015
28C6	–	–	–	–	–	–	–
17Mn5Cr4	–	590H17 590M17	17 MnCr 5	18ХГ	10	–	–
20Ni7Mo2	4620 4620H	665H20 665M20	–	20H2M	11	–	4620 4620H
20Ni2Cr2Mo2	8620 8620H	805A20 805H20 805M20	–	20ХГНМ	12	SNCM 220 SNCM 220 H	8620 8620H

5 REFERENCES

Books

1. *ASM Handbook,* vol. 1, ASM International, Materials Park, Ohio, U.S.A., 1990.
2. M. J. Neale (Ed.), *Tribology Handbook,* Newnes-Butterworths, London, U.K., 1973.

Standards

3. IS 4398 – 1994: *Carbon-Chromium Steel for the Manufacture of Balls, Rollers and Bearing Races—Specification.*
4. IS 5489 – 1975: *Specification for Carburizing Steels for Use in Bearing Industry.*

3.22

TOOL STEELS

1 INTRODUCTION

Tool steels are unalloyed and alloy steels which are used to make tools for cutting, forming or shaping of materials, and for handling and measuring workpieces. Many alloy tool steels are also widely used for machinery components and structural applications where particularly severe requirements must be met, such as special purpose valves and bearings for elevated temperature service. These steels are characterized by high hardness, high wear resistance and toughness.

A distinction is made between the following groups of tool steels:

Cold-work tool steels

Cold-work tool steels are unalloyed and alloy steels whose use is restricted to applications in which the surface temperature is generally below 200 °C. They possess high hardness, high wear resistance and toughness, and high resistance to pressure and impact.

Hot-work tool steels

Hot-work tool steels are alloy steels which are used for applications in which the surface temperature is generally greater than 200 °C, such as those encountered in hot forging, stamping, extrusion, etc. These steels are characterized by good red-hardness, good wear resistance, high shock resistance, good resistance to heat checking and an ability to withstand drastic cooling after having been heated to relatively high temperatures.

High-speed tool steels

High-speed tool steels are alloy steels which are primarily used for cutting applications at temperatures of up to 600 °C. These steels exhibit maximum red-hardness, good wear resistance and adequate shock resistance.

2 SPECIFICATIONS

2.1 Chemical composition

TABLE 3.139 Chemical composition (cast analysis) of tool steels, conforming to IS 3748 – 1990 and IS 7291 – 1981

Steel designation	Chemical composition[1)] [% (m/m)]										
	C	Si	Mn	P max.	S max.	Cr	Mo	Ni	Co	V	W
Cold-work tool steels (IS 3748 – 1990)											
50T8	0.45–0.55	0.10–0.35	0.60–0.90	0.035	0.035	–	–	–	–	–	–
55T8	0.50–0.60	0.10–0.35	0.60–0.90	0.035	0.035	–	–	–	–	–	–
60T6	0.55–0.65	0.10–0.35	0.50–0.80	0.035	0.035	–	–	–	–	–	–
65T6	0.60–0.70	0.10–0.35	0.50–0.80	0.035	0.035	–	–	–	–	–	–
70T6	0.65–0.75	0.10–0.35	0.50–0.80	0.035	0.035	–	–	–	–	–	–
75T6	0.70–0.80	0.10–0.35	0.50–0.80	0.035	0.035	–	–	–	–	–	–
80T6	0.75–0.85	0.10–0.35	0.50–0.80	0.035	0.035	–	–	–	–	–	–
85T6	0.80–0.90	0.10–0.35	0.50–0.80	0.035	0.035	–	–	–	–	–	–

(Contd.)

TABLE 3.139 Chemical composition (cast analysis) of tool steels, conforming to IS 3748 – 1990 and IS 7291 – 1981 (continued)

Steel designation	Chemical composition[1)] [% (m/m)]										
	C	Si	Mn	P max.	S max.	Cr	Mo	Ni	Co	V	W
Cold-work tool steels (IS 3748 – 1990)											
70T3	0.65–0.75	0.10–0.30	≤ 0.40	0.035	0.035	–	–	–	–	–	–
80T3	0.75–0.85	0.10–0.30	≤ 0.40	0.035	0.035	–	–	–	–	–	–
90T3	0.85–0.95	0.10–0.30	≤ 0.40	0.035	0.035	–	–	–	–	–	–
103T3	0.95–1.10	0.10–0.30	≤ 0.40	0.035	0.035	–	–	–	–	–	–
118T3	1.10–1.25	0.10–0.30	≤ 0.40	0.035	0.035	–	–	–	–	–	–
133T3	1.25–1.40	0.10–0.30	≤ 0.40	0.035	0.035	–	–	–	–	–	–
T80V2	0.75–0.85	0.10–0.30	≤ 0.40	0.035	0.035	–	–	–	–	0.15–0.30	–
T90V2	0.85–0.95	0.10–0.30	≤ 0.40	0.035	0.035	–	–	–	–	0.15–0.30	–
T103V2	0.95–1.10	0.10–0.30	≤ 0.40	0.035	0.035	–	–	–	–	0.15–0.30	–

(Contd.)

TABLE 3.139 Chemical composition (cast analysis) of tool steels, conforming to IS 3748 – 1990 and IS 7291 – 1981 (continued)

Steel designation	Chemical composition[1)] [% (m/m)]										
	C	Si	Mn	P max.	S max.	Cr	Mo	Ni	Co	V	W
Cold-work tool steels (IS 3748 – 1990)											
T118Cr2	1.10–1.25	0.10–0.30	≤ 0.40	0.035	0.035	0.30–0.60	–	–	–	≤ 0.30	–
T135Cr2	1.25–1.40	0.10–0.30	≤ 0.40	0.035	0.035	0.30–0.60	–	–	–	≤ 0.30	
T105Cr5	0.90–1.20	0.10–0.35	0.20–0.40	0.035	0.035	1.00–1.60	–	–	–	–	–
T105Cr5Mn2	0.90–1.20	0.10–0.35	0.40–0.80	0.035	0.035	1.00–1.60	–	–	–	–	–
T140W15Cr2	1.30–1.50	0.10–0.35	0.25–0.50	0.035	0.035	0.30–0.70	–	–	–	–	3.50–4.20
T60Ni5	0.55–0.65	0.10–0.35	0.50–0.80	0.035	0.035	≤ 0.30	–	1.00–1.50	–	–	–
T40Ni14	0.35–0.45	0.10–0.35	0.50–0.80	0.035	0.035	≤ 0.30	–	3.20–3.60	–	–	–
T30Ni16Cr5	0.26–0.34	0.10–0.35	0.40–0.70	0.035	0.035	1.10–1.40	–	3.90–4.30	–	–	–

(Contd.)

TABLE 3.139 Chemical composition (cast analysis) of tool steels, conforming to IS 3748 – 1990 and IS 7291 – 1981 (continued)

Steel designation	Chemical composition[1)] [% (*m*/*m*)]										
	C	Si	Mn	P max.	S max.	Cr	Mo	Ni	Co	V	W
Cold-work tool steels (IS 3748 – 1990)											
T55Ni6CrMo3	0.50–0.60	0.10–0.35	0.50–0.80	0.035	0.035	0.50–0.80	0.25–0.35	1.25–1.75	–	–	–
T40Ni6Cr4Mo3	0.35–0.45	0.10–0.35	0.40–0.70	0.035	0.035	0.90–1.30	0.20–0.35	1.25–1.75	–	–	–
T31Ni10Cr3Mo6	0.27–0.35	0.10–0.35	0.40–0.70	0.035	0.035	0.50–0.80	0.40–0.70	2.25–2.75	–	–	–
T40Ni10Cr3Mo6	0.36–0.44	0.10–0.35	0.40–0.70	0.035	0.035	0.50–0.80	0.40–0.70	2.25–2.75	–	–	–
T105W6CrV2	0.90–1.20	0.10–0.35	≤ 0.40	0.035	0.035	0.40–0.80	≤ 0.25	–	–	0.20–0.30	1.25–1.75
T110W6Cr4	1.00–1.20	0.10–0.35	0.25–0.50	0.035	0.035	0.90–1.30	–	–	–	–	1.25–1.75
T90Mn6WCr2	0.85–0.95	0.10–0.35	1.25–1.75	0.035	0.035	0.30–0.60	–	–	–	≤ 0.25[2)]	0.40–0.60
XT160Cr12	1.50–1.70	0.10–0.35	0.25–0.55	0.035	0.035	11.0–13.0	≤ 0.80	–	–	≤ 0.80[2)]	≤ 0.60

(Contd.)

TABLE 3.139 Chemical composition (cast analysis) of tool steels, conforming to IS 3748 – 1990 and IS 7291 – 1981 (continued)

Steel designation	Chemical composition[1)] [% (m/m)]										
	C	Si	Mn	P max.	S max.	Cr	Mo	Ni	Co	V	W
Cold-work tool steels (IS 3748 – 1990)											
XT215Cr12	2.00–2.30	0.10–0.35	0.25–0.50	0.035	0.035	11.0–13.0	≤ 0.80	–	–	≤ 0.80[2)]	–
T55Cr3	0.50–0.60	0.10–0.35	0.60–0.80	0.035	0.035	0.60–0.80	–	–	–	–	–
T45Cr5Si3	0.40–0.50	0.80–1.10	0.55–0.75	0.035	0.035	1.20–1.60	–	–	–	–	–
T55Cr3V2	0.50–0.60	0.10–0.35	0.60–0.80	0.035	0.035	0.60–0.80	–	–	–	0.10–0.20	–
T50Cr4V2	0.45–0.55	0.10–0.35	0.50–0.80	0.035	0.035	0.90–1.20	–	–	–	0.15–0.30	–
T55Si7	0.50–0.60	1.50–2.00	0.80–1.00	0.035	0.035	–	–	–	–	–	–
T55Si7Mo3	0.50–0.60	1.50–2.00	0.80–1.00	0.035	0.035	–	0.25–0.40	–	–	0.12–0.20[2)]	–
T40W8Cr5V2	0.35–0.45	0.50–1.00	0.20–0.40	0.035	0.035	1.00–1.50	–	–	–	0.10–0.25	1.75–2.25
T50W8Cr5V2	0.45–0.55	0.50–1.00	0.20–0.40	0.035	0.035	1.00–1.50	–	–	–	0.10–0.25	1.75–2.25

(Contd.)

TABLE 3.139 Chemical composition (cast analysis) of tool steels, conforming to IS 3748 – 1990 and IS 7291 – 1981 (continued)

Steel designation	Chemical composition[1)] [% (m/m)]										
	C	Si	Mn	P max.	S max.	Cr	Mo	Ni	Co	V	W
Hot-work tool steels (IS 3748 – 1990)											
XT33W9Cr3V4	0.25–0.40	0.10–0.35	0.20–0.40	0.035	0.035	2.80–3.30	–	–	–	0.25–0.50	8.00–10.0
XT35Cr5Mo1V3	0.30–0.40	0.80–1.20	0.25–0.50	0.035	0.035	4.75–5.50	1.20–1.60	–	–	0.20–0.40	–
XT35Cr5MoV1	0.30–0.40	0.80–1.20	0.25–0.50	0.035	0.035	4.75–5.50	1.20–1.60	–	–	1.00–1.20	–
XT35Cr5MoW1V3	0.30–0.40	0.80–1.20	0.25–0.50	0.035	0.035	4.75–5.50	1.20–1.60	–	–	0.20–0.40	1.20–1.60
XT55W14Cr3V4	0.50–0.60	0.10–0.35	0.20–0.40	0.035	0.035	2.80–3.30	–	–	–	0.30–0.60	13.0–15.0
High-speed tool steels (IS 7291 – 1981)[3)]											
XT72W18Cr4V1	0.65–0.80	0.15–0.40	0.20–0.40	0.030	0.030	3.75–4.50	–	–	–	1.00–1.25	17.5–19.0
XT75W18Co5Cr4MoV1	0.70–0.80	0.15–0.40	0.20–0.40	0.030	0.030	3.75–4.50	0.40–1.00	–	4.50–5.50	1.00–1.25	17.5–19.0
XT80W20Co12Cr4V2Mo1	0.75–0.85	0.15–0.40	0.20–0.40	0.030	0.030	4.00–4.75	0.40–1.00	–	11.0–12.5	1.25–1.75	19.5–21.0
XT125WCo10CrMo4V3	1.20–1.30	0.15–0.40	0.20–0.40	0.030	0.030	3.75–4.50	3.00–4.00	–	8.80–10.7	2.80–3.50	8.80–10.7

(Contd.)

TABLE 3.139 Chemical composition (cast analysis) of tool steels, conforming to IS 3748 – 1990 and IS 7291 – 1981 (continued)

Steel designation	Chemical composition[1)] [% (*m*/*m*)]										
	C	Si	Mn	P max.	S max.	Cr	Mo	Ni	Co	V	W
High-speed tool steels (IS 7291 – 1981)[3)]											
XT87W6Mo5Cr4V2	0.82–0.92	0.15–0.40	0.15–0.40	0.030	0.030	3.75–4.50	4.75–5.50	–	–	1.75–2.05	5.75–6.75
XT90W6CoMo5Cr4V2	0.85–0.95	0.15–0.40	0.20–0.40	0.030	0.030	3.75–4.50	4.75–5.50	–	4.75–5.25	1.70–2.20	5.75–6.75
XT110Mo10Co8Cr4W2	1.05–1.15	0.15–0.40	0.15–0.40	0.030	0.030	3.50–4.50	9.00–10.0	–	7.75–8.75	0.95–1.35	1.15–1.85

1) Residual elements up to the following limits are acceptable:
(a) for cold-work and hot-work tool steels: Cr ≤ 0.25 %; Mo ≤ 0.25 %; Ni ≤ 0.25 %; Co ≤ 0.10 %; Cu ≤ 0.35 %; Sn ≤ 0.05 %; V ≤ 0.05 %; W ≤ 0.25 %.
(b) for high-speed tool steels: Mo ≤ 0.40 %, Ni ≤ 0.40 % Cu ≤ 0.20 %, Sn ≤ 0.05 %.
2) Optional.
3) For improved machinability, high-speed tool steels can be ordered with a S content of 0.09–0.15 %.

Note: All steels should be fully killed.

2.2 Hardness

TABLE 3.140 Hardness requirements for tool steels, conforming to IS 3748 – 1990 and IS 7291–1981, in the annealed condition

Steel designation	Maximum Brinell hardness HB	Steel designation	Maximum Brinell hardness HB
Cold-work tool steels (IS 3748 – 1990)		**Cold-work tool steels (IS 3748 – 1990)**	
50T8	210	**T135Cr2**	220
55T8	220	**T105Cr5Mn2**	230
60T6	220	**T140W15Cr2**	250
65T6	220	**T60Ni5**	255
70T6	220	**T40Ni14**	255
75T6	220	**T30Ni16Cr5**	255
80T6	220	**T55Ni6CrMo3**	255
85T6	220	**T40Ni6Cr4Mo3**	255
70T3	220	**T31Ni10Cr4Mo6**	255
80T3	220	**T40Ni10Cr3Mo6**	255
90T3	220	**T105W6CrV2**	230
103T3	220	**T110W6Cr4**	230
118T3	220	**T90Mn6WCr2**	230
133T3	220	**XT160Cr12**	230
T80V2	220	**XT215Cr12**	255
T90V2	220	**T55Cr3**	230
T103V2	220	**T45Cr5Si3**	230
T118Cr2	220	**T55Cr3V2**	230

(Contd.)

TABLE 3.140 Hardness requirements for tool steels, conforming to IS 3748–1990 and IS 7291–1981, in the annealed condition (continued)

Steel designation	Maximum Brinell hardness HB
Cold-work tool steels (IS 3748 – 1990)	
T50Cr4V2	230
T55Si7	230
T55Si7Mo3	230
T40W8Cr5V2	230
T50W8Cr5V2	230
Hot-work tool steels (IS 3748 – 1990)	
XT33W9Cr3V4	245
XT35Cr5Mo1V3	235
XT35Cr5MoV1	235
XT35Cr5MoW1V3	235
Hot-work tool steels (IS 3748 – 1990)	
XT55W14Cr3V4	248
High-speed tool steels (IS 7291 – 1981)	
XT72W18Cr4V1	255
XT75W18Co5Cr4MoV1	269
XT80W20Co12Cr4V2Mo1	302
XT125WCo10CrMo4V3	269
XT87W6Mo5Cr4V2	248
XT90W6CoMo5Cr4V2	269
XT110Mo10Co8Cr4W2	269

TABLE 3.141 Hardness requirements for high-speed tool steels, conforming to IS 7291 – 1981, in the hardened and tempered condition

Steel designation	Minimum hardness	
	HRC	HV
XT72W18Cr4V1	63	772
XT75W18Co5Cr4MoV1	63	772
XT80W20Co12Cr4V2Mo1	64	800
XT125WCo10CrMo4V3	66	862
XT87W6Mo5Cr4V2	63	772
XT90W6CoMo5Cr4V2	64	800
XT110Mo10Co8Cr4W2	66	862

3 RECOMMENDED HOT WORKING AND HEAT TREATING TEMPERATURES

TABLE 3.142 Recommended hot working and heat treating temperatures for hot-work and high-speed tool steels specified in Indian standards

Steel designation	Hot working temperature °C	Annealing temperature °C	Hardening			Tempering temperature °C
			Preheat temperature °C	Hardening temperature °C	Quenching agent[1)]	
Hot-work tool steels (IS 3748 – 1990)						
XT33W9Cr3V4	1150–850	850	800	1100	o, a	600–650
XT35Cr5Mo1V3	1150–850	850	800	1030	o, a	550–650
XT35Cr5MoV1	1150–850	850	800	1000	o, a	550–650
XT35Cr5MoW1V3	1150–850	850	800	1010	o, a	550–650
XT55W14Cr3V4	1150–950	850	850	1260	o, a	570[2)]
High-speed tool steel (IS 7291 – 1981)						
XT72W18Cr4V1	1150–950	860	850	1250–1290	o, a, s	530–570
XT75W18Co5Cr4MoV1	1150–950	860	850	1270–1310	o, a, s	530–570
XT80W20Co12Cr4V2Mo1	1150–950	860	850	1280–1320	o, a, s	530–570
XT125WCo10CrMo4V3	1150–950	860	850	1230–1250	o, a, s	530–570
XT87W6Mo5Cr4V2	1150–950	860	850	1190–1230	o, a, s	530–570
XT90W6CoMo5Cr4V2	1150–950	860	850	1200–1240	o, a, s	530–570
XT110Mo10Co8Cr4W2	1150–950	860	850	1180–1230	o, a, s	530–570

1) a = air; o = oil; s = salt bath.
2) Should be double tempered.

Note: Double tempering is recommended for high-speed steels except for cobalt bearing steels which should be triple tempered.

4 EQUIVALENT GRADES

TABLE 3.143 Comparison of grades of cold-work tool steels specified in IS 3748 – 1990 with nearest equivalent grades specified in other national and international standards

IS 3748–90	ASTM A 681–94 ASTM A 686–92	BS 4659–89	DIN 17350–80	GOST 1435–90 5950–73	ISO 4957–80	JIS G 4401–83 JIS G 4404–83	SAE J438b–70
50T8	–	–	–	–	–	–	–
55T8	–	–	–	–	–	–	–
60T6	–	–	C 60 W	–	–	–	–
65T6	–	–	–	–	–	–	–
70T6	–	–	–	–	–	–	–
75T6	–	–	–	–	–	–	–
80T6	–	–	–	–	–	–	–
85T6	–	–	C 85 W	–	–	–	–
70T3	–	–	C 70 W 2	Y7, Y7A	TC 70	–	–
80T3	–	–	C 80 W 1	Y8, Y8A	TC 80	–	W108
90T3	W1-8 ½	–	–	Y9, Y9A	TC 90	–	W109
103T3	W1-9 ½	–	C 105 W 1	Y10, Y10A	TC 105	SK 3	W110
118T3	W1-11	–	–	Y12, Y12A	TC 120	SK 2	W112
133T3	–	–	–	Y13, Y13A	TC 140	SK 1	–

(Contd.)

TABLE 3.143 Comparison of grades of cold-work tool steels specified in IS 3748 – 1990 with nearest equivalent grades specified in other national and international standards (continued)

IS 3748–90	ASTM A 681–94 ASTM A 686–92	BS 4659–89	DIN 17350–80	GOST 1435–90 5950–73	ISO 4957–80	JIS G 4401–83 JIS G 4404–83	SAE J438b–70
T80V2	–	–	–	–	–	–	–
T90V2	W2-8 ½	–	–	–	–	SKS 44	W209
T103V2	W2-9 ½	BW2	–	–	TCV 105	SKS 43	W210
T118Cr2	W5	–	115 CrV 3	11XΦ	–	–	–
T135Cr2	–	–	–	13X	–	SKS 8	–
T105Cr5	L3	–	100 Cr 6	X	100 Cr 2	–	–
T105Cr5Mn2	–	BL1	–	12X1	–	–	–
T140W15Cr2	F2	–	–	XB4Φ	–	–	–
T60Ni5	–	–	–	–	–	–	–
T40Ni14	–	–	–	–	–	–	–
T30Ni16Cr5	–	–	–	–	–	–	–
T55Ni6CrMo3	–	–	–	5XHM	–	–	–
T40Ni6Cr4Mo3	–	–	–	–	–	–	–
T31Ni10Cr3Mo6	–	–	–	–	–	–	–
T40Ni10Cr3Mo6	–	–	–	–	–	–	–

(Contd.)

TABLE 3.143 Comparison of grades of cold-work tool steels specified in IS 3748 – 1990 with nearest equivalent grades specified in other national and international standards (continued)

IS 3748–90	ASTM A 681–94 ASTM A 686–92	BS 4659–89	DIN 17350–80	GOST 1435–90 5950–73	ISO 4957–80	JIS G 4401–83 JIS G 4404–83	SAE J438b–70
T105W6CrV2	O7	–	–	B2Φ	–	SKS 2	–
T110W6Cr4	–	–	105 WCr 6	XBΓ	105 WCr 1	SKS 31	–
T90Mn6WCr2	O1	BO1	–	9XBΓ	95 MnCrW 1	SKS 3	O1
XT160Cr12	D2	BD2	X 165 CrMoV 12	X12MΦ	160 CrMoV 12	SKD 11	D2
XT215Cr12	D3	BD3	X 210 Cr 12	X12	210 Cr 12	SKD 1	D3
T55Cr3	–	–	–	–	–	–	–
T45Cr5Si3	–	–	–	–	–	–	–
T55Cr3V2	–	–	–	–	–	–	–
T50Cr4V2	–	–	51 CrV 4	–	51 CrMnV 1	–	–
T55Si7	–	–	–	–	60 SiMn 2	–	–
T55Si7Mo3	S5	–	–	–	–	–	S5
T40W8Cr5V2	–		–	4XB2C	45 WCrV 2	–	–
T50W8Cr5V2	S1	BS1	–	5XB2CΦ	50 WCrV 2	–	S1

(Contd.)

TABLE 3.144 Comparison of grades of hot-work tool steels specified in IS 3748 – 1990 with nearest equivalent grades specified in other national and international standards

IS 3748–90	ASTM A 681–94	BS 4659–89	DIN 17350–80	GOST 5950–73	ISO 4957–80	JIS G 4404–83	SAE J438b–70
XT33W9Cr3V4	H21	BH21	–	3X2B8Φ	30 WCrV 9	SKD 5	H21
XT35Cr5Mo1V3	H11	BH11	X 38 CrMoV 5 1	4X5MΦC	35 CrMoV 5	SKD 6	H11
XT35Cr5MoV1	H13	BH13	X 40 CrMoV 5 1	4X5MΦ1C	40 CrMoV 5	SKD 61	H13
XT35Cr5MoW1V3	H12	BH12	–	–	–	SKD 62	H12
XT55W14Cr3V4	–	–	–	–	–	–	–

TABLE 3.145 Comparison of grades of high-speed tool steels specified in IS 7291 – 1981 with nearest equivalent grades specified in other national and international standards

IS 7291–81	ASTM A 600–92a	BS 4659–89	DIN 17350–80	GOST 19265–73	ISO 4957–80	JIS G 4403–83	SAE J438b–70
XT72W18Cr4V1	T1	BT1	–	P18	HS 18-0-1	SKH 2	T1
XT75W18Co5Cr4MoV1	T4	BT4	S 18-1-2-5	–	HS 18-1-1-5	SKH 3	T4
XT80W20Co12Cr4V2Mo1	T6	BT6	–	–	–	–	–
XT125WCo10CrMo4V3	–	BT42	S 10-4-3-10	–	HS 10-4-3-10	SKH 57	–
XT87W6Mo5Cr4V2	M2, regular C	BM2	S 6-5-2	P6M5	HS 6-5-2	SKH 51	M2
XT90W6CoMo5Cr4V2	–	BM35	S 6-5-2-5	P6M5K5	HS 6-5-2-5	SKH 55	–
XT110Mo10Co8Cr4W2	M42	BM42	S 2-10-1-8	–	HS 2-9-1-8	SKH 59	–

5 REFERENCES

Book

1. *Metals Handbook*, vol. 3, 9th ed., American Society for Metals, Metals Park, Ohio, U.S.A., 1980.

Standards

2. IS 3748 – 1990: *Tool and Die Steels—Specification.*
3. IS 7291 – 1981: *Specification for High-Speed Tool Steels.*
4. DIN 17350 – 1980: *Tool Steels.*

3.23

CAST STEELS

1 INTRODUCTION

Cast steels are unalloyed and alloy steels which are commonly classified on the basis of application into cast steels for structural purposes, and cast steels for special applications, such as those requiring resistance to abrasion, corrosion and elevated temperature.

1.1 Cast steels for structural purposes

Unalloyed cast steels: The mechanical properties of these steels are primarily controlled by the carbon content and heat treatment. Strength and hardness increase with increasing carbon content, but at the expense of ductility and toughness.

Low-alloy cast steels: These steels possess higher strength, toughness and abrasion resistance than can be achieved in unalloyed steels. They are also more resistant to atmospheric corrosion than unalloyed steels.

1.2 Cast steels for special applications

Abrasion-resistant cast steels: These are unalloyed, low-alloy and austenitic manganese steels which are used in applications requiring resistance to wear and good toughness.

The unalloyed and low-alloy steels may be used in the case hardened, surface hardened, normalized or hardened and tempered condition.

Cast steels for low temperature service: These steels are ferritic hardenable steels and austenitic non-hardenable steels which are especially suited for service in cold environments, and for storing and handling liquefied gases such as argon, nitrogen and oxygen.

Corrosion-resistant cast steels: These steels are iron-chromium and iron-chromium-nickel alloys which resist attack by aqueous solutions at or near ambient temperature, and by hot gases and high boiling point liquids at temperatures up to 650 °C.

Cast steels for high temperature service: These steels are unalloyed and alloy steels intended for service at elevated temperatures, possibly under reactive environments, as in gas turbines and chemical processing equipment.

The use of unalloyed, low-alloy and medium-alloy steels is generally restricted to 620 °C. However, the high-alloy ferritic and austenitic steels are recommended for service above 620 °C as they possess excellent corrosion and oxidation resistance alongwith high strength. They are capable of sustained operation while exposed either continuously or intermittently to high temperatures.

2 SPECIFICATIONS

2.1 Chemical composition

TABLE 3.146 Chemical composition (cast analysis) of steel castings specified in Indian Standards

Grade	Chemical composition [% (m/m)]								
	C	Si	Mn	P max.	S max.	Cr	Mo	Ni	Other elements
Austenitic manganese steel castings (IS 276 – 1992)									
1	1.05–1.35	≤ 1.00	11.0–14.0	0.090	0.050	–	–	–	–
2	0.90–1.05	≤ 1.00	11.5–14.0	0.090	0.050	–	–	–	–
3	1.05–1.35	≤ 1.00	11.5–14.0	0.090	0.050	1.50–2.50	–	–	–
4	0.70–1.30	≤ 1.00	11.5–14.0	0.090	0.050	–	–	3.00–5.00	–
5	1.05–1.45	≤ 1.00	11.5–14.0	0.090	0.050	–	1.80–2.10	–	–
Carbon steel castings for general engineering purposes (IS 1030 – 1989)									
200–400N	–	–	–	0.045	0.040	–	–	–	–
200–400W	≤ 0.25	≤ 0.60	≤ 1.00	0.040	0.035	–	–	–	–
230–450N	–	–	–	0.045	0.040	–	–	–	–
230–450W	≤ 0.25	≤ 0.60	≤ 1.20	0.040	0.035	–	–	–	–
280–520N	–	–	–	0.045	0.040	–	–	–	–
280–520W	≤ 0.25	≤ 0.60	≤ 1.20	0.040	0.035	–	–	–	–
340–570N	–	–	–	0.045	0.040	–	–	–	–
340–570W	≤ 0.25	≤ 0.60	≤ 1.50	0.040	0.035	–	–	–	–

(Contd.)

TABLE 3.146 Chemical composition (cast analysis) of steel castings specified in Indian Standards (continued)

Grade	Chemical composition [% (m/m)]								
	C	Si	Mn	P max.	S max.	Cr	Mo	Ni	Other elements
High strength steel castings for general engineering and structural purposes (IS 2644 – 1994)									
1	–	≤ 0.60	–	0.035	0.035	–	–	–	–
2	–	≤ 0.60	–	0.035	0.035	–	–	–	–
3	–	≤ 0.60	–	0.035	0.035	–	–	–	–
4	–	≤ 0.60	–	0.035	0.035	–	–	–	–
5	–	≤ 0.60	–	0.035	0.035	–	–	–	–
Carbon steel castings for surface hardening (IS 2707 – 1989)									
1	0.42–0.48	≤ 0.60	≤ 1.00	0.040	0.040	–	–	–	–
2	0.39–0.45	0.30–0.80	1.00–1.40	0.040	0.040	–	–	–	–
3	0.52–0.58	≤ 0.60	≤ 1.00	0.040	0.040	–	–	–	–
1.5 % Manganese steel castings for general engineering purposes (IS 2708 – 1993)									
1	0.18–0.25	≤ 0.60	1.20–1.60	0.050	0.050	–	–	–	–
2	0.25–0.33	≤ 0.60	1.20–1.60	0.050	0.050	–	–	–	–
3	0.25–0.33	≤ 0.60	1.20–1.60	0.050	0.050	–	–	–	–

(Contd.)

TABLE 3.146 Chemical composition (cast analysis) of steel castings specified in Indian Standards (continued)

Grade	Chemical composition [% (m/m)]								
	C	Si	Mn	P max.	S max.	Cr	Mo	Ni	Other elements
Carbon steel castings for pressure containing parts suitable for high temperature service (fusion welding quality) (IS 2856 – 1987)									
1	≤ 0.25	≤ 0.60	≤ 1.20	0.040	0.045	–	–	–	–
2	≤ 0.30	≤ 0.60	≤ 1.00	0.040	0.045	–	–	–	–
3	≤ 0.25	≤ 0.60	≤ 0.70	0.040	0.045	–	–	–	–
Alloy steel castings for pressure containing parts suitable for high temperature service (IS 3038 – 1983)									
1	0.18–0.25	≤ 0.60	1.20–1.60	0.040	0.045	–	–	–	–
2	≤ 0.20	0.20–0.60	0.50–1.00	0.040	0.045	–	0.45–0.65	–	–
3	0.10–0.15	0.20–0.45	0.40–0.70	0.030	0.030	0.30–0.50	0.40–0.60	–	Al ≤ 0.025; V = 0.22–0.30
4	≤ 0.20	≤ 0.60	0.50–0.80	0.040	0.045	1.00–1.50	0.45–0.65	–	–
4A	0.12–0.16	0.30–0.50	0.55–0.75	0.030	0.030	0.85–1.15	0.90–1.10	–	Al ≤ 0.025; V = 0.20–0.30
5	≤ 0.15	0.20–0.45	0.40–0.70	0.030	0.030	2.00–2.50	0.90–1.20	–	Al ≤ 0.025
5A	≤ 0.18	≤ 0.60	0.40–0.70	0.040	0.045	2.00–2.75	0.90–1.20	–	–
6	≤ 0.20	≤ 0.75	0.40–0.70	0.040	0.045	4.00–6.50	0.45–0.65	–	–
7	≤ 0.20	≤ 1.00	0.35–0.65	0.040	0.045	8.00–10.0	0.90–1.20	–	–

(Contd.)

TABLE 3.146 Chemical composition (cast analysis) of steel castings specified in Indian Standards (continued)

Grade	Chemical composition [% (m/m)]								
	C	Si	Mn	P max.	S max.	Cr	Mo	Ni	Other elements
Alloy steel castings for pressure containing parts suitable for high temperature service (IS 3038–1983)									
8	≤ 0.15	≤ 1.50	≤ 1.00	0.040	0.040	11.5–14.0	≤ 0.50	–	–
Corrosion-resistant alloy steel and nickel-based castings for general applications (IS 3444–1978)									
1	≤ 0.08	≤ 2.00	≤ 1.50	0.040	0.040	18.0–21.0	–	8.00–11.0	–
2	≤ 0.12	≤ 2.00	≤ 1.50	0.040	0.040	20.0–23.0	–	10.0–13.0	–
3	≤ 0.20	≤ 2.00	≤ 1.50	0.040	0.040	18.0–21.0	–	8.00–11.0	–
4	≤ 0.08	≤ 2.00	≤ 1.50	0.040	0.040	18.0–21.0	2.00–3.00	9.00–12.0	–
4A	≤ 0.12	≤ 2.00	≤ 1.50	0.040	0.040	18.0–21.0	2.00–3.00	9.00–12.0	–
5	≤ 0.08	≤ 2.00	≤ 1.50	0.040	0.040	18.0–21.0	–	9.00–12.0	Nb ≥ 8 x %C ≤ 1.00
6	≤ 0.16	≤ 2.00	≤ 1.50	0.040	0.040	18.0–21.0	–	9.00–12.0	–
7	≤ 0.20	≤ 2.00	≤ 1.50	0.040	0.040	22.0–26.0	–	12.0–15.0	–
8	≤ 0.20	≤ 2.00	≤ 2.00	0.040	0.040	23.0–27.0	–	19.0–22.0	–
9	≤ 0.30	≤ 2.00	≤ 1.50	0.040	0.040	26.0–30.0	–	8.00–11.0	–
10	≤ 0.15	≤ 1.50	≤ 1.00	0.040	0.040	11.5–14.0	≤ 0.50	≤ 1.00	–
11	≤ 0.15	≤ 0.65	≤ 1.00	0.040	0.040	11.5–14.0	0.15–1.00	≤ 1.00	–

(Contd.)

TABLE 3.146 Chemical composition (cast analysis) of steel castings specified in Indian Standards (continued)

Grade	Chemical composition [% (m/m)]								
	C	Si	Mn	P max.	S max.	Cr	Mo	Ni	Other elements
Corrosion-resistant alloy steel and nickel-based castings for general applications (IS 3444 – 1978)									
12	≤ 0.30	≤ 1.50	≤ 1.00	0.040	0.040	18.0–21.0	–	≤ 2.00	–
13	≤ 0.50	≤ 1.50	≤ 1.00	0.040	0.040	26.0–30.0	–	≤ 4.00	–
14	0.20–0.40	â 1.50	≤ 1.00	0.040	0.040	11.5–14.0	≤ 0.50	≤ 1.00	–
15	≤ 0.030	≤ 2.00	≤ 1.50	0.040	0.040	17.0–21.0	–	8.00–12.0	–
16	≤ 0.030	≤ 1.50	≤ 1.50	0.040	0.040	17.0–21.0	2.00–3.00	9.00–13.0	–
17	≤ 0.08	≤ 1.50	≤ 1.50	0.040	0.040	18.0–21.0	3.00–4.00	9.00–13.0	–
18	≤ 0.07	≤ 1.50	≤ 1.50	0.040	0.040	19.0–22.0	2.00–3.00	27.5–30.5	Cu = 3.00–4.00
19	≤ 0.12	≤ 1.50	≤ 1.00	0.040	0.030	15.5–20.0	16.0–20.0	Rem.	Co ≤ 2.50; Fe ≤ 7.50; V ≤ 0.40; W ≤ 5.25
20	≤ 0.40	≤ 3.00	≤ 1.50	0.030	0.030	14.0–17.0	–	Rem.	Fe ≤ 11.0
21	≤ 1.00	≤ 2.00	≤ 1.50	0.030	0.030	–	–	Rem.	Cu ≤ 1.25; Fe ≤ 3.00
22	≤ 0.35	≤ 2.00	≤ 1.50	0.030	0.030	–	–	Rem.	Cu = 26.0–33.0; Fe ≤ 3.50
23	≤ 0.12	≤ 1.00	≤ 1.00	0.040	0.030	≤ 1.00	26.0–33.0	Rem.	Co ≤ 2.50; Fe ≤ 6.00; V ≤ 0.60
24	≤ 0.06	≤ 1.00	≤ 1.00	0.040	0.040	11.5–14.0	0.40–1.00	3.50–4.50	–

(Contd.)

TABLE 3.146 Chemical composition (cast analysis) of steel castings specified in Indian Standards (continued)

Grade	Chemical composition [% (m/m)]								
	C	Si	Mn	P max.	S max.	Cr	Mo	Ni	Other elements
Steel castings of high magnetic permeability (IS 4491 – 1994)									
1	≤ 0.15	≤ 0.60	≤ 0.50	0.040	0.040	–	–	–	–
2	≤ 0.25	≤ 0.60	≤ 0.50	0.040	0.040	–	–	–	–
Heat-resistant alloy steel castings (IS 4522 – 1979)									
1	≤ 0.40	≤ 2.00	≤ 1.00	0.050	0.050	12.0–14.0	≤ 0.50	≤ 1.00	–
2	0.30–0.60	≤ 2.00	≤ 1.00	0.050	0.050	27.0–30.0	≤ 0.50	–	–
3	1.20–1.40	≤ 2.00	≤ 1.00	0.050	0.050	27.0–30.0	≤ 0.50	–	–
4	0.20–0.50	≤ 2.00	≤ 1.00	0.050	0.050	26.0–30.0	≤ 0.50	4.00–7.00	–
5	0.20–0.50	≤ 2.00	≤ 2.00	0.050	0.050	18.0–20.0	≤ 0.50	8.00–10.0	–
6	0.20–0.50	≤ 2.00	≤ 2.00	0.050	0.050	26.0–30.0	≤ 0.50	8.00–10.0	–
7	0.20–0.50	≤ 2.00	≤ 2.00	0.050	0.050	23.0–27.0	≤ 0.50	11.0–14.0	–
8	0.15–0.35	≤ 2.50	≤ 1.50	0.050	0.050	19.0–21.0	≤ 0.50	13.0–15.0	–
9	0.20–0.50	≤ 2.00	≤ 2.00	0.050	0.050	23.0–27.0	≤ 0.50	18.0–22.0	–
10	0.20–0.60	≤ 2.50	≤ 2.00	0.050	0.050	28.0–32.0	≤ 0.50	18.0–22.0	–
11	0.20–0.50	≤ 2.00	≤ 2.00	0.050	0.050	19.0–23.0	≤ 0.50	23.0–27.0	–

(Contd.)

TABLE 3.146 Chemical composition (cast analysis) of steel castings specified in Indian Standards (continued)

Grade	Chemical composition [% (m/m)]								
	C	Si	Mn	P max.	S max.	Cr	Mo	Ni	Other elements
Heat-resistant alloy steel castings (IS 4522 – 1979)									
12	0.35–0.75	≤ 2.50	≤ 2.00	0.050	0.050	13.0–17.0	≤ 0.50	33.0–37.0	–
13	0.35–0.75	≤ 2.50	≤ 2.00	0.050	0.050	17.0–21.0	≤ 0.50	37.0–41.0	–
14	0.35–0.75	≤ 2.50	≤ 2.00	0.050	0.050	15.0–19.0	≤ 0.50	64.0–68.0	–
1 % Chromium steel castings for resistance to abrasion (IS 4896 – 1992)									
1	0.45–0.55	≤ 0.75	0.50–1.00	0.040	0.040	0.80–1.20	–	–	–
2	0.55–0.70	≤ 0.75	0.50–1.00	0.040	0.040	0.80–1.50	0.20–0.40	–	–
3	0.90–1.20	≤ 0.75	0.50–1.00	0.040	0.040	0.80–1.50	–	–	–
Steel castings for case carburizing (IS 4898 – 1989)									
1	0.10–0.18	≤ 0.60	0.60–1.00	0.040	0.040	–	–	–	–
2	0.12–0.18	≤ 0.60	0.30–0.60	0.040	0.040	0.60–1.10	0.15–0.25	3.00–3.75	–
Ferritic and martensitic steel castings for use at low temperature (IS 4899 – 1991)									
1	≤ 0.25	≤ 0.60	≤ 0.70	0.040	0.045	–	–	–	–
2	≤ 0.30	≤ 0.60	≤ 1.00	0.040	0.045	–	–	–	–
3	≤ 0.25	≤ 0.60	≤ 1.20	0.040	0.045	–	–	–	–

(Contd.)

TABLE 3.146 Chemical composition (cast analysis) of steel castings specified in Indian Standards (continued)

Grade	Chemical composition [% (m/m)]								
	C	Si	Mn	P max.	S max.	Cr	Mo	Ni	Other elements
Ferritic and martensitic steel castings for use at low temperature (IS 4899 – 1991)									
4	≤ 0.25	≤ 0.60	0.50–0.80	0.040	0.045	–	0.45–0.65	–	–
5	≤ 0.25	≤ 0.60	0.50–0.80	0.040	0.045	–	–	2.00–3.00	–
6	≤ 0.22	≤ 0.50	0.55–0.75	0.040	0.045	1.35–1.85	0.30–0.60	2.50–3.50	–
7	≤ 0.15	≤ 0.60	0.50–0.80	0.040	0.045	–	–	3.00–4.00	–
8	≤ 0.15	≤ 0.60	0.50–0.90	0.040	0.045	–	–	4.00–5.00	–
9	≤ 0.13	≤ 0.45	≤ 0.90	0.040	0.045	–	–	8.50–10.0	–
10	≤ 0.06	≤ 1.00	≤ 1.00	0.040	0.030	11.5–14.0	0.40–1.00	3.50–4.50	–
Martensitic and austenitic high alloy steel castings for high temperature corrosion services (IS 7806 – 1993)									
1	≤ 0.15	≤ 1.50	≤ 1.00	0.040	0.040	11.5–14.0	≤ 0.50	≤ 1.00	–
1A	≤ 0.15	≤ 0.65	≤ 1.00	0.040	0.040	11.5–14.0	0.75–1.00	≤ 1.00	–
2, 2A	≤ 0.030	≤ 2.00	≤ 1.50	0.040	0.040	17.0–21.0	–	8.00–12.0	–
3, 3A	≤ 0.08	≤ 2.00	≤ 1.50	0.040	0.040	18.0–21.0	–	8.00–11.0	–
4, 4A	≤ 0.030	≤ 1.50	≤ 1.50	0.040	0.040	17.0–21.0	2.00–3.00	9.00–13.0	–
5	≤ 0.08	≤ 1.50	≤ 1.50	0.040	0.040	18.0–21.0	2.00–3.00	9.00–12.0	–

(Contd.)

TABLE 3.146 Chemical composition (cast analysis) of steel castings specified in Indian Standards (continued)

Grade	Chemical composition [% (m/m)]								
	C	Si	Mn	P max.	S max.	Cr	Mo	Ni	Other elements
Martensitic and austenitic high alloy steel castings for high temperature corrosion services (IS 7806 – 1993)									
5A	≤ 0.08	≤ 2.00	≤ 1.50	0.040	0.040	18.0–21.0	–	9.00–12.0	Nb ≥ 8 x %C ≤ 1.00
6	≤ 0.08	≤ 1.50	≤ 1.50	0.040	0.040	22.0–26.0	–	12.0–15.0	–
7	≤ 0.10	≤ 2.00	≤ 1.50	0.040	0.040	22.0–26.0	–	12.0–15.0	–
8	≤ 0.20	≤ 2.00	≤ 1.50	0.040	0.040	22.0–26.0	–	12.0–15.0	–
9	≤ 0.20	≤ 1.75	≤ 1.50	0.040	0.040	23.0–27.0	–	19.0–22.0	–
10	0.25–0.35	≤ 1.75	≤ 1.50	0.040	0.040	23.0–27.0	–	19.0–22.0	–
11	0.35–0.45	≤ 1.75	≤ 1.50	0.040	0.040	23.0–27.0	–	19.0–22.0	–
12	0.25–0.35	≤ 2.50	≤ 2.00	0.040	0.040	13.0–17.0	≤ 0.50	33.0–37.0	–
13	≤ 0.10	≤ 1.50	≤ 1.50	0.040	0.040	15.0–18.0	1.75–2.25	13.0–16.0	Nb ≥ 10 x %C ≤ 1.20
14	≤ 0.07	≤ 1.50	≤ 1.50	0.040	0.040	19.0–22.0	2.00–3.00	27.5–30.5	Cu = 3.00–4.00
15	≤ 0.08	≤ 1.50	≤ 1.50	0.040	0.040	18.0–21.0	3.00–4.00	9.00–13.0	–
16	≤ 0.04	≤ 1.00	≤ 1.00	0.040	0.040	24.5–26.5	1.75–2.75	4.75–6.00	Cu = 2.75–3.25
17	0.05–0.15	0.50–1.50	0.15–1.50	0.040	0.040	19.0–21.0	–	31.0–34.0	Cu = 0.15–1.50

(Contd.)

TABLE 3.146 Chemical composition (cast analysis) of steel castings specified in Indian Standards (continued)

Grade	Chemical composition [% (m/m)]								
	C	Si	Mn	P max.	S max.	Cr	Mo	Ni	Other elements
Alloy steel castings suitable for pressure service (IS 7899–1992)									
1	≤ 0.30	≤ 0.80	≤ 1.00	0.040	0.045	–	–	–	V = 0.07–0.15
2	≤ 0.30	≤ 0.80	1.00–1.40	0.040	0.045	–	0.10–0.30	–	–
4	≤ 0.30	≤ 0.80	≤ 1.00	0.040	0.045	0.40–0.80	0.15–0.30	0.40–0.80	–
6	≤ 0.38	≤ 0.80	1.30–1.70	0.040	0.045	0.40–0.80	0.30–0.40	0.40–0.80	–
7	≤ 0.20	≤ 0.80	0.60–1.00	0.040	0.045	0.40–0.80	0.40–0.60	0.70–1.00	B = 0.002–0.006; Cu = 0.15–0.50; V = 0.03–0.10
8	≤ 0.20	≤ 0.80	0.50–0.90	0.040	0.045	2.00–2.75	0.90–1.10	–	–
9	≤ 0.33	≤ 0.80	0.60–1.00	0.040	0.045	0.75–1.10	0.15–0.30	–	–
10	≤ 0.30	≤ 0.80	0.60–1.00	0.040	0.045	0.55–0.90	0.20–0.40	1.40–2.00	–
11	≤ 0.15	≤ 1.50	≤ 1.00	0.040	0.040	11.5–14.0	≤ 0.50	≤ 1.00	–
12	≤ 0.15	≤ 0.65	≤ 1.00	0.040	0.040	11.5–14.0	0.15–1.00	≤ 1.00	–
13	≤ 0.06	≤ 1.00	≤ 1.00	0.040	0.040	11.5–14.0	0.40–1.00	3.50–4.50	–

2.2 Mechanical properties

TABLE 3.147 Mechanical properties[1] of steel castings specified in Indian Standards

Grade	Treatment condition[2]	R_m	R_e	*A* min.	*Z* min.	*KV* min.		Brinell hardness	Bend test	
							Test temperature		Angle of bend	Bend mandrel diameter
		MPa	MPa	%	%	J	°C	HB	°	mm
Carbon steel castings for general engineering purposes (IS 1030 – 1989)										
200–400N	A, N or N + T	≥ 400	≥ 200	25	40	30	–	–	90	50
200–400W	A, N or N + T	≥ 400	≥ 200	25	40	40	–	–	90	50
230–450N	A, N or N + T	≥ 450	≥ 230	22	31	25	–	–	90	50
230–450W	A, N or N + T	≥ 450	≥ 230	22	31	35	–	–	90	50
280–520N	A, N or N + T	≥ 520	≥ 280	18	25	22	–	–	60	50
280–520W	A, N or N + T	≥ 520	≥ 280	18	25	22	–	–	60	50
340–570N	A, N or N + T	≥ 570	≥ 340	15	21	20	–	–	60	50
340–570W	A, N or N + T	≥ 570	≥ 340	15	21	20	–	–	60	50
High strength steel castings for general engineering and structural purposes (IS 2644 – 1986)										
1	HT	≥ 640	≥ 390	15	35	25[4]	–	≥ 190	–	–
2	HT	≥ 700	≥ 580	14	30	25[4]	–	≥ 207	–	–
3	HT	≥ 840	≥ 700	12	28	20[4]	–	≥ 248	–	–

(Contd.)

TABLE 3.147 Mechanical properties[1] of steel castings specified in Indian Standards (continued)

Grade	Treatment condition[2]	R_m	R_e	*A* min.	*Z* min.	*KV* min.		Brinell hardness	Bend test	
							Test temperature		Angle of bend	Bend mandrel diameter
		MPa	MPa	%	%	J	°C	HB	°	mm
High tensile steel castings (IS 2644 – 1986)										
4	HT	≥ 1030	≥ 850	8	20	15[4]	–	≥ 305	–	–
5	HT	≥ 1230	≥ 1000	5	12	–	–	≥ 355	–	–
Carbon steel castings for surface hardening (IS 2707–1989)										
1	HT	≥ 620	≥ 320	12	–	–	–	–	–	–
2	HT	≥ 690	≥ 490	12	–	–	–	–	–	–
3	HT	≥ 700	≥ 370	8						
1.5 % Manganese steel castings for general engineering purposes (IS 2708–1993)										
1	N	≥ 520	≥ 320	16	–	30		152–207	–	–
2	N + T	≥ 620	≥ 370	13	–	25		179–229	–	–
3	Q + T	≥ 690	≥ 495	13	–	22		201–255	–	–
Carbon steel castings for pressure containing parts suitable for high temperature service (fusion welding quality) (IS 2856 – 1987)										
1	HT	485–655	≥ 275	20	35	20	–	–	90	25
2	HT	485–655	≥ 250	20	35	20	–	–	90	25
3	HT	415–585	≥ 205	22	35	22	–	–	90	25

(Contd.)

TABLE 3.147 Mechanical properties[1)] of steel castings specified in Indian Standards (continued)

Grade	Treatment condition[2)]	R_m	R_e	A min.	Z min.	KV min.		Brinell hardness	Bend test	
							Test temperature		Angle of bend	Bend mandrel diameter
		MPa	MPa	%	%	J	°C	HB	°	mm
Alloy steel castings for pressure containing parts suitable for high temperature service (IS 3038 – 1983)										
1	N + T	≥ 540	≥ 320	16	–	20	–	–	90	50
2	N + T	≥ 460	≥ 260	18	–	20	–	–	90	50
3	N + T	≥ 510	≥ 295	17	–	20	–	–	90	50
4	N + T	≥ 480	≥ 280	17	–	30	–	–	90	50
4A	N + T	≥ 590	≥ 440	15	–	25	–	–	90	50
5	N + T	≥ 510	≥ 310	17	–	25	–	–	90	50
5A	N + T	≥ 540	≥ 325	17	–	25	–	–	90	50
6	N + T	≥ 620	≥ 420	15	–	25	–	–	90	50
7	N + T	≥ 620	≥ 420	15	–	–	–	–	90	50
8	N + T	≥ 620	≥ 450	15	–	25	–	–	90	50
Corrosion-resistant steel and nickel-based castings for general engineering applications (IS 3444 – 1978)										
1	HT	≥ 450	≥ 190	31	–	–	–	–	–	–
2	HT	≥ 480	≥ 190	31	–	–	–	–	–	–
3	HT	≥ 480	≥ 210	26	–	–	–	–	–	–

(Contd.)

TABLE 3.147 Mechanical properties[1)] of steel castings specified in Indian Standards (continued)

Grade	Treatment condition[2)]	R_m	R_e	A min.	Z min.	KV min.		Brinell hardness	Bend test	
							Test temperature		Angle of bend	Bend mandrel diameter
		MPa	MPa	%	%	J	°C	HB	°	mm
Corrosion-resistant steel and nickel-based castings for general engineering applications (IS 3444 – 1978)										
4	HT	≥ 480	≥ 210	26	–	–	–	–	–	–
4A	HT	≥ 480	≥ 210	26	–	–	–	–	–	–
5	HT	≥ 480	≥ 210	26	–	–	–	–	–	–
6	HT	≥ 480	≥ 210	22	–	–	–	–	–	–
6A	HT	≥ 480	≥ 210	22	–	–	–	–	–	–
7	HT	≥ 480	≥ 210	26	–	–	–	–	–	–
7A	HT	≥ 480	≥ 210	26	–	–	–	–	–	–
8	HT	≥ 450	≥ 190	26	–	–	–	–	–	–
9	HT	≥ 550	≥ 280	9	–	–	–	–	–	–
10	HT	≥ 620	≥ 450	16	30	–	–	–	–	–
11	HT	≥ 620	≥ 450	16	30	–	–	–	–	–
12	HT	≥ 450	≥ 210	–	–	–	–	–	–	–
13	HT	≥ 380	–	–	–	–	–	–	–	–
14	HT	≥ 690	≥ 480	14	25	–	–	–	–	–

(Contd.)

TABLE 3.147 Mechanical properties[1] of steel castings specified in Indian Standards (continued)

Grade	Treatment condition[2]	R_m	R_e	A min.	Z min.	KV min.		Brinell hardness	Bend test	
							Test temperature		Angle of bend	Bend mandrel diameter
		MPa	MPa	%	%	J	°C	HB	°	mm
Corrosion-resistant steel and nickel-based castings for general engineering applications (IS 3444 – 1978)										
15	HT	≥ 450	≥ 190	31	–	–	–	–	–	–
16	HT	≥ 480	≥ 210	26	–	–	–	–	–	–
17	HT	≥ 520	≥ 240	22	–	–	–	–	–	–
18	HT	≥ 430	≥ 170	31	–	–	–	–	–	–
19	HT	≥ 500	≥ 320	4	–	–	–	–	–	–
20	HT	≥ 480	≥ 190	26	–	–	–	–	–	–
21	HT	≥ 350	≥ 120	9	–	–	–	–	–	–
22	HT	≥ 450	≥ 210	22	–	–	–	–	–	–
23	HT	≥ 500	≥ 320	6	–	–	–	–	–	–
24	HT	≥ 760	≥ 550	14	35	–	–	–	–	–
Steel castings of high magnetic permeability (IS 4491 – 1994)										
1	A	340–430	≥ 190	22	40	–	–	–	120[5]	$3a$[3), 5)]
2	A	400–490	≥ 220	22	30	–	–	–	120[5]	$3a$[3), 5)]

(Contd.)

TABLE 3.147 Mechanical properties[1)] of steel castings specified in Indian Standards (continued)

Grade	Treatment condition[2)]	R_m	R_e	*A* min.	*Z* min.	*KV* min.		Brinell hardness	Bend test	
							Test temperature		Angle of bend	Bend mandrel diameter
		MPa	MPa	%	%	J	°C	HB	°	mm
1 % Chromium steel castings for resistance to abrasion (IS 4896 – 1992)										
1	A + N + T	≥ 690	10	–	–	–	200–250	–	–	
2	N + T, Q + T or A + Q + T	–	–	–	–	–	–	330–380	–	–
3		–	–	–	–	–	–	≥ 340	–	–
Steel castings for case carburizing (IS 4898 – 1989)										
1	Simulated case hardened	≥ 490	–	12	–	25	–	–	–	–
2		≥ 1000	–	7	–	20	–	–	–	–
Ferritic and martensitic steel castings for use at low temperature (IS 4899 – 1991)										
1	HT	415–585	≥ 205	23	35	18	–32	–	–	–
2	HT	450–620	≥ 240	23	35	18	–46	–	–	–
3	HT	485–655	≥ 275	21	35	20	–46	–	–	–
4	HT	450–620	≥ 240	23	35	18	–59	–	–	–
5	HT	485–655	≥ 275	23	35	20	–73	–	–	–
6	HT	725–895	≥ 550	17	30	41	–73	–	–	–
7	HT	485–655	≥ 275	23	35	20	–101	–	–	–

(Contd.)

TABLE 3.147 Mechanical properties[1)] of steel castings specified in Indian Standards (continued)

Grade	Treatment condition[2)]	R_m	R_e	*A* min.	*Z* min.	*KV* min.		Brinell hardness	Bend test	
							Test temperature		Angle of bend	Bend mandrel diameter
		MPa	MPa	%	%	J	°C	HB	°	mm
Ferritic and martensitic steel castings for use at low temperature (IS 4899 – 1991)										
8	HT	485–655	≥ 275	23	35	20	–115	–	–	–
9	HT	≥ 585	≥ 515	19	30	27	–196	–	–	–
10	HT	760–930	≥ 550	14	75	27	–73	–	–	–
Martensitic and austenitic high-alloy steel castings for high temperature corrosion services (IS 7806 – 1993)										
1	N + T	≥ 620	≥ 450	16	30	–	–	–	–	–
1A	N + T	≥ 755	≥ 550	15	35	–	–	–	–	–
2	ST	≥ 480	≥ 210	33	–	–	–	–	–	–
2A	ST	≥ 530	≥ 240	33	–	–	–	–	–	–
3	ST	≥ 480	≥ 210	33	–	–	–	–	–	–
3A	ST	≥ 530	≥ 240	33	–	–	–	–	–	–
4	ST	≥ 480	≥ 210	29	–	–	–	–	–	–
4A	ST	≥ 550	≥ 260	29	–	–	–	–	–	–
5	ST	≥ 480	≥ 210	29	–	–	–	–	–	–
5A	ST	≥ 480	≥ 210	29	–	–	–	–	–	–

(Contd.)

TABLE 3.147 Mechanical properties[1)] of steel castings specified in Indian Standards (continued)

Grade	Treatment condition[2)]	R_m	R_e	*A* min.	*Z* min.	*KV* min.		Brinell hardness	Bend test	
							Test temperature		Angle of bend	Bend mandrel diameter
		MPa	MPa	%	%	J	°C	HB	°	mm
Martensitic and austenitic high-alloy steel castings for high temperature corrosion services (IS 7806 – 1993)										
6	ST	≥ 450	≥ 190	29	–	–	–	–	–	–
7	ST	≥ 480	≥ 210	29	–	–	–	–	–	–
8	ST	≥ 480	≥ 210	29	–	–	–	–	–	–
9	ST	≥ 450	≥ 190	29	–	–	–	–	–	–
10	As-cast	≥ 450	≥ 240	10	–	–	–	–	–	–
11	As-cast	≥ 450	≥ 240	10	–	–	–	–	–	–
12	As-cast	≥ 450	≥ 190	15	–	–	–	–	–	–
13	ST	≥ 480	≥ 210	19	–	–	–	–	–	–
14	ST	≥ 430	≥ 170	33	–	–	–	–	–	–
15	ST	≥ 515	≥ 240	23	–	–	–	–	–	–
16	ST	≥ 690	≥ 485	15	–	–	–	–	–	–
17	As-cast	≥ 435	≥ 170	19	–	–	–	–	–	–
Alloy steel castings suitable for pressure service (IS 7899 – 1992)										
1A	HT	585–760	≥ 380	20	40	–	–	–	–	–

(Contd.)

TABLE 3.147 Mechanical properties[1)] of steel castings specified in Indian Standards (continued)

Grade	Treatment condition[2)]	R_m	R_e	*A* min.	*Z* min.	*KV* min.		Brinell hardness	Bend test	
							Test temperature		Angle of bend	Bend mandrel diameter
		MPa	MPa	%	%	J	°C	HB	°	mm
Alloy steel castings suitable for pressure service (IS 7899 – 1992)										
1B	HT	620–795	≥ 450	20	45	–	–	–	–	–
1C	HT	≥ 620	≥ 450	20	45	–	–	–	–	–
2A	HT	585–760	≥ 365	20	35	–	–	–	–	–
2B	HT	620–795	≥ 450	20	40	–	–	–	–	–
2C	HT	≥ 620	≥ 450	20	40	–	–	–	–	–
4A	HT	620–795	≥ 415	17	40	–	–	–	–	–
4B	HT	725–895	≥ 585	16	35	–	–	–	–	–
4C	HT	≥ 620	≥ 415	17	35	–	–	–	–	–
4D	HT	≥ 690	≥ 515	16	35	–	–	–	–	–
4E	HT	≥ 795	≥ 655	14	35	–	–	–	–	–
6A	HT	≥ 795	≥ 550	17	30	–	–	–	–	–
6B	HT	≥ 825	≥ 655	11	25	–	–	–	–	–
7A	HT	≥ 795	≥ 690	14	30	–	–	–	–	–
8A	HT	585–760	≥ 380	18	35	–	–	–	–	–

(Contd.)

TABLE 3.147 Mechanical properties[1)] of steel castings specified in Indian Standards (continued)

Grade	Treatment condition[2)]	R_m	R_e	A min.	Z min.	KV min.		Brinell hardness	Bend test	
							Test temperature		Angle of bend	Bend mandrel diameter
		MPa	MPa	%	%	J	°C	HB	°	mm
Alloy steel castings suitable for pressure service (IS 7899 – 1992)										
8B	HT	≥ 725	≥ 585	16	30	–	–	–	–	–
8C	HT	≥ 690	≥ 515	16	35	–	–	–	–	–
9A	HT	≥ 620	≥ 415	17	35	–	–	–	–	–
9B	HT	≥ 725	≥ 585	15	35	–	–	–	–	–
9C	HT	≥ 620	≥ 415	17	35	–	–	–	–	–
9D	HT	≥ 690	≥ 515	16	35	–	–	–	–	–
10A	HT	≥ 690	≥ 485	17	35	–	–	–	–	–
10B	HT	≥ 860	≥ 690	14	35	–	–	–	–	–
11A	HT	620–795	≥ 450	17	30	–	–	–	–	–
11B	HT	965–1170	760–895	9	25	–	–	–	–	–
11C	HT	≥ 620	≥ 415	17	35	–	–	–	–	–
11D	HT	≥ 690	≥ 515	16	35	–	–	–	–	–
12A	HT	620–795	≥ 450	17	30	–	–	–	–	–

(Contd.)

TABLE 3.147 Mechanical properties[1] of steel castings specified in Indian Standards (continued)

Grade	Treatment condition[2]	R_m	R_e	*A* min.	*Z* min.	*KV* min.		Brinell hardness	Bend test	
							Test temperature		Angle of bend	Bend mandrel diameter
		MPa	MPa	%	%	J	°C	HB	°	mm
Alloy steel castings suitable for pressure service (IS 7899 – 1992)										
13A	HT	760–930	≥ 515	14	35	–	–	–	–	–
13B	HT	≥ 690	≥ 520	16	35	–	–	–	–	–

1) Measured on test bars cast either separately from or attached to the castings they represent.
2) A = annealed; N = normalized; Q = quenched; T = tempered; HT = heat treated, ST = solution treated.
3) *a* = diameter or thickness of the bend test piece.
4) The Charpy (V–notch) impact test optional.
5) The bend test is optional.

3 REFERENCES

Books

1. *ASM Handbook*, vol. 1, ASM International, Materials Park, Ohio, U.S.A., 1990.
2. P.F. Wieser (Ed.), *Steel Castings Handbook*, 5th ed., Steel Founder's Society of America, Rocky River, Ohio, U.S.A., 1980.

Standards

3. IS 276 – 1992: *Austenitic Manganese Steel Castings—Specification.*
4. IS 1030 – 1989: *Carbon Steel Castings for General Engineering Purposes—Specification.*
5. IS 2644 – 1994: *High Strength Steel Castings for General Engineering and Structural Purposes—Specification.*
6. IS 2707 – 1989: *Carbon Steel Castings for Surface Hardening—Specification.*
7. IS 2708 – 1993: *1.5 Per Cent Manganese Steel Castings for General Engineering Purposes—Specification.*
8. IS 2856 – 1987: *Specification for Carbon Steel Castings for Pressure Containing Parts Suitable for High Temperature Service (Fusion Welding Quality).*
9. IS 3038 – 1983: *Specification for Alloy Steel Castings for Pressure Containing Parts Suitable for High Temperature Service.*
10. IS 3444 – 1978: *Specification for Corrosion Resistant Alloy Steel and Nickel Based Castings for General Applications.*
11. IS 4491 – 1994: *Steel Castings of High Magnetic Permeability—Specification.*
12. IS 4522 – 1979: *Specification for Heat Resistant Alloy Steel Castings.*
13. IS 4896 – 1992: *One Per Cent Chromium Steel Castings for Resistance to Abrasion—Specification.*
14. IS 4898 – 1989: *Steel Castings for Case Carburizing—Specification.*
15. IS 4899 – 1991: *Ferritic and Martensitic Steel Castings for Use at Low Temperature—Specification.*
16. IS 7806 – 1993: *Martensitic Steel and Austenitic High Alloy Steel Castings for High Temperature Corrosion Services—Specification.*
17. IS 7899 – 1992: *Alloy Steel Castings Suitable for Pressure Service—Specification.*

4

ALUMINIUM AND ITS ALLOYS

CHAPTER **4**

Aluminium and Its Alloys

4.1	Aluminium	4.5
4.2	Designation system for aluminium and aluminium alloys	4.8
4.3	Temper designation system for aluminium and aluminium alloys	4.12
4.4	Cast aluminium and aluminium alloys	4.16
4.5	Wrought aluminium and aluminium alloys	4.35
4.6	Heat treatment of aluminium and aluminium alloys	4.77

4.1

ALUMINIUM

1 INTRODUCTION

Pure aluminium is a silvery-white metal with a density of 2698.9 kg/m^3 at 20 °C, which is approximately one-third that of ferrous alloys, copper or brass. It is an excellent conductor of heat and electricity, and is also highly reflective to radiant energy. It exhibits excellent corrosion resistance in most environments—including atmosphere, water (including sea water), oils, and many chemicals. It is non-toxic, non-ferromagnetic, and also has non-sparking characteristics. A summary of some of the important physical properties of pure aluminium is given in Table 1.1.

Aluminium is a soft metal with a low tensile strength of 40–50 MPa in the annealed condition. However, by alloying, cold working or heat treatment, it can be made much stronger and harder.

Aluminium is used extensively:

a) in cast and wrought aluminium and aluminium alloys for use in building and construction applications, consumer durables, containers and packaging, electrical applications, machinery and equipment, transportation, etc;
b) as an alloying element in copper-base, magnesium-base and zinc-base alloys, and in nitriding and heat-resisting steels;
c) as a reducing agent;
d) as a deoxidizer in the manufacture of steel;
e) in aluminium coatings;
f) as sacrificial anodes for cathodic protection of metals and alloys;
g) in powders, paints and chemicals; and
h) in metal-matrix composites.

2 SPECIFICATIONS

TABLE 4.1 Chemical composition of grades of primary aluminium ingots for remelting specified in Indian Standards

Designation	Chemical composition [% (*m*/*m*)]								
	Al	Cu	Si	Fe	Mn	Mg	Zn	Ti + V	Ga	Other elements (each)
	min.	max.	max.	max.	max.	max.	max.	max.	max.	max.
Primary aluminium ingots for remelting (IS 2590 – 1987)										
19000	99.00	0.10	0.50	0.60	0.10	–	0.10	0.07	–	0.10
19500	99.50	0.05	0.30	0.40	0.05	–	0.10	0.07	–	0.07
19600	99.60	0.04	0.25	0.35	0.03	–	0.07	0.05	–	0.05
19700	99.70	0.03	0.20	0.25	0.03	–	0.06	0.05	–	0.05
High purity primary aluminium ingots for remelting (IS 11890 – 1987)										
19800	99.80	0.03	0.15	0.15	0.03	0.02	0.03	0.05	0.03	0.02
19850	99.85	0.03	0.10	0.12	0.02	0.02	0.03	0.05	0.03	0.01
19900	99.90	0.02	0.07	0.07	0.01	0.01	0.03	0.05	0.03	0.01
19950	99.95	0.010	0.030	0.040	0.010	0.010	0.010	0.010	0.010	0.005

3 EQUIVALENT GRADES

TABLE 4.2 Comparison of grades of primary aluminium ingots for remelting specified in Indian Standards with nearest equivalent grades specified in other national and international standards

IS 2590–87 IS 11890–87	ASTM	BS 1490–88	DIN 1712 (1)–76	GOST 11069–74	ISO R 115–68	JIS H 2102–68 JIS H 2111–68
Primary aluminium ingots for remelting (IS 2590 – 1987)						
19000	–	–	–	*A0*	Al 99.0	H 2102–Class 3
19500	–	LM0	Al99.5H	*A5*	Al 99.5	H 2102–Class 2
19600	–	–	–	*A6*	–	–
19700	–	–	Al99.7H	*A7*	Al 99.7	H 2102–Class 1
High purity aluminium ingots for remelting (IS 11890 – 1987)						
19800	–	–	Al99.8H	*A8*	Al 99.8	–
19850	–	–	–	*A85*	–	H 2102–Special Class 2
19900	–	–	Al99.9H	–	–	H 2102–Special Class 1
19950	–	–	–	*A95*	–	H 2111–Class 2

4 REFERENCES

Books

1. *ASM Handbook,* vol. 2, ASM International, Ohio, U.S.A., 1990.
2. A.S. Russell, 'Aluminium', *McGraw-Hill Encyclopedia of Science & Technology.* vol. 1, 5th ed., New York, U.S.A., 1982.

Standards

3. IS 2590 – 1987: *Specification for Primary Aluminium Ingots for Remelting for General Engineering Purposes.*
4. IS 11890 – 1987: *Specification for High Purity Primary Aluminium Ingots for Remelting for Special Applications.*

4.2

DESIGNATION SYSTEM FOR ALUMINIUM AND ALUMINIUM ALLOYS

1 DESIGNATION SYSTEM

The grades of wrought and cast aluminium and aluminium alloys specified in the Indian Standards are designated on the basis of their chemical composition.

1.1 Designation system for wrought aluminium and aluminium alloys

A five-digit numerical designation system is used for identifying wrought aluminium and aluminium alloys. The first digit in the designation indicates the alloy group and the major alloying element(s), as follows:

• Aluminium, ≥ 99.00 %	1xxxx
• Aluminium alloys grouped by major alloying elements	
– Copper	2xxxx
– Manganese	3xxxx
– Silicon	4xxxx
– Magnesium	5xxxx
– Magnesium and silicon	6xxxx
– Zinc	7xxxx
– Other elements	8xxxx
• Unused series	9xxxx

In the 1xxxx group, the second, third and fourth digits in the designation indicate the minimum aluminium percentage. The second digit in the designation is always 9, which is the digit to the left of the decimal point in the minimum aluminium percentage. The third and fourth digits are the same as the two digits to the right of the decimal point when it is expressed to the nearest 0.01%. The fifth digit, if other than zero, indicates modifications in impurity limits.

In the alloy groups 2xxxx through 8xxxx, the second digit in the designation indicates:

For the 2xxxx, 3xxxx, 5xxxx, 7xxxx and 8xxxx alloy groups: The specified mean percentage content of the major alloying element rounded to the nearest whole number;

For the 4xxxx alloy group: 0.5 times the specified mean percentage silicon content rounded to the nearest whole number; and

For the 6xxxx alloy group: Five times the specified mean percentage magnesium content rounded to the nearest whole number.

The third, fourth and fifth digits in the alloy designation indicate the minor alloying elements (see Table 4.3 for element identification numbers). The sequence of the numbers indicating the minor alloying elements are in decreasing order of the value of their content. Where the values of contents are the same for two or more elements, the corresponding element identification numbers are indicated in the same order as given in Table 4.3. For the alloy group 6xxxx, however, the third digit refers to either magnesium or silicon which is in excess of that required for the formation of magnesium silicide; where the alloy composition is balanced, the third digit is zero.

EXAMPLES

1. Wrought unalloyed aluminium containing Al $\geq$ 99.50 % is designated as 19500.
2. A wrought aluminium alloy containing Mg_{mean} = 0.65 % and Si_{mean} = 0.50 % is designated as 63400.

TABLE 4.3 Element identification numbers

Alloying element	Element identification number
Copper	2
Manganese	3
Silicon	4
Magnesium	5
Zinc	7
Other elements (such as bismuth, chromium, iron, lead, nickel, titanium, etc.)	8
None	0

1.2 Designation system for cast aluminium and aluminium alloys

A four-digit numerical designation system is used for identifying cast aluminium and aluminium alloys. The first digit in the designation indicates the alloy group and the major alloying element(s), as follows:

• Aluminium, ≥ 99.0 %	1xxx
• Aluminium alloys grouped by major alloying elements	
– Copper	2xxx
– Manganese	3xxx
– Silicon	4xxx
– Magnesium	5xxx
– Magnesium and silicon	6xxx
– Zinc	7xxx
– Other elements	8xxx
• Unused series	9xxx

In the 1xxx group, the second and third digits in the designation indicate the minimum aluminium percentage. The second digit in the designation is always 9, which is the digit to the left of the decimal point in the minimum aluminium percentage. The third digit is the same as the digit to the right of the decimal point when it is expressed to the nearest 0.1 %. The fourth digit, if other than zero, indicates modifications in impurity limits.

In the alloy groups 2xxx through 8xxx, the second digit in the designation indicates 0.5 times the specified mean percentage content of the major alloying element rounded to the nearest whole number. The third and fourth digits in the alloy designation indicate the minor alloying elements (see Table 4.3 for element identification numbers). The sequence of the numbers indicating the minor alloying elements are in decreasing order of the value of their content. Where the values of contents are the same for the two elements, the corresponding element identification numbers are indicated in the same order as given in Table 4.3.

EXAMPLES

1. Cast unalloyed aluminium containing Al ≥ 99.5 % is designated as 1950.
2. A cast aluminium alloy containing Si_{mean} = 10.25 % and Cu_{mean} = 1.6 % is designated as 4520.

2 REFERENCES

Standards

1. IS 6051 – 1970: *Code for Designation of Aluminium and its Alloys.*
2. ANSI H35.1(M) – 1993: *Alloy and Temper Designation Systems for Aluminium.*

4.3

Temper Designation System for Aluminium and Aluminium Alloys

1 TEMPER DESIGNATION SYSTEM

The temper designation system used in India for wrought and cast aluminium and aluminium alloy products, except ingots, is described in IS 5052 – 1993. The temper designations define the sequence of basic treatments used to produce the various tempers. The temper designation follows the material designation, and is separated from it by a hyphen.

Basic temper designations consist of individual capital letters. Subdivisions of the basic tempers, where required, are indicated by one or more digits following the letter of the basic temper. These digits designate specific sequence of basic treatments, but only those treatments or operations which significantly influence the product characteristics are recognized.

1.1 Basic temper designations

M—*As manufactured:* Applies to products which acquire some temper from hot shaping processes for which mechanical property limits apply.

F—*As fabricated:* Applies to products of shaping processes without special control over thermal conditions or strain-hardening and for which no mechanical property limits apply for wrought products.

O—*Annealed:* Applies to wrought products which are fully annealed to obtain the lowest strength condition, and to cast products which are annealed to improve ductility and dimensional stability.

H—*Strain-hardened (wrought products only):* Applies to products subjected to the application of cold work after annealing (or hot forming), or to a combination of cold work and partial annealing or stabilizing in order to secure the specified mechanical properties. The letter H is always followed by two or more digits.

T—*Thermally treated to produce tempers other than* M, F, O *or* H: Applies to products which have their strength increased by thermal treatment, with or without supplementary strain-hardening. The letter T is always followed by one or more digits.

1.2 Subdivisions of basic temper designations

1.2.1 Subdivisions of H temper: Strain-hardened

The first digit following the letter H indicates the specific combination of basic operations, as follows:

H1—*Strain-hardened only:* Applies to products which are strain-hardened to obtain the desired strength without supplementary thermal treatment. The digit following this designation indicates the degree of strain-hardening.

H2—*Strain-hardened and partially annealed:* Applies to products which are strain-hardened more than the desired final amount and then reduced in strength to the desired level by partial annealing. The digit following this designation indicates the degree of strain-hardening remaining after the product has been partially annealed.

H3—*Strain-hardened and stabilized:* Applies to products which are strain-hardened and whose mechanical properties are stabilized either by a low temperature thermal treatment or as a result of heat introduced during fabrication. Stabilization usually improves ductility. This designation is applicable only to those alloys which, unless stabilized, gradually age-soften at room temperature. The digit following this designation indicates the degree of strain-hardening after the stabilization treatment.

The final degree of strain-hardening is indicated by the following designation (the letter X represents 1, 2 or 3 as appropriate):

HX8 – full hard temper;
HX4 – tensile strength approximately midway between that of the O temper and that of the HX8 temper;
HX2 – tensile strength approximately midway between that of the O temper and that of the HX4 temper; and
HX6 – tensile strength approximately midway between that of the HX4 temper and that of the HX8 temper.

1.2.2 Subdivisions of T temper: Thermally treated

Numerals one through ten following the letter T indicate specific sequence of basic treatments, as follows:

T1—*Cooled from an elevated temperature shaping process and naturally aged to a substantially stable condition:* Applies to products which are not cold worked after cooling from an elevated temperature shaping process, or in which the effect of cold work in flattening or straightening may not be recognized in mechanical property limits.

T2—*Cooled from an elevated temperature shaping process, cold worked and naturally aged to a substantially stable condition:* Applies to products which are cold worked to improve strength after cooling from an elevated temperature shaping process, or in which the effect of cold work in flattening or straightening is recognized in mechanical property limits.

T3—*Solution heat-treated, cold worked and naturally aged to a substantially stable condition:* Applies to products which are cold worked to improve strength after solution heat-treatment, or in which the effect of cold work in flattening or straightening is recognized in mechanical property limits.

T4—*Solution heat-treated and naturally aged to a substantially stable condition:* Applies to products which are not cold worked after solution heat treatment, or in which the effect of cold work in flattening or straightening may not be recognized in mechanical property limits.

T5—*Cooled from an elevated temperature shaping process and then artificially aged:* Applies to products which are not cold worked after cooling from an elevated temperature shaping process, or in which the effect of cold work in flattening or straightening may not be recognized in mechanical property limits.

T6 —*Solution heat-treated and then artificially aged:* Applies to products which are not cold worked after solution heat treatment, or in which the effect of cold work in flattening or straightening may not be recognized in mechanical property limits.

T7—*Solution heat-treated and stabilized:* Applies to wrought products which are artificially aged after solution heat-treatment to carry them beyond the point of maximum strength to provide control of some special characteristics. Also applies to cast products that are artificially aged after solution heat treatment to provide dimensional and strength stability.

T8—*Solution heat-treated, cold worked and then artificially aged:* Applies to products which are cold worked to improve strength, or in which the effect of cold work in flattening or straightening is recognized in mechanical property limits.

T9—*Solution heat-treated, artificially aged and then cold worked:* Applies to products which are cold worked to improve strength after solution heat treatment and artificial ageing.

T10—*Cooled from an elevated temperature shaping process, cold worked and then artificially aged:* Applies to products which are cold worked to improve strength after cooling from an elevated temperature shaping process, or in which the effect of cold work in flattening or straightening is recognized in mechanical property limits.

2 EQUIVALENT TEMPER DESIGNATIONS

TABLE 4.4 Comparison of temper designations specified in Indian Standards with equivalent temper designations specified by American National Standards Institute (ANSI) and International Organization for Standardization (ISO)

IS	ANSI	ISO
M	H112	M
F	F	F

TABLE 4.4 Comparison of temper designations specified in Indian Standards with equivalent temper designations specified by American National Standards Institute (ANSI) and International Organization for Standardization (ISO) (continued)

IS	ANSI	ISO
O	O	O
H12, H22, H32	H12, H22, H32	H1B, H2B, H3B
H14, H24, H34	H14, H24, H34	H1D, H2D, H3D
H16, H26, H36	H16, H26, H36	H1F, H2F, H3F
H18, H28, H38	H18, H28, H38	H1H, H2H, H3H
–	H19, H29, H39	H1J, H2J, H3J
T1	T1	TA
T2	T2	TC
T3	T3	TD
T4	T4	TB
T5	T5	TE
T6	T6	TF
T7	T7	TM
T8	T8	TH
T9	T9	TL
T10	T10	TG

2 REFERENCES

Standards

1. IS 5052 – 1993: *Aluminium and Its Alloys—Temper Designations.*
2. ISO 2107 – 1983: *Aluminium, Magnesium and Their Alloys—Temper designations.*
3. ANSI H35.1 (M) – 1993: *Alloy and Temper Designation Systems for Aluminium.*

4.4

CAST ALUMINIUM AND ALUMINIUM ALLOYS

1 INTRODUCTION

1.1 Unalloyed aluminium (1xxx, series)

Unalloyed aluminium castings (Al ≥ 99.0 %) are characterized by low strength, high ductility, poor castability, excellent corrosion resistance and high electrical conductivity.

1.2 Aluminium-copper alloys (2xxx, series)

Aluminium-copper alloys are heat-treatable alloys which are characterized by medium to high strength, good elevated temperature properties and fair castability. The corrosion resistance of these alloys is lower than those of other casting alloys and protection by surface coatings is required in critical applications. Addition of magnesium to aluminium-copper alloys improves their response to natural ageing and also increases their maximum strength. Addition of nickel to aluminium-copper-magnesium alloys improves their strength and hardness at elevated temperatures.

1.3 Aluminium-silicon alloys (4xxx, series)

1.3.1 Hypoeutectic and eutectic aluminium-silicon alloys

Hypoeutectic and eutectic binary aluminium-silicon alloys are non-heat-treatable alloys, which are characterized by excellent casting characteristics, good pressure tightness, good corrosion resistance and good weldability. These alloys are more difficult to machine than the aluminium-copper and aluminium-magnesium alloys. The strength and ductility of these alloys can be improved through modification of the eutectic, which is generally accomplished by the addition of sodium or strontium. Addition of copper to aluminium-silicon alloys increases their strength and machinability, but reduces their ductility and corrosion resistance. Addition of magnesium to aluminium-silicon alloys makes these alloys heat treatable.

1.3.2 Hypereutectic aluminium-silicon alloys (Si >12 %)

Hypereutectic aluminium-silicon alloys are characterized by outstanding wear resistance, a low coefficient of thermal expansion, good elevated temperature strength and very good casting characteristics. The high silicon content in these alloys can give rise to machining problems and, therefore, special techniques with diamond tools should be used. These alloys are used in a condition in which the silicon phase is refined with phosphorus.

1.4 Aluminium-magnesium alloys (5xxx, series)

Aluminium-magnesium alloys are characterized by moderate-to-high strength and toughness, excellent corrosion resistance especially to seawater and marine atmospheres, good machinability, good weldability, relatively poor castability and an attractive appearance when anodized. Aluminium-magnesium alloys containing less than about 7% magnesium are not heat treated; however, the Al-10Mg cast alloy is used only in the heat treated condition. Addition of up to 0.75 % manganese to aluminium-magnesium alloys increases hardness, decreases ductility but has little effect on their resistance to corrosion. Addition of zinc or silicon to aluminium-magnesium alloys moderately improves the casting characteristics.

1.5 Aluminium-zinc-magnesium alloys (7xxx, series)

Aluminium-zinc-magnesium alloys are heat-treatable alloys which are characterized by moderately high strength, poor castability, good corrosion resistance despite some susceptibility to stress-corrosion cracking, good machinability and good finishing characteristics. These alloys are not recommended for service at elevated temperatures. Addition of copper, together with small amounts of chromium and manganese to aluminium-zinc-magnesium alloys helps achieve the highest strength aluminium casting alloys.

2 SPECIFICATION

2.1 Chemical Composition

TABLE 4.5 Chemical composition of aluminium and aluminium alloy ingots and castings, conforming to IS 617 – 1994

Designation	Chemical composition, [% (m/m)]										
	Al	Cu	Si	Mg	Fe	Mn	Ni	Zn	Pb	Sn	Ti
General purpose alloys											
Sand, gravity die, and investment casting alloys											
4223	Rem.	2.0–4.0	4.0–6.0	≤ 0.15	≤ 0.8	0.2–0.6	≤ 0.3	≤ 0.5	≤ 0.1	≤ 0.1	≤ 0.2
4423	Rem.	1.5–2.5	6.0–8.0	≤ 0.3	≤ 0.8	0.2–0.6	≤ 0.3	≤ 1.0	≤ 0.2	≤ 0.1	≤ 0.2
4450	Rem.	≤ 0.1	6.5–7.5	0.20–0.45	≤ 0.5	≤ 0.3	≤ 0.1	≤ 0.1	≤ 0.1	≤ 0.05	≤ 0.2[1)]
4528	Rem.	1.75–2.5	8.5–9.5	≤ 0.15	0.4–0.6	≤ 0.8	≤ 0.8	≤ 0.5	≤ 0.1	≤ 0.1	≤ 0.2
4600	Rem.	≤ 0.1	10.0–13.0	≤ 0.1	≤ 0.6	≤ 0.5	≤ 0.1	≤ 0.1	≤ 0.1	≤ 0.05	≤ 0.2
4600A	Rem.	≤ 0.4	10.0–13.0	≤ 0.2	≤ 1.0	≤ 0.5	≤ 0.1	≤ 0.2	≤ 0.1	≤ 0.01	≤ 0.2
Pressure die casting alloys											
4420	Rem.	3.0–4.0	7.5–9.5	≤ 0.3	≤ 1.3	≤ 0.5	≤ 0.5	≤ 3.0	≤ 0.3	≤ 0.2	≤ 0.2
4520	Rem.	0.7–2.5	9.0–11.5	≤ 0.3	≤ 1.0	≤ 0.5	≤ 0.5	≤ 2.0	≤ 0.3	≤ 0.2	≤ 0.2
4600	Rem.	≤ 0.1	10.0–13.0	≤ 0.1	≤ 0.6	≤ 0.5	≤ 0.1	≤ 0.1	≤ 0.1	≤ 0.05	≤ 0.2
4600A	Rem.	≤ 0.4	10.0–13.0	≤ 0.2	≤ 1.0	≤ 0.5	≤ 0.1	≤ 0.2	≤ 0.1	≤ 0.1	≤ 0.2
4628	Rem.	1.75–2.5	11.0–12.5	≤ 0.3	0.7–1.1	≤ 0.5	≤ 0.3	≤ 1.5	≤ 0.0.5	≤ 0.1	≤ 0.2

(Contd.)

TABLE 4.5 Chemical composition of aluminium and aluminium alloy ingots and castings, conforming to IS 617 – 1994 (continued)

Designation	Chemical composition, [% (m/m)]										
	Al	Cu	Si	Mg	Fe	Mn	Ni	Zn	Pb	Sn	Ti
Special purpose alloys											
1900	≥ 99.0	≤ 0.2	≤ 0.5	≤ 0.05	≤ 0.6	≤ 0.2	≤ 0.1	≤ 0.1	≤ 0.05	≤ 0.05	–
1950	≥ 99.5	≤ 0.03	≤ 0.3	≤ 0.03	≤ 0.4	≤ 0.03	≤ 0.03	≤ 0.07	≤ 0.03	≤ 0.03	–
2280	Rem.	4.0–5.0	≤ 0.25	≤ 0.1	≤ 0.25	≤ 0.1	≤ 0.1	≤ 0.1	≤ 0.05	≤ 0.05	0.2–0.3
2285[2]	Rem.	3.5–4.5	≤ 0.7	1.2–1.8	≤ 0.7	≤ 0.6	1.7–2.3	≤ 0.1	≤ 0.05	≤ 0.05	≤ 0.2
2550	Rem.	9.0–11.0	≤ 2.5	0.2–0.4	≤ 1.0	≤ 0.6	≤ 0.5	≤ 0.8	≤ 0.1	≤ 0.1	≤ 0.2
4223A	Rem.	2.8–3.8	4.0–6.0	≤ 0.05	≤ 0.6	0.2–0.6	≤ 0.2	≤ 0.15	≤ 0.1	≤ 0.05	≤ 0.2
4225	Rem.	1.0–1.5	4.5–6.0	0.3–0.6	≤ 0.8	≤ 0.5	≤ 0.3	≤ 0.5	≤ 0.2	≤ 0.1	≤ 0.2[1]
4300	Rem.	≤ 0.1	4.5–6.0	≤ 0.1	≤ 0.6	≤ 0.5	≤ 0.1	≤ 0.1	≤ 0.1	≤ 0.05	≤ 0.2
4323	Rem.	3.0–5.0	5.0–7.0	0.1–0.3	≤ 1.0	0.2–0.6	≤ 0.3	≤ 2.0	≤ 0.2	≤ 0.1	≤ 0.2
4525	Rem.	2.0–4.0	8.5–10.5	0.5–1.5	≤ 1.2	≤ 0.5	≤ 1.0	≤ 1.0	≤ 0.2	≤ 0.1	≤ 0.2
4635	Rem.	≤ 0.1	10.0–13.0	0.2–0.6	≤ 0.6	0.3–0.7	≤ 0.1	≤ 0.1	≤ 0.1	≤ 0.05	≤ 0.2
4652	Rem.	0.7–1.5	10.0–12.0	0.8–1.5	≤ 1.0	≤ 0.5	0.7–1.5	≤ 0.5	≤ 0.1	≤ 0.1	≤ 0.2
5230	Rem.	≤ 0.1	≤ 0.3	3.0–6.0	≤ 0.6	0.3–0.7	≤ 0.1	≤ 0.1	≤ 0.05	≤ 0.05	≤ 0.2
5500	Rem.	≤ 0.1	≤ 0.25	9.5–11.0	≤ 0.4	≤ 0.1	≤ 0.1	≤ 0.1	≤ 0.05	≤ 0.05	≤ 0.2

1) If Ti alone is used for grain refining, the amount present should be ≥ 0.05 %
2) The Cr content in this alloy should be ≤ 0.02 %.

2.2 Mechanical properties

TABLE 4.6 Mechanical properties[1] of cast aluminium alloys, conforming to IS 617 – 1994

Designation	Temper	R_m min.		$A^{2)}$ min.	
		Sand cast MPa	Chill cast MPa	Sand cast %	Chill cast %
General purpose alloys					
Sand, gravity die, and investment casting alloys					
4223	M	140	160	2	2
	T6	225	280	–	–
4423	M	140	160	1	2
4450	M	135	160	2	3
	T5	160	190	1	2
	T6	225	275	–	2
	T7	160	225	2.5	5
4528	M	150	220	1	1.5
	T5	140	200	1.5	3
	T6	–	320	–	2
4600	M	165	190	5	7
4600A	M	165	190	5	5
Pressure die casting alloys					
4420	M	–	180	–	1.5
4520	M	125	150	–	–
4600	M	165	190	5	7
4600A	M	165	190	5	5
4628	M	–	270	–	1.5

(Contd.)

TABLE 4.6 Mechanical properties[1] of cast aluminium alloys, conforming to IS 617–1994 (continued)

Designation	Temper	R_m min.		$A^{2)}$ min.	
		Sand cast MPa	Chill cast MPa	Sand cast %	Chill cast %
Special purpose alloys					
1900	M	–	–	–	–
1950	M	–	–	–	–
2280	T4	215	265	7	13
	T6	275	310	4	9
2285	T6	215	280	–	–
2550	M	–	170	–	–
4223A	T4	–	245	–	8
4225	T4	175	230	2	3
	T6	230	280	–	–
4300	M	120	140	3	4
4323	M	160	175	1	1
4525	T5	–	210	–	–
4635	M	–	190	–	3
	T5	170	230	1.5	2
	T6	240	295	–	–
4652	T5	–	210	–	–
	T6	175	280	–	–
	T7	140	200	–	–
5230	M	140	170	3	5

(Contd.)

TABLE 4.6 Mechanical properties[1] of cast aluminium alloys, conforming to IS 617 – 1994 (continued)

Designation	Temper	R_m min.		$A^{2)}$ min.	
		Sand cast MPa	Chill cast MPa	Sand cast %	Chill cast %
Special purpose alloys					
5500	T4	275	310	8	12

[1] Measured on separately cast test samples.
[2] $L_o = 5.65 \sqrt{S_o}$ or 50 mm.

3 TYPICAL PHYSICAL AND MECHANICAL PROPERTIES

TABLE 4.7 Typical physical and mechanical properties of cast aluminium and aluminium alloys, conforming to IS 617–1994

Designation	Temper	Density	Approximate melting range	Coefficient of thermal expansion between 20 °C and 100 °C	Thermal conductivity at 25 °C	Electrical conductivity at 20 °C	Electrical resistivity at 20 °C	Modulus of elasticity
		kg/m^3	°C	μm/(m K)	W/(m K)	% IACS	μΩ m	GPa
1900	–	–	–	–	–	–	–	–
1950	M	2700	643–657	24	209	57	0.0302	69
2280	–	2800	540–650	–	–	–	–	–
2285	–	2820	530–640	–	–	–	–	–
2550	M	2940	525–625	22	130	33	0.523	71
4223	M	2750	525–625	21	121	32	0.0539	71
4223A	T4	2770	525–625	21	121	32	0.039	71
4225	T4	2700	550–620	23	142	36	0.0478	71
4300	M	2690	565–625	22	142	37	0.0465	71
4323	M	2810	520–615	21	121	32	0.0539	71
4420	M	2790	520–580	21	96.3	24	0.0718	71
4423	M	2750	525–605	21	155	27	0.0639	71

(Contd.)

TABLE 4.7 Typical physical and mechanical properties of cast aluminium and aluminium alloys, conforming to IS 617–1994 (continued)

Designation	Temper	Density	Approximate melting range	Coefficient of thermal expansion between 20 °C and 100 °C	Thermal conductivity at 25 °C	Electrical conductivity at 20 °C	Electrical resistivity at 20 °C	Modulus of elasticity
		kg/m^3	°C	µm/(m K)	W/(m K)	% IACS	µΩ m	GPa
4450	M	2680	550–615	22	151	39	0.044	71
4520	M	2740	525–570	20	100	26	0.0663	71
4525	T5	2760	520–580	21	105	26	0.0663	71
4528	–	–	–	–	–	–	–	–
4600	M	2650	565–575	20	142	37	0.0465	71
4600A	M	2680	565–575	20	155	37	0.0465	71
4628	–	–	–	–	–	–	–	–
4635	M	2680	550–575	22	147	38	0.0455	71
4652	T5	2700	525–560	19	117	29	0.0595	71
5230	M	2650	580–642	23	138	31	0.0556	71
5500	T5	2570	450–620	25	87.9	20	0.0862	71

Source: References 3 and 4

4 RECOMMENDED CASTING TEMPERATURE

TABLE 4.8 Recommended casting temperature range for cast aluminium and aluminium alloys, conforming to IS 617 – 1994

Designation	Casting temperature °C	Designation	Casting temperature °C
1900	–	**4423**	680–740
1950	–	**4450**	680–740
2280	675–750	**4520**	–
2285	700–750	**4525**	670–740
2550	700–760	**4528**	–
4223	700–760	**4600**	710–740
4223A	700–740	**4600A**	680–740
4225	690–760	**4628**	–
4300	700–740	**4635**	690–740
4323	680–760	**4652**	680–760
4420	–	**5230**	680–740
		5500	680–720

Source: Reference 4

5 COMPARISON RATINGS OF CAST ALUMINIUM AND ALUMINIUM ALLOYS

TABLE 4.9 Comparison of the characteristics of cast aluminium and aluminium alloys, conforming to IS 617 – 1994

Designation	Suitability for					Fluidity	Resistance to hot tearing	Pressure tightness	Machin-ability	Corros-ion resistance
	Sand casting	Permanent mould casting	Gravity die casting	Low pressure die casting	High pressure die casting					
1900	–	–	–	–	–	–	–	–	–	–
1950	F	F	F	F	F	F	F	F	F	E
2280	F	P	–	–	–	F	P	P	G	F
2285	F	G	–	–	–	G	G	E	G	F
2550	F	G	G	P	1)	F	G	G	E	P
4223	E	G	E	G	G 1)	G	G	E	G	F
4223A	G 1)	G	G	G	1)	G	G	G	G	F
4225	G	G	G	G	1)	G	G	G	G	G
4300	G	G	–	–	–	G	E	E	F	E
4323	G	G	G	G	1)	G	G	G	G	F
4420	F 1)	F 1)	F	F	E	G	E	E	G	F
4423	E	E	E	G 1)	G 1)	G	E	E	G	G
4450	E	E	E	E	G 1)	G	E	E	G	E

(Contd.)

TABLE 4.9 Comparison of the characteristics of cast aluminium and aluminium alloys, conforming to IS 617 – 1994 (continued)

Designation	Suitability for					Fluidity	Resistance to hot tearing	Pressure tightness	Machin-ability	Corros-ion resistance
	Sand casting	Permanent mould casting	Gravity die casting	Low pressure die casting	High pressure die casting					
4520	G 1)	G 1)	G	G	E	G	E	E	G	F
4525	G	G	G	G 1)	1)	G	E	F	G	F
4528	–	–	–	–	–	–	–	–	–	–
4600	E	E	E	E	G	E	E	E	P	E
4600A	E 1)	E	E	E	E	E	E	E	P	G
4628	–	–	–	–	–	–	–	–	–	–
4635	G	E	G	E	1)	G	E	G	F	G
4652	G	G	G	G	1)	G	E	F	F	G
5230	F	F	F	F 1)	F	F	F	P	E	E
5500	F	F	–	–	–	F	G	P	G	E

1) Not usually cast by this method.

Key: E = excellent; G = good; F = fair; P = poor.

Source: References 3 and 4

6 TYPICAL APPLICATIONS

TABLE 4.10. Typical applications of cast aluminium and aluminium alloys specified in IS 617 – 1994

Designation	Typical applications
1900	Fittings for electrical transmission, food and chemical plant fittings.
1950	Fittings for electrical transmission, food and chemical plant fittings.
2280	Artificial limbs, flywheel housings, moulding boxes, propellers, reciprocating parts of engines.
2285	Air-cooled cylinder heads, aircraft generator housings, motocycle, diesel and aircraft pistons, other applications where excellent high-temperature strength is required.
2550	Air-cooled cylinder heads, castings for hydraulic equipment, pistons, valve tappet guides.
4223	Automobile engine and transmission components, domestic and office equipment, household fittings, electrical tools and switchgear.
4223A	Castings for heavy duty service in road transport vehicles, structural components.
4225	Air-compressor pistons, aircraft supercharger covers, fuel-pump bodies, liquid-cooled cylinder heads, master brake cylinders.
4300	Architectural castings, cooking utensils, electric floor polishers, food-handling equipment, marine fittings, steam-irons, vacuum cleaners, waffle irons.
4323	Heavy duty automotive parts.
4420	Floor polishers, parts for automotive and electrical industries such as motor frames and housings, vacuum cleaners.
4423	Automobile engine and transmission components, domestic and office equipment, household fittings, electrical tools and switchgear.

(Contd.)

TABLE 4.10 Typical applications of cast aluminium and aluminium alloys specified in IS 617–1994 (continued)

Designation	Typical applications
4450	Aircraft fittings and control parts, aircraft pump parts, automotive transmission cases, automotive wheels, cylinder heads, water-cooled cylinder blocks, radar and electronic housings.
4520	Barbeque covers, computer bases, escalator steps, lawn-mower decks, oil pumps, transmission cases.
4525	Automotive and compressor pistons.
4528	Automotive castings.
4600	'On-deck' castings and other marine applications, water-cooled manifolds and jackets, thin-walled and intricate instrument cases, switch boxes and motor housings, very large castings such as doors and panels, pumps and other equipment in the chemical and dye industries, castings used in the manufacture of paint and food and for a wide range of domestic tools and kitchen equipment.
4600A	The applications of this alloy are similar to those of 4600, except that the alloy has a corrosion resistance that is not quite as good as 4600. However, the alloy has a slightly better castability than 4600 and its higher iron content reduces the problem of molten metal die interaction known as 'welding-on'.
4628	Thin-wall automotive die castings.
4635	Low pressure die castings such as scooter parts.
4652	Pistons, mainly for internal combustion engines.
5230	Castings for marine, food processing and decorative applications.
5500	Aircraft fittings, rail-road passenger-car frames, miscellaneous castings requiring strength and shock resistance.

7 EQUIVALENT GRADES

TABLE 4.11 Comparison of grades of cast aluminium and aluminium alloys ingots specified in IS 617 – 1994 with nearest equivalent grades specified in other national and international standards

IS 617–94	ASTM B 179–95a	BS 1490–88	DIN 1725 (5)–86	GOST	ISO	JIS H 2211–92 JIS H 2118–90
General purpose alloys						
Sand, gravity die, and investment casting alloys						
4223	–	LM4	–	–	–	–
4423	–	LM27	–	–	–	–
4450	356.1	LM25	GB–AlSi7Mg	–	–	AC4C.1
4528	–	–	–	–	–	–
4600	–	LM6	GB–AlSi12	–	–	AC3A.1
4600A	–	LM20	GB–AlSi12(Cu)	–	–	–
Pressure die casting alloys						
4420	A380.1	LM24	–	–	–	AD 10Z.1
4520	383.1	LM2	–	–	–	AD 12Z.1
4600	A413.2	LM6	GBD–AlSi12	–	–	AD 1.2
4600A	A413.1	LM20	GBD–AlSi12(Cu)	–	–	AD 1.1
4628	–	–	–	–	–	–
Special purpose alloys						
1900	–	–	–	–	–	–
1950	–	LM0	–	–	–	–
2280	–	–	GB–AlCu4Ti	–	–	–
2285	242.1	–	–	–	–	AC5A.1
2550	222.1	LM12	–	–	–	–

(Contd.)

TABLE 4.11 Comparison of grades of cast aluminium and aluminium alloys ingots specified in IS 617–1994 with nearest equivalent grades specified in other national and international standards (continued)

IS 617–94	ASTM B 179–95a	BS 1490–88	DIN 1725 (5)–86	GOST	ISO	JIS H 2211–92 JIS H 2118–90
Special purpose alloys						
4223A	–	LM22	–	–	–	–
4225	355.1	LM16	–	–	–	AC4D.1
4300	B443.1	–	–	–	–	–
4323	–	LM21	GB–AlSi6Cu4	–	–	–
4525	332.1	LM26	–	–	–	AC8C.1
4635	–	LM9	–	–	–	–
4652	–	LM13	–	–	–	AC8A.1
5230	–	LM5	–	–	–	–
5500	520.2	–	–	–	–	–

TABLE 4.12 Comparison of grades of aluminium and aluminium alloy castings specified in IS 617–1994 with nearest equivalent grades specified in other national and international standards

IS 617–94	ASTM B 26/B 26M–95 ASTM B 85–95 ASTM B 108–95	BS 1490–88	DIN 1725 (2)–86	GOST 2685–75	ISO 3522–85	JIS H 5202–92 H 5302–90	SAE J452–89
General purpose alloys							
Sand, gravity die, and investment casting alloys							
4223	–	LM4	–	–	Al-Si5 Cu3	–	–
4423	–	LM27	–	–	–	–	–
4450	356.0	LM25	G-/GK-/GF-AlSi7Mg	AЛ9	Al-Si7 Mg (Fe)	AC 4 C	356.0
4528	–	–	–	–	–	–	–
4600	–	LM6	G-/GK-AlSi12	–	Al-Si12	AC 3 A	–
4600A	–	LM20	G-/GK-AlSi12(Cu)	AЛ2	Al-Si12 Cu	–	–
Pressure die casting alloys							
4420	A380.0	LM24	–	–	Al-Si8 Cu3 Fe	ADC 10 Z	A380.0
4520	383.0	LM2	–	–	–	ADC 12 Z	383.0
4600	–	LM6	GD-AiSi12	–	Al-Si12 Fe	–	–
4600A	A413.0	LM20	GD-AlSi12 (Cu)	AЛ2	Al-Si12 Cu Fe	ADC 1	A413.0
4628	–	–	–	–	–	–	–

(Contd.)

TABLE 4.12 Comparison of grades of aluminium and aluminium alloy castings specified in IS 617–1994 with nearest equivalent grades specified in other national and international standards (continued)

IS 617–94	ASTM B 26/B 26M–95 ASTM B 85–95 ASTM B 108–95	BS 1490–88	DIN 1725 (2)–86	GOST 2685–75	ISO 3522–85	JIS H 5202–92 H 5302–90	SAE J452–89
Special purpose alloys							
1900	–	–	–	–	–	–	–
1950	–	LM0	–	–	–	–	–
2280	–	–	G-/GK-AlCu4Ti	–	Al-Cu4 Ti	–	–
2285	242.0	–	–	АЛ1	Al-Cu4 Ni2 Mg2	AC 5 A	242.0
2550	222.0	LM12	–	–	–	–	222.0
4223A	–	LM22	–	–	Al-Si5 Cu3	–	–
4225	355.0	LM16	–	АЛ5	Al-Si5 Cu1 Mg	AC 4 D	355.0
4300	B443.0	–	–	–	Al-Si5	–	B443.0
4323	–	LM21	G-/GK-AlSi6Cu4	–	Al-Si6 Cu4	–	–
4525	332.0	LM26	–	–	–	AC 8 C	332.0
4635	–	LM9	–	–	–	–	–
4652	–	LM13	–	АЛ30	–	AC 8 A	–
5230	–	LM5	–	–	–	–	–
5500	520.0	–	–	АЛ8	Al-Mg10	–	520.0

8 REFERENCES

Books

1. *ASM Handbook*, vol. 2, ASM International, Materials Park, Ohio, U.S.A., 1990.
2. J.E. Hatch (Ed.), *Aluminum: Properties and Physical Metallurgy*, American Society for Metals, Metals Park, Ohio, U.S.A., 1984.
3. *The Properties of Aluminium and Its Alloys*, 9th ed., The Aluminium Federation, Birmingham, U.K., 1993.
4. C.J. Smithells (Ed.), *Metals Reference Book*, 5th ed., Butterworths, London, U.K., 1976.

Standards

5. IS 617 – 1994: *Cast Aluminium and Its Alloys—Ingots And Castings For General Engineering Purposes—Specification.*
6. BS 1490 – 1988: Specification for Aluminium and aluminium alloy ingots and castings for general engineering purposes.

4.5

WROUGHT ALUMINIUM AND ALUMINIUM ALLOYS

1 INTRODUCTION

1.1 Unalloyed aluminium (1xxxx series)

Wrought unalloyed aluminium (Al ≥ 99.00 %) is characterized by low strength, excellent workability, excellent corrosion resistance, and high thermal and electrical conductivity. It can be strengthened moderately by cold working.

1.2 Aluminium-copper alloys (2xxxx series)

Aluminium–copper alloys are heat-treatable alloys which are characterized by high strength, good elevated temperature properties and limited weldability. Their corrosion resistance is lower than that of most other wrought aluminium alloys and under certain conditions are prone to intergranular corrosion. Therefore, these alloys in the form of plate, sheet or strip are clad with high-purity aluminium or with an aluminium-magnesium-silicon alloy (6xxxx series) to improve their corrosion resistance. Addition of magnesium to aluminium-copper alloys improves their response to natural ageing and also increases their maximum strength. Addition of nickel to these alloys improves their strength and hardness at elevated temperatures.

1.3 Aluminium-manganese and aluminium-manganese-magnesium alloys (3xxxx series)

Aluminium-manganese and aluminium–manganese–magnesium alloys are non-heat treatable alloys which are characterized by moderate strength (about 20% higher than that offered by unalloyed aluminium), good workability and good corrosion resistance.

1.4 Aluminium-silicon alloys (4xxxx series)

Aluminium–silicon alloys are non-heat-treatable alloys which are used in welding wire and as brazing alloys where a lower melting point than that of the parent metal is required.

1.5 Aluminium-magnesium alloys (5xxxx series)

Aluminium–magnesium alloys are non-heat treatable alloys which are characterized by medium to high strength, good forming and welding characteristics, good finishing characteristics, and good corrosion resistance in marine atmospheres. Addition of small amounts of chromium, manganese or titanium to these alloys improves their strength.

1.6 Aluminium-magnesium-silicon alloys (6xxxx series)

Aluminium–magnesium–silicon alloys are heat-treatable alloys which are characterized by medium strength (although not as strong as the aluminium–copper and aluminium–zinc alloys) good formability, good weldability, good machinability, and good corrosion resistance. Addition of chromium, copper and manganese, singly or in combination, further increases their strength.

1.7 Aluminium-zinc-magnesium and aluminium-zinc-magnesium-copper alloys (7xxxx series)

Aluminium–zinc–magnesium and aluminium–zinc–magnesium–copper alloys are heat-treatable alloys which are characterized by medium to high strength, good fabrication characteristics but unsatisfactory resistance to stress-corrosion cracking. Addition of small amounts of chromium or overageing decreases their susceptibility to stress-corrosion cracking.

2 SPECIFICATIONS

2.1 Plate

2.1.1 Chemical composition

TABLE 4.13 Chemical composition of wrought aluminium and aluminium alloy plate, conforming to IS 736 – 1986

Designation	Chemical composition [% (*m*/*m*)]								
	Al	Cu	Mg	Si	Fe	Mn	Zn	Ti[1)]	Cr	Remarks
19000	≥ 99.00	≤ 0.1	≤ 0.2	≤ 0.5	≤ 0.7	≤ 0.1	≤ 0.1	–	–	2)
19500	≥ 99.50	≤ 0.05	–	≤ 0.3	≤ 0.4	≤ 0.05	≤ 0.1	–	–	3)
19600	≥ 99.60	≤ 0.05	–	≤ 0.25	≤ 0.35	≤ 0.03	≤ 0.06	–	–	4)
19700	≥ 99.70	≤ 0.03	–	≤ 0.2	≤ 0.25	≤ 0.03	≤ 0.06	–	–	5)
19800	≥ 99.80	≤ 0.03	–	≤ 0.15	≤ 0.15	≤ 0.03	≤ 0.06	–	–	6)

(Contd.)

TABLE 4.13 Chemical composition of wrought aluminium and aluminium alloy plate, conforming to IS 736 – 1986 (continued)

Designation	Chemical composition [% (*m*/*m*)]								
	Al	Cu	Mg	Si	Fe	Mn	Zn	Ti[1)]	Cr	Remarks
24345	Rem.	3.8–5.0	0.2–0.8	0.5–1.2	≤0.7	0.3–1.2	≤0.2	≤0.3[7)]	≤0.3[7)]	–
Alclad 24345	Rem.	3.8–5.0	0.2–0.8	0.5–1.2	≤0.7	0.3–1.2	≤0.2	≤0.3[7)]	≤0.3[7)]	–
31000	Rem.	≤0.1	≤0.1	≤0.6	≤0.7	0.8–1.5	≤0.2	≤0.2	≤0.2	–
40800	≥98.00	≤0.2	≤0.1	0.6–0.95	0.6–0.95	≤0.1	≤0.2	≤0.2	–	–
51000-A	Rem.	≤0.2	0.5–1.1	≤0.6	≤0.7	≤0.2	≤0.25	–	≤0.1	–
51000-B	Rem.	≤0.2	1.1–1.8	≤0.6	≤0.7	≤0.2	≤0.25	–	≤0.1	–
52000	Rem.	≤0.1	1.7–2.6	≤0.6	≤0.7	≤0.5	≤0.2	–	≤0.25	8)
53000	Rem.	≤0.1	2.8–4.0	≤0.6	≤0.7	≤0.5	≤0.2	≤0.2	≤0.25	8)
54300	Rem.	≤0.1	4.0–4.9	≤0.4	≤0.7	0.5–1.0	≤0.2	≤0.2	≤0.25	–
55000	Rem.	≤0.1	4.5–5.5	≤0.6	≤0.7	≤0.5	≤0.2	≤0.2	≤0.25	8)
64430	Rem.	≤0.1	0.4–1.2	0.6–1.3	≤0.6	0.4–1.0	≤0.1	≤0.2	≤0.25	–
65032	Rem.	0.15–0.4	0.7–1.2	0.4–0.8	≤0.7	0.2–0.8	≤0.2	≤0.2	0.15–0.35	9)
74530	Rem.	≤0.2	1.0–1.5	≤0.4	≤0.7	0.2–0.7	4.0–5.0	≤0.2	≤0.2	–

1) Ti and/or other grain refining elements.
2) Cu + Mg + Si + Fe + Mn + Zn ≤ 1.0 %.
3) Cu + Si + Fe + Mn + Zn ≤ 0.5 %.
4) Cu + Si + Fe + Mn + Zn ≤ 0.4 %.
5) Cu + Si + Fe + Mn + Zn ≤ 0.3 %.
6) Cu + Si + Fe + Mn + Zn ≤ 0.2 %.
7) Ti[1)] + Cr ≤ 0.3 %.
8) Mn + Cr ≤ 0.5 %.
9) Either Mn or Cr should be present.

2.1.2 Mechanical properties

TABLE 4.14 Mechanical properties of wrought aluminium and aluminium alloy plate, conforming to IS 736 – 1986

Designation	Temper	R_m	$R_{p0.2}$ min.	$A_{50\,mm}$ min. for a thickness, in mm, of	
				> 6.0 ≤ 12.5	> 12.5 ≤ 25.0
		MPa	MPa	%	%
19000	F	≥ 70	–	28	28
	O	70–110	–	28	28
	HX4	110–140	–	7	–
19500	F	≥ 65	–	28	28
	O	≤ 100	–	30	30
	HX4	100–135	–	7	–
19600	F	≥ 60	–	28	28
	O	≤ 100	–	30	30
	HX4	95–125	–	7	–
19700	F	≥ 60	–	30	30
	O	≤ 95	–	34	34
	HX4	90–115	–	8	–
19800	F	≥ 55	–	30	30
	O	≤ 90	–	34	34
	HX4	85–110	–	8	–
24345	F	≥ 150	–	10	10
	O	≤ 240	–	12	10
	T4	≥ 390	225	12	10
	T6	≥ 405	310	7	5

(Contd.)

TABLE 4.14 Mechanical properties of wrought aluminium and aluminium alloy plate, conforming to IS 736 – 1986 (continued)

Designation	Temper	R_m	$R_{p0.2}$ min.	A_{50mm} min. for a thickness, in mm, of	
				>6.0≤12.5	>12.5≤25.0
		MPa	MPa	%	%
Alclad 24345	F	≥150	–	10	10
	O	≤240	–	12	10
	T4	≥375	230	12	10
	T6	≥420	345	7	5
31000	F	≥ 95	–	23	23
	O	90–130	–	22	23
	HX4	130–180	–	5	–
40800	F	≥ 90	–	28	28
	O	85–120	–	30	30
	HX4	120–160	–	7	–
51000-A	F	≥105	–	20	20
	O	95–145	–	22	22
	HX4	140–190	–	4	–
51000-B	F	≥135	–	17	17
	O	125–170	–	19	19
	HX4	≥170	–	4	–
52000	F	≥190	–	12	12
	O	175–215	60	18	18
	HX4	200–255	160	5	–

(Contd.)

TABLE 4.14 Mechanical properties of wrought aluminium and aluminium alloy plate, conforming to IS 736 – 1986 (continued)

Designation	Temper	R_m	$R_{p0.2}$ min.	A_{50mm} min. for a thickness, in mm, of	
				> 6.0 ≤ 12.5	> 12.5 ≤ 25.0
		MPa	MPa	%	%
53000	F	≥ 215	–	12	12
	O	205–265	85	16	16
	HX4	280–325	190	5	–
54300	F	≥ 280	125	12	12
	O	270–340	115	16	16
	HX4	≥ 355	275	5	–
55000	F	≥ 275	125	–	–
	O	265–355	100	16	16
	HX4	≥ 355	275	5	–
64430	F	≥ 110	–	12	12
	O	≤ 150	–	15	15
	T4	≥ 200	115	15	15
	T6	≥ 285	240	8	8
65032	F	≥ 110	–	12	12
	O	≤ 150	–	15	15
	T4	≥ 200	110	15	15
	T6	≥ 280	235	8	8
74530	T4[1)]	≥ 265	160	8	8
	T6	≥ 305	255	7	7

1) Naturally aged for 30 days.

Note: The mechanical properties in the F temper are for information only.

2.2 Sheet and strip

2.2.1 Chemical composition

TABLE 4.15 Chemical composition of wrought aluminium and aluminium alloy sheet and strip, conforming to IS 737 – 1986

Designation	Chemical composition [% (*m*/*m*)]								
	Al	Cu	Mg	Si	Fe	Mn	Zn	Ti[1)]	Cr	Remarks
19000	≥ 99.00	≤0.1	≤0.2	≤0.5	≤0.7	≤0.1	≤0.1	–	–	2)
19500	≥ 99.50	≤0.05	–	≤0.3	≤0.4	≤0.05	≤0.1	–	–	3)
19600	≥ 99.60	≤0.05	–	≤0.25	≤0.35	≤0.03	≤0.06	–	–	4)
19700	≥ 99.70	≤0.03	–	≤0.2	≤0.25	≤0.03	≤0.06	–	–	5)
19800	≥ 99.80	≤0.03	–	≤0.15	≤0.15	≤0.03	≤0.06	–	–	6)
19990	≥ 99.99	–	–	–	–	–	–	–	–	7)
24345	Rem.	3.8–5.0	0.2–0.8	0.5–1.2	≤0.7	0.3–1.2	≤0.2	≤0.3[8)]	≤0.3[8)]	–
Alclad 24345	Rem.	3.8–5.0	0.2–0.8	0.5–1.2	≤0.7	0.3–1.2	≤0.2	≤0.3[8)]	≤0.3[8)]	–
31000	Rem.	≤0.1	≤0.1	≤0.6	≤0.7	0.8–1.5	≤0.2	≤0.2	≤0.2	–
31500	Rem.	≤0.2	0.6–1.3	≤0.4	≤0.7	1.0–1.5	≤0.2	≤0.2	–	–
40800	≥ 98.00	≤0.2	≤0.1	0.6–0.95	0.6–0.95	≤0.1	≤0.2	≤0.2	–	–
51000-A	Rem.	≤0.2	0.5–1.1	≤0.6	≤0.7	≤0.2	≤0.25	–	≤0.1	–
51000-B	Rem.	≤0.2	1.1–1.8	≤0.6	≤0.7	≤0.2	≤0.25	–	≤0.1	–
51300	Rem.	≤0.3	0.2–0.9	≤0.6	≤0.9	0.2–0.7	≤0.4	≤0.2	≤0.2	–
52000	Rem.	≤0.1	1.7–2.6	≤0.6	≤0.7	≤0.5	≤0.2	≤0.2	≤0.25	9)
53000	Rem.	≤0.1	2.8–4.0	≤0.6	≤0.7	≤0.5	≤0.2	≤0.2	≤0.25	9)
54300	Rem.	≤0.1	4.0–4.9	≤0.4	≤0.7	0.5–1.0	≤0.2	≤0.2	≤0.25	–
55000	Rem.	≤0.1	4.5–5.5	≤0.6	≤0.7	≤0.5	≤0.2	≤0.2	≤0.25	9)
64430	Rem.	≤0.1	0.4–1.2	0.6–1.3	≤0.6	0.4–1.0	≤0.1	≤0.2	≤0.25	–

(Contd.)

TABLE 4.15 Chemical composition of wrought aluminium and aluminium alloy sheet and strip, conforming to IS 737 – 1986 (continued)

Designation	Chemical composition [% (m/m)]								
	Al	Cu	Mg	Si	Fe	Mn	Zn	Ti[1)]	Cr	Remarks
65032	Rem.	0.15–0.4	0.7–1.2	0.4–0.8	≤0.7	0.2–0.8	≤0.2	≤0.2	0.15–0.35	10)
74530	Rem.	≤0.2	1.0–1.5	≤0.4	≤0.7	0.2–0.7	4.0–5.0	≤0.2	≤0.2	–

1) Ti and/or other grain refining elements.
2) Cu + Mg + Si + Fe + Mn + Zn ≤ 1.0 %.
3) Cu + Si + Fe + Mn + Zn ≤ 0.5 %.
4) Cu + Si + Fe + Mn + Zn ≤ 0.4 %.
5) Cu + Si + Fe + Mn + Zn ≤ 0.3 %.
6) Cu + Si + Fe + Mn + Zn ≤ 0.2 %.
7) Cu + Si + Fe ≤ 0.01 %
8) Ti[1)] + Cr ≤ 0.3 %.
9) Mn + Cr ≤ 0.5 %.
10) Either Mn or Cr should be present.

2.2.2 Mechanical properties

TABLE 4.16 Mechanical properties of wrought aluminium and aluminium alloy sheet and strip, conforming to IS 737 – 1986

Designation	Temper	R_m	$R_{p0.2}$	$A_{50\,mm}$[1)] min. for a thickness, in mm, of				Bend mandrel diameter (180° bend)[2), 3), 4)]
				> 0.5 ≤ 0.8	> 0.8 ≤ 1.3	> 1.3 ≤ 2.6	> 2.6 ≤ 6.3	
		MPa	MPa	%	%	%	%	
19000	O	70–110	–	20	25	29	30	0a
	HX2	90–130	–	5	6	8	8	0a
	HX4	105–140	–	3	4	5	5	1a
	HX6	125–150	–	2	3	4	4	1a
	HX8	≥ 140	–	2	2	3	3	2a
19500	O	55–95	–	22	25	29	30	0a
	HX4	95–135	–	4	5	6	6	1a
	HX8	≥ 135	–	3	3	4	4	2a

(Contd.)

TABLE 4.16 Mechanical properties of wrought aluminium and aluminium alloy sheet and strip, conforming to IS 737 – 1986 (continued)

Designation	Temper	R_m	$R_{p0.2}$	$A_{50\,mm}$ [1)] min. for a thickness, in mm, of				Bend mandrel diameter (180° bend)[2), 3), 4)]
				$>0.5 \le 0.8$	$>0.8 \le 1.3$	$>1.3 \le 2.6$	$>2.6 \le 6.3$	
		MPa	MPa	%	%	%	%	
19600	O	≤95	–	25	25	29	32	0a
	HX4	95–125	–	4	5	6	6	1a
	HX8	≥125	–	3	3	4	4	2a
19700	O	≤95	–	27	27	29	34	0a
	HX4	95–120	–	4	5	6	7	1a
	HX8	≥120	–	3	3	4	4	2a
19800	O	≤90	–	29	29	29	34	0a
	HX4	90–120	–	5	6	7	8	1a
	HX8	≥120	–	3	4	4	5	2a
19990	O	≤65	–	30	35	40	45	0a
	HX4	80–100	–	7	8	10	12	1a
	HX8	≥100	–	3	4	5	6	2a
24345	O	≤240		14	14	14	14	0a
	T4	≥380	≥ 240	13	14	14	14	6a
	T6	≥425	≥ 345	6	6	6	6	10a
Alclad 24345	O	≤240	–	14	14	14	14	0a
	T4	≥370	≥ 225	13	14	14	14	6a
	T6	≥395	≥ 320	6	6	6	7	10a

TABLE 4.16 Mechanical properties of wrought aluminium and aluminium alloy sheet and strip, conforming to IS 737 – 1986 (continued)

Designation	Temper	R_m	$R_{p0.2}$	$A_{50\,mm}$ [1)] min. for a thickness, in mm, of				Bend mandrel diameter (180° bend)[2), 3), 4)]
				> 0.5 ≤ 0.8	> 0.8 ≤ 1.3	> 1.3 ≤ 2.6	> 2.6 ≤ 6.3	
		MPa	MPa	%	%	%	%	
31000	O	90–130	–	20	23	24	24	0a
	HX2	115–150	–	5	6	7	8	0a
	HX4	130–180	–	3	4	5	5	1a
	HX6	150–195	–	2	3	4	4	2a
	HX8	≥ 170	–	2	2	3	3	6a
31500	O	125–165	–	16	16	18	20	0a
	HX2	150–210	–	5	5	6	8	1a
	HX4	190–245	–	3	4	5	5	2a
	HX6	215–275	–	2	2	3	4	4a
	HX8	≥ 245	–	1	1	1	2	8a
40800	O	85–120	–	20	23	25	30	0a
	HX2	105–140	–	5	6	7	8	0a
	HX4	125–160	–	3	4	5	5	1a
	HX6	150–180	–	2	3	4	4	2a
	HX8	≥ 175	–	2	2	3	3	6a
51000-A	O	105–150	–	18	18	18	22	0a
	HX2	120–160	≥ 85	3	4	5	6	0a
	HX4	140–180	≥ 105	2	3	4	4	1a
	HX6	160–200	≥ 125	1	2	2	2	2a
	HX8	≥ 185	–	1	2	2	2	6a

(Contd.)

TABLE 4.16 Mechanical properties of wrought aluminium and aluminium alloy sheet and strip, conforming to IS 737 – 1986 (continued)

Designation	Temper	R_m	$R_{p0.2}$	$A_{50\,mm}$[1)] min. for a thickness, in mm, of				Bend mandrel diameter (180° bend)[2), 3), 4)]
				$>0.5 \le 0.8$	$>0.8 \le 1.3$	$>1.3 \le 2.6$	$>2.6 \le 6.3$	
		MPa	MPa	%	%	%	%	
51000-B	O	125–170	–	18	18	18	19	0a
	HX2	155–195	≥ 110	3	4	5	6	1a
	HX4	175–215	≥ 140	2	3	4	4	2a
	HX6	190–225	≥ 155	1	2	2	2	–
	HX8	≥ 200	–	1	2	2	2	–
51300	O	95–145	–	14	14	15	16	0a
	HX2	130–180	–	6	6	7	8	1a
	HX4	150–200	–	4	4	5	6	2a
	HX6	175–215	–	2	2	3	4	4a
	HX8	≥ 195	–	1	1	1	2	8a
52000	O	175–215	≥ 60	16	16	16	18	0a
	HX2	200–240	≥ 125	3	4	5	6	1a
	HX4	230–275	≥ 175	2	2	3	4	2a
	HX6	235–295	≥ 190	2	2	3	–	–
	HX8	≥ 265	≥ 215	1	2	3	–	–
53000	O	210–270	≥ 85	12	14	16	18	0a
	HX2	240–290	≥ 160	4	5	6	7	2a
	HX4	270–320	≥ 220	3	3	5	5	4a
	HX6	290–340	≥ 225	2	2	4	–	–
	HX8	≥ 310	≥ 235	2	2	3	–	–

(Contd.)

TABLE 4.16 Mechanical properties of wrought aluminium and aluminium alloy sheet and strip, conforming to IS 737 – 1986 (continued)

Designation	Temper	R_m	$R_{p0.2}$	$A_{50\,mm}$[1] min. for a thickness, in mm, of				Bend mandrel diameter (180° bend)[2), 3), 4)]
				> 0.5 ≤ 0.8	> 0.8 ≤ 1.3	> 1.3 ≤ 2.6	> 2.6 ≤ 6.3	
		MPa	MPa	%	%	%	%	
54300	O	265–365	≥ 130	12	14	16	16	0a
	HX2	315–395	≥ 235	5	6	7	7	4a
	HX4	≥ 355	≥ 275	4	4	5	5	6a
55000	O	265–365	≥ 130	12	14	16	16	0a
	HX2	310–395	≥ 220	5	6	7	7	4a
64430	O	≤ 175	–	14	16	16	17	0a
	T4	≥ 200	≥ 115	12	15	15	15	4a
	T6	≥ 295	≥ 250	5	5	5	6	6a
65032	O	≤ 175	–	14	16	16	18	0a
	T4	≥ 200	≥ 110	12	15	15	15	4a
	T6	≥ 280	≥ 235	5	5	5	6	6a
74530	T4[5)]	≥ 280	≥ 175	8	9	9	10	10a
	T6	≥ 315	≥ 270	6	7	7	8	10a

1) The values of $A_{50\,mm}$ for thickness ≤ 2.6 mm are for information only.
2) The bend test is applicable for thicknesses ≤ 2.6 mm.
3) a = thickness of the bend test piece; 0a means flat on itself.
4) The axis of the bend should be parallel to the direction of rolling.
5) Naturally aged for 30 days.

2.3 Rods, Bars and Sections

2.3.1 Chemical composition

TABLE 4.17 Chemical composition of wrought aluminium and aluminium alloy rods, bars and sections, conforming to IS 733 – 1983

Designation	Chemical composition [% (*m*/*m*)]								
	Al	Cu	Mg	Si	Fe	Mn	Zn	Ti[1]	Cr	Remarks
19000	≥ 99.00	≤ 0.1	–	≤ 0.5	≤ 0.6	≤ 0.1	–	–	–	2)
19500	≥ 99.50	≤ 0.05	–	≤ 0.3	≤ 0.4	≤ 0.05	–	–	–	3)
19600	≥ 99.60	≤ 0.05	–	≤ 0.25	≤ 0.35	≤ 0.03	–	–	–	4)
24345	Rem.	3.8–5.0	0.2–0.8	0.5–1.2	≤ 0.7	0.3–1.2	≤ 0.2	≤ 0.3[5]	≤ 0.3[5]	–
24534	Rem.	3.5–4.7	0.4–1.2	0.2–0.7	≤ 0.7	0.4–1.2	≤ 0.2	≤ 0.3	–	–
43000	Rem.	≤ 0.1	≤ 0.2	4.5–6.0	≤ 0.6	≤ 0.5	≤ 0.2	–	–	–
46000	Rem.	≤ 0.1	≤ 0.2	10.0–13.0	≤ 0.6	≤ 0.5	≤ 0.2	–	–	–
52000	Rem.	≤ 0.1	1.7–2.6	≤ 0.6	≤ 0.5	≤ 0.5	≤ 0.2	≤ 0.2	≤ 0.25	6)
53000	Rem.	≤ 0.1	2.8–4.0	≤ 0.6	≤ 0.5	≤ 0.5	≤ 0.2	≤ 0.2	≤ 0.25	6)
54300	Rem.	≤ 0.1	4.0–4.9	≤ 0.4	≤ 0.7	0.5–1.0	≤ 0.2	≤ 0.2	≤ 0.25	–
63400	Rem.	≤ 0.1	0.4–0.9	0.3–0.7	≤ 0.6	≤ 0.3	≤ 0.2	≤ 0.2	≤ 0.1	–
64423	Rem.	0.5–1.0	0.5–1.3	0.7–1.3	≤ 0.8	≤ 1.0	–	–	–	–
64430	Rem.	≤ 0.1	0.4–1.2	0.6–1.3	≤ 0.6	0.4–1.0	≤ 0.1	≤ 0.2	≤ 0.25	–
65032	Rem.	0.15–0.4	0.7–1.2	0.4–0.8	≤ 0.7	0.2–0.8	≤ 0.2	≤ 0.2	0.15–0.35	7)
74530	Rem.	≤ 0.2	1.0–1.5	≤ 0.4	≤ 0.7	0.2–0.7	4.0–5.0	≤ 0.2	≤ 0.2	–
76528	Rem.	1.2–2.0	2.1–2.9	≤ 0.5	≤ 0.7	≤ 0.3	5.1–6.1	≤ 0.2	0.20–0.28	–

1) Ti and/or other grain refining elements.
2) Ti + V ≤ 0.07 %; total impurities ≤ 1.00 %.
3) Ti + V ≤ 0.07 %; total impurities ≤ 0.50 %.
4) Ti + V ≤ 0.05 %; total impurities ≤ 0.40 %.
5) Ti[1] + Cr ≤ 0.3 %.
6) Mn + Cr ≤ 0.5 %.
7) Either Mn or Cr should be present.

2.3.2 Mechanical properties

TABLE 4.18 Mechanical properties of wrought aluminium and aluminium alloy rods, bars and sections, conforming to IS 733 – 1983

Designation	Temper	Diameter, width across flats or thickness mm	R_m MPa	$R_{p0.2}$ MPa	$A^{2)}$ min %
19000	F	–	≥65	≥20	18
	O	–	≤110	–	25
19500	F	–	≥65	≥18	23
	O	–	≤100	–	25
19600	F	–	≥65	≥17	23
	O	–	≤95	–	25
24345	F	–	≥150	≥90	12
	O	–	≤240	≤175	12
	T4	≤ 10	≥375	≥225	10
		> 10 ≤ 75	≥385	≥235	10
		> 75 ≤ 150	≥385	≥235	8
		> 150 ≤ 200	≥375	≥225	8
	T6	≤ 10	≥430	≥375	6
		> 10 ≤ 25	≥460	≥400	6
		> 25 ≤ 75	≥480	≥420	6
		> 75 ≤ 150	≥460	≥405	6
		> 150 ≤ 200	≥430	≥380	6

(Contd.)

TABLE 4.18 Mechanical properties of wrought aluminium and aluminium alloy rods, bars and sections, conforming to IS 733 – 1983 (continued)

Designation	Temper	Diameter, width across flats or thickness	R_m	$R_{p0.2}$	$A^{2)}$ min
		mm	MPa	MPa	%
24534	F	–	≥ 150	≥ 90	12
	O	–	≤ 240	≤ 175	12
	T4	≤ 10	≥ 375	≥ 220	10
		> 10 ≤ 75	≥ 385	≥ 235	10
		> 75 ≤ 150	≥ 385	≥ 235	8
		> 150 ≤ 200	≥ 375	≥ 225	8
43000	F	≤ 15	≥ 90	–	18
	O	≤ 15	≤ 130	–	18
46000	F	≤ 15	≥ 100	–	10
	O	≤ 15	≤ 150	–	12
52000	F	≤ 150	≥ 160	≥ 70	14
	O	≤ 150	≤ 240	–	18
53000	F	≤ 50	≥ 215	≥ 100	14
		> 50 ≤ 150	≥ 200	≥ 100	14
	O	≤ 150	≤ 260	–	16
54300	F	≤ 150	≥ 265	≥ 130	11
	O	≤ 150	≤ 350	≥ 125	13

(Contd.)

TABLE 4.18 Mechanical properties of wrought aluminium and aluminium alloy rods, bars and sections, conforming to IS 733 – 1983 (continued)

Designation	Temper	Diameter, width across flats or thickness mm	R_m MPa	$R_{p0.2}$ MPa	$A^{2)}$ min %
63400	F	–	≥ 110	–	13
	O	–	≤ 130	–	18
	T4	≤ 150	≥ 140	≥ 80	14
		> 150 ≤ 200	≥ 125	≥ 80	13
	T5	≤ 25	≥ 150	≥ 110	7
	T6	≤ 150	≥ 185	≥ 150	7
		> 150 ≤ 200	≥ 150	≥ 130	6
64423	F	–	≥ 120	–	10
	O	–	≤ 215	≤ 125	15
	T4	–	≥ 266	≥ 155	13
	T6	–	≥ 330	≥ 265	7
64430	F	–	≥ 110	≥ 80	12
	O	–	≤ 150	–	16
	T4	≤ 150	≥ 185	≥ 120	14
		> 150 ≤ 200	≥ 170	≥ 100	12
	T6	≤ 5	≥ 295	≥ 255	7
		> 5 ≤ 75	≥ 310	≥ 270	7
		> 75 ≤ 150	≥ 295	≥ 270	7
		> 150 ≤ 200	≥ 280	≥ 240	6

(Contd.)

TABLE 4.18 Mechanical properties of wrought aluminium and aluminium alloy rods, bars and sections, conforming to IS 733 – 1983 (continued)

Designation	Temper	Diameter, width across flats or thickness mm	R_m MPa	$R_{p0.2}$ MPa	A[2] min %
65032	F	–	≥ 110	≥ 50	12
	O	–	≤ 150	≤ 115	16
	T4	≤ 150	≥ 185	≥ 115	14
		> 150 ≤ 200	≥ 170	≥ 100	12
	T6	≤ 150	≥ 280	≥ 235	7
		> 150 ≤ 200	≥ 245	≥ 200	6
74530	T4[1]	≤ 6	ô 255	≥ 220	9
		> 6 ≤ 75	≥ 275	≥ 230	9
		> 75 ≤ 150	≥ 265	≥ 220	9
	T6	≤ 6	≥ 285	≥ 245	7
		> 6 ≤ 75	≥ 310	≥ 260	7
		> 75 ≤ 150	≥ 290	≥ 245	7
76528	O	All sizes	≤ 290	–	10
	T6	≤ 6	ô 500	≥ 430	6
		> 6 ≤ 75	≥ 530	≥ 455	6
		> 75 ≤ 150	≥ 500	≥ 430	6

1) Naturally aged for 30 days.
2) $L_o = 5.65\sqrt{S_o}$ or 50 mm.

Note: The mechanical properties in the F temper are for information only.

2.4 Wire

2.4.1 Chemical composition

TABLE 4.19 Chemical composition of wrought aluminium and aluminium alloy wire, conforming to IS 739 – 1992

Designation	Chemical composition [% (*m*/*m*)]								
	Al	Cu	Mg	Si	Fe	Mn	Zn	Ti[1]	Cr	Remarks
19000	≥ 99.00	≤0.1	–	≤0.5	≤0.6	≤0.1	–	–	–	[2]
19500	≥ 99.50	≤0.05	–	≤0.3	≤0.4	≤0.05	–	–	–	[3]
24345	Rem.	3.8–5.0	0.2–0.8	0.5–1.2	≤0.7	0.3–1.2	≤0.2	≤0.3	≤0.3	4)
31000	Rem.	≤0.1	≤0.1	≤0.6	≤0.7	0.8–1.5	≤0.2	≤0.2	≤0.2	[5]
43000	Rem.	≤0.1	≤0.2	4.5–6.0	≤0.6	≤0.5	≤0.2	–	–	–
46000	Rem.	≤0.1	≤0.2	10.0–13.0	≤0.6	≤0.5	≤0.2	–	–	–
52000	Rem.	≤0.1	1.7–2.6	≤0.6	≤0.5	≤0.5	≤0.2	≤0.2	≤0.25	[6]
53000	Rem.	≤0.1	2.8–4.0	≤0.6	≤0.5	≤0.5	≤0.2	≤0.2	≤0.25	[6]
55000	Rem.	≤0.1	4.5–5.5	≤0.6	≤0.7	≤0.5	≤0.2	≤0.2	≤0.25	[6]
55380	Rem.	≤0.1	5.0–5.5	–	–	0.6–1.0	≤0.2	0.05–0.20[7]	0.05–0.20	[8]
63400	Rem.	≤0.1	0.4–0.9	0.3–0.7	≤0.6	≤0.3	≤0.2	≤0.2	≤0.1	–
64430	Rem.	≤0.1	0.4–1.2	0.6–1.3	≤0.6	0.4–1.0	≤0.1	≤0.2	≤0.25	–
65032	Rem.	0.15–0.4	0.7–1.2	0.4–0.8	≤0.7	0.2–0.8	≤0.2	≤0.2	0.15–0.35	[9]

[1] Ti and/or other grain refining elements.
[2] Ti[1] + V ≤ 0.07 %; total impurities ≤ 1.0 0%.
[3] Ti[1] + V ≤ 00 7 %; total impurities ≤ 0.5 0%.
[4] Ti[1] + Cr ≤ 0 3 %.
[5] Ti[1] + Cr ≤ 0.2 %.
[6] Mn + Cr ≤ 0.5 %.
[7] Ti only.
[8] Si + Fe ≤ 0.40 %.
[9] Either Mn or Cr should be present.

2.4.2 Mechanical properties

TABLE 4.20 Mechanical properties of wrought aluminium and aluminium alloy wire, conforming to IS 739 – 1992

Designation	Temper	Diameter mm	R_m MPa
19000	F	≤6	–
	O	≤6	≤100
	HX8	≤6	≥140
19500	F	≤6	–
	O	≤6	≤95
	HX8	≤6	≥135
24345	T4	≤6	≥380
	T6	≤6	≥425
31000	F	≤6	–
	O	≤6	≤125
	HX8	≤6	≥170
43000	F	≤6	–
46000	F	≤6	–
52000	F	≤6	–
	O	≤6	165–215
	HX8	≤6	≥260
53000	F	≤6	–

(Contd.)

TABLE 4.20 Mechanical properties of wrought aluminium and aluminium alloy wire, conforming to IS 739–1992 (continued)

Designation	Temper	Diameter mm	R_m MPa
55000	F	≤6	–
	O	≤6	245–310
	HX4	≤6	310–355
	HX8	≤6	≥380
55380	F	≤6	–
63400	T3	≤6	≥280
	T4	≤6	≥140
	T6	≤6	≥180
64430	T4	≤6	≥200
	T6	≤6	≥295
	T8	≤6	≥355
65032	T8	≤6	≥355

2.5 Drawn tube

2.5.1 Chemical composition

TABLE 4.21 Chemical composition of wrought aluminium and aluminium alloy drawn tube, conforming to IS 738 – 1994

Designation	Chemical composition [% (*m*/*m*)]									
	Al	Cu	Mg	Si	Fe	Mn	Zn	Ti[1)]	Cr	Remarks
19000	≥ 99.00	≤0.1	–	≤0.5	≤0.6	≤0.1	–	–	–	2)
19500	≥ 99.50	≤0.05	–	≤0.3	≤0.4	≤0.05	–	–	–	3)

(Contd.)

TABLE 4.21 Chemical composition of wrought aluminium and aluminium alloy drawn tube, conforming to IS 738–1994 (continued)

Designation	Chemical composition [% (m/m)]									
	Al	Cu	Mg	Si	Fe	Mn	Zn	Ti[1)]	Cr	Remarks
24345	Rem.	3.8–5.0	0.2–0.8	0.5–1.2	≤0.7	0.3–1.2	≤0.2	≤0.3	≤0.3	4)
31000	Rem.	≤0.1	≤0.1	≤0.6	≤0.7	0.8–1.5	≤0.2	≤0.2	≤0.2	–
52000	Rem.	≤0.1	1.7–2.6	≤0.6	≤0.5	≤0.5	≤0.2	≤0.2	≤0.25	5)
63400	Rem.	≤0.1	0.4–0.9	0.3–0.7	≤0.6	≤0.3	≤0.2	≤0.2	≤0.1	–
64430	Rem.	≤0.1	0.4–1.2	0.6–1.3	≤0.6	0.4–1.0	≤0.1	≤0.2	≤0.25	–
65028	Rem.	0.15–0.4	0.7–1.2	0.4–0.8	≤0.6	≤0.2	≤0.2	≤0.2	0.15–0.35	
65032	Rem.	0.15–0.4	0.7–1.2	0.4–0.8	≤0.6	0.2–0.8	≤0.2	≤0.2	≤0.2	–

1) Ti and/or other grain refining elements.
2) Ti + V ≤ 0.07 %; total impurities ≤ 1.00 %.
3) Ti + V ≤ 0.07 %; total impurities ≤ 0.50 %.
4) Ti[1)] + Cr ≤ 0.3 %.
5) Mn + Cr ≤ 0.5 %.

2.5.2 Mechanical properties

TABLE 4.22 Mechanical properties of wrought aluminium and aluminium alloy drawn tube, conforming to IS 738 – 1994

Designation	Temper	Wall thickness mm	R_m MPa	$R_{p0.2}$ min. MPa	$A_{50\,mm}$ min. %
19000	O	≤ 12	≤ 105	–	–
	HX8	≤ 12	≥ 120	–	–
19500	O	≤ 12	≤ 95	–	–
	HX8	≤ 12	≥ 110	–	–
24345	T4[1)]	≤ 10	≥ 395	290	8

(Contd.)

TABLE 4.22 Mechanical properties of wrought aluminium and aluminium alloy drawn tube, conforming to IS 738–1994 (continued)

Designation	Temper	Wall thickness mm	R_m MPa	$R_{p0.2}$ min. MPa	$A_{50\,mm}$ min. %
31000	O	≤ 10	≤ 110	35	24
	HX4	≤ 10	≥ 140	120	5
52000	O	≤ 10	≤ 200	–	18
	HX4	≤ 10	≥ 225	170	5
63400	O	≤ 10	≤ 155	–	–
	T4	≤ 10	≥ 150	95	15
	T6	≤ 10	≥ 200	170	8
64430	T4	≤ 10	≥ 215	110	12
	T6	≤ 6	≥ 310	245	7
		> 6 ≤ 10	≥ 310	230	9
65028	T4	≤ 6	≥ 215	110	12
		> 6 ≤ 10	≥ 215	110	14
	T6	≤ 6	≥ 295	230	7
		> 6 ≤ 10	≥ 295	215	9
65032	T4	≤ 6	≥ 215	110	12
		> 6 ≤ 10	≥ 215	110	14
	T6	≤ 6	≥ 295	230	7
		> 6 ≤ 10	≥ 295	215	9

[1)] If the user reheat-treats this material, the minimum R_m and $R_{p0.2}$ values may be reduced to 385 MPa and 230 MPa, respectively. Similar properties may also be obtained when the tubes are supplied in the annealed condition and subsequently heat-treated.

2.6 Extruded round tube and hollow sections

2.6.1 Chemical composition

TABLE 4.23 Chemical composition of wrought aluminium and aluminium alloy extruded round tube and hollow sections, conforming to IS 1285 – 1975

Designation	Chemical composition [% (m/m)]								
	Al	Cu	Mg	Si	Fe	Mn	Zn	Ti[1]	Cr	Remarks
19000	≥ 99.00	≤ 0.1	≤ 0.2	≤ 0.5	≤ 0.7	≤ 0.1	≤ 0.1	–	–	[2]
19500	≥ 99.50	≤ 0.05	–	≤ 0.3	≤ 0.4	≤ 0.05	≤ 0.1	–	–	[3]
24345	Rem.	3.8–5.0	0.2–0.8	0.5–1.2	≤ 0.7	0.3–1.2	≤ 0.2	≤ 0.3[4]	≤ 0.3[4]	–
52000	Rem.	≤ 0.1	1.7–2.6	≤ 0.6	≤ 0.5	≤ 0.5	≤ 0.2	≤ 0.2	≤ 0.25	[5]
53000	Rem.	≤ 0.1	2.8–4.0	≤ 0.6	≤ 0.5	≤ 0.5	≤ 0.2	≤ 0.2	≤ 0.25	[5]
54300	Rem.	≤ 0.1	4.0–4.9	≤ 0.4	≤ 0.5	0.5–1.0	≤ 0.2	≤ 0.2	≤ 0.25	–
62400	Rem.	≤ 0.1	0.4–0.6	0.5–0.9	≤ 0.7	≤ 0.1	–	–	–	–
63400	Rem.	≤ 0.1	0.4–0.9	0.3–0.7	≤ 0.6	≤ 0.3	≤ 0.2	≤ 0.2	≤ 0.1	–
64423	Rem.	0.5–1.0	0.5–1.3	0.7–1.3	≤ 0.8	≤ 1.0	–	–	–	–
64430	Rem.	≤ 0.1	0.4–1.2	0.6–1.3	≤ 0.6	0.4–1.0	≤ 0.1	≤ 0.2	≤ 0.25	–
65032	Rem.	0.15–0.4	0.7–1.2	0.4–0.8	≤ 0.7	0.2–0.8	≤ 0.2	≤ 0.2	0.15–0.35	[6]
74530	Rem.	≤ 0.2	1.0–1.5	≤ 0.4	≤ 0.7	0.2–0.7	4.0–5.0	≤ 0.2	≤ 0.2	–

1) Ti and/or other grain refining elements.
2) Cu + Mg + Si + Fe + Mn + Zn ≤ 1.0 %.
3) Cu + Si + Fe + Mn + Zn ≤ 0.5 %.
4) Ti[1] + Cr ≤ 0.3 %.
5) Mn + Cr ≤ 0.5 %.
6) Either Mn or Cr should be present.

2.6.2 Mechanical properties

TABLE 4.24 Mechanical properties of wrought aluminium and aluminium alloy extruded round tube and hollow sections, conforming to IS 1285 – 1975

Designation	Temper	Wall thickness mm	R_m min. MPa	$R_{p0.2}$ min. MPa	$A_{50\,mm}$ min. %
19000	F	–	65	–	18
19500	F	–	65	–	23
24345	T4	–	375	225	10
	T6	–	430	275	6
52000	F	–	170	–	14
53000	F	–	215	100	14
54300	F	–	275	130	11
62400	T6	–	255	200	15
63400	F	–	110	–	13
	T4	–	140	80	14
	T5	≤ 3.15	170	140	7
		> 3.15 ≤ 12.5	150	110	7
	T6	–	185	150	7
65032	F	–	110	–	–
	T4	–	185	115	14
	T6	–	280	235	7
64423	F	–	120	–	11
	T4	–	265	155	14
	T6	–	330	265	9

(Contd.)

TABLE 4.24 Mechanical properties of wrought aluminium and aluminium alloy extruded round tube and hollow sections, conforming to IS 1285–1975 (continued)

Designation	Temper	Wall thickness mm	R_m min. MPa	$R_{p0.2}$ min. MPa	$A_{50\,mm}$ min. %
64430	F	–	110	–	14
	T4	–	185	120	14
	T6	≤ 6.3	295	255	7
		> 6.3	310	270	7
74530	T4[1)]	≤ 6.3	255	220	9
		> 6.3 ≤ 76	275	230	9
		> 76 ≤ 152	260	230	9
	T6	≤ 6.3	285	245	7
		> 6.3 ≤ 76	310	260	7
		> 76 ≤ 152	290	245	7

[1)] Naturally aged for 30 days.

Note: The mechanical properties in the F temper are for information only.

2.7 Forging stock and forgings

2.7.1 Chemical composition

TABLE 4.25 Chemical composition of wrought aluminium and aluminium alloy forging stock and forgings, conforming to IS 734 – 1975

Designation	Chemical composition [% (*m*/*m*)]									
	Al	Cu	Mg	Si	Fe	Mn	Zn	Ti[1)]	Cr	Remarks
19000	≥ 99.00	≤ 0.1	≤ 0.2	≤ 0.5	≤ 0.7	≤ 0.1	≤ 0.1	–	–	2)
19500	≥ 99.50	≤ 0.05	–	≤ 0.3	≤ 0.4	≤ 0.05	≤ 0.1	–	–	3)

(Contd.)

TABLE 4.25 Chemical composition of wrought aluminium and aluminium alloy forging stock and forgings, conforming to IS 734–1975 (continued)

Designation	Chemical composition [% (*m*/*m*)]									
	Al	Cu	Mg	Si	Fe	Mn	Zn	Ti[1]	Cr	Remarks
22588	Rem.	1.8–2.7	1.2–1.6	≤1.3	0.6–1.2	≤0.2	≤0.2	≤0.3	–	4)
24345	Rem.	3.8–5.0	0.2–0.8	0.5–1.2	≤0.7	0.3–1.2	≤0.2	≤0.3[5]	≤0.3[5]	–
24534	Rem.	3.5–4.7	0.4–1.2	0.2–0.7	≤0.7	0.4–1.0	≤0.2	≤0.3	–	–
31000	Rem.	≤0.1	≤0.1	≤0.6	≤0.7	0.8–1.5	≤0.2	≤0.2	≤0.2	–
52000	Rem.	≤0.1	1.7–2.6	≤0.6	≤0.5	≤0.5	≤0.2	≤0.2	≤0.25	6)
53000	Rem.	≤0.1	2.8–4.0	≤0.6	≤0.5	≤0.5	≤0.2	≤0.2	≤0.25	6)
54300	Rem.	≤0.1	4.0–4.9	≤0.4	≤0.5	0.5–1.0	≤0.2	≤0.2	≤0.25	–
63400	Rem.	≤0.1	0.4–0.9	0.3–0.7	≤0.6	≤0.3	≤0.2	≤0.2	≤0.1	–
64423	Rem.	0.5–1.0	0.5–1.3	0.7–1.3	≤0.8	≤1.0	–	–	–	–
64430	Rem.	≤0.1	0.4–1.2	0.6–1.3	≤0.6	0.4–1.0	≤0.1	≤0.2	≤0.25	–
65032	Rem.	0.15–0.4	0.7–1.2	0.4–0.8	≤0.7	0.2–0.8	≤ 0.2	≤0.2	0.15–0.35	7)
74530	Rem.	≤0.2	1.0–1.5	≤0.4	≤0.7	0.2–0.7	4.0–5.0	≤0.2	≤0.2	–

1) Ti and/or other grain refining elements.
2) Cu + Mg + Si + Fe + Mn + Zn ≤ 1.0 %.
3) Cu + Si + Fe + Mn + Zn ≤ 0.5 %.
4) Ni = 0.6–1.4 %.
5) Ti[1] + Cr ≤ 0.3 %.
6) Mn + Cr ≤ 0.5 %.
7) Either Mn or Cr should be present.

2.7.2 Mechanical properties

TABLE 4.26 Mechanical properties of wrought aluminium and aluminium alloy forging stock and forgings, conforming to IS 734 – 1975

Designation	Temper	Condition of test sample	Size of bar mm	R_m min. MPa	$R_{p0.2}$ min. MPa	$A_{50\,mm}$ min. %
19000	F/O	Forged or extruded	–	65	–	18
19500	F/O	Forged or extruded	–	65	–	23
22588	T6	Forged	–	375	265	–
		Extruded	–	385	265	–
24345	T4	Forged	–	375	225	–
		Extruded	≤ 10	375	225	10
			$> 10 \le 75$	385	235	10
			$> 75 \le 150$	385	235	–
			$> 150 \le 200$	375	225	–
	T6	Forged	–	445	385	6
		Extruded	≤ 10	430	375	6
			$> 10 \le 25$	460	400	6
			$> 25 \le 75$	480	420	–
			$> 75 \le 150$	460	405	–
			$> 150 \le 200$	430	380	–

(Contd.)

TABLE 4.26 Mechanical properties of wrought aluminium and aluminium alloy forging stock and forgings, conforming to IS 734 – 1975 (continued)

Designation	Temper	Condition of test sample	Size of bar mm	R_m min. MPa	$R_{p0.2}$ min. MPa	$A_{50\,mm}$ min. %
24534	T4	Extruded	≤ 10	375	220	10
			> 10 ≤ 75	385	235	10
			> 75 ≤ 150	385	235	8
			> 150 ≤ 200	375	225	8
31000	F/O	Forged or extruded	–	105	–	–
52000	F/O	Forged or extruded	–	170	–	14
53000	F/O	Forged	–	215	–	14
		Extruded	≤ 50	215	–	14
			> 50	200	–	14
54300	O	Forged	–	265	–	13
	F	Extruded	–	275	–	11
63400	T4	Forged	–	140	80	14
		Extruded	≤ 150	140	80	14
			> 150 ≤ 200	125	80	–
	T6	Forged	–	185	150	7
		Extruded	≤ 150	185	150	7
			> 150 ≤ 200	150	130	–

(Contd.)

TABLE 4.26 Mechanical properties of wrought aluminium and aluminium alloy forging stock and forgings, conforming to IS 734 – 1975 (continued)

Designation	Temper	Condition of test sample	Size of bar mm	R_m min. MPa	$R_{p0.2}$ min. MPa	$A_{50\,mm}$ min. %
64423	T4	Forged	–	265	155	14
		Extruded	–	265	155	14
	T6	Forged	–	330	265	9
		Extruded	–	330	265	9
64430	T4	Forged	–	185	120	14
		Extruded	≤150	185	120	14
			>150 ≤200	170	100	–
	T6	Forged	–	295	255	7
		Extruded	≤5	295	255	7
			>5 ≤ 75	310	270	7
			>75 ≤150	295	255	–
			>150 ≤200	280	240	–
65032	T4	Forged	–	185	115	14
		Extruded	≤150	185	115	14
			>150 ≤200	170	100	14
	T5	Forged	–	280	235	7
		Extruded	≤150	280	235	7
			>150 ≤200	245	195	7

TABLE 4.26 Mechanical properties of wrought aluminium and aluminium alloy forging stock and forgings, conforming to IS 734 – 1975 (continued)

Designation	Temper	Condition of test sample	Size of bar mm	R_m min. MPa	$R_{p0.2}$ min. MPa	$A_{50\,mm}$ min. %
74530	T4[1]	Forged	–	255	220	9
		Extruded	≤6.3	255	220	9
			>6.3≤75	275	230	9
			>75≤150	260	220	9
	T6	Forged	–	285	245	7
		Extruded	≤ 6.3	285	245	7
			>6.3≤75	310	260	7
			>75≤150	290	245	7

[1] Naturally aged for 30 days.

Note: The mechanical properties in the F temper are for information only.

3 TYPICAL PHYSICAL AND MECHANICAL PROPERTIES

TABLE 4.27 Typical physical and mechanical properties of wrought aluminium and aluminium alloys specified in Indian Standards

Designation	Temper	Density at 20 °C kg/m^3	Approximate melting range °C	Coefficient of thermal expansion between 20 °C and 100 °C μm/(m K)	Thermal conductivity at 20 °C W/(m K)	Electrical conductivity at 20 °C % IACS	Electrical resistivity at 20 °C μΩ m	Modulus of elasticity GPa
19000	O	2710	–	24	226	59.5	0.029	–
19500	O	2705	646–657	23.6	231	61.3	0.0281	–
19600	O	2705	646–657	23.6	234[1)]	62	0.0278	69
19700	–	–	–	–	–	–	–	–
19800	O	2700	–	24	230[2)]	61.6	0.028	69
19990	O	2700	660[3)]	23.6	243	64.5	0.0267	62
22588	–	–	–	–	–	–	–	–
24345	O	2800	507–638	22.5	192	50	0.034	72.4
24534	O	2800	513–640	23.6	193[1)]	50	–	72.4
31000	O	2730	645–655	23	172[2)]	43.1	0.040	69
31500	O	2720	629–654	23.2	162	42	0.041	70
40800	–	–	–	–	–	–	–	–

(Contd.)

TABLE 4.27 Typical physical and mechanical properties of wrought aluminium and aluminium alloys specified in Indian Standards (continued)

Designation	Temper	Density at 20 °C kg/m³	Approximate melting range °C	Coefficient of thermal expansion between 20 °C and 100 °C μm/(m K)	Thermal conductivity at 20 °C W/(m K)	Electrical conductivity at 20 °C % IACS	Electrical resistivity at 20 °C μΩ m	Modulus of elasticity GPa
43000	O	2680	575–630	22.0	–	42	0.041	–
46000	–	–	–	–	–	–	–	–
51000-A	O	2700	632–652	23.7	205	52	0.0332	68.2
51000-B	O	2690	627–652	23.8	191	50	0.034	68.9
51300	O	2710	638–657	23.6	173	45	0.0383	69
52000	O	2690	595–650	24	155[2)]	36.7	0.047	71
53000	–	2660	593–643	23.9	127	32	0.0539	69.3
54300	–	2660	574–638	24.2	120	20	0.0595	70.3
55000	O	2640	568–638	24.1	120	29	0.059	71.7
55380	–	–	–	–	–	–	–	–
62400	T5	2700	607–654	23.4	167[1)]	49	0.035	69
63400	O	2690	615–655	23.4	218[1)]	58	0.030	68.3
64423	–	–	–	–	–	–	–	–
64430	T4	2700	570–660	23	172[2)]	42.1	0.041	69

(Contd.)

TABLE 4.27 Typical physical and mechanical properties of wrought aluminium and aluminium alloys specified in Indian Standards (continued)

Designation	Temper	Density at 20 °C	Approximate melting range	Coefficient of thermal expansion between 20 °C and 100 °C	Thermal conductivity at 20 °C	Electrical conductivity at 20 °C	Electrical resistivity at 20 °C	Modulus of elasticity
		kg/m^3	°C	μm/(m K)	W/(m K)	% IACS	μΩ m	GPa
65032	O	2700	582–652	23.6	180[2)]	47	0.037	68.9
74530	T4	2780	–	–	134[2)]	37.5	0.046	–
76528	T6	2800	477–635	23.4	130	33	0.0522	71

1) At 25 °C.
2) Between 0°C and 100 °C.
3) Melting point.

Source: References 1 and 3.

4 COMPARISON RATINGS OF WROUGHT ALUMINIUM AND ALUMINIUM ALLOYS

TABLE 4.28 Comparison of the characteristics of wrought aluminium and aluminium alloys specified in Indian Standards

Designation	Temper	Resistance to corrosion		Work ability[3] (cold)	Machinability[3]	Weldability[4]			Brazeability[4]	Solderability[5]
		General[1]	Stress-corrosion cracking[2]			Gas	Arc	Resistance, spot and seam		
19000	–	–	–	–	–	–	–	–	–	–
19500	O	A	A	A	E	A	A	B	A	A
19600	O	A	A	A	E	A	A	B	A	A
19700	–	–	–	–	–	–	–	–	–	–
19800	–	–	–	–	–	–	–	–	–	–
19990	O	A	A	A	E	A	A	B	A	A
22588	–	–	–	–	–	–	–	–	–	–
24345	O	–	–	–	D	D	D	B	D	C
24534	T4	D	C	C	B	B	D	B	D	–
31000	–	–	–	–	–	–	–	–	–	–
31500	O	A	A	A	D	B	A	B	B	B
40800	–	–	–	–	–	–	–	–	–	–
43000	–	B	A	NA	C	NA	NA	NA	NA	NA
46000	–	–	–	–	–	–	–	–	–	–

(Contd.)

TABLE 4.28 Comparison of the characteristics of wrought aluminium and aluminium alloys specified in Indian Standards (continued)

Design-ation	Temper	Resistance to corrosion		Work ability[3] (cold)	Machin-ability[3]	Weldability[4]			Braze-ability[4]	Solder-ability[5]
		General[1]	Stress-corrosion cracking[2]			Gas	Arc	Resistance, spot and seam		
51000-A	O	A	A	A	E	A	A	B	B	B
51000-B	O	A	A	A	E	A	A	B	B	C
51300	O	A	A	A	E	B	A	B	B	B
52000	–	–	–	–	–	–	–	–	–	–
53000	O	A[6]	A[6]	A	D	C	A	B	D	D
54300	O	A[6]	B6)	B	D	C	A	B	D	D
55000	O	A[6]	B6)	A	D	C	A	B	D	D
55380	–	–	–	–	–	–	–	–	–	–
62400	T5	B	A	C	C	A	A	A	A	NA
63400	T5	A	A	B	C	A	A	A	A	B
64423	–	–	–	–	–	–	–	–	–	–
64430	–	–	–	–	–	–	–	–	–	–
65028	O	B	A	A	D	A	A	B	A	B
65032	–	–	–	–	–	–	–	–	–	–

(Contd.)

TABLE 4.28 Comparison of the characteristics of wrought aluminium and aluminium alloys specified in Indian Standards (continued)

Designation	Temper	Resistance to corrosion		Work ability[3] (cold)	Machinability[3]	Weldability[4]			Brazeability[4]	Solderability[5]
		General[1]	Stress-corrosion cracking[2]			Gas	Arc	Resistance, spot and seam		
74530	–	–	–	–	–	–	–	–	–	–
76528	T6	C[7]	C	D	B	D	C	B	D	D

[1] Ratings A through E are relative ratings in decreasing order of merit, based on exposures to sodium chloride solution by intermittent spraying or immersion. Alloys with A and B ratings can be used in industrial and sea coast atmospheres without protection. Alloys with C, D and E ratings should generally be protected, at least on faying surfaces.

[2] Stress-corrosion cracking ratings are based on service experience and on laboratory tests of specimens exposed to the 3.5 % sodium chloride alternate immersion test.
A = no known instance of failure in service or in laboratory tests.
B = no known instance of failure in service; limited failures in laboratory tests of short transverse specimens.
C = service failures with sustained tension stress acting in short transverse direction relative to grain structure; limited failures in laboratory tests of long transverse specimens.
D = limited service failures with sustained longitudinal or long transverse stress.

[3] Ratings A through D for workability (cold), and A through E for machinability, are relative ratings in decreasing order of merit.

[4] Ratings A through D for weldability and brazeability are relative ratings defined as follows:
A = generally weldable by all commercial procedures and methods.
B = weldable with special techniques or for specific applications which justify preliminary trials or testing to develop welding procedure and weld performance.
C = limited weldability because of crack sensitivity or loss in resistance to corrosion and mechanical properties.
D = no commonly used welding methods have been developed.

[5] Ratings A through D and NA for solderability are relative ratings defined as follows:
A = excellent; B = good; C = fair; D = poor; NA = not applicable.

[6] This rating may be different for materials held at elevated temperature for long periods.

[7] In relatively thick sections the rating would be E.

Source: Reference 1

5 TYPICAL APPLICATIONS

TABLE 4.29 Typical applications of wrought aluminium and aluminium alloys specified in Indian Standards

Designation	Typical applications
19000	Food and chemical handling and storage equipment, drawn or spun hollowware, sheet metal work.
19500	Extruded coiled tube for equipment and containers for food, chemical and brewing industries, collapsible tubes, pyrotechnic powders.
19600	Chemical process equipment.
19990	Electrolytic capacitor foil, vapor deposited coatings for optically reflecting surfaces.
22588	Aircraft undercarriage wheels, cylinder heads, forged aircraft and automobile pistons.
24345	Heavy-duty forgings, plate and extrusions for aircraft fittings, space booster tankage and structure, truck frame and suspension components.
24534	Components for general engineering purposes, fittings, structural applications in construction and transportation, screw and machine products.
31000	Air-conditioning ducting, cooking utensils, detonator caps, fan blades, heat exchangers, irrigation tubing, pilfer-proof caps, pressure vessels, vehicle panelling.
31500	Builder's hardware, chemical handling and storage equipment, drawn and ironed rigid containers, incandescent and fluorescent lamp bases, sheet metal work.
40800	Fan blades, vehicle panelling.
43000	General purpose weld filler alloy (rod or wire) for welding all wrought and foundry alloys except those rich in magnesium.
46000	Filler wire for brazing.
51000-A	Appliances, architectural applications, cooking utensils, electrical conductor wire.
51000-B	Sheet used as trim in refrigerator applications, tube for automotive gas and oil lines, welded irrigation pipe.

(Contd.)

TABLE 4.29 Typical applications of wrought aluminium and aluminium alloys specified in Indian Standards (continued)

Designation	Typical applications
51300	Bottle caps and closures, gutters and downspouts, mobile home sheet, residential siding, sheet metal work.
52000	Boat and ship building, vehicle construction.
53000	Marine structures, pressure vessels, storage tanks, transportation trailer tanks, welded structures.
54300	Marine, auto and aircraft applications, cryogenics, drilling rigs, missile components, TV towers, transportation equipment, unfired welded pressure vessels.
55000	Nails, rivets for use with magnesium alloy and cable sheathing, zipper stock.
62400	Ladders and TV antennas.
63400	Architectural extrusions, doors, furniture, irrigation pipes, railings, truck and trailer flooring, windows.
64423	Textile machinery components.
64430	Bridges, cargo containers, cranes, deep-drawn containers, flooring, milk containers, road and rail transport vehicles, roof trusses.
65028	Canoes, furniture, pipelines, rail-road cars, towers, trucks.
65032	Canoes, furniture, pipelines, rail-road cars, towers, trucks.
74530	Bridges, chequered plates, dump-truck bodies, pressure vessels, rail coaches.
76528	Aircraft structural parts.

Source: References 1, and 4–10

9 EQUIVALENT GRADES

TABLE 4.30 Comparison of grades of wrought aluminium and aluminium alloys specified in Indian Standards with nearest equivalent grades specified in other national and international standards.

IS 733–83 IS 734–75 IS 736–86 IS 737–87 IS 738–94 IS 739–92 IS 1285–75	ASTM B 209M–95 ASTM B 210M–95 ASTM B 211M–95 ASTM B 221M–95a ASTM B 241 / B 241M–95a ASTM B 247M–95a ASTM B 483M–95	BS 1470–87 BS 1471–72 BS 1472–72 BS 1474–87 BS 1475–72	DIN 1712 (3)–76 DIN 1725 (1)–83	GOST 4784–74 GOST 11069–74	ISO 209 (1)–89	JIS H 4000–88 JIS H 4040–88 JIS H 4080–88 JIS H 4100–88 JIS H 4140–88	SAE J454–91
19000	–	1200	Al99	A0	Al 99.0	1200	–
19500	–	1050A	Al99.5	A5	Al 99.5	1050	–
19600	1060	–	–	A6	Al 99.6	–	–
19700	–	–	Al99.7	A7	Al 99.7	1070	–
19800	–	1080A	Al99.8	A8	Al 99.8(A)	1080	–
19990	–	–	Al99.98R	A99	–	–	–
22588	–	–	–	–	–	–	–
24345	2014	2014A	AlCuSiMn	AK8	Al Cu4SiMg	2014	–
Alclad 24345	Alclad 2014	Clad 2014A	–	–	–	2014 Clad plate	–
24534	2017	–		–	Al Cu4MgSi	2017	2017

TABLE 4.30 Comparison of grades of wrought aluminium and aluminium alloys specified in Indian Standards with nearest equivalent grades specified in other national and international standards. (continued)

IS733–83 IS734–75 IS736–86 IS737–87 IS738–94 IS739–92 IS1285–75	ASTM B209M–95 ASTM B210M–95 ASTM B211M–95 ASTM B221M–95a ASTM B241/B241M–95a ASTM B247M–95a ASTM B483M–95	BS1470–87 BS1471–72 BS1472–72 BS1474–87 BS1475–72	DIN 1712 (3)–76 DIN 1725 (1)–83	GOST 4784–74 GOST 11069–74	ISO 209 (1)–89	JIS H 4000–88 JIS H 4040–88 JIS H 4080–88 JIS H 4100–88 JIS H 4140–88	SAE J454–91
31000	–	3103	AlMn1	–	Al Mn1	–	–
31500	3004	–	–	–	Al Mn1Mg1	3004	3004
40800	–	–	–	–	–	–	–
43000	–	4043A	–	–	Al Si5(A)	–	–
46000	–	4047A	–	–	Al Si12(A)	–	–
51000-A	5005	5005	AlMg1	AMr1	Al Mg1(B)	5005	5005
51000-B	5050	–	–	–	Al Mg1.5(C)	–	–
51300	3105	3105	–	–	Al Mn0.5Mg0.5	3105	–
52000	–	5251	–	–	Al Mg2	–	–
53000	5154	5154A	–	–	Al Mg3.5(A)	5154	–
54300	5083	5083	AlMg4.5Mn	–	Al Mg4.5 Mn0.7	5083	5083
55000	5056	5056A	AlMg5	–	Al Mg5	5056	–
55380	–	5556A	–	–	–	–	–

(Contd.)

TABLE 4.30 Comparison of grades of wrought aluminium and aluminium alloys specified in Indian Standards with nearest equivalent grades specified in other national and international standards. (continued)

IS 733–83 IS 734–75 IS 736–86 IS 737–87 IS 738–94 IS 739–92 IS 1285–75	ASTM B 209M–95 ASTM B 210M–95 ASTM B 211M–95 ASTM B 221M–95a ASTM B 241 / B 241M–95a ASTM B 247M–95a ASTM B 483M–95	BS 1470–87 BS 1471–72 BS 1472–72 BS 1474–87 BS 1475–72	DIN 1712 (3)–76 DIN 1725 (1)–83	GOST 4784–74 GOST 11069–74	ISO 209 (1)–89	JIS H 4000–88 JIS H 4040–88 JIS H 4080–88 JIS H 4100–88 JIS H 4140–88	SAE J454–91
62400	–	–	AlMgSi0.7	–	Al SiMg	–	–
63400	6063	6063	AlMgSi0.5	АД31	Al Mg0.7Si	6063	6063
64423	–	–	–	–	–	–	–
64430	–	6082	AlMgSi1	–	Al Si1MgMn	–	–
65028	6061	6061	–	АД33	Al Mg1SiCu	6061	6061
65032	–	–	–	–	–	–	–
74530	–	–	AlZn4.5Mg1	–	Al Zn4.5 Mg1.5Mn	7 N 01	–
76528	7075	–	AlZnMgCu1.5	–	Al Zn5.5MgCu	7075	–

7 REFERENCES

Books

1. *ASM Handbook*, vol. 2, ASM International, Materials Park, Ohio, U.S.A., 1990.
2. J.E. Hatch (Ed.), *Aluminium: Properties, and Physical Metallurgy*, American Society for Metals, Metals Park, Ohio, U.S.A., 1984.
3. *The Properties of Aluminium and Its Alloys*, 9th ed., The Aluminium Federation, Birmingham, U.K., 1993.

Standards

4. IS 733 – 1983: *Specification for Wrought Aluminium and Aluminium Alloy Bars, Rods and Sections (for General Engineering Purposes).*
5. IS 734 – 1975: *Specification for Wrought Aluminium and Aluminium Alloy Forging Stock and Forgings (for General Engineering Purposes).*
6. IS 736 – 1986: *Specification for Wrought Aluminium and Aluminium Alloy Plate for General Engineering Purposes.*
7. IS 737 – 1986: *Specification for Wrought Aluminium and Aluminium Alloy Sheet and Strip for General Engineering Purposes.*
8. IS 738 – 1994: *Wrought Aluminium and its Alloys—Drawn Tubes for General Engineering Purposes —Specification.*
9. IS 739 – 1992: *Wrought Aluminium and Aluminium Alloys – Wire for General Engineering Purposes —Specification.*
10. IS 1285 – 1975: *Specification for Wrought Aluminium and Aluminium Alloy Extruded Round Tube and Hollow Sections (for General Engineering Purposes).*

4.6

HEAT TREATMENT OF ALUMINIUM AND ALUMINIUM ALLOYS

1 DEFINITIONS

Heat-treatable alloy: An alloy capable of being strengthened by suitable thermal treatment.

Non-heat-treatable alloy: An alloy strengthened by cold working only and incapable of being substantially strengthened by thermal treatment.

2 ANNEALING

2.1 Full annealing of precipitation-hardened wrought aluminium alloys

Purpose: To eliminate the strength increase produced by precipitation hardening.

Recommended heat treatment cycle: Heat to 415 ± 10 °C. Hold at this temperature for a sufficient length of time until a uniform temperature is achieved (minimum holding time = 1 h). Furnace cool to 230 °C at a rate not exceeding 30 °C/h, and continue cooling to ambient temperature in still air.

2.2 Full annealing of strain hardened wrought aluminium and aluminium alloys

Purpose: To eliminate the strength increase produced by strain hardening.

Recommended heat treatment cycle: Heat to 345 °C. Hold at this temperature for a sufficient length of time until a uniform temperature is achieved. Cool to ambient temperature in still air.

2.3 Partial annealing of wrought aluminium and aluminium alloys

Purpose: To reduce the strength properties of a cold worked metal or alloy to a controlled level.

Recommended heat treatment cycle: Heat to 340 ± 10 °C. Hold at this temperature for a sufficient length of time until a uniform temperature is achieved (minimum holding time = 0.5 h). Cool to ambient temperature in still air.

Note: In annealing thin-gage clad products, the heating time at the annealing temperature should be limited to avoid excessive diffusion from core to cladding.

2.4 Annealing of cast aluminium alloys

Purpose: To improve ductility and shock resistance and to improve dimensional stability, particularly in products intended for elevated temperature service.

Recommended heat treatment cycle: Heat to 315–345 °C. Hold at this temperature for 2–4 h. Cool to ambient temperature in still air.

3 PRECIPITATION HARDENING

The purpose of precipitation hardening is to increase strength and hardness of heat-treatable aluminium alloys. It is achieved through a sequence of solution heat treatment, quenching and natural/artificial ageing. However, certain alloys which are relatively insensitive to cooling rates during quenching can be precipitation hardened either by air cooling or by water quenching directly from the elevated temperature shaping process followed by a ageing treatment.

3.1 Solution heat treatment

Purpose: To take into solid solution the maximum practical amounts of the solute elements, such as copper, magnesium, silicon or zinc, that participate in the subsequent ageing process.

Recommended heat treatment cycle: Heat to a temperature in the single-phase, solid-solution range (see Table 4.31). Hold at this temperature for a sufficient length of time (see Tables 4.32 and 4.33) to achieve a homogeneous structure. The holding time required can vary from less than a minute for thin sheets to as much as 20 h for large sand castings.

Notes:

1) The solution heat treating temperature is determined by the chemical composition of the alloy.
2) Overheating of the alloy, resulting in incipient melting of the low temperature eutectic and grain boundary phases, should be avoided, as such a melting would result in quench cracks and, low strength and ductility. The alloy would thus become brittle and nonsalvageable.
3) Underheating of the alloy to a temperature below the solvus temperature should be avoided, as this would result in a significant decrease in the final strength.
4) The holding time at the solution heat treating temperature is a function of the microstructure before heat treatment.
5) The holding time for alclad sheets and for parts made from alclad sheets should be kept to a minimum to avoid excessive diffusion of alloying elements from the core into the cladding.
6) Heating of aluminium alloys in an atmosphere containing water vapour should be avoided as this leads to blistering of the surface of the material and, occassionally, formation of internal discontinuities and voids. This phenomenon which is termed as high temperature oxidation is aggravated in the presence of sulphur compounds.

TABLE 4.31 Recommended solution and precipitation heat treatment for wrought and cast aluminium alloys specified in Indian Standards

Alloy designation	Product form	Solution heat treatment temperature	Precipitation heat treatment	
			Temperature[1)]	Time[1)]
		°C	°C	h
22500	Cold forged rivet stock	490–500	room temperature	≥96
22588	Forgings for use at elevated temperatures	520–530	200 ± 5	20
24345	Sheet, extruded rods, bars and sections, forging stock and forgings	495–505	160	12–20
			175	9–12
			190	3–6
24435	Cold forged rivet stock	500–510	room temperature	≥48
24530	Cold forged rivet stock	490–500	room temperature	≥48
24534	Sheet, rods, bars, wire	495–510	room temperature	≥96
62400	Drawn tube	500–540	160–175	8–20
63400	Extruded rods, bars and shapes, tubes	500–540	160–175	8–20
64423	Die forgings	515–543	room temperature	≥96
			171–182	≥8
64430	Plate, sheet, extrusions, forgings	515–550	165–185	8–12
65032	Plate, sheet, rods, bars, wire, drawn tube, forgings	515–552	room temperature	≥96
			160 ± 5	6–20
			177 ± 5	6–10
65400	Die forgings	510–525	room temperature	≥96
			165–177	≥10

(Contd.)

TABLE 4.31 Recommended solution and precipitation heat treatment for wrought and cast aluminium alloys specified in Indian Standards (continued)

Alloy designation	Product form	Solution heat treatment temperature °C	Precipitation heat treatment Temperature[1)] °C	Time[1)] h
74530	Plate, sheet	400–470	room temperature	≥72
	Extrusions (rods, bars, shapes, tubes)	430–470	85–95	≥8
	Forgings	450–470	145–155	≥16
76528	Clad sheet	450–470	130–140	≥12
2280	Casting	500–521	149–160	12–20 (S) 1–8 (D)
				1–8 (D)
2285	Casting	510–527	160–177 (S)	22–26 (S)
			205–232 (D)	1–3 (D)
2550	Casting	500–515	193–204 (S)	10–12 (S)
			165–177 (D)	7–9 (D)
4223	Casting	493–510	149–166	1–6
4225	Casting	515–532	221–232 (S)	7–9 (S)
			149–160 (D)	1–6 (D)
4450	Casting	527–543	149–160	1–6
4685	Casting	505–521	149–177	14–18
5500	Casting	427–438	room temperature	≥96

1) S = sand casting; D = die casting.

Source: Reference 4

TABLE 4.32 Recommended soaking time for solution heat treatment of wrought aluminium alloys specified in Indian Standards

Product form	Thickness	Soaking time	
		Air furnace	Salt bath
	mm	min	min
Annealed clad sheets	≤ 1.4	10–12	5
	> 1.4 ≤ 2.0	15–20	7
	> 2.0 ≤ 4.0	20–25	10
	> 4.0 ≤ 10	35–40	20
Annealed non-clad sheets	≤ 1.2	10–20	5
	> 1.2 ≤ 3.0	15–30	10
Annealed cold formed parts, hot-rolled plates, hot-pressed sections, rods, strips and bushes	> 3.0 ≤ 5.0	20–45	15
	> 5.0 ≤ 10	30–60	20
	> 10 ≤ 20	35–75	25
	> 20 ≤ 30	45–90	30
	> 30 ≤ 50	60–120	40
	> 50 ≤ 75	100–150	50
	> 75 ≤ 100	120–180	70
	> 100 ≤ 150	150–210	80

TABLE 4.32 Recommended soaking time for solution heat treatment of wrought aluminium alloys specified in Indian Standards (continued)

Product form	Thickness	Soaking time	
		Air furnace	Salt bath
	mm	min	min
Stampings and forgings	≤2.5	15–30	10
	>2.5 ≤5.0	20–45	15
	>5.0 ≤15	30–50	25
	>15 ≤30	40–60	40
	>30 ≤50	60–150	50
	>50 ≤75	150–210	60
	>75 ≤100	180–240	90–180
	>100 ≤150	210–360	120–140

Source: Reference 4

TABLE 4.33 Recommended soaking time for solution heat treatment of cast aluminium alloys specified in Indian Standards

Alloy designation	Soaking time h	Alloy designation	Soaking time h
Sand casting alloys		**Permanent mould casting alloys**	
2280	6–18	**2250**	6–18
2285	2–10	**2285**	2–10
2350	6–18	**4225**	4–12
4223	6–18	**4450**	6–18
4225	6–18	**4685**	6–18
4450	6–18	Source: Reference 4	
5500	12–24		

3.2 Quenching

Purpose: To produce a solid solution supersaturated with solute atoms and vacancies for the subsequent ageing treatment.

Recommended heat treatment cycle: Quench rapidly from the solution heat treatment temperature to room temperature. Most frequently, the parts are quenched in water maintained at a temperature of 40 °C. Parts of complex shape and other parts prone to distortion or warpage are commonly quenched in a medium that provides slower cooling. This medium may be hot water, boiling water, an aqueous solution of polyalkylene glycol, oil or some other fluid medium such as forced air or mist.

For optimum results precipitation during cooling should be avoided. This can be accomplished by minimizing the quench delay time (see Table 4.34) and cooling rapidly through the critical temperature range, particularly in the temperature range 400–260 °C.

Notes:

1) The highest strength, and most often the greatest corrosion resistance, is obtained with the fastest quenching rate.
2) Cold working of the natural ageing alloys should be completed within 2 h of quenching, or if severe forming is intended within 0.5 h. If this is impractical then natural ageing may be delayed or suppressed by holding at a temperature of –18 °C or lower.

TABLE 4.34 Maximum quench delay for wrought and cast aluminium alloys specified in Indian Standards

Thickness mm	Maximum quench delay[1] s
Wrought alloys	
≤ 0.4	5
> 0.4 ≤ 0.8	7
> 0.8 ≤ 2.4	10
> 2.4	15
Cast alloys	
–	10

[1] Quench delay time begins when the furnace door begins to open or when the first corner of a load emerges from a salt bath, and ends when the last corner of the load is immersed in the quenching agent.

Source: References 4 and 5

3.3 Ageing

After solution heat treatment and quenching, hardening can be achieved either by natural ageing or artificial ageing.

Purpose: To produce an optimum size and distribution of precipitates.

Natural ageing

In some alloys, sufficient precipitation takes place by holding at room temperature for a few days. This treatment yields products with properties that are adequate for many applications.

3.3.3 Artificial ageing

Alloys with slow precipitation reactions at room temperature are heated to a moderately high temperature (about 95-205 °C) and held for a sufficient length of time (see Table 4.31 for recommended temperatures and time) to achieve the desired properties.

Notes:
1) The response to artificial ageing of several aluminium-copper alloys is accelerated by strain hardening the solution heat treated material in excess of levels normally required for flattening and straightening.
2) Maximum strength is generally achieved by prolonged ageing at low temperature rather than rapid ageing at high temperature.

3.4 Re-heat treatment

Alloys which have been incorrectly heat treated for example by solution heat treatment at a temperature lower than that recommended can be re-solution treated and then precipitation treated to achieve the optimum properties. Reheat treatment of alclad products should, however, be limited to a minimum (see Table 4.35). Overheated structures are nonsalvageable.

TABLE 4.35 Re-heat treatment of alclad alloys

Thickness mm	Maximum number of re-heat treatments permissible
Wrought alloys	
≤ 0.5	0
> 0.5 ≤ 3.0	1
> 3.0	2
Source: Reference 4	

4 REFERENCES

Books

1. *ASM Handbook*, vol. 4, ASM International, Ohio, U.S.A., 1991.
2. J.E. Hatch (Ed.), *Aluminium: Properties and Physical* Metallurgy, American Society for Metals, Metals Park, Ohio, U.S.A., 1984.

3. *The Properties of Aluminium and Its Alloys*, 9th ed., The Aluminium Federation, Birmingham, U.K. 1993.

Standards

4. IS 8860 – 1978: *Code of Practice for Heat Treatment of Aluminium Alloys.*
5. ASTM B 597 – 1992: *Standard Practice for Heat Treatment of Aluminium Alloys.*
6. DIN 29850 – 1989: *Aerospace; Heat Treatment of Wrought Aluminium Alloys.*
7. ISO 3134 (1) – 1985: *Light Metals and their Alloys—Terms and Definitions—Part 1: Materials.*
8. ISO 3134 (5) – 1981: *Light Metals and their Alloys—Terms and Definitions—Part 5: Methods of Processing and Treatment.*

5

COPPER AND ITS ALLOYS

CHAPTER 5

Copper and Its Alloys

5.1	Copper	5.5
5.2	Designation system for copper and copper alloys	5.10
5.3	Temper designation system for copper and copper alloys	5.12
5.4	Cast copper and copper alloys	5.15
5.5	Wrought copper and copper alloys	5.41
5.6	Heat treatment of copper and copper alloys	5.98

5.1

COPPER

1 INTRODUCTION

Pure copper is a yellowish-red metal with a density of 8930 kg/m^3 at 20 °C. It is characterized by low strength, very high ductility , excellent electrical and thermal conductivity, outstanding corrosion resistance, and good forming and joining characteristics. It is slightly paramagnetic. A summary of some of the important physical properties of pure copper is given in Table 1.1.

Copper is used extensively:

a) in wrought and cast copper-base alloys for use in building construction, electrical and electronic products, industrial machinery and equipment, transportation applications, and consumer and general products;
b) as an alloying element in aluminium-base, nickel-base, zinc-base, tin-base and lead-base alloys, steels and cast irons;
c) in platings;
d) in claddings; and
e) in chemicals.

Types of commercially pure copper

Cathode copper (CuCATH): It is the direct product of electrolytic refining, and is used for the production of:

a) electrolytic high-conductivity coppers, both tough-pitch and oxygen-free;
b) high conductivity copper castings; and
c) high-copper alloys, such as cadmium–copper, silver–copper and tellurium–copper.

Electrolytic tough-pitch copper (CuETP): Commercially pure high-conductivity copper which has been refined by electrolytic deposition, then melted and oxidized to the tough-pitch condition with a controlled low oxygen content (generally, 0.02–0.04 %), and finally cast into shapes.

Fire-refined tough-pitch high-conductivity copper (CuFRHC): Commercially pure high-conductivity copper which is furnace refined without, at any stage, having been electrolytically refined. It is melted and oxidized to the tough-pitch condition with a controlled oxygen content, and then cast into various shapes. The maximum impurity limits in this type of copper is slightly higher than in CuETP, but the same minimum conductivity requirements apply.

Fire-refined tough-pitch copper (CuFRTP): Commercially pure copper which is fire-refined without, at any stage, having been electrolytically refined. It is melted and oxidized to the tough-pitch condition with a controlled oxygen content, and then cast into shapes. The conductivity of this type of copper may be appreciably below that of the high-conductivity coppers (CuETP CuFRHC and CuOF).

Arsenical tough-pitch copper (CuATP): Tough-pitch copper containing 0.20–0.50 % arsenic in solid solution. This type of copper has a relatively low electrical conductivity, but its tensile strength is slightly higher than that of the non-arsenical varieties. Arsenic also raises the softening temperature and enhances corrosion resistance in specific environments.

Oxygen-free copper (CuOF): Commercially pure high-conductivity copper made by remelting and pouring electrodeposited copper in a protective gas atmosphere or in vacuum. No oxygen is absorbed by the copper, which is thus free from oxide or deoxidants. It is not susceptible to embrittlement when heated in a reducing atmosphere.

Phosphorus-deoxidized copper, high residual phosphorus (CuDHP): Phosphorus-deoxidized copper containing 0.015–0.10 % phosphorus. This type of copper has low electrical conductivity due to the high amount of phosphorus present. It is not susceptible to hydrogen embrittlement.

Phosphorus-deoxidized arsenical copper (CuDPA): Phosphorus-deoxidized copper containing 0.20–0.50 % arsenic in solid solution. This type of copper has relatively low electrical conductivity, but its tensile strength is slightly higher than that of the non-arsenical varities. Arsenic also raises the softening temperature and enhances corrosion resistance in specific environments.

2 SPECIFICATION

TABLE 5.1 Chemical composition of grades of commercially pure copper, conforming to IS 191 (IV to X) – 1980

Designation	Chemical composition [(% (m/m)]											
	Cu + Ag	P	As	Sb	Bi	Fe	Pb	Ni	O	Te	Se + Te	Sn	Total impurities
CuCATH	≥99.90	–	–	–	≤0.001	–	≤0.005	–		–	–	–	≤0.03[1)]
CuETP	≥99.90	–	–	–	≤0.001	–	≤0.005	–	–	–	–	–	≤0.03[2)]
CuFRHC	≥99.90	–	–	–	≤0.0025	–	≤0.005	–	–	–	–	–	≤0.04[2)]
CuFRTP-1	≥99.85	–	≤0.02	≤0.005	≤0.003	≤0.010	≤0.005	≤0.05	≤0.10	–	≤0.025	≤0.01	≤0.05[2)]
CuFRTP-2	≥99.50	–	≤0.10	≤0.05	≤0.02	≤0.03	≤0.10	≤0.50	≤0.10	–	≤0.07	≤0.05	–
CuDHP	≥99.80	0.015–0.10	≤0.05	≤0.005	≤0.003	≤0.03	≤0.01	≤0.10	–	≤0.01	≤0.02	≤0.01	≤0.06[3)]
CuATP	≥99.20	–	0.20–0.50	≤0.01	≤0.005	≤0.02	≤0.02	≤0.15	≤0.10	–	≤0.03	≤0.03	–
CuDPA	≥99.20	0.015–0.10	0.20–0.50	≤0.01	≤0.003	≤0.03	≤0.010	≤0.15	–	≤0.01	≤0.02	≤0.01	≤0.07[3)]

1) Excluding Ag.
2) Excluding Ag and O.
3) Excluding Ag, P, As and Ni.

3 ELECTRICAL PROPERTIES

TABLE 5.2 Electrical properties of grades of commercially pure copper, conforming to IS 191 (IV to X) – 1980

Designation	Mandatory value	Equivalent values for guidance only		
	Mass resistivity	Volume resistivity	Electrical conductivity	
	Ω g/m²	Ω mm²/m	MS/m	% IACS
CuCATH	≤0.15328	≤0.01724	≥58.00	≥100.0
CuETP	≤0.15328[1)]	≤0.01724[1)]	≥58.00[1)]	≥100.0[1)]
	≤0.15596[2)]	≤0.01754[2)]	≥57.00[2)]	≥ 98.3[2)]
CuFRHC	≤0.15328[1)]	≤0.01724[1)]	≥58.00[1)]	≥100.0[1)]
	≤0.15596[2)]	≤0.01754[2)]	≥57.00[2)]	≥ 98.3[2)]

1) When copper is used for electrical purposes.
2) When copper is used for non-electrical purposes.

Source: Reference 5

4 EQUIVALENT GRADES

TABLE 5.3 Comparison of grades of commercially pure copper specified in IS 191 (IV to X) – 1980 with nearest equivalent grades specified in other national and international standards

IS 191 (IV to X)–80	ASTM B 5–89 ASTM B 115–91 ASTM B 216–89 ASTM B 379–80 ASTM B 623–80	BS 6017–81	DIN 1708–73	ISO 431–81	JIS H 2121–61 JIS H 2123–90	SAE
CuCATH	B 115 - CATH	Cu-CATH-2	KE-Cu	Cu-CA TH	H 2121 – Electrolytic cathode copper	–
CuETP	B 5 - ETP (C11000)	Cu-ETP-2	E1-Cu58	Cu-ETP	H 2123 – C 1100	–
CuFRHC	B 623 - FRHC (C11020)	Cu-FRHC	E2-Cu58	Cu-FRHC	–	–

(Contd.)

TABLE 5.3 Comparison of grades of commercially pure copper specified in IS 191 (IV to X) – 1980 with nearest equivalent grades specified in other national and international standards (continued)

IS 191 (IV to X)–80	ASTM B 5–89 ASTM B 115–91 ASTM B 216–89 ASTM B 379–80 ASTM B 623–80	BS 6017–81	DIN 1708–73	ISO 431–81	JIS H 2121–61 JIS H 2123–90	SAE
CuFRTP-1	B 216 - FRTP (C12500)	Cu-FRTP	F-Cu	Cu-FRTP	–	–
CuFRTP-2	–	–	–	–	–	–
CuDHP	B 379 - DHP (C12200)	Cu-DHP	SF-Cu	Cu-DHP	H 2123 – C 1220	–
CuATP	–	–	–	–	–	–
CuDPA	B 379 - DPA (C14200)	–	–	–	–	–

5 REFERENCES

Books

1. *ASM Handbook,* vol. 2, ASM International, Materials Park, Ohio, U.S.A., 1990.
2. *Copper and its Alloys in Engineering and Technology,* C.D.A. Publication No. 43, Copper Development Association, Herts, U.K., 1955.

Standards

3. IS 191 (I to X) – 1980: *Specification for Copper.*
4. ISO 197 (1) – 1983: *Copper and Copper Alloys—Terms and Definitions—Part 1: Materials.*
5. ISO 431 – 1981: Copper Refinery Shapes.
6. ASTM B 224 – 1992: *Classification of Coppers.*

5.2

DESIGNATION SYSTEM FOR COPPER AND COPPER ALLOYS

1 DESIGNATION SYSTEM

The grades of wrought and cast copper and copper alloys specified in the Indian Standards are designated on the basis of their chemical composition.

1.1 Unalloyed copper

The grades of unalloyed copper are designated as follows, in the order given:

a) by the chemical symbol Cu of the base element copper;

b) by the symbol indicating the type of copper,as follows;

Symbol	Type of copper
CATH	Cathode copper
ETP	Electrolytic tough-pitch copper
FRHC	Fire-refined tough-pitch high-conductivity copper
FRTP	Fire-refined tough-pitch copper
DHP	Phosphorus-deoxidized copper, high residual phosphorus
DPA	Phosphorus-deoxidised arsenical copper
ATP	Arsenical tough-pitch copper
OF	Oxygen-free copper

EXAMPLE CuETP

1.2 Wrought and cast copper alloys

The grades of wrought and cast copper alloys are designated as follows, in the order given:

a) by the chemical symbol Cu of the base element copper;

b) by the chemical symbol(s) of the alloying element(s). The sequence of symbols are in decreasing order of the value of their content; where the values of contents are the same for two or more elements, the corresponding symbols are indicated in alphabetical order, provided that the principal alloying element is listed first irrespective of its content.
In cases where more than two alloying elements are present, it is not necessary to list all of the minor constituents in the designation, except where they are essential for the proper designation of the alloy.
Each alloying element is followed by a whole number indicating its content, if present in nominal amounts of 1 % or more. When a range is specified for an alloying element, the rounded off mean is used in the designation. When only a minimum percentage is specified for the alloying element, the rounded off minimum percentage is used in the designation. When the mean of the range is halfway between two whole numbers, it is rounded off to the nearest even number; and

c) by the symbol K to distinguish between two alloys having the same composition but differing only in the limits of an impurity element.

Notes:

1) Due to similar composition limits wrought and cast alloys may have the same designation. Thus cast alloys should have any one the following prefixes for identification:
   a) G for sand casting;
   b) GC for chill casting;
   c) GD for die casting;
   d) GW for centrifugal casting; and
   e) GX for investment casting ;

2) The designation of a casting alloy in ingot form is derived from the composition specified for the corresponding alloy in the form of casting.

EXAMPLE CuZn30.

2 REFERENCES

Standards

1. IS 2378 – 1974: *Code for Designation of Copper and Copper Alloys.*
2. ISO 1190 (1) – 1982: *Copper and Copper Alloys—Code of Designation—Part 1: Designation of Materials.*

5.3

TEMPER DESIGNATION SYSTEM FOR COPPER AND COPPER ALLOYS

1 TEMPER DESIGNATION SYSTEM

The temper designation system used in India for wrought and cast copper and copper alloy products, except ingots, is described in IS 2378 – 1974. The temper designations define the sequence of basic treatments used to produce the various tempers. The temper designation follows the material designation and is separated from it by a hyphen.

Basic temper designations consist of individual capital letters. Subdivisions of the basic tempers, where required, are indicated by a second letter following the letter of the basic temper. These second letters designate specific sequence of basic treatments, but only those treatments or operations which significantly influence the product characteristics are recognized. Special subdivisions are indicated by a third letter where necessary.

1.1 Basic temper designations

M— *As manufactured:* Applies to products of shaping processes in which no special control over thermal condition or strain-hardening is exercised.

O—*Annealed:* Applies to wrought products which are fully annealed and to cast products which are annealed to improve ductility and dimensional stability.

H— *Strain-hardened (wrought products only):* Applies to products subjected to the application of cold work after annealing or to a combination of cold work and partial annealing/stabilizing in order to secure the specified mechanical properties. The capital letter H is always followed by a second letter indicating the various stages of strain-hardening.

T—*Thermally treated to produce tempers other than* M, O *or H:* Applies to products which have their tensile strength increased by thermal treatment, with or without supplementary strain-hardening. The capital letter T is always followed by a second letter indicating the specific sequence of treatments.

1.2 Subdivisions of basic temper designations

1.2.1 Subdivisions of O *temper: Annealed*

The basic temper designation O applies to products which have no special requirement for grain size. A second letter following the capital letter O indicates a product in the annealed condition having special characteristics.

OS*:* Applies to copper and copper alloy products required to be specially annealed to obtain restricted grain size ranges. The designation OS should be followed by figures indicating the nominal grain size, the maximum and minimum limits of which are listed for each alloy and product in the relevant document on mechanical properties.

1.2.2 Subdivisions of H *temper: Strain-hardened*

Subdivisions for various stages of strain hardening are HA, HB, HC, etc. The designations are in alphabetical order and in ascending order of tensile strength as indicated in the relevant mechanical property documents.

Note: If copper and copper alloys should be stress relieved after strain-hardening to improve stress-corrosion characteristics or dimensional stability after machining, this temper is indicated, as a special subdivision, by the capital letter *R* in the third place, for example HAR, HCR, etc.

1.2.3 Subdivisions of T *temper: Thermally treated*

The letter following the capital letter T indicate specific sequence of basic treatments, as follows:

TA—*Cooled from an elevated temperature shaping process and naturally aged:* Applies to products for which the rate of cooling from an elevated temperature shaping process, such as casting or extrusion, is controlled, or is such that the product is subject to natural ageing. Properties of some alloys in this temper are unstable.

TB—*Solution heat-treated and naturally aged:* Applies to products which receive no cold work after solution heat treatment except as may be required to flatten or straighten them. Properties of some alloys in this temper are unstable.

TC—*Cooled from an elevated temperature shaping process, cold worked and naturally aged:* Applies to products which are subject to a controlled amount of cold working following controlled cooling from an elevated temperature shaping process, such as forging or extrusion, to improve strength or reduce internal stresses. Properties of some alloys in this temper are unstable.

TD—*Solution heat-treated, cold worked and naturally aged:* Applies to products which are subjected to a controlled amount of cold working following solution heat treatment to improve strength or reduce internal stresses. Properties of some alloys in this temper are unstable.

TE—*Cooled from an elevated temperature shaping process and precipitation treated:* Applies to products which are precipitation-treated following cooling from an elevated temperature shaping process such as casting or extrusion. May be achieved by precipitation treatment of TA temper products or, in some cases, by precipitation treatment of M temper products.

TF—*Solution heat-treated and precipitation-treated:* Applies to products which are precipitation treated following TB treatment.

TG—*Cooled from an elevated temperature shaping process, cold worked and precipitation-treated:* Applies to products which are precipitation-treated following TC treatment.

TH—*Solution heat-treated, cold worked and precipitation treated:* Applies to products which are precipitation-treated following TD treatment.

TK—*Cooled from an elevated temperature shaping process, precipitation treated and cold worked:* Applies to products which are subjected to a controlled amount of cold working following TE treatment.

TL—*Solution heat-treated, precipitation-treated and cold worked:* Applies to products which are subjected to a controlled amount of cold working following TF treatment.

1.3 Further variations of temper designations

If necessary, a third letter (or a digit) may be used to identify two or more variations of a subdivision of basic tempers H and T.

2 REFERENCES

Standards

1. IS 2378 – 1974: *Code for Designation of Copper and Copper Alloys.*
2. ISO 1190 (2) – 1982: *Copper and Copper Alloys—Code of Designation—Part 2: Designation of Tempers.*

5.4

CAST COPPER AND COPPER ALLOYS

1 INTRODUCTION

1.1 High conductivity coppers

High conductivity coppers are unalloyed coppers (Cu ≥ 99.80 %) which are characterized by low strength and hardness, good corrosion and oxidation resistance, and high electrical and thermal conductivity. However, to produce a sound copper casting care must be exercised during melting and casting.

1.2 Brasses

Brasses are copper–zinc alloys, with or without relatively small quantities of other elements, namely aluminium, iron, lead, manganese, nickel, silicon and tin. These alloys are characterized by moderate to high strength (depending on grade), good casting characteristics, good corrosion resistance, good electrical and thermal conductivity, and an attractive colour.

Lead is frequently added to brass in amounts of 0.5–3.0 % to improve machinability and pressure-tightness. Addition of tin to the extent of 1.0–1.5 % helps improve corrosion resistance. Aluminium in amounts of 0.2– 0.3 % is sometimes added to increase fluidity and to give a smooth surface.

High tensile brasses: These are copper–zinc alloys whose strength is considerably increased by the addition of elements such as aluminium, iron, manganese, nickel and tin. These alloys are characterized by high strength, good wear resistance and reasonably good corrosion resistance.

Silicon brasses: These are copper-zinc-silicon alloys which possess attractive mechanical properties (comparable to those of certain of the $\alpha + \beta$ high tensile brasses), good bearing characteristics, good corrosion resistance (superior to that of the high tensile brasses), and good casting characteristics.

1.3 Gunmetals

Gunmetals are copper–tin–zinc alloys which may additionally contain lead (for improved pressure tightness and machinability), and nickel. These alloys have moderate strength, good casting characteristics, good corrosion resistance, good wear resistance and low coefficient of friction.

1.4 Leaded tin bronzes

Leaded tin bronzes are copper–tin–lead alloys which may additionally contain zinc. They are used for bearing applications where a combination of wear resistance and good anti-friction properties are desired. These alloys also possess good casting characteristics, good corrosion resistance, good machinability and good pressure tightness.

1.5 Phosphor bronzes

Phosphor bronzes are copper–tin–phosphorus alloys which may sometimes also contain lead. These alloys are characterized by high hardness, good wear resistance, good toughness, good bearing properties, and good corrosion resistance.

1.6 Aluminium bronzes

Aluminium bronzes are copper–aluminium alloys which normally contain iron, often with nickel and manganese. These alloys are characterized by high strength, good ductility, good resistance to cavitation erosion and wear, excellent resistance to corrosion and oxidation, good bearing properties, and good casting and welding characteristics. These alloys also retain a useful proportion of their strength at elevated temperatures.

1.7 Silicon bronzes

Silicon bronzes are copper–silicon alloys which are characterized by high strength, very good aqueous and atmospheric corrosion resistance, and excellent weldability.

2 SPECIFICATIONS

2.1 *Chemical composition*

2.1.1 *Ingots*

TABLE 5.4 Chemical composition of copper alloy ingots specified in Indian Standards

Designation/ Grade	Chemical composition [% (*m*/*m*)]														
	Cu	Sn	Zn	Pb	P	Ni	Fe	Al	Mn	Sb	As	Si	Bi	Mg	Other elements	Total impurities
Leaded brass (IS 292 – 1983)																
LCB 1	70.0–77.0[1)]	1.0–3.0	Rem.	2.0–5.0	–	–	≤ 0.50	≤ 0.01	–	–	–	–	–	–	–	–
LCB 2	63.0–67.0[1)]	≤1.5	Rem.	1.0–3.0	–	–	≤ 0.50	≤ 0.01	–	–	–	–	–	–	–	–
Brass for gravity die castings (IS 1264 – 1989)																
DCB 1	59.0–62.0	–	Rem.	≤0.25	–	–	–	≤ 0.25	–	–	–	–	–	–	–	≤ 0.75[2)]
DCB 2	58.0–63.0	≤1.0	Rem.	0.5–2.5	–	≤ 1.0	≤ 0.5	0.2–0.8	≤ 0.5	–	–	–	–	–	–	≤ 2.00[2)]
High tensile brass (IS 304 – 1981)																
HTB 1	≥55.0[1)]	≤1.0	Rem.	≤0.50	–	–	0.7–2.0	0.5–2.5	≤3.0	–	–	≤0.10	–	–	Total: ≤0.20	–
HTB 2	≥55.0[1)]	≤0.20	Rem	≤0.20	–	–	1.5–3.25	3.0–6.0	≤4.0	–	–	≤0.10	–	–	Total: ≤0.20	–

(Contd.)

TABLE 5.4 Chemical composition of copper alloy ingots specified in Indian Standards (continued)

Designation/ Grade	Chemical composition [% (m/m)]														
	Cu	Sn	Zn	Pb	P	Ni	Fe	Al	Mn	Sb	As	Si	Bi	Mg	Other elements	Total impurities
Silicon brass (IS 11109 – 1984)																
1	≥79.0	–	12.5–16.0	≤0.50	–	–	≤0.30	≤0.50	–	–	–	3.2–5.0	–	–	Total: ≤0.50	–
2	≥88.0	–	4.5–7.0	≤0.50	–	–	≤0.30	–	–	–	–	3.7–5.5	–	–	Total: ≤0.50	–
3	80.0–83.0	–	Rem.	≤0.40	–	–	≤0.30	≤0.05	–	–	–	4.1–4.7	–	–	Total: ≤0.50	–
Tin bronze (IS 306 – 1983)																
–	Rem.	9.5–10.5	1.75–3.25	≤1.5	–	≤1.0	≤0.15	≤0.01	–	–	–	≤0.02	≤0.03	–	–	≤0.50[3)]
Leaded tin bronze (IS 318 – 1981)																
LTB 1	Rem.	6.0–8.0	1.5–3.0	2.5–3.5	–	≤2.0	≤0.30	≤0.01	–	≤0.3	–	≤0.01	–	–	–	≤0.70[4)]
LTB 2	Rem.	4.0–6.0	4.0–6.0	4.0–6.0	–	≤2.0	≤0.35	≤0.01	–	≤0.4	–	≤0.02	–	–	–	≤0.80[4)]
LTB 3	Rem.	6.0–8.0	≤0.75	9.0–11.0	–	≤2.0	≤0.35	≤0.01	–	≤0.5	–	≤0.02	–	–	–	≤0.80[5)]
LTB 4	Rem.	6.0–8.0	≤0.75	14.0–16.0	–	≤2.0	≤0.35	–	–	≤0.5	–	≤0.02	–	–	–	≤0.80[5)]
LTB 5	Rem.	9.0–11.0	≤1.0	8.5–11.0	–	≤2.0	≤0.35	≤0.01	–	≤0.5	–	≤0.02	–	–	–	≤0.80[5)]
LTB 6	Rem.	4.0–6.0	≤1.0	18.0–23.0	–	≤2.0	≤0.35	–	–	≤0.5	–	≤0.01	–	–	–	≤0.80[5)]

(Contd.)

TABLE 5.4 Chemical composition of copper alloy ingots specified in Indian Standards (continued)

Designation/ Grade	Chemical composition [% (m/m)]														
	Cu	Sn	Zn	Pb	P	Ni	Fe	Al	Mn	Sb	As	Si	Bi	Mg	Other elements	Total impurities
Phosphor bronze (IS 28 – 1985)																
1	Rem.	6.0–8.0	≤0.50	≤0.25	0.30–0.50	≤0.70	≤0.30	≤0.01	–	≤0.10	–	≤0.02	–	–	–	≤1.20
2	Rem.	≥10.0	≤0.05	≤0.25	≥0.50	≤0.10	≤0.10	≤0.01	–	–	–	≤0.02	–	–	–	≤0.60
3	Rem.	6.5–8.5	≤2.0	2.0–5.0	≥0.30	≤1.0	–	–	–	–	–	–	–	–	–	≤0.50
4	Rem.	9.0–11.0	≤0.05	≤0.25	≤0.15	≤0.25	–	–	–	–	–	–	–	–	–	≤0.80
5	Rem.	11.0–13.1	≤0.30	≤0.50	≥0.15	≤0.50	≤0.15	≤0.01	–	–	–	≤0.02	–	–	–	≤0.20
Aluminium bronze (IS 305 – 1981)																
AB 1	Rem.	≤0.10	≤0.50	≤0.05	–	≤1.0	1.5–3.5	8.5–10.5	≤1.0	–	–	≤0.25	–	≤0.05	≤0.30[6)]	–
AB 2	Rem.	≤0.10	≤0.50	≤0.05	–	4.0–5.5	4.0–5.5	8.8–10.0	≤1.5	–	–	≤0.10	–	≤0.05	≤0.30[6)]	–

(Contd.)

TABLE 5.4 Chemical composition of copper alloy ingots specified in Indian Standards (continued)

Designation/Grade	Chemical composition [% (m/m)]														
	Cu	Sn	Zn	Pb	P	Ni	Fe	Al	Mn	Sb	As	Si	Bi	Mg	Other elements	Total impurities
Silicon bronze (IS 1028 – 1987)																
–	≥ 89.0	≤ 1.0	≤ 5.0	≤ 0.5	–	–	≤ 2.5	≤ 1.5	≤ 1.5	–	–	1.0–5.0	–	–	–	–

1) Including Ni.
2) Excluding Pb, Ni and Al.
3) Including Fe, Al, Sb, As, Si, and Bi.
4) Excluding Ni.
5) Excluding Zn and Ni.
6) Total of Sn, Pb, Si, and Mg.

TABLE 5.5 Chemical composition of copper and copper alloy castings specified in Indian Standards

Designation/Grade	Chemical composition [% (m/m)]															
	Cu	Sn	Zn	Pb	P	Ni	Fe	Al	Mn	Sb	As	Si	Bi	Mg	Other elements	Total impurities
High conductivity copper (IS 9805 – 1981)																
1	≥ 99.90	–	–	≤ 0.005	–	–	–	–	–	–	–	–	≤ 0.0025	–	–	≤ 0.04 1)
2	≥ 99.80 2)	≤ 0.01	–	≤ 0.01	≤ 0.03	≤ 0.10	≤ 0.03	–	–	≤ 0.005	≤ 0.05	–	≤ 0.001	–	Te ≤ 0.01, Se + Te ≤ 0.02	≤ 0.06 3)

(Contd.)

TABLE 5.5 Chemical composition of copper and copper alloy castings specified in Indian Standards (continued)

Designation/ Grade	Chemical composition [% (m/m)]														
	Cu	Sn	Zn	Pb	P	Ni	Fe	Al	Mn	Sb	As	Si	Bi	Mg	Other elements	Total impurities
Leaded brass (IS 292 – 1983)																
LCB 1	70.0–80.0[4)]	1.0–3.0	Rem.	2.0–5.0	–	–	≤ 0.75	≤ 0.01	–	–	–	–	–	–	–	–
LCB 2	63.0–70.0[4)]	≤ 1.5	Rem.	1.0–3.0	–	–	≤ 0.75	≤ 0.01	–	–	–	–	–	–	–	–
Brass for gravity die castings (IS 1264 – 1989)																
DCB 1	59.0–63.0	–	Rem.	≤ 0.25	–	–	–	≤ 0.5	–	–	–	–	–	–	–	≤ 0.75[5)]
DCB 2	58.0–63.0	≤ 1.0	Rem.	0.5–2.5	–	≤ 1.0	≤ 0.8	0.2–0.8	≤ 0.5	–	–	–	–	–	–	≤ 2.00[5)]
High tensile brass (IS 304 – 1981)																
HTB 1	≥ 55.0[4)]	≤ 1.0	Rem.	≤ 0.50	–	–	0.7–2.0	0.5–2.5	≤ 3.0	–	–	≤ 0.10	–	–	Total ≤ 0.20	–
HTB 2	≥ 55.0[4)]	≤ 0.20	Rem.	≤ 0.20	–	–	1.5–3.25	3.0–6.0	≤ 4.0	–	–	≤ 0.10	–	–	Total ≤ 0.20	–
Silicon brass (IS 11109 – 1984)																
1	≥ 79.0	–	12.0–16.0	≤ 0.50	–	–	≤ 0.30	≤ 0.50	–	–	–	3.0–5.0	–	–	Total ≤ 0.50	–

(Contd.)

TABLE 5.5 Chemical composition of copper and copper alloy castings specified in Indian Standards (continued)

Designation/ Grade	Chemical composition [% (m/m)]															
	Cu	Sn	Zn	Pb	P	Ni	Fe	Al	Mn	Sb	As	Si	Bi	Mg	Other elements	Total impurities
Silicon brass (IS 11109 – 1984)																
2	≥88.0	–	4.0–7.0	≤0.50	–	–	≤0.30	–	–	–	–	3.5–5.5	–	–	Total ≤ 0.50	–
3	80.0–83.0	–	Rem.	≤0.40	–	–	≤0.30	≤0.05	–	–	–	3.9–4.7	–	–	Total ≤ 0.50	–
Tin bronze (IS 306–1983)																
–	Rem.	9.5–10.5	1.5–3.0	≤1.5	–	≤1.0	≤0.15	≤0.01	–	–	–	≤0.02	≤0.03	–	–	≤0.50[6)]
Leaded tin bronze (IS 318 – 1981)																
LTB 1	Rem.	6.0–8.0	1.5–3.0	2.5–3.5	–	≤2.0	≤0.30	≤0.01	–	≤0.3	–	≤0.01	–	–	–	≤0.70[7)]
LTB 2	Rem.	4.0–6.0	4.0–6.0	4.0–6.0	–	≤2.0	≤0.35	≤0.01	–	≤0.4	–	≤0.02	–	–	–	≤0.80[7)]
LTB 3	Rem.	6.0–8.0	≤0.75	9.0–11.0	–	≤2.0	≤0.35	≤0.01	–	≤0.5	–	≤0.02	–	–	–	≤0.80[8)]
LTB 4	Rem.	6.0–8.0	≤0.75	14.0–16.0	–	≤2.0	≤0.35	–	–	≤0.5	–	≤0.02	–	–	–	≤0.80[8)]
LTB 5	Rem.	9.0–11.0	≤1.0	8.5–11.0	–	≤2.0	≤0.35	≤0.01	–	≤0.5	–	≤0.02	–	–	–	≤0.80[8)]
LTB 6	Rem.	4.0–6.0	≤1.0	18.0–23.0	–	≤2.0	≤0.35	–	–	≤0.5	–	≤0.01	–	–	–	≤0.80[8)]

(Contd.)

TABLE 5.5 Chemical composition of copper and copper alloy castings specified in Indian Standards (continued)

Designation/ Grade	Chemical composition [% (m/m)]															
	Cu	Sn	Zn	Pb	P	Ni	Fe	Al	Mn	Sb	As	Si	Bi	Mg	Other elements	Total impurities
Phosphor bronze (IS 28 – 1985)																
1	Rem.	6.0–8.0	≤0.50	≤0.25	0.30–0.50	≤0.70	≤0.30	≤0.01	–	≤0.10	–	≤0.02	–	–	–	≤1.20
2	Rem.	≥10.0	≤0.05	≤0.25	≥0.50	≤0.10	≤0.10	≤0.01	–	–	–	≤0.02	–	–	–	≤0.60
3	Rem.	6.5–8.5	≤2.0	2.0–5.0	≥0.30	≤1.0	–	–	–	–	–	–	–	–	–	≤0.50
4	Rem.	9.0–11.0	≤0.05	≤0.25	≤ 0.15	≤0.25	–	–	–	–	–	–	–	–	–	≤0.80
5	Rem.	11.0–13.1	≤0.30	≤0.50	≥0.15	≤0.50	≤0.15	≤0.01	–	–	–	≤0.02	–	–	–	≤0.20
Aluminium bronze (IS 305 – 1981)																
AB 1	Rem.	≤0.10	≤0.50	≤0.05	–	≤1.0	1.5–3.5	8.5–10.5	≤1.0	–	–	≤0.25	–	≤0.05	≤0.30[9)]	≤0.30[9)]
AB 2	Rem.	≤0.10	≤0.50	≤0.05	–	4.0–5.5	4.0–5.5	8.8–10.0	≤1.5	–	–	≤0.10	–	≤0.05	≤0.30[9)]	≤0.30[9)]
Silicon bronze (IS 1028 – 1987)																
–	≥89.0	≤1.0	≤5.0	≤0.5	–	–	≤2.5	≤1.5	≤1.5	–	–	1.0–5.0	–	–	–	–

1) Excluding Ag and O.
2) Including Ag.
3) Excluding Ag, P, Ni, and As.
4) Including Ni.
5) Excluding Pb, Ni and Al.
6) Including Fe, Al, Sb, As, Si, and Bi.
7) Excluding Ni
8) Excluding Zn and Ni.
9) Total of Sn, Pb, Si, and Mg.

2.2 Mechanical properties

TABLE 5.6 Mechanical properties[1)] of cast copper alloys specified in Indian Standards

Designation/ Grade	Sand cast			Chill cast			Continuously cast		
	R_m min. MPa	$R_{p0.2}$ min. MPa	A min. %	R_m min. MPa	$R_{p0.2}$ min. MPa	A min. %	R_m min. MPa	$R_{p0.2}$ min. MPa	A min. %
Brass for gravity die castings (IS 1264 – 1989)									
DCB 1	–	–	–	275	–	23	–	–	–
DCB 2	–	–	–	315	–	20	–	–	–
High tensile brass (IS 304 – 1981)									
HTB 1	470	170[2)]	18	500	210[2)]	18	–	–	–
HTB 2	740	400[2)]	11	–	–	–	–	–	–
Silicon brass (IS 11109 – 1984)									
1	414	165[2)]	16[3)]	–	–	–	–	–	–
2	414	207[2)]	16[3)]	–	–	–	–	–	–
3	390	175[2)]	25[3)]	–	–	–	–	–	–
Tin Bronze (IS 306 – 1983)									
–	260	120	13	210	120	3	–	–	–
Leaded tin bronze (IS 318 – 1981)									
LTB 1	250	130[2)]	16	250	130[2)]	5	–	–	–
LTB 2	190	100[2)]	13	190	100[2)]	6	–	–	–
LTB 3	175	75[2)]	4	200	75[2)]	3	–	–	–
LTB 4	160	70[2)]	4	190	70[2)]	3	–	–	–
LTB 5	190	80[2)]	5	220	140[2)]	3	–	–	–
LTB 6	140	60[2)]	5	150	60[2)]	5	–	–	–

(Contd.)

TABLE 5.6 Mechanical properties[1] of cast copper alloys specified in Indian Standards (continued)

Designation/ Grade	Sand cast			Chill cast			Continuously cast		
	R_m min. MPa	$R_{p0.2}$ min. MPa	A min. %	R_m min. MPa	$R_{p0.2}$ min. MPa	A min. %	R_m min. MPa	$R_{p0.2}$ min. MPa	A min. %
Phosphor bronze (IS 28 – 1985)									
1	190	–	3	205	–	5	275	–	8
2	220	–	3	310	–	2	360	–	7
3	190	–	3	220	–	2	270	–	5
4	230	–	6	270	–	5	310	–	9
5	220	–	5	270	–	3	310	–	5
Aluminium bronze (IS 305 – 1981)									
AB 1	500	170[2]	18	540	200[2]	18	–	–	–
AB 2	640	250[2]	13	650	250[2]	13	–	–	–
Silicon bronze (IS 1028 – 1987)									
–	310	–	20	–	–	–	–	–	–

1) Measured on separately cast test bars.
2) For information only.
3) $L_o = 4\sqrt{S_o}$.

3 TYPICAL PHYSICAL PROPERTIES

TABLE 5.7 Typical physical properties of cast copper and copper alloys specified in Indian Standards

Designation/ Grade	Density kg/m^3	Coefficient of thermal expansion between 0 °C and 250 °C μm/(m K)	Thermal conductivity at 15 °C W/(m K)	Electrical conductivity at 15 °C % IACS	Electrical resistivity at 15 °C μΩ m
High conductivity copper (IS 9805 – 1981)					
1	8900	17	372	90	0.019
2	–	–	–	80	–
Leaded brass (IS 292 – 1983)					
LCB 1	8500	19	81	18	0.09
LCB 2	8400	20	90	20	0.08
Brass for gravity die castings (IS 1264 – 1989)					
DCB 1	8300	21	81	18	0.09
DCB 2	8300	21	81	18	0.09
High tensile brass (IS 304 – 1981)					
HTB 1	8300	21	87	22	0.08
HTB 2	7900	21	42	8	0.22
Silicon brass (IS 11109 – 1984)					
1	8280	19.6[1)]	28[2)]	6.7[2)]	0.284[2)]
2	8300	–	–	6.0[2)]	–
3	8280	19.6[1)]	28[2)]	6.7[2)]	0.284[2)]

(Contd.)

TABLE 5.7 Typical physical properties of cast copper and copper alloys specified in Indian Standards (continued)

Designation/ Grade	Density kg/m^3	Coefficient of thermal expansion between 0 °C and 250 °C µm/(m K)	Thermal conductivity at 15 °C W/(m K)	Electrical conductivity at 15 °C % IACS	Electrical resistivity at 15 °C µΩ m
Tin bronze (IS 306 – 1983)					
–	8800	18	47	11	0.16
Leaded tin bronze (IS 318 – 1981)					
LTB 1	8800	18	61	13	0.13
LTB 2	8800	18	71	15	0.11
LTB 3	–	–	–	–	–
LTB 4	9250	18.5[3)]	52[2)]	11.5[2)]	–
LTB 5	9000	19	47	10	0.17
LTB 6	9200	19	71	14	0.11
Phosphor bronze (IS 28 – 1985)					
1	–	–	–	–	–
2	8800	18	47	9	0.17
3	8800	18	47	11	0.16
4	8800	18	50	11	0.16
5	8800	19	45	9	0.19
Aluminium bronze (IS 305 – 1981)					
AB 1	7600	17	61	13	0.13
AB 2	7600	17	42	8	0.22

TABLE 5.7 Typical physical properties of cast copper and copper alloys specified in Indian Standards (continued)

Designation/ Grade	Density kg/m³	Coefficient of thermal expansion between 0 °C and 250 °C µm/(m K)	Thermal conductivity at 15 °C W/(m K)	Electrical conductivity at 15 °C % IACS	Electrical resistivity at 15 °C µΩ m
Silicon bronze (IS 1028 –1987)					
–	8360	19.6[1)]	28[2)]	6.7[2)]	–

1) Between 20 °C and 260 °C.
2) At 20 °C.
3) Between 20 °C and 205 °C.

Source: References 1 and 15

4 RECOMMENDED CASTING TEMPERATURE

TABLE 5.8 Recommended casting temperatures for cast copper and copper alloys specified in Indian Standards

Designation/ Grade	Casting temperature, in °C, for a section thickness, in mm, of		
	< 13	≥ 13 ≤ 39	> 40
High conductivity copper (IS 9805 – 1981)			
1	1200	1170	1130
2	–	–	–
Leaded brass (IS 292 – 1983)			
LCB 1	1150	1100	1070
LCB 2	1100	1050	1020
Brass for gravity die castings (IS 1264 – 1989)			
DCB 1	1050	1050	1050
DCB 2	1050	1050	1050
High tensile brass (IS 304 – 1981)			
HTB 1	1060	1020	980
HTB 2	1060	1020	980
Silicon brass (IS 11109 – 1984)			
1	–	–	–
2	–	–	–
3	–	–	–
Tin bronze (IS 306 – 1983)			
–	1200	1170	1130

Designation/ Grade	Casting temperature, in °C, for a section thickness, in mm, of		
	< 13	≥ 13 ≤ 39	> 40
Leaded tin bronze (IS 318 – 1981)			
LTB 1	1200	1160	1120
LTB 2	1200	1150	1120
LTB 3	–	–	–
LTB 4	–	–	–
LTB 5	1130	1080	1030
LTB 6	1090	1030	1010
Phosphor bronze (IS 28 – 1985)			
1	–	–	–
2	1120	1100	1040
3	1130	1050	1030
4	1100	1070	1040
5	1170	1120	1070
Aluminium bronze (IS 305 – 1981)			
AB 1	1250	1200	1150
AB 2	1240	1170	1120
Silicon bronze (IS 1028 – 1987)			
–	1200	1160	1120

Source: Reference 4

5 COMPARISON RATINGS OF CAST COPPER AND COPPER ALLOYS

TABLE 5.9 Comparison of the characteristics of cast copper and copper alloys specified in Indian Standards

Designation/ Grade	Suitability for					Pressure tightness[b]		Machinability[c]
	sand casting[a]	gravity die casting[a]	continuous casting[a]	centrifugal casting[a]	chill casting[a]	Thin section	Thick section	
High conductivity copper (IS 9805 – 1981)								
1	2	3	3	2	–	1	2	3
2	–	–	–	–	–	–	–	–
Leaded brass (IS 292 – 1983)								
LCB 1	1	–	2	3	–	1	1	1
LCB 2	1	–	2	3	–	1[d]	1[d]	1
Brass for gravity die castings (IS 1264 – 1989)								
DCB 1	–	1	–	2	–	–	–	2
DCB 2	–	1	3	2	–	–	–	1
High tensile brass (IS 304 – 1981)								
HTB 1	2	2	–	2	–	1	1	3
HTB 2	2	4	–	2	–	1	1	3
Silicon brass (IS 11109 – 1984)								
1	–	–	–	–	–	–	–	–
2	–	–	–	–	–	–	–	–
3	–	–	–	–	–	–	–	–

(Contd.)

TABLE 5.9 Comparison of the characteristics of cast copper and copper alloys specified in Indian Standards (continued)

Designation/ Grade	Suitability for					Pressure tightness[b)]		Machinability[c)]
	sand casting[a)]	gravity die casting[a)]	continuous casting[a)]	centrifugal casting[a)]	chill casting[a)]	Thin section	Thick section	
Tin bronze (IS 306 – 1983)								
–	2	3	1	2	3	2	2	2
Leaded tin bronze (IS 318 – 1981)								
LTB 1	1	3	1	1	2	2	1	1
LTB 2	1	3	1	1	2	1	2	1
LTB 3	–	–	–	–	–	–	–	–
LTB 4	–	–	–	–	–	–	–	–
LTB 5	2	4	1	2	1	2	2	1
LTB 6	3	4	3	3	2	2	3	1
Phosphor bronze (IS 28 – 1985)								
1	–	–	–	–	–	–	–	–
2	2	3	1	1	1	3	3	2
3	2	3	1	2	1	2	2	1
4	2	–	–	–	–	2	3	2
5	2	3	1	1	1	3	3	2
Aluminium bronze (IS 305 – 1981)								
AB 1	2	1	3	2	–	1	1	3
AB 2	2	2	3	2	2	1	1	3

(Contd.)

TABLE 5.9 Comparison of the characteristics of cast copper and copper alloys specified in Indian Standards (continued)

Designation/ Grade	Suitability for					Pressure tightness[b)]		Machinability[c)]
	sand casting[a)]	gravity die casting[a)]	continuous casting[a)]	centrifugal casting[a)]	chill casting[a)]	Thin section	Thick section	
Silicon bronze (IS 1028 – 1987)								
–	–	–	–	–	–	–	–	–

a) 1 = excellent; 2 = satisfactory; 3 = possible with special techniques; 4 = unsuitable.
b) 1 = suitable; 2 = less suitable; 3 = unsuitable.
c) 1 = excellent; 2 = good; 3 = satisfactory with special techniques.
d) For pressure tight castings, aluminium should not be greater than 0.02 %.

Source: Reference 15

6 TYPICAL APPLICATIONS

TABLE 5.10 Typical applications of cast copper and copper alloys specified in Indian Standards

Designation/Grade	Typical applications
High conductivity copper (IS 9805 – 1981)	
1	Electrode clamps for arc furnaces, cooling rings for blast furnaces, lance nozzle tips for L.D. converters.
2	
Leaded brass (IS 292 – 1983)	
LCB 1	Cocks, hardware fittings, ornamental castings, plumbing fittings and fixtures, low pressure valves.
LCB 2	Door and furniture fittings, gas fittings, ornamental castings, plumbing and pipe fittings, switch-gear brush holders.
Brass for gravity die castings (IS 1264 – 1989)	
DCB 1	Gas fittings, ornamental castings, plumbing fittings.
DCB 2	Bushings, hardware fittings, lock hardware, ornamental castings.

(Contd.)

TABLE 5.10 Typical applications of cast copper and copper alloys specified in Indian Standards (continued)

Designation/Grade	Typical applications
High tensile brass (IS 304 – 1981)	
HTB 1	Gun mountings, hydraulic equipment, locomotive axle boxes, marine castings and fittings, marine propellers and cones, pump casings, rudders and rudder posts.
HTB 2	Rolling mill castings, spur and gear wheels which are heavily loaded and slow moving.
Silicon brass (IS 11109 – 1984)	
1	Bearings, gears, impellers, rocker arms, valve stems, small boat propellers.
2	Valve stems.
3	Brackets, brush holders, clamps, hexagonal nuts, high-strength thin-wall die castings, lever arms.
Tin bronze (IS 306 – 1983)	
–	Bearings, bushings, gears, pumps and pump fittings, valves, valve bodies, valve guides.
Leaded tin bronze (IS 318 – 1981)	
LTB 1	Bearings, elbows, engine components, pipes, pressure components, pumps, valve parts.
LTB 2	Gasoline-fittings and oil-line fittings, general plumbing hardware, fire-equipment fittings, general plumbing hardware, low-pressure valves, pipe fittings, small gears, small pump parts.
LTB 3	Bearings for high speed and heavy pressure.
LTB 4	Backs for lined journal bearings for locomotive tenders and passenger cars, freight car bearings, general purpose wearing metal for rod bushings, locomotive engine castings, pump impellers, and bodies for use in mine water.
LTB 5	Applications requiring corrosion resistance, bearings for high speed and heavy pressure, impellers, pressure-tight castings, pumps.
LTB 6	Steel backed bearings for car and aero engines. It is also used for its self-lubricating properties under conditions where lubrication is difficult.

(Contd.)

TABLE 5.10 Typical applications of cast copper and copper alloys specified in Indian Standards (continued)

Designation/Grade	Typical applications
Phosphor bronze (IS 28 – 1985)	
1	Connecting rod small-end bushes, locomotive slide valves, slippers for bridge girders.
2	Bearings and bushings for heavy loads and heavy duty with adequate lubrication, such as bearings for aero engines, diesel engines, electrical generators and rolling mills, gears and worm wheels, pump parts particularly in mine water and for marine work.
3	Bearings and bushings for lighter duties for use with limited lubrication.
4	Gears, general purpose castings.
5	Gears and worm wheels resistant to shock loads, heavily loaded bearings.
Aluminium bronze (IS 305 – 1981)	
AB 1	Acid-resisting pumps, bearings, bushings, gears, guides, pickling hooks, plungers, pump rods, non-sparking hardware, valve seats.
AB 2	Gears, fittings, propeller blades and hubs for fresh- and salt-water service, structural applications, valve guides and seals, worm wheels.
Silicon bronze (IS 1028 – 1987)	
–	Bearings, corrosion-resistant castings, impellers, marine fittings, pump and valve components.
Source: References 1 and 2	

7 EQUIVALENT GRADES

TABLE 5.11 Comparison of grades of copper alloy ingots for castings specified in Indian Standards with nearest equivalent grades specified in other national and international standards

IS	ASTM B 30–95	BS 1400–85	DIN 17656–73	GOST 614–73 GOST 1020–77	ISO	JIS H 2202–85 JIS H 2203–85 JIS H 2204–85 JIS H 2205–85 JIS H 2206–85 JIS H 2207–85	SAE
Leaded brass (IS 292 – 1983)							
LCB 1	C85200	SCB1	–	–	–	–	–
LCB 2	–	SCB3	GB-CuZn33Pb	–	–	YBsCIn 2	–
Brass for gravity die castings (IS 1264 – 1989)							
DCB 1	–	DCB1	–	–	–	–	—
DCB 2	C85710	DCB3	GB-CuZn37Pb	–	–	–	–
High tensile brass (IS 304 – 1981)							
HTB 1	C86500	HTB1	GB-CuZn35Al1	–	–	HBsCIn 2	–
HTB 2	–	HTB3	–	лАжМu	–	–	–
Silicon brass (IS 11109 – 1984)							
1	C87500	–	GB-CuZn15Si4	лK1	–	SzBCIn 2	–
2	C87600	–	–	–	–	–	–
3	C87800	–	GB-CuZn15Si4	–	–	SzBCIn 3	–
Tin bronze (IS 306 – 1983)							
–	C90500	G1	GB-CuSn10Zn	-	-	BCIn 3	–
Leaded tin bronze (IS 318 – 1981)							
LTB 1	–	LG4	–	–	–	–	–
LTB 2	C83600	LG2	GB-CuSn5ZnPb	БрО5ц6С5	–	BCIn 6	–

(Contd.)

TABLE 5.11 Comparison of grades of copper alloy ingots for castings specified in Indian Standards with nearest equivalent grades specified in other national and international standards (continued)

IS	ASTM B 30–95	BS 1400–85	DIN 17656–73	GOST 614–73 GOST 1020–77	ISO	JIS H 2202–85 JIS H 2203–85 JIS H 2204–85 JIS H 2205–85 JIS H 2206–85 JIS H 2207–85	SAE
Leaded tin bronze (IS 318 – 1981)							
LTB 3	–	–	–	–	–	–	–
LTB 4	C93800	–	GB-CuPb15Sn	–	–	LBCIn 4	–
LTB 5	C93700	LB2	GB-CuPb10Sn	–	–	LBCIn 3	–
LTB 6	C94100	LB5	GB-CuPb20Sn	–	–	–	–
Phosphor bronze (IS 28 – 1985)							
1	–	–	–	–	–	–	–
2	–	PB1	–	–	–	–	–
3	–	LPB1	–	–	–	–	–
4	–	CT1	GB-CuSn10	–	–	PBCIn 2	–
5	–	PB2	–	–	–	PBCIn 3	–
Aluminium bronze (IS 305 – 1981)							
AB 1	C95200	AB1	GB-CuAl10Fe	–	–	AlBCIn 1	–
AB 2	C95800	AB2	GB-CuAl10Ni	–	–	AlBCIn 3	–
Silicon bronze (IS 1028 – 1987)							
–	–	–	–	–	–	–	–

TABLE 5.12 Comparison of grades of copper and copper alloy castings specified in Indian Standards with nearest equivalent grades specified in other national and international standards

IS	ASTM B 148–93a ASTM B 176–95 ASTM B 271–93a ASTM B 505–95 ASTM B 584–93b ASTM B 806–93a	BS 1400–85	DIN 1705–81 DIN 1709–81 DIN 1714–81 DIN 1716–81 DIN 17655–81	GOST 493–79 GOST 613–79 GOST 17711–80	ISO	JIS H 5100–90 JIS H 5101–88 JIS H 5102–88 JIS H 5111–88 JIS H 5112–88 JIS H 5113–88 JIS H 5114–88 JIS H 5115–88	SAE J462–81
High conductivity copper (IS 9805 – 1981)							
1	–	HCC1	–	–	–	CuC 3	–
2	–	–	G-/GK-Cu L50 G-/GK-SCu L50	– –	– –	– –	– –
Leaded brass (IS 292 – 1983)							
LCB 1	C85200	SCB1	–	–	–	–	C85200
LCB 2	–	SCB3	G-CuZn33Pb	–	–	YBsC 2	–
Brass for gravity die castings (IS 1264 – 1989)							
DCB 1	–	DCB1	–	–	–	–	–
DCB 2	–	DCB3	GD-/GK-CuZn37Pb	–	–	–	–
High tensile brass (IS 304 – 1981)							
HTB 1	C86500	HTB1	G-/GZ/GK-CuZn35Al1	лц23А6ж3Мц2	–	HBsC 2/HBsC 2 C	C86500
HTB 2	–	HTB3	–	–	–	–	–

(Contd.)

TABLE 5.12 Comparison of grades of copper and copper alloy castings specified in Indian Standards with nearest equivalent grades specified in other national and international standards (continued)

IS	ASTM B 148–93a ASTM B 176–95 ASTM B 271–93a ASTM B 505–95 ASTM B 584–93b ASTM B 806–93a	BS 1400–85	DIN 1705–81 DIN 1709–81 DIN 1714–81 DIN 1716–81 DIN 17655–81	GOST 493–79 GOST 613–79 GOST 17711–80	ISO	JIS H 5100–90 JIS H 5101–88 JIS H 5102–88 JIS H 5111–88 JIS H 5112–88 JIS H 5113–88 JIS H 5114–88 JIS H 5115–88	SAE J462–81
Silicon brass (IS 11109 – 1984)							
1	C87500	–	G-/GD-/GK-CuZn15Si4	лц16К4	–	SzBC 2	C87500
2	C87600	–	–	–	–	–	–
3	C87800	–	G-/GD-/GK-CuZn15Si4	–	–	SzBC 3	C87800
Tin bronze (IS 306 – 1983)							
–	C90500	G1	G-CuSn10Zn	БрО10ц2	–	BC 3/BC 3 C	C90500
Leaded tin bronze (IS 318 – 1981)							
LTB 1	–	LG4	–	–	–	–	–
LTB 2	C83600	LG2	G-CuSn5ZnPb	БрО5ц5С5	–	BC 6 /BC 6 C	C83600
LTB 3	–	–	–	–	–	–	–
LTB 4	C93800	–	G-/GZ-/GC-CuPb15Sn	–	–	LBC 4/LBC 4 C	C93800
LTB 5	C93700	LB2	G-/GZ-/GC-CuPb10Sn	БрО10С10	–	LBC 3/LBC 3 C	C93700
LTB 6	C94100	LB5	G-CuPb20Sn	–	–	–	–

(Contd.)

TABLE 5.12 Comparison of grades of copper and copper alloy castings specified in Indian Standards with nearest equivalent grades specified in other national and international standards (continued)

IS	ASTM B 148–93a ASTM B 176–95 ASTM B 271–93a ASTM B 505–95 ASTM B 584–93b ASTM B 806–93a	BS 1400–85	DIN 1705–81 DIN 1709–81 DIN 1714–81 DIN 1716–81 DIN 17655–81	GOST 493–79 GOST 613–79 GOST 17711–80	ISO	JIS H 5100–90 JIS H 5101–88 JIS H 5102–88 JIS H 5111–88 JIS H 5112–88 JIS H 5113–88 JIS H 5114–88 JIS H 5115–88	SAE J462–81
Phosphor bronze (IS 28 – 1985)							
1	–	–	–	–	–	–	–
2	–	PB1	–	БрО10ф1	–	–	–
3	–	LPB1	–	–	–	–	–
4	–	CT1	G-CuSn10	–	–	PBC 2	–
5	–	PB2	–	–	–	PBC 3 B	–
Aluminium bronze (IS 305 – 1981)							
AB 1	C95200	AB1	G-/GK-/GZ-CuAl10Fe	БрА9ж3л	–	AlBC 1/AlBC 1 C	C95200
AB 2	C95800	AB2	G-/GK-/GZ-/GC-CuAl10Ni	БрА10ж4Н4л	–	AlBC 3/AlBC 3 C	C95800
Silicon bronze (IS 1028–1987)							
–	–	–	–	–	–	–	C87200

8 REFERENCES

Books

1. *ASM Handbook,* vol. 2, ASM International, Materials Park, Ohio, U.S.A., 1990.
2. *Copper and Copper Alloy Castings—Companion Volume to BS 1400–1961,* The Association of Bronze & Brass Founders, Birmingham, U.K., 1961.
3. *Copper and its Alloys in Engineering and Technology,* C.D.A. Publication No. 43, Copper Development Association, Herts, U.K., 1955.
4. C.J. Smithells (Ed.), *Metals Reference Book,* 5th ed., Butterworths, London, U.K., 1976.

Standards/article

5. IS 28 – 1985: *Specification for Phosphor Bronze Ingots and Castings.*
6. IS 292 – 1983: *Specification for Leaded Brass Ingots and Castings.*
7. IS 304 – 1981: *Specification for High Tensile Brass Ingots and Castings.*
8. IS 305 – 1981: *Specification for Aluminium Bronze Ingots and Castings.*
9. IS 306 – 1983: *Specification for Tin Bronze Ingots and Castings.*
10. IS 318 – 1981: *Specification for Leaded Tin Bronze Ingots and Castings.*
11. IS 1028 – 1987: *Specification for Silicon Bronze Ingots and Castings.*
12. IS 1264 – 1989: *Brass Gravity Die Castings (Ingots and Castings)—Specification.*
13. IS 9805 – 1981: *Specification for High Conductivity Copper Castings.*
14. IS 11109 – 1984: *Specification for Silicon Brass Ingots and Castings.*
15. BS 1400 – 1985: *Specification for Copper Alloy Ingots and Copper Alloy and High Conductivity Copper Castings.*
16. D.T. Peters and K.J.A. Kundig, Selecting Copper and Copper Alloys—Part II: Cast Products, Advanced Materials & Processes, 145(6), 1994.

5.5

WROUGHT COPPER AND COPPER ALLOYS

1 INTRODUCTION

1.1 Unalloyed coppers and high-copper alloys

Unalloyed coppers are characterized by low strength, high ductility, excellent corrosion resistance, and high electrical and thermal conductivity. They can be moderately strengthened by cold working.

High-copper alloys, containing small amounts of alloying elements, such as beryllium, chromium, or zirconium, are preferred for applications where higher strength and hardness, higher resistance to softening at elevated temperatures or other mechanical attributes are needed along with good electrical and thermal conductivity.

1.2 Brasses

Brasses are copper–zinc alloys, with or without relatively small quantities of other elements, namely aluminium, arsenic, iron, lead, manganese, nickel, silicon, and tin. These alloys can be produced to a wide range of mechanical properties and are characterized by good corrosion resistance, good electrical and thermal conductivity, good forming and joining characteristics, and an attractive colour.

Brasses are often classified on the basis of their microstructure into:

a) alpha brasses: These are single phase alloys which contain a minimum of 63 % copper. They are characterized by excellent cold-working properties; and
b) alpha-beta brasses: These are duplex alloys which usually contain between 38 % and 42 % zinc. These brasses possess excellent hot-working characteristics; however, their ability to be deformed at room temperature is limited.

Alloying elements are frequently added to brasses to:

a) Improve strength: Elements such as aluminium, manganese, nickel, silicon, and tin may be added in small amounts to improve strength, hardness and wear resistance;
b) Improve machinability: Machinability increases with lead content but a reduction in ductility restricts the amount of lead which can be added; and

c) Improve corrosion resistance: Addition of small amounts of aluminium, arsenic, nickel, and tin are made to improve corrosion resistance.

1.3 Phosphor bronzes

Phosphor bronzes are copper–tin alloys which contain a small amount of phosphorus (0.02–0.40%). These alloys are stronger than the brasses, and possess good cold-working properties, good wear resistance and good corrosion resistance.

1.4 Aluminium bronzes

Aluminium bronzes are copper–aluminium alloys which normally contain small amounts of iron, often with nickel, manganese or silicon. These alloys are characterized by high strength, good toughness, good elevated temperature properties, good resistance to cavitation erosion and wear, excellent resistance to corrosion and oxidation, and good bearing properties.

Aluminium silicon bronzes are copper–aluminium–silicon alloys which are characterized by high strength, good ductility, good corrosion resistance, good wear resistance, and low coefficient of friction.

1.5 Nickel silvers

Nickel silvers are copper–nickel–zinc alloys, which owe their name to their silvery appearance and their ability to take on a high polish. These alloys are somewhat stronger than most brasses and have a good tarnish resistance.

1.6 Copper–nickel alloys

Copper and nickel form a continuous series of solid solution alloys which may be hot or cold worked over the whole range. These alloys possess excellent resistance to salt water corrosion, erosion, impingement, and biological fouling.

2 SPECIFICATIONS

2.1 *Plate, sheet, strip and foil*

2.1.1 *Coppers*

1 *Chemical composition*

TABLE 5.13 Chemical composition of copper plate, sheet and strip, conforming to IS 1972 – 1989

Designation	Chemical composition [% (m/m)]												
	Cu + Ag	P	As	Sb	Bi	Fe	Pb	Ni	O	Te	Se + Te	Sn	Total impurities
FRTP-1	≥ 99.85	–	≤ 0.02	≤ 0.005	≤ 0.003	≤ 0.010	≤ 0.005	≤ 0.05	≤ 0.10	–	≤ 0.025	≤ 0.01	≤ 0.05[1)]
DHP	≥ 99.80	0.015–0.10	≤ 0.05	≤ 0.005	≤ 0.003	≤ 0.03	≤ 0.01	≤ 0.10	–	≤ 0.01	≤ 0.02	≤ 0.01	≤ 0.06[2)]
ATP	≥ 99.20	–	0.20–0.50	≤ 0.01	≤ 0.005	≤ 0.02	≤ 0.02	≤ 0.15	≤ 0.10	–	≤ 0.03	≤ 0.03	–
DPA	≥ 99.20	0.015–0.10	0.20–0.50	≤ 0.01	≤ 0.003	≤ 0.03	≤ 0.010	≤ 0.15	–	≤ 0.01	≤ 0.02	≤ 0.01	≤ 0.07[2)]

1) Excluding Ag and O.
2) Excluding Ag, P, As, and Ni.

2 Mechanical properties

TABLE 5.14 Mechanical properties of copper plate, sheet and strip, conforming to IS 1972 – 1989

Designation	Temper	R_m min. MPa	$A^{1)}$ min. %	Vickers hardness HV	Bend test[2)] Bend mandrel diameter: Transverse bend, Bend mandrel diameter[2)] (180°)	Longitudinal bend, Bend mandrel diameter[2)] (180°)
FRTP-1, DHP, ATP, DPA	O	205	35	≤60	0*a*	0*a*
	HB	245	–	≥75	2*a*	2*a*
	HD	295	–	≥90	4*a*	4*a*

1) $L_o = 5.65\sqrt{S_o}$ or $L_o = 50$ mm.
2) *a* = thickness of the bend test piece; 0*a* means flat on itself.

Notes:
1) The tensile test is not applicable for thicknesses ≤ 0.50 mm.
2) The elongation test is not applicable for thicknesses < 0.80 mm when widths are < 12.5 mm.

2.1.2 Brass

1 Chemical composition

TABLE 5.15 Chemical composition of brass plate, sheet, strip and foil specified in Indian Standards

Alloy designation	Chemical composition [% (*m*/*m*)]: Cu	Zn	Al	As	Fe	Pb	Sn	Total impurities[1)]
Cold-rolled brass sheet, strip and foil (IS 410 – 1977)								
CuZn30	68.5–71.5	Rem.	–	–	≤0.05	≤0.05	–	≤0.3
CuZn37	61.5–64.5	Rem.	–	–	≤0.075	≤0.3	–	≤0.6
CuZn40	58.5–61.5	Rem.	–	–	≤0.1	≤0.3	–	≤0.75

(Contd.)

TABLE 5.15 Chemical composition of brass plate, sheet, strip and foil specified in Indian Standards (continued)

Alloy designation	Chemical composition [% (*m*/*m*)]							
	Cu	Zn	Al	As	Fe	Pb	Sn	Total impurities[1)]
	Rolled brass plate (IS 12443–1988)							
CuZn30	68.5–71.5	Rem.	–	–	≤0.05	≤0.05	–	≤0.3
CuZn20Al2As	76.0–79.0	Rem.	1.8–2.5	0.02–0.06	≤0.06	≤0.07	–	≤0.3
CuZn28As	70.0–73.0	Rem.	–	0.02–0.06	≤0.06	≤0.075	–	≤0.3
CuZn36Sn1	61.0–64.0	Rem.	–	–	–	–	1.0–1.5	≤0.75
CuZn39Sn	59.0–62.0	Rem.	–	–	≤0.1	≤0.02	0.5–1.0	≤0.5
CuZn40Pb	58.0–61.0	Rem.	–	–	≤0.15	0.4–1.0	≤0.25	≤0.5

1) Including Fe.

2 Mechanical properties

TABLE 5.16 Mechanical properties of cold-rolled brass sheet, strip and foil, conforming to IS 410 – 1977

Alloy designation	Temper	Thickness	R_m min. for a width, in mm, of		$A_{50\,mm}$ min.	Vickers hardness for a width, in mm, of		Bend test			
								Transverse bend		Longitudinal bend	
			≤ 450	> 450		≤ 450	> 450	Angle of bend	Bend mandrel diameter[1)]	Angle of bend	Bend mandrel diameter[1)]
		mm	MPa	MPa	%	HV	HV	°		°	
CuZn30	O	≤ 10	275	275	50	≤ 80	≤ 80	180	0*a*	180	0*a*
	HA	≤ 10	320	320	35	≥ 75	≥ 75	180	0*a*	180	0*a*
	HB	≤ 3.5	345	345	20	ô 100	≥ 95	180	0*a*	180	0*a*
		> 3.5 ≤ 10	345	345	20	≥ 100	≥ 95	180	2*a*	180	2*a*
	HD	≤ 10	405	380	5	≥ 125	≥ 120	90	4*a*	90	2*a*
CuZn37	O	≤ 10	275	275	40	≤ 80	≤ 80	180	0*a*	180	0a
	HA	≤ 10	335	320	30	≥ 75	≥ 75	180	0*a*	180	0a
	HB	≤ 3.5	380	345	15	≥ 110	≥ 100	180	0*a*	180	0a
		> 3.5 ≤ 10	380	345	15	≥ 110	≥ 100	180	2*a*	180	2a

(Contd.)

TABLE 5.16 Mechanical properties of cold-rolled brass sheet, strip and foil, conforming to IS 410 – 1977 (continued)

Alloy designation	Temper	Thickness	R_m min. for a width, in mm, of		$A_{50\,mm}$ min.	Vickers hardness for a width, in mm, of		Bend test			
								Transverse bend		Longitudinal bend	
			≤ 450	> 450		≤ 450	> 450	Angle of bend	Bend mandrel diameter[1)]	Angle of bend	Bend mandrel diameter[1)]
		mm	MPa	MPa	%	HV	HV	°		°	
CuZn37	HD	≤ 10	450	405	5	≥ 135	≥ 125	90	4*a*	90	2a
	HE	≤ 10	515	–	–	≥ 165	–	–	–	90	4*a*
	HS	≤ 5	660	–	–	≥ 185	–	–	–	90	4*a*
CuZn40	O	≤ 10	275	275	30	≤ 85	≤ 85	180	0*a*	180	0a
	HB	≤ 10	420	420	12	≥ 100	≥ 100	180	2*a*	180	2a
	HD	≤ 10	490	490	5	≥ 125	≥ 125	90	4*a*	90	2a

[1)] *a* = thickness of the bend test piece; 0*a* means flat on itself.

TABLE 5.17 Mechanical properties of rolled brass plate, conforming to IS 12443 – 1988

Alloy designation	Temper	Thickness mm	R_m min. MPa	A min. %
CuZn30	M/O	> 10	280	40
	H	> 10 ≤ 15	360	18
		> 15 ≤ 25	340	20
CuZn20Al2As	M	> 10	280	36
	O	> 10	280	40
CuZn28As	M/O	> 10	280	40
	H	> 10 ≤ 15	360	18
		> 15 ≤ 25	340	20
CuZn36Sn1	M	> 10 ≤ 25	360	18
		> 25 ≤ 125	340	18
	H	> 125	310	18
		> 10 ≤ 12.5	400	18
CuZn39Sn	M	> 10 ≤ 20	315	20
		> 20	305	20
CuZn40Pb	M	> 10 ≤ 20	345	16
		> 20 ≤ 50	315	16
		> 50	275	16

2.1.3 Phosphor bronze

1 Chemical composition

TABLE 5.18 Chemical composition of phosphor bronze sheet and strip, conforming to IS 7814 – 1985

Grade	Chemical composition [% (*m*/*m*)]					
	Cu	P	Pb	Sn	Zn	Total impurities
I	Rem.	0.02–0.40	≤ 0.02	3.0–4.2	≤ 0.3	≤ 0.5
II	Rem.	0.02–0.40	≤ 0.02	4.2–5.5	≤ 0.3	≤ 0.5
III	Rem.	0.02–0.40	≤ 0.02	5.5–7.5	≤ 0.3	≤ 0.5

2 *Mechanical properties*

TABLE 5.19 Mechanical properties of phosphor bronze sheet and strip, conforming to IS 7814 – 1985

Grade	Temper	Thickness	R_m min. for a width, in mm, of		$R_{p0.2}$[1] min. for a width, in mm, of		$A_{50\,mm}$	Vickers hardness for a width, in mm, of		Bend test			
										Transverse bend		Longitudinal bend	
			≤ 450	> 450	≤ 450	> 450		≤ 450	> 450	Angle of bend	Bend mandrel diameter[2]	Angle of bend	Bend mandrel diameter[2]
		mm	MPa	MPa	MPa	MPa	%	HV	HV	°		°	
I	O	≤ 10.0	295	295	–	–	≥ 40	≤ 80	≤ 80	180	0*a*	180	0*a*
	HA	≤ 10.0	340	340	125	125	≥ 30	≥ 100	≥ 100	180	0*a*	180	0*a*
	HB	≤ 10.0	460	400	390	340	≥ 8	≥ 150	≥ 130	90	2*a*	180	2*a*
	HD	≤ 6.0	540	495	480	435	≥ 4	≥ 180	≥ 150	–	–	90	2*a*
	HE	≤ 6.0	620	–	580	–	–	≥ 190	–	–	–	90	2*a*
II	O	≤ 10.0	310	310	–	–	≥ 45	≤ 85	≤ 85	180	0*a*	180	0*a*
	HA	≤ 10.0	350	350	140	140	≥ 35	≥ 110	≥ 110	180	0*a*	180	0*a*
	HB	≤ 10.0	495	460	420	385	≥ 10	≥ 160	≥ 140	90	2*a*	180	2*a*

(Contd.)

TABLE 5.19 Mechanical properties of cold-rolled phosphor bronze sheet, strip and foil, conforming to IS 7814 – 1985 (continued)

Grade	Temper	Thickness	R_m min. for a width in mm, of		$R_{p0.2}$[1] min. for a width, in mm, of		$A_{50\,mm}$	Vickers hardness for a width, in mm, of		Bend test			
										Transverse bend		Longitudinal bend	
			≤ 450	> 450	≤ 450	> 450		≤ 450	> 450	Angle of bend	Bend mandrel diameter[2]	Angle of bend	Bend mandrel diameter[2]
		mm	MPa	MPa	MPa	MPa	%	HV	HV	°		°	
II	HD	≤ 6.0	570	525	520	480	≥ 4	≥ 180	≥ 160	–	–	90	2*a*
	HE	≤ 6.0	645	–	615	–	–	≥ 200	–	–	–	90	2*a*
	HS	≤ 0.9	–	–	–	–	–	215–230	–	–	–	90	2*a*
III	O	≤ 10.0	340	340	–	–	≥ 50	≤ 90	≤ 90	180	0*a*	180	0a
	HA	≤ 10.0	385	385	200	200	≥ 40	≥ 115	≥ 115	180	0*a*	180	0a
	HB	≤ 10.0	525	460	440	380	≥ 12	≥ 170	≥ 150	90	2*a*	180	2a
	HD	≤ 6.0	620	540	550	480	≥ 6	≥ 200	≥ 165	–	–	90	2*a*
	HE	≤ 6.0	695	–	650	–	–	≥ 215	–	–	–	90	2*a*
	HS	≤ 0.9	–	–	–	–	–	220–240[3]	–	–	–	90	2*a*
	HES	≤ 0.6	–	–	–	–	–	≥ 240[3]	–	–	–	–	–

1) For information only.
2) *a* = thickness of the bend test piece; 0*a* means flat on itself.
3) For widths ≤ 150 mm.

2.1.4 Nickel silver

1 Chemical composition

TABLE 5.20 Chemical composition of nickel silver sheet, strip and foil, conforming to IS 2283 – 1981

Alloy designation	Chemical composition [% (*m*/*m*)]					
	Cu	Fe	Mn	Ni	Pb	Zn	Other elements (total)
NS 10	60.0–65.0	≤0.3	≤0.5	9.0–11.0	≤0.04	Rem.	≤0.5
NS 12	60.0–65.0	≤0.3	≤0.5	11.0–13.0	≤0.04	Rem.	≤0.5
NS 15	60.0–65.0	≤0.3	≤0.5	14.0–16.0	≤0.04	Rem.	≤0.5
NS 18	60.0–65.0	≤0.3	≤0.5	16.5–19.5	≤0.04	Rem.	≤0.5
NS 18T[1)]	53.5–56.5	≤0.3	≤0.5	16.5–19.5	≤0.04	Rem.	≤0.1
NS 20	60.0–65.0	≤0.3	≤0.5	18.5–21.5	≤0.04	Rem.	≤0.5

1) For telecommunication purposes only.

2 Mechanical properties

TABLE 5.21 Mechanical properties of nickel silver sheet and strip, conforming to IS 2283 – 1981

Alloy designation	Temper	Vickers hardness	Bend test			
			Transverse bend		Longitudinal bend	
			Angle of bend	Bend mandrel diameter[1)]	Angle of bend	Bend mandrel diameter[1)]
		HV	°		°	
NS 10	O	≤ 100	180	2*a*	180	2*a*
	HB	≥ 125	180	2*a*	180	2*a*
	HC	–	90	2*a*	120	2*a*
	HD	≥ 160	90	2*a*	90	2*a*
	HE	≥ 185	–	–	90	2*a*
NS 12	O	≤ 100	180	2*a*	180	2*a*
	HB	≥ 130	180	2*a*	180	2*a*
	HC	–	90	2*a*	120	2*a*
	HD	≥ 160	90	2*a*	90	2*a*
	HE	≥ 190	–	–	90	2*a*
NS 15	O	≤ 105	180	2*a*	180	2*a*
	HB	≥ 135	180	2*a*	180	2*a*
	HC	–	90	2*a*	120	2*a*
	HD	≥ 165	90	2*a*	90	2*a*
	HE	≥ 195	–	–	90	2*a*

(Contd.)

TABLE 5.21 Mechanical properties of nickel silver sheet and strip, conforming to IS 2283 – 1981 (continued)

Alloy designation	Temper	Vickers hardness	Bend test			
			Transverse bend		Longitudinal bend	
			Angle of bend	Bend mandrel diameter[1)]	Angle of bend	Bend mandrel diameter[1)]
		HV	°		°	
NS 18	O	≤110	180	2*a*	180	2*a*
	HB	≥135	180	2*a*	180	2*a*
	HC	–	90	2*a*	120	2*a*
	HD	≥170	90	2*a*	90	2*a*
	HE	≥200	–	–	90	2*a*
NS 18T	O	≤115	180	2*a*	180	2*a*
	HB	140–170	180	2*a*	180	2*a*
	HC	170–195	90	2*a*	120	2*a*
	HD	195–220	90	2*a*	90	2*a*
	HE	≥220	–	–	90	2*a*
NS 20	O	≤110	180	2*a*	180	2*a*
	HB	≥140	180	2*a*	180	2*a*
	HC	–	90	2*a*	120	2*a*
	HD	≥175	90	2*a*	90	2*a*
	HE	≥205	–	–	90	2*a*

1) *a* = thickness of the bend test piece.

2.2 Rods, bars and sections

2.2.1 Coppers

1 Chemical composition

TABLE 5.22 Chemical composition of copper and free-cutting copper rods, bars and sections specified in Indian Standards

Designation	Chemical composition [% (m/m)]											
	Cu + Ag	P	As	Sb	Bi	Fe	Pb	Ni	Te	Se + Te	Sn	Total impurities
Copper rods and bars (IS 4171 – 1983)												
–	≥ 99.50	0.015–0.06	≤ 0.05	≤ 0.005	≤ 0.003	≤ 0.030	≤ 0.01	≤ 0.10	–	≤ 0.02	≤ 0.01	≤ 0.06[1)]
Free-cutting copper rods, bars and sections (IS 8328 – 1977)												
Cu-Te	≥ 99.90	–	–	–	–	–	–	–	0.3–0.7	–	–	–
Cu-Pb	≥ 99.90	–	–	–	–	–	0.8–1.5	–	–	–	–	–

1) Excluding Ag, P and Ni.

2 Mechanical properties

TABLE 5.23 Mechanical properties of copper and free-cutting copper rods, bars and sections specified in Indian Standards

Designation	Temper	Product form	Diameter or width across flats mm	R_m MPa	*A* min. %
Copper rods and bars (IS 4171 – 1983)					
–	M	Rods and bars	>6	≥230	13
	O	Rods and bars	>6	≤260	33
Free-cutting copper rods, bars and sections (IS 8328 – 1977)					
Cu-Te, Cu-Pb	HB	Rods, bars and sections	≤ 65	≥260	12
	HD	Rods	≤ 32	≥305	8
			>32≤ 50	≥275	8
		Bars and sections	>10≤ 13	≥275	10
			>13≤ 50	≥225	12
			>50≤100	≥220	12

2.2.2 Brass

1 Chemical composition

TABLE 5.24 Chemical composition of brass rods, bars and sections specified in Indian Standards

Alloy designation/ Grade	Chemical composition [% (*m*/*m*)]								
	Cu	Zn	Al	Fe	Mn	Pb	Sb	Sn	Other elements (total)
Brass rods (IS 4170 – 1967)									
CuZn20	79.0–81.0	Rem.	–	≤0.05	–	≤0.1	–	–	≤0.3[1)]

(Contd.)

TABLE 5.24 Chemical composition of brass rods, bars and sections specified in Indian Standards (continued)

Alloy designation/ Grade	Chemical composition [% (*m*/*m*)]								
	Cu	Zn	Al	Fe	Mn	Pb	Sb	Sn	Other elements (total)
Brass rods (IS 4170 – 1967)									
CuZn30	68.0–72.0	Rem.	–	≤0.05	–	≤0.03	–	–	≤0.3[1)]
CuZn40	59.0–62.0	Rem.	–	≤0.1	–	≤0.75[2)]	3)	–	≤0.3[1)]
Naval brass rods and sections (IS 291 – 1989)									
1	61.0–64.0[4)]	Rem.	–	≤0.1	–	≤0.2	–	1.0–1.5	≤0.2
2	59.0–62.0[4)]	Rem.	–	≤0.1	–	0.5–1.0	–	0.5–1.0	≤0.2
Free-cutting leaded brass rods, bars and sections (IS 319 – 1989)									
1	56.0–59.0[5)]	Rem.	–	≤0.35	–	2.0–3.5	3)	–	≤0.7[6)]
2	60.0–63.0[5)]	Rem.	–	≤0.35	–	2.5–3.7	–	–	≤0.5[6)]
3	60.0–63.0[5)]	Rem.	–	≤0.2	–	0.5–1.5	–	–	≤0.5[6)]
High tensile brass rods and sections (IS 320 – 1980)									
HT 1	56.0–60.0[4)]	Rem.	≤0.2	0.2–1.25	0.25–2.0	0.2–1.5[7)]	3)	0.2–1.0	≤0.5
HT 2	56.0–61.0[4)]	Rem.	0.3–2.0	0.2–1.5	0.5–2.0	0.5–1.5[7)]	3)	≤1.0	≤0.5

1) Excluding Pb.
2) If a Pb-free material is required, it may be ordered with a Pb content of ≤0.1%.
3) If required, the material may be ordered with a Sb content of ≤0.02%.
4) Including Ni.
5) Including Ag and Ni.
6) Excluding Fe.
7) If the material is required with a lower Pb content, it may be ordered with a Pb content of either ≤0.1% or ≤0.5%.

2 *Mechanical properties*

TABLE 5.25 Mechanical properties of brass rods, bars and sections specified in Indian Standards

Alloy designation/ Grade	Temper	Diameter or width across flats mm	R_m min. MPa	$R_{p0.2}$ min. MPa	A min. %	Vickers hardness HV
Brass rods (IS 4170 – 1967)						
CuZn20	M	≥5	315	–	25	–
	O	≥5	245	–	50	≤90
CuZn30	M	≥5	345	–	25	≤100
	O	≥5	275	–	50	≤90
CuZn40	M	≥5	345	–	25	–
	O	≥5	275	–	30	≤90
Naval brass rods and sections (IS 291 – 1989)						
1	O	All sizes	325	–	20	–
	HB	≤ 12.5	390	–	18	–
		> 12.5 ≤ 50	380	–	18	–
		> 50 ≤ 100	345	–	18	–
	HD	≤ 12.5	430	–	10	–
		> 12.5 ≤ 25	410	–	10	–
		> 25 ≤ 50	390	–	10	–
2	O	All sizes	355	–	20	–
	HB	All sizes	390	–	15	–

(Contd.)

TABLE 5.25 Mechanical properties of brass rods, bars and sections specified in Indian Standards (continued)

Alloy designation/ Grade	Temper	Diameter or width across flats mm	R_m min. MPa	$R_{p0.2}$ min. MPa	A min. %	Vickers hardness HV
Free-cutting leaded brass rods, bars and sections (IS 319–1989)						
1	O	>6 ≤ 25	345	–	12	–
		>25 ≤ 50	315	–	17	–
		>50	285	–	22	–
	HB	>66 ≤ 12	405	–	4	–
		>12 ≤ 25	395	–	6	–
		>25 ≤ 50	355	–	12	–
		>50	325	–	17	–
	HD	>6≤ 12	550	–	–	–
		>12 ≤ 25	490	–	4	–
2	O	> 6 ≤25	355	–	15	–
		>25 ≤ 50	305	–	20	–
		>50	275	–	25	–
	HB	>6 ≤ 12	395	–	7	–
		>12≤ 25	385	–	10	–
		>25≤ 50	345	–	15	–
		>50	315	–	20	–
	HD	>6 ≤ 12	550	–	–	–
		>12 ≤ 25	485	–	4	–

(Contd.)

TABLE 5.25 Mechanical properties of brass rods, bars and sections specified in Indian Standards (continued)

Alloy designation/ Grade	Temper	Diameter or width across flats mm	R_m min. MPa	$R_{p0.2}$ min. MPa	A min. %	Vickers hardness HV
Free-cutting leaded brass rods, bars and sections (IS 319–1989)						
3	O	> 6 ≤ 25	315	–	22	–
		> 25 ≤ 50	285	–	27	–
		> 50	255	–	32	–
	HB	> 6 ≤ 12	355	–	8	–
		> 12 ≤ 25	345	–	12	–
		> 25 ≤ 50	305	–	22	–
		> 50	285	–	27	–
	HD	> 6 ≤ 12	460	–	–	–
		> 12 ≤ 25	400	–	14	–
High tensile brass rods and sections (IS 320–1980)						
HT 1	M	All sizes	430	240	20	–
	Cold worked and stress relieved	> 10 ≤ 40	480	–	12	–
		> 40	460	–	15	–
HT 2	M	All sizes	460	280	20	–
	Cold worked and stress relieved	> 10 ≤ 40	520	–	12	–
		> 40	500	–	15	–

2.2.3 Phosphor bronze

1 Chemical composition

TABLE 5.26 Chemical composition of phosphor bronze rods and bars, conforming to IS 7811 – 1985

Grade	Chemical composition [% (m/m)]					
	Cu	P	Pb	Sn	Zn	Total impurities
–	Rem.	0.02–0.40	≤ 0.02	4.0–5.5	≤ 0.03	≤ 0.05[1)]

1) Including Zn.

2 Mechanical properties

TABLE 5.27 Mechanical properties of phosphor bronze rods and bars, conforming to IS 7811 – 1985

Grade	Temper	Diameter or width across flats mm	R_m min. MPa	$R_{p0.2}$ min. MPa	A min. %
–	M	> 6 ≤ 18	500	410[1)]	12
		> 18 ≤ 38	460	380[1)]	12
		> 38 ≤ 70	380	315[1)]	16
		> 70 ≤ 100	315	235[1)]	20
		> 100 ≤ 120	275	118[1)]	22
		> 120	255	80[1)]	25

1) For information only.

2.2.4 Aluminium bronze

1 Chemical composition

TABLE 5.28 Chemical composition of aluminium bronze rods and bars specified in Indian Standards

Alloy designation	Chemical composition [% (m/m)]											
	Cu	Al	As	Fe	Mg	Mn	Ni	Pb	Si	Sn	Zn	Total impurities
Aluminium bronze rods and bars (IS 10569 – 1983)												
CuAl8Fe3	Rem.	6.5–8.0	–	2.0–3.5	≤0.05	≤0.5	≤0.5	≤0.05	≤0.15	≤0.1	≤0.4	≤0.5[1]
CuAl10Fe3	Rem.	8.5–10.0	–	≤4.0[2]	≤0.05	≤0.5	–	≤0.05	≤0.1	≤0.1	≤0.4	≤0.5[3]
CuAl10Fe5Ni5	Rem.	8.5–11.0	–	4.0–6.0	≤0.05	≤0.05	4.0–6.0	≤0.05	≤0.1	≤0.1	≤0.4	≤0.5[3]
Aluminium-silicon bronze rods and bars (IS 10723 – 1983)												
–	Rem.	6.3–7.6	≤0.15	≤0.3	–	≤0.1	≤0.25	≤0.05	1.5–2.2	≤0.2	≤0.5	≤0.5[4]

1) Excluding Mn, Ni and Zn.
2) Fe + Ni.
3) Excluding Mn.
4) Excluding Fe, Mn and Ni.

2 Mechanical properties

TABLE 5.29 Mechanical properties of aluminium bronze rods and bars specified in Indian Standards

Alloy designation	Temper	Diameter or width across flats mm	R_m min. MPa	$R_{p0.2}$ min. MPa	A_4 min. %
Aluminium bronze rods and bars (IS 10569 – 1983)					
CuAl8Fe3	M	$>6 \leq 70$	520	220	30
		>70	460	190	30
	O	>6	460	190	30
CuAl10Fe3	M	>6	520	215	22
CuAl10Fe5Ni5	M	$>6 \leq 70$	700	400	12
		$>70 \leq 120$	700	350	14
		>120	650	320	12
Aluminium-silicon bronze rods and bars (IS 10723 – 1983)					
–	M	$>6 \leq 10$	620	310	9
		$>10 \leq 25$	585	310	12
		$>25 \leq 50$	550	290	12
		$>50 \leq 75$	515	240	15

2.2.5 Nickel silver

1 Chemical composition

TABLE 5.30 Chemical composition of nickel silver rods and bars, conforming to IS 10757 – 1983

Alloy designation	Chemical composition [% (m/m)]						
	Cu	Fe	Mn	Ni	Pb	Zn	Other elements (total)
CuNi10Zn27	61.0–65.0	≤0.3	≤0.5	9.0–11.0	≤0.05	Rem.	≤0.3
CuNi18Zn20	60.0–64.0	≤0.3	≤0.7	17.0–19.0	≤0.03	Rem.	≤0.3
CuNi12Zn24	62.0–66.0	≤0.3	≤0.5	11.0–13.0	≤0.05	Rem.	≤0.3
CuNi18Zn27	53.0–56.0	≤0.3	≤0.5	17.0–19.0	≤0.05	Rem.	≤0.3
CuNi10Zn42Pb2	44.0–48.0	≤0.5	≤0.5	9.0–11.0	1.0–2.5	Rem.	≤0.5

2 Mechanical properties

TABLE 5.31 Mechanical properties of nickel silver rods and bars, conforming to IS 10757 – 1983

Alloy designation	Temper	Cross-section	Diameter or width across flats mm	R_m MPa	A min. %	Minimum Vickers hardness HV
CuNi10Zn27	O	All sections	–	≥360	50	80
CuNi10Zn27	HD	Round, hexagonal and octagonal	>6≤10	550–690	8	–
			>10≤25	515–655	–	–
			>25	485–620	–	–
		Square and rectangular	All sizes	515–650	5	–

(Contd.)

TABLE 5.31 Mechanical properties of nickel silver rods and bars, conforming to IS 10757–1983 (continued)

Alloy designation	Temper	Cross-section	Diameter or width across flats	R_m	*A* min.	Minimum Vickers hardness
			mm	MPa	%	HV
CuNi18Zn20	O	All sections	–	≥400	43	90
	HD	Round, hexagonal and octagonal	>6≤10	485–620	12	–
			>10≤25	450–590	12	–
			>25	415–550	8	–
		Square and rectangular	All sizes	470–605	8	–
CuNi12Zn24	O	All sections	–	≥370	50	90
	HD	Round, hexagonal and octagonal	>6≤10	550–690	8	–
			>10≤25	515–655	–	–
			>25	485–620	–	–
		Square and rectangular	All sizes	515–650	5	–
CuNi18Zn27	O	All sections	–	≥400	45	95
	HD	Round, hexagonal and octagonal	>6≤10	550–690	8	–
			>10≤25	515–655	–	–
			>25	485–620	–	–
		Square and rectangular	All sizes	515–650	5	–
CuNi10Zn42Pb2	O	All sections	–	≥540	22	150
	HD	Round, hexagonal and octagonal	>6≤10	485–620	12	–
			>10≤25	450–590	12	–
			>25	415–550	8	–
		Square and rectangular	All sizes	470–605	8	–

2.3 Wire

2.3.1 Coppers

1 Chemical composition

TABLE 5.32 Chemical composition of copper wire, conforming to IS 4412 – 1981

Designation	Chemical composition [% (*m*/*m*)]										
	Cu + Ag	P	As	Sb	Bi	Fe	Pb	Ni	Te	Se + Te	Sn	Total impurities
DHP	≥99.80	0.015–0.10	≤0.05	≤0.005	≤0.003	≤0.03	≤0.01	≤0.10	≤0.01	≤0.02	≤0.01	≤0.06[1)]

1) Excluding Ag, P, As, and Ni.

2 Mechanical properties

TABLE 5.33 Mechanical properties of copper wire, conforming to IS 4412 – 1981

Designation	Temper	Diameter or width across flats	R_m min	$A_{100\,mm}$ min.
		mm	MPa	%
DHP	O	>1 ≤6	225	35
	HB	>1 ≤6	280	–
	HD	>1 ≤6	340	–

2.3.2 Brass

1 Chemical composition

TABLE 5.34 Chemical composition of brass wire specified in Indian Standards

Alloy designation/ Grade	Chemical composition [% (*m*/*m*)]						
	Cu	Zn	Fe	Pb	Sb	Other elements (total)	Total impurities
Brass wire (IS 4413 – 1981)							
CuZn30	68.5–71.5	Rem.	≤0.05	≤0.05[1)]	–	–	≤0.3
CuZn37	62.0–65.0	Rem.	≤0.1	≤0.3[1)]	–	–	≤0.6
Brass wire for cold-headed and machined parts (IS 2704 – 1983)							
CuZn35	63.0–68.0	Rem.	≤0.05	≤0.02	–	–	≤0.3
CuZn35Pb1	62.0–65.0	Rem.	≤0.1	0.75–1.5	–	–	≤0.5

(Contd.)

TABLE 5.34 Chemical composition of brass wire specified in Indian Standards (continued)

Alloy designation/ Grade	Chemical composition [% (m/m)]						
	Cu	Zn	Fe	Pb	Sb	Other elements (total)	Total impurities
Free-cutting brass wire (IS 8364 – 1989)							
1	56.0–59.0[2)]	Rem.	≤0.35	2.0–3.5	[3)]	≤0.7[4)]	–
2	60.0–63.0[2)]	Rem.	≤0.35	2.5–3.7	–	≤0.5[4)]	–
3	60.0–63.0[2)]	Rem.	≤0.2	0.5–1.5	–	≤0.5[4)]	–

1) If the material is required for hot-rolling, it may be ordered with a Pb content of ≤ 0.015%.
2) Including Ag and Ni.
3) If required, the material may be ordered with a Sb content of ≤ 0.02%.
4) Excluding Fe.

2 Mechanical properties

TABLE 5.35 Mechanical properties of brass wire specified in Indian Standards

Alloy designation	Temper	Diameter or width across flats mm	R_m MPa	$A_{100\,mm}$ min. %	Vickers hardness HV
Brass wire (IS 4413 – 1981)					
CuZn30	O	≤6	≥315	45	–
	HB	≤6	460–620	–	–
	HD	≤6	≥620	–	–
CuZn37	O	≤6	≥325	35	–
	HB	≤6	460–620	–	–
	HD	≤6	620–775	–	–

(Contd.)

TABLE 5.35 Mechanical properties of brass wire specified in Indian Standards (continued)

Alloy designation	Temper	Diameter or width across flats mm	R_m MPa	$A_{100\,mm}$ min. %	Vickers hardness HV
Brass wire for cold-headed and machined parts (IS 2704 – 1983)					
CuZn35	HA	≤6	345–410	30	110–130
	HB	≤6	420–510	15	131–160
CuZn35Pb1	HA	≤6	325–390	30	110–130
	HB	≤6	400–490	20	131–150
	HD	≤6	500–685	15	151–175
Free-cutting brass wire (IS 8364 – 1989)					
1	O	≤6	≥355	10	–
	HB	≤6	≥410	4	–
	HD	≤6	≥590	–	–
2	O	≤6	≥345	12	–
	HB	≤6	≥400	6	–
	HD	≤6	≥590	–	–
3	O	≤6	≥325	20	–
	HB	≤6	≥400	8	–
	HD	≤6	≥500	–	–

2.3.3 Phosphor bronze

1 Chemical composition

TABLE 5.36 Chemical composition of phosphor bronze wire, conforming to IS 7608 – 1987

Grade	Chemical composition [% (*m*/*m*)]				
	Cu	P	Pb	Sn	Total impurities
I	Rem.	0.02–0.40	≤0.05	4.2–5.5	≤0.2
II	Rem.	0.02–0.40	≤0.05	5.5–7.5	≤0.2

2 Mechanical properties

TABLE 5.37 Mechanical properties of phosphor bronze wire, conforming to IS 7608 – 1987

Grade	Temper	Diameter mm	R_m MPa	$A_{100\ mm}$ min. %
I	O	>0.45 ≤6.0	≥340	40
	HB	>0.45 ≤6.0	540–700	–
	HD	>0.45 ≤6.0	700–850	–
	HE	>0.45 ≤2.5	≥850	–
		>2.5 ≤6.0	≥800	–
II	O	>0.45 ≤6.0	≥370	50
	HB	>0.45 ≤6.0	590–740	–
	HD	>0.45 ≤6.0	740–900	–
	HE	>0.45 ≤2.5	≥900	–
		>2.5 ≤6.0	≥850	–

2.4 Tubes

2.4.1 Coppers

1 Chemical composition

TABLE 5.38 Chemical composition of copper tubes specified in Indian Standards

Designation	Chemical composition [% (m/m)]												
	Cu + Ag	P	As	Sb	Bi	Fe	Pb	Ni	O	Te	Se + Te	Sn	Total impurities
Copper tubes (IS 2501 – 1995)													
ETP	≥ 99.90	–	–	–	≤ 0.001	–	≤ 0.005	–	–	–	–	–	≤ 0.03[1)]
FRTP-1	≥ 99.85	–	≤ 0.02	≤ 0.005	≤ 0.003	≤ 0.010	≤ 0.005	≤ 0.05	≤ 0.10	–	≤ 0.025	≤ 0.01	≤ 0.05[1)]
FRTP-2	≥ 99.50	–	≤ 0.10	≤ 0.05	≤ 0.02	≤ 0.03	≤ 0.10	≤ 0.50	≤ 0.10	–	≤ 0.07	≤ 0.05	–
DHP	≥ 99.80	0.015–0.10	≤ 0.05	≤ 0.005	≤ 0.003	≤ 0.03	≤ 0.01	≤ 0.10	–	≤ 0.01	≤ 0.02	≤ 0.01	≤ 0.06[2)]
ATP	≥ 99.20	–	0.20–0.50	≤ 0.01	≤ 0.005	≤ 0.02	≤ 0.02	≤ 0.15	≤ 0.10	–	≤ 0.03	≤ 0.03	–
DPA	≥ 99.20	0.015–0.10	0.20–0.50	≤ 0.01	≤ 0.003	≤ 0.03	≤ 0.010	≤ 0.15	–	≤ 0.01	≤ 0.02	≤ 0.01	≤ 0.07[2)]
Copper tubes for condensers and heat exchangers (IS 1545 – 1994)													
DLP	≥ 99.90	0.004–0.015	–	–	–	–	–	–	–	–	–	–	–
DHP	≥ 99.85	0.014–0.15	≤ 0.05	–	–	≤ 0.03	≤ 0.01	≤ 0.1	–	–	–	≤ 0.01	≤ 0.05[2)]

1) Excluding Ag and O.
2) Excluding Ag, P, As, and Ni.

2 *Mechanical properties*

TABLE 5.39 Mechanical properties of copper tubes specified in Indian Standards

Designation	Temper	Type of test piece	R_m MPa	*A* min. %	Vickers hardness HV
Copper tubes for condensers and heat exchangers (IS 1545 – 1994)					
DLP, DHP	O	Tube (in full section) or longitudinal strip	≥ 205	40	≤ 60
	HB		245–325	–	80–100
	HD		≥ 315	–	≥ 105
Copper tubes (IS 2501–1995)					
ETP, FRTP-1, FRTP-2, DHP, ATP, DPA	O	Tube (in full section)	≥ 205	40	–
		Longitudinal strip	≥ 195	45	–
	HB	Tube (in full section)	≥ 235	25	–
		Longitudinal strip	≥ 225	25	–
	HD	Tube (in full section)	≥ 280	–	–
		Longitudinal strip	≥ 250	–	–

2.4.2 Brass

1 Chemical composition

TABLE 5.40 Chemical composition of brass tubes specified in Indian Standards

Alloy designation	Chemical composition [% (m/m)]							
	Cu	Zn	Al	As	Fe	Pb	Sn	Total impurities
Brass tubes for general purposes (IS 407 – 1981)								
CuZn37	62.0–65.0[1)]	Rem.	–	≤0.06	≤0.1	≤0.3	–	≤0.6
CuZn30As	68.5–71.5[1)]	Rem.	–	0.02–0.06	≤0.06	≤0.07	–	≤0.3
Brass tubes for condensers and heat exchangers (IS 1545 – 1994)								
CuZn21Al2As	76.0–79.0	Rem.	1.8–2.3	0.02–0.06	≤0.06	≤0.07	–	≤0.3
CuZn29Sn1As	70.0–73.0	Rem.	–	0.02–0.06	≤0.06	≤0.07	1.0–1.5	≤0.3
CuZn30As	69.0–71.0	Rem.	–	0.02–0.06	≤0.06	≤0.07	–	≤0.3

1) Including Ni.

2 Mechanical properties

TABLE 5.41 Mechanical properties of brass tubes specified in Indian Standards

Alloy designation	Temper	R_m MPa	*A* min. MPa	Vickers hardness HV
Brass tubes for general purposes (IS 407 – 1981)				
CuZn37	O	≥285	–	≤80
	Temper annealed	≥320	–	80–110
	HD	≥400	–	ô 130

(Contd.)

TABLE 5.41 Mechanical properties of brass tubes specified in Indian Standards (continued)

Alloy designation	Temper	R_m MPa	A min. MPa	Vickers hardness HV
Brass tubes for condensers and heat exchangers (IS 1545 – 1982)				
CuZn30As	O	≥285	–	≤75
	Temper annealed	≥300	–	80–110
	HD	≥400	–	≥ 135
CuZn21Al2As	O	≤400	–	≤85
	Temper annealed	≥355	50	80–110
	HD	≥415	–	≥130
CuZn29Sn1As	O	≤375	–	≤80
	Temper annealed	≥340	35	80–105
	HD	≥385	–	≥130
CuZn30As	O	≤375	≤375	≤80
	Temper annealed	≥340	≥340	80–105
	HD	≥385	≥385	≥130

2.4.3 Copper-nickel alloys

1 Chemical composition

TABLE 5.42 Chemical composition of copper-nickel alloy tubes for condensers and heat exchangers, conforming to IS 1545 – 1994

Alloy designation	Chemical composition [% (*m*/*m*)]						
	Cu	Fe	Mn	Ni	Pb	Zn	Other elements (total)
CuNi10Fe1Mn	Rem.	1.0–1.8	0.5–1.0	9.0–11.0[1)]	≤0.05	≤0.5	≤0.3
CuNi30Mn1Fe	Rem.	0.4–1.0	0.5–1.0	29.0–33.0[1)]	≤0.05	≤0.5	≤0.3

2 Mechanical properties

TABLE 5.43 Mechanical properties of copper-nickel alloy tubes for condensers and heat exchangers, conforming to IS 1545 – 1994

Alloy designation	Temper	R_m MPa	*A* min. %	Vickers hardness HV 5
CuNi10Fe1Mn	O	≤295	35	≤110
	HD	≥385	–	≥130
CuNi30Mn1Fe	O	≤360	35	≤115
	HD	≥480	–	≥140

2.5 Forging stock and forgings

2.5.1 Coppers

1 Chemical composition

TABLE 5.44 Chemical composition of copper forging stock and forgings, conforming to IS 6912 – 1985

Designation	Chemical composition [% (m/m)]										
	Cu + Ag	P	As	Sb	Bi	Fe	Pb	Ni	Te	Se + Te	Sn	Total impurities
ETP	≥ 99.90	–	–	–	≤ 0.001	–	≤ 0.005	–	–	–	–	≤ 0.03[1]
FRHC	≥ 99.90	–	–	–	≤ 0.0025	–	≤ 0.005	–	–	–	–	≤ 0.04[1]
DPA	≥ 99.20	0.02–0.10	0.20–0.50	≤ 0.01	≤ 0.003	≤ 0.03	≤ 0.010	≤ 0.15	≤ 0.01	≤ 0.02	≤ 0.01	≤ 0.07[2]

1) Excluding Ag and O.
2) Excluding Ag, P, As, and Ni.

2 Mechanical properties

TABLE 5.45 Mechanical properties of copper forging stock and forgings, conforming to IS 6912 – 1985

Designation	Temper	Diameter or thickness of test sample mm	R_m min. MPa	$R_{p0.2}$ min. MPa	A min. %
ETP	O	≥6	210	–	45
FRHC	O	≥6	210	–	45
DPA	O	≥6	215	–	55

2.5.2 Brass

1 Chemical composition

TABLE 5.46 Chemical composition of brass forging stock and forgings, conforming to IS 6912 – 1985

Alloy designation	Chemical composition [% (m/m)]								
	Cu	Zn	Al	Fe	Mn	Pb	Sb	Sn	Total impurities
Naval brass									
FNB	61.0–64.0[1)]	Rem.	–	≤0.1	–	≤0.2	–	1.0–1.5	≤0.2
Leaded brass									
FLB	56.5–60.0	Rem.	–	≤0.3	≤0.2	0.6–2.0	2)	–	≤0.25[3)]
High tensile brass									
FHTB1	56.0–60.0	Rem.	≤0.2	0.2–1.25	0.25–2.0	0.2–1.5[4)]	2)	0.2–1.0	≤0.5[3)]
FHTB2	56.0–61.0	Rem.	0.3–2.0	0.2–1.5	0.5–2.0	0.5–1.5[4)]	2)	≤1.0	≤0.5[3)]

1) Including Ni.
2) If required, the material may be ordered with a Sb content of < 0.02 %.
3) Including Sb.
4) If the material is required with a lower Pb content than that specified, it may be ordered with a Pb content of either ≤ 0.1% or ≤ 0.5%.

2 *Mechanical properties*

TABLE 5.47 Mechanical properties of brass forging stock and forgings, conforming to IS 6912 – 1985

Alloy designation	Temper	Diameter or thickness of test sample mm	R_m min. MPa	$R_{p0.2}$ min. MPa	A min. %
FNB	M	≥6	340	–	15
FLB	M	≥6	310	–	20
FHTB1	M	≥6	430	240[1)]	15
FHTB2	M	≥6	460	280[1)]	12

[1)] For information only.

2.5.3 Aluminium bronze

1 Chemical composition

TABLE 5.48 Chemical composition of aluminium bronze forging stock and forgings, conforming to IS 6912 – 1985

Alloy designation	Chemical composition [% (*m*/*m*)]										
	Cu	Al	As	Fe	Mg	Mn	Ni	Pb	Si	Sn	Zn	Total impurities
Aluminium bronze												
FAB1	Rem.	6.5–8.0	–	2.0–3.5	≤0.05	≤0.5	≤0.50	≤0.05	≤0.15	≤0.1	≤0.4	≤0.5[1)]
FAB2	Rem.	8.5–10.0	–	≤4.0[2)]	≤0.05	≤0.5	–	≤0.05	≤0.1	≤0.1	≤0.4	≤0.5[3)]
FAB3	Rem.	8.5–11.0	–	4.0–6.0	≤0.05	≤0.5	4.0–6.0	≤0.05	≤0.1	≤0.1	≤0.4	≤0.5[3)]
Aluminium silicon bronze												
FABS	Rem.	6.3–7.6	≤0.15	≤0.3	–	≤0.1	≤0.25	≤0.05	1.5–2.2	≤0.2	≤0.5	≤0.5[4)]

1) Excluding Mn, Ni and Zn.
2) Fe + Ni.
3) Excluding Mn.
4) Excluding Fe, Mn and Ni.

2 *Mechanical properties*

TABLE 5.49 Mechanical properties of aluminium bronze forging stock and forgings, conforming to IS 6912 – 1985

Alloy designation	Temper	Diameter or thickness of test sample mm	R_m min. MPa	$R_{p0.2}$ min. MPa	A min. %
FAB1	M	≥ 6 ≤ 10	540	240[1)]	30
		> 110≤ 70	520	220[1)]	30
		> 70	460	190[1)]	30
FAB2	M	≥6	520	215[1)]	20
FAB3	M	≥6≤ 10	700	400[1)]	10
		> 10≤10	700	400[1)]	12
		> 70≤120	700	350[1)]	12
		> 120	650	320[1)]	12
FABS	M	≥ 6≤10	620	310[1)]	9
		> 10≤ 25	585	310[1)]	12
		> 25≤ 50	550	290[1)]	12
		> 50≤ 75	515	240[1)]	15

[1)] For information only.

2.5.4 Nickel silver

1 Chemical composition

TABLE 5.50 Chemical composition of nickel silver forging stock and forgings, conforming to IS 6912 – 1985

Alloy designation	Chemical composition [% (*m*/*m*)]						
	Cu	Fe	Mn	Ni	Pb	Zn	Other elements (total)
FNS	58.0–63.0	≤0.5	0.2–0.5	9.0–11.0	1.0–2.5	Rem.	≤0.5

2 Mechanical properties

TABLE 5.51 Mechanical properties of nickel silver forging stock and forgings, conforming to IS 6912 – 1985

Alloy designation	Temper	Diameter or thickness of test sample mm	R_m min. MPa	$R_{p0.2}$ min. MPa	A min. %
FNS	M	≥6	460	–	8

3 TYPICAL PHYSICAL AND MECHANICAL PROPERTIES

TABLE 5.52 Typical physical and mechanical properties of wrought copper and copper alloys specified in Indian Standards

Designation/ Grade	Density at 20 °C kg/m³	Liquidus temperature °C	Solidus temperature °C	Coefficient of thermal expansion between 20 °C and 300 °C µm/(m K)	Thermal conductivity at 20 °C W/(m K)	Electrical conductivity at 20 °C % IACS	Electrical resistivity at 20 °C µΩ m	Modulus of elasticity GPa
Coppers								
CuETP	8890	1083	1065[1)]	17.7	388	100–101.5[2)]	0.01700–0.01724[2)]	115[2)]
CuFRTP-1	8890	1085	–	17.7	377	98[3)]	0.0176[3)]	–
CuFRTP-2	–	–	–	–	–	–	–	–
CuFRHC	–	–	–	–	–	–	–	–
CuDHP	8940	1083	–	17.7	338	85[3)]	0.0203[3)]	118
CuATP	8940	1080	–	17.7	194	45[3)]	0.0382[3)]	118
CuATP	–	–	–	–	–	–	–	–
Cu-Te	8940	1075	1051	17.8	355	93	0.0186	115
Cu-Pb	8940	1080	950	17.6	377	96	0.0179	115

(Contd.)

TABLE 5.52 Typical physical and mechanical properties of wrought copper and copper alloys specified in Indian Standards (continued)

Designation/ Grade	Density at 20 °C	Liquidus temperature	Solidus temperature	Coefficient of thermal expansion between 20 °C and 300 °C	Thermal conductivity at 20 °C	Electrical conductivity at 20 °C	Electrical resistivity at 20 °C	Modulus of elasticity
	kg/m^3	°C	°C	μm/(m K)	W/(m K)	% IACS	μΩ m	GPa
Brass								
CuZn20	8670	1000	965	19.1	140	32[3)]	0.054[3)]	110
CuZn30	8530	955	915	19.9	120	28[3)]	0.062[3)]	110
CuZn35	8470	930	905	20.3	116	27[3)]	0.064[3)]	105
CuZn37	8440	916	–	20.5	117	27.6[3)]	–	103
CuZn40	8390	905	900	20.8	123	28	0.0616	105
CuZn20Al2As	8330	970	935	18.5	100	23[3)]	0.075[3)]	110
CuZn28As	–	–	–	–	–	–	–	–
CuZn29Sn1As	8530	935	900	20.2	110	25	0.069	110
CuZn30As	8530	955	915	19.9	121	28[3)]	0.0616[3)]	110
CuZn36Sn1	–	–	–	–	–	–	–	–
CuZn37Sn1Pb1	8440	900	885	21.2	116	26	0.0663	100
CuZn39Sn	8410	900	885	21.2	116	26[3)]	0.0663[3)]	100

(Contd.)

TABLE 5.52 Typical physical and mechanical properties of wrought copper and copper alloys specified in Indian Standards (continued)

Designation/ Grade	Density at 20 °C	Liquidus temperature	Solidus temperature	Coefficient of thermal expansion between 20 °C and 300 °C	Thermal conductivity at 20 °C	Electrical conductivity at 20 °C	Electrical resistivity at 20 °C	Modulus of elasticity
	kg/m^3	°C	°C	µm/(m K)	W/(m K)	% IACS	µΩ m	GPa
Brass								
HT 1	8360	890	865	21.2	109	24[3)]	0.0718[3)]	103
HT 2	–	–	–	–	–	–	–	–
CuZn35Pb1	8470	925	885	20.3	115	26[3)]	0.066	105
CuZn37Pb1	8470	915	895	20.3	115	26	0.066	105
CuZn36Pb3	8500	900	885	20.5	115	26[3)]	0.066	97
CuZn40Pb	8410	900	885	20.8	123	28[3)]	0.062	105
CuZn39Pb1	–	–	–	–	–	–	–	–
CuZn40Pb2	–	–	–	–	–	–	–	–
CuZn39Pb3	8470	890	875	20.9	123	28[3)]	0.062	97
Phosphor bronze								
I (CuSn4)	8860	1060	975	17.8	84	20	0.087	110
II (CuSn5)	8860	1060	975	17.8	84	20	0.087	110
III (CuSn6)	–	–	–	–	–	–	–	–

(Contd.)

TABLE 5.52 Typical physical and mechanical properties of wrought copper and copper alloys specified in Indian Standards (continued)

Designation/ Grade	Density at 20 °C	Liquidus temperature	Solidus temperature	Coefficient of thermal expansion between 20 °C and 300 °C	Thermal conductivity at 20 °C	Electrical conductivity at 20 °C	Electrical resistivity at 20 °C	Modulus of elasticity
	kg/m^3	°C	°C	µm/(m K)	W/(m K)	% IACS	µΩ m	GPa
Aluminium bronze								
CuAl8Fe3	7890	1045	1040	16.2	56.5	14	0.123	115
CuAl10Fe3	7650	1045	1040	16.2	54.4	12	0.144	115
CuAl10Fe5Ni5	7640	1060	1040	16.2	36	7	0.246	115
Aluminium-silicon bronze								
CuAl7Si2	7690	1005	985	18.1	45	8	0.186	110
Nickel silver								
CuNi10Zn27	8690	1020	–	16.4	45	9	0.192	120
CuNi10Zn42Pb2	–	–	–	–	–	–	–	–
CuNi12Zn24	8690	1040	–	16.2	40	8	0.216	125
CuNi15Zn21	8700	1075	1040	16.2	36	7	0.246	125
CuNi18Zn20	–	–	–	–	–	–	–	–
CuNi18Zn27	8700	1055	–	16.7	29	5.5	0.314	125

(Contd.)

TABLE 5.52 Typical physical and mechanical properties of wrought copper and copper alloys specified in Indian Standards (continued)

Designation/ Grade	Density at 20 °C	Liquidus temperature	Solidus temperature	Coefficient of thermal expansion between 20 °C and 300 °C	Thermal conductivity at 20 °C	Electrical conductivity at 20 °C	Electrical resistivity at 20 °C	Modulus of elas-ticity
	kg/m^3	°C	°C	μm/(m K)	W/(m K)	% IACS	μΩ m	GPa
Copper-nickel alloy								
CuNi10Fe1Mn	8940	1150	1100	17.1	40	9.1	0.19[3)]	140
CuNi30Mn1Fe	8940	1240	1170	16.2	29	4.6[3)]	0.375[3)]	150

1) Eutectic point.
2) Soft annealed.
3) Annealed.

Source: References 1 and 3

4 TYPICAL APPLICATION

TABLE 5.53 Typical applications of wrought copper and copper alloys specified in Indian Standards

Designation/ Grade	Typical applications
Coppers	
CuETP	Busbars, chemical process equipment, contacts, cotter pins, electrical wire, flashing gaskets, kettles, printing rolls, radiators, radio parts, rivets, soldering copper, stranded conductors, switches.
CuFRTP–1	Anodes, busbars, chemical process equipment, commutator segments, contacts, gaskets, kettles, printing rolls, radiators, radio parts, switches, terminals.
CuFRHC	Busbars, chemical process equipment, cables, heat exchangers, electrical wiring.
CuDHP	Air, gasoline, hydraulic and oil lines, air conditioners, brewery tubes, condensers, evaporator and heat exchanger tubes, distiller tubes, kettles, steam and water lines.
CuATP	Heat exchanger and condenser tubes, plates for locomotive fireboxes.
CuATP	Electrical items of lower conductivity.
Cu–Te	Electrical connectors, motor parts, plumbing fixtures, soldering tips, transistor bases and parts that are assembled by furnace brazing, welding torch tips.
Cu–Pb	Electrical connectors, motor parts, switch parts, screw machine parts requiring high conductivity.
Brass	
CuZn20	Bellows and musical instruments, clock dials, flexible hose, ornamental metalwork, medallions, pump lines.
CuZn30	Ammunition components, bead chain, etched articles, eyelets, fasteners, flashlight reflectors, hinges, lamp fixtures, locks, pins, plumbing accessories, radiator cores and tubes, screw shells, springs, socket shells, stampings, tubes.
CuZn35	Bead chain, chain, eyelets, fasteners, flashlight shells, grillwork, grommets, lamp fixtures, pins, radiator cores and tanks, reflectors, rivets, screw shells, screws, sink strainers, springs.
CuZn37	Plumbing accessories, traps.

(Contd.)

TABLE 5.53 Typical applications of wrought copper and copper alloys specified in Indian Standards (continued)

Designation/ Grade	Typical applications
Brass	
CuZn40	Architectural panel sheets, bolting, brazing rod, heavy plates for structural applications, hot forgings, valve stems.
CuZn20A12As	Condenser, evaporator and heat exchanger tubes, distiller tubes, ferrules.
CuZn29Sn1As	Condenser, condenser–tube plates, distiller and heat–exchanger tubes, ferrules, strainers.
CuZn37Sn1Pb1	Marine hardware, screw machine products, valve stems.
CuZn39Sn	Balls bolts, condenser plates, fittings, marine hardware, nuts, propeller shafts, rivets, valve stems.
HT1	Forged parts, heat exchangers, marine equipment, underwater items.
HT2	Architectural sections, forgings, pump parts, pressure tubes.
CuZn35Pb1	Couplings, dials, engravings, free–machining screws and rivets, gears, tire valve stems.
CuZn37Pb1	Bearing cages, book dies, clock plates, engraving plates, gears, hinges, hose couplings, keys, lock parts, lock tumblers, meter parts, nuts, sink strainers, templates, type characters, washers, wear plates.
CuZn36Pb3	Automatic high–speed screw–machine parts, gears, pinions.
CuZn39Pb3	Architectural extrusions, hardware, butts, hinges, industrial forgings, lock bodies, thresholds and trim.
Phosphor bronze	
I (CuSn4)	Bridge bearing plates, beater bars, bellows, chemical hardware, clutch disks, connectors, diaphragms, fasteners, fuse clips, lock washers, sleeve bushings, springs, switch parts, terminals.
II (CuSn5)	Bridge bearing plates, beater bars, bellows, bourdon tubing, chemical hardware, clutch disks, cotter pins, diaphragms, fasteners, fuse clips, sleeve bushings, springs, switch parts, welding rods, wire brushes.
III (CuSn6)	Anti–friction components, clips, pinions and gears, pump parts, springs, tubes to resist corrosion and erosion, washers.

(Contd.)

TABLE 5.53 Typical applications of wrought copper and copper alloys specified in Indian Standards (continued)

Designation/ Grade	Typical applications
	Aluminium bronze
CuAl8Fe3	Condenser and heat exchanger tubes, fasteners, corrosion resistant vessels.
CuA110Fe3	Bearings, bushings, bolts, cams, gears, nuts, valve guides, pump rods.
CuA110Fe5Ni5	Bolting, nuts, pump parts, shafts, miscellaneous uses for corrosion resistant and spark–resistant parts for industrial, marine and submarine applications.
	Copper-nickel-zinc alloys
CuNi10Zn27	Etching stock, hollowware, nameplates, optical parts, rivets, screws, slide fasteners.
CuNi10Zn42Pb2	Architectural fittings and trim, clock and watch components, hot stamped and extruded shapes, machined parts, small hinges.
CuNi12Zn24	Camera parts, etching stock, nameplates, optical parts, slide fasteners.
CuNi15Zn21	Camera parts,etching stock, jewelery, optical equipment.
CuNi18Zn20	Camera parts, costume jewelery, etching stock, hollowware, nameplates.
CuNi18Zn27	Optical goods, springs, resistance wire.
	Copper-nickel alloys
CuNi10Fe 1Mn	Boat hulls, condensers, condenser plates, distiller tubes, evaporator and heat exchanger tubes, ferrules, salt water piping.
CuNi30Mn1Fe	Feed water heaters and fittings, mechanical fasteners, tubes and plates for marine condenser and desalination plants.

5 EQUIVALENT GRADES

TABLE 5.54 Comparison of grades of wrought copper and copper alloy plate, sheet, strip, and foil specified in Indian Standards with nearest equivalent grades specified in other national and international standards

IS	ASTM B 36/B 36M–95 ASTM B 103/B 103M–93 ASTM B 121/B 121M–91 ASTM B 122–92a ASTM B 152M–94	BS 2870–80 BS 2875–69	DIN 1787–73 DIN 17660–83 DIN 17662–83 DIN 17663–83	ISO 426 (1)–83 ISO 426 (2)–83 ISO 427–83 ISO 430–83 ISO 1337–80	JIS H 3100–92 JIS H 3110–92	SAE J463–81
			Copper (IS 1972 – 1989)			
FRTP-1	C12500	C 104	–	Cu-FRTP	–	–
DHP	C12200	C 106	SF-Cu	Cu-DHP	C 1220 P/C 1220 R	C12200
DPA	C14200	C 107	–	–	–	–
ATP	–	C 105	–	–	–	–
			Brass (IS 410 – 1977 and IS 12443 – 1988)			
CuZn30	C26000	CZ 106	CuZn30	CuZn30	C 2600 P/C 2600 R	C26000
CuZn37	C27200	CZ 108	CuZn37	CuZn37	C 2720 P/C 2720 R	–
CuZn40	C28000	–	CuZn40	CuZn40	C 2801 P/C 2801 R	–
CuZn20Al2As	–	CZ 110	CuZn20Al2	CuZn20Al2	–	–
CuZn28As	–	CZ 105	–	–	C 4430 P/C 4430 R	–
CuZn36Sn1	–	CZ 112	–	–	C 4621 P	

(Contd.)

TABLE 5.54 Comparison of grades of wrought copper and copper alloy plate, sheet, strip, and foil specified in Indian Standards with nearest equivalent grades specified in other national and international standards (continued)

IS	ASTM B 36/B 36M–95 ASTM B 103/B 103M–93 ASTM B 121/B 121M–91 ASTM B 122–92a ASTM B 152M–94	BS 2870–80 BS 2875–69	DIN 1787–73 DIN 17660–83 DIN 17662–83 DIN 17663–83	ISO 426 (1)–83 ISO 426 (2)–83 ISO 427–83 ISO 430–83 ISO 1337–80	JIS H 3100–92 JIS H 3110–92	SAE J463–81
Brass (IS 410 – 1977 and IS 12443 – 1988)						
CuZn39Sn	–	–	CuZn38Sn1	CuZn38Sn1	C 4640 P	–
CuZn40Pb	–	CZ 123	CuZn39Pb0.5	CuZn40Pb	C 3710 P/C 3710 R	–
Phosphor bronze (IS 7814 – 1985)						
I	C51100	PB 101	CuSn4	CuSn4	C 5111 P/C 5111 R	C51100
II	C51000	PB 102	–	CuSn5	C 5102 P/C 5102 R	C51000
III	C51900	PB 103	CuSn6	CuSn6	C 5191 P/C 5191 R	-
Nickel silver (IS 2283 – 1981)						
NS 10	C74500	NS 103	–	CuNi10Zn27	C 7451 P/C 7451 R	–
NS 12	–	NS 104	CuNi12Zn24	CuNi12Zn24	–	–
NS 15	–	NS 105	–	CuNi15Zn21	C 7541 P/C 7541 R	–
NS 18	C75200	NS 106	CuNi18Zn20	CuNi18Zn20	C 7521 P/C 7521 R	C75200
NS 18T	C77000	NS 107	–	CuNi18Zn27	–	C77000
NS 20	–	–	–	–	–	–

TABLE 5.55 Comparison of grades of wrought copper and copper alloy rods, bars, and sections specified in Indian Standards with nearest equivalent grades specified in other national and international standards

IS	ASTM B 16M–92 ASTM B 21M–90a ASTM B 138M–92 ASTM B 139M–95 ASTM B 150M–95 ASTM B 151M–94 ASTM B 187M–94 ASTM B 301M–93 ASTM B 453M–88 ASTM B 455–91	BS 2874–86	DIN 1787–73 DIN 17660–83 DIN 17662–83 DIN 17663–83 DIN 17665–83 DIN 17666–83	ISO 426 (1)–83 ISO 426 (2)–83 ISO 427–83 ISO 428–83 ISO 430–83 ISO 1187–83 ISO 1336–80	JIS H 3250–92 JIS H 3270–92	SAE J463–81
Phosphor bronze (IS 7811 – 1985)						
–	C51000	PB 102	–	CuSn5	C 5102 B	C51000
Aluminium bronze (IS 10569 – 1983)						
CuAl8Fe3	C61400	–	CuAl8Fe3	CuAl8Fe3	–	C61400
CuAl10Fe3	C62300	–	–	CuAl10Fe3	–	C62300
CuAl10Fe5Ni5	C63200	CA 104	CuAl10Ni5Fe4	CuAl10Ni5Fe4	–	–
Aluminium silicon bronze (IS 10723 – 1983)						
–	C64200	–	–	CuAl7Si2	–	C64200
Nickel silver (IS 10757 – 1983)						
CiNi10Zn27	C74500	–	–	–	C 7451 B	–
CuNi18Zn20	C75200	–	CuNi18Zn20	CuNi18Zn20	C 7521 B	C75200
CuNi12Zn24	C75700	–	CuNi12Zn24	CuNi12Zn24	–	–

(Contd.)

TABLE 5.55 Comparison of grades of wrought copper and copper alloy rods, bars, and sections specified in Indian Standards with nearest equivalent grades specified in other national and international standards (continued)

IS	ASTM B 16M–92 ASTM B 21M–90a ASTM B 138M–92 ASTM B 139M–95 ASTM B 150M–95 ASTM B 151M–94 ASTM B 187M–94 ASTM B 301M–93 ASTM B 453M–88 ASTM B 455–91	BS 2874–86	DIN 1787–73 DIN 17660–83 DIN 17662–83 DIN 17663–83 DIN 17665–83 DIN 17666–83	ISO 426 (1)–83 ISO 426 (2)–83 ISO 427–83 ISO 428–83 ISO 430–83 ISO 1187–83 ISO 1336–80	JIS H 3250–92 JIS H 3270–92	SAE J463–81
Nickel silver (IS 10757 – 1983)						
CuNi18Zn27	C77000	–	–	CuNi18Zn27	C 7701 B	C77000
CuNi10Zn42Pb2	–	NS 101	–	–	–	–

TABLE 5.56 Comparison of grades of wrought copper and copper alloy wire specified in Indian Standards with nearest equivalent grades specified in other national and international standards

IS	ASTM B 134–93 ASTM B 159M–94	BS 2873–69	DIN 1787–73 DIN 17660–83 DIN 17662–83	ISO 426 (1)–83 ISO 426 (2)–83 ISO 427–83 ISO 1337–80	JIS H 3260–92 JIS H 3270–92	SAE J463–81
Copper wire (IS 4412 – 1981)						
DHP	–	C 106	SF-Cu	Cu-DHP	C 1220 W	–

(Contd.)

TABLE 5.56 Comparison of grades of wrought copper and copper alloy wire specified in Indian Standards with nearest equivalent grades specified in other national and international standards (continued)

IS	ASTM B 134–93 ASTM B 159M–94	BS 2873–69	DIN 1787–73 DIN 17660–83 DIN 17662–83	ISO 426 (1)–83 ISO 426 (2)–83 ISO 427–83 ISO 1337–80	JIS H 3260–92 JIS H 3270–92	SAE J463–81
Brass wire (IS 4413 – 1981)						
CuZn30	C26000	CZ 106	CuZn30	CuZn30	C 2600 W	C26000
CuZn37	C27400	CZ 108	CuZn37	CuZn37	C 2720 W	–
Brass wire for cold-headed and machined parts (IS 2704 – 1983)						
CuZn35	C27000	CZ 107	CuZn36	CuZn35	C 2700 W	C27000
CuZn35Pb1	–	–	CuZn36Pb1.5	CuZn35Pb1	–	–
Free cutting brass wire (IS 8364 – 1989)						
1	–	–	CuZn39Pb3	CuZn39Pb3	C 3603 W	–
2	–	–	CuZn36Pb3	CuZn36Pb3	C 3601 W	–
3	–	–	–	CuZn37Pb1	C 3501 W	–
Phosphor bronze wire (IS 7608 – 1987)						
I	C51000	PB 102	–	CuSn5	C 5102 W	C51000
II	–	PB 103	CuSn6	CuSn6	C 5191 W	–

TABLE 5.57 Comparison of grades of wrought copper and copper alloy tubes specified in Indian Standards with nearest equivalent grades specified in other national and international standards

IS	ASTM B 68M–92a ASTM B 75M–93 ASTM B 111M–93 ASTM B 135M–91 ASTM B 359M–94 ASTM B 466M–92	BS 2871 (2)–72 BS 2871 (3)–72	DIN 1787–73 DIN 17660–83 DIN 17664–83	ISO 426 (1)–83 ISO 429–83 ISO 1337–80	JIS H 3300–92	SAE J463–81
Copper tubes (IS 2501 – 1995)						
ETP	–	C 101	E-Cu58	Cu-ETP	C 1100 T	–
FRTP-1	–	–	–	–	–	–
FRTP-2	–	–	–	–	–	–
DHP	C12200	C 106	SF-Cu	Cu-DHP	C1220 T	C12200
ATP	–	–	–	–	–	–
Brass tubes (IS 407 – 1981)						
CuZn37	C27200	CZ 108	CuZn37	CuZn37	–	–
CuZn21Al2As	C68700	CZ 110	CuZn20Al2	CuZn20Al2	C 6870 T	–
CuZn29Sn1As	C44300	CZ 111	CuZn28Sn1	CuZn28Sn1	C 4430 T	–
CuZn30As	–	CZ 126	–	CuZn30As	–	–
Copper-nickel alloy tubes (IS 1545 – 1994)						
CuNi10Fe1Mn	C70600	CN 102	CuNi10Fe1Mn	CuNi10Fe1Mn	C 7060 T	C70600
CuNi30Mn1Fe	C71500	CN 107	CuNi30Mn1Fe	CuNi30Mn1Fe	C 7150 T	C71500

(Contd.)

TABLE 5.58 Comparison of grades of wrought copper alloy forging stock and forgings specified in IS 6912 – 1985 with nearest equivalent grades specified in other national and international standards

IS 6912–85	ASTM B 124M–94 ASTM B 283–94	BS 2872–89	DIN 1787–73 DIN 17660–83 DIN 17663–83 DIN 17665–83	ISO 426 (1)–83 ISO 426 (2)–83 ISO 428–83 ISO 430–83 ISO 1337–80	JIS	SAE J463–81
			Copper			
ETP	C11000	C 101	E-Cu57	Cu-ETP	–	–
FRHC	–	C 102	–	Cu-FRHC	–	–
DPA	–	–	–	–	–	–
			Naval brass			
FLB	–	CZ 112	–	CuZn38Sn1	–	–
			Leaded brass			
FNB	–	CZ 129	CuZn39Pb2	CuZn39Pb1	–	–
			High tensile brass			
FHTB1	C67500	CZ 115	–	CuZn39AlFeMn	–	–
FHTB2	–	CZ 114	–	CuZn39AlFeMn	–	–

(Contd.)

TABLE 5.58 Comparison of grades of wrought copper alloy forging stock and forgings specified in IS 6912 – 1985 with nearest equivalent grades specified in other national and international standards (continued)

IS 6912–85	ASTM B 124M–94 ASTM B 283–94	BS 2872–89	DIN 1787–73 DIN 17660–83 DIN 17663–83 DIN 17665–83	ISO 426 (1)–83 ISO 426 (2)–83 ISO 428–83 ISO 430–83 ISO 1337–80	JIS	SAE J463–81
Aluminium bronze						
FAB1	–	–	CuAl8Fe3	–	–	–
FAB2	C62300	–	–	CuAl10Fe3	–	C62300
FAB3	C63200	CA 104	CuAl10Ni5Fe4	CuAl10Ni5Fe4	–	–
Aluminium-silicon bronze						
FABS	C64200	–	–	–	–	–
Nickel silver						
FNS	–	–	–	–	–	–

8 REFERENCES

Books

1. *ASM Handbook*, vol. 2, ASM International, Materials Park, Ohio, U.S.A., 1990.
2. *Copper and its Alloys in Engineering and Technology*, C.D.A. Publication No. 43, Copper Development Association, Herts, U.K., 1955.
3. *Standards Handbook—Wrought Copper and Copper Alloy Mill Products — Part 2: Alloy Data*, 8th ed., Copper Development Association, Greenwich, Connecticut, U.S.A., 1985.

Standards/article

4. IS 291 – 1989: *Naval Brass Rods and Sections for Machining Purposes – Specification.*
5. IS 319 – 1989: *Free-Cutting Leaded Brass Bars, Rods and Sections — Specification.*
6. IS 320 – 1980: *Specification for High Tensile Brass Rods and Sections (other than Forging Stock).*
7. IS 407 – 1981: *Specification for Brass Tubes for General Purposes.*
8. IS 410 – 1977: *Specification for Cold Rolled Brass Sheet, Strip and Foil.*
9. IS 1545 – 1994: *Solid Drawn Copper and Copper Alloy Tubes for Condensers and Heat Exchangers—Specification.*
10. IS 1972 – 1989: *Copper Plate, Sheet and Strip for Industrial Purposes—Specification.*
11. IS 2283 – 1981: *Specification for Nickel Silver Sheet, Strip and Foil.*
12. IS 2501 – 1995: *Solid drawn Copper Tubes for General Engineering Purposes—Specification.*
13. IS 2704 – 1983: *Specification for Brass Wires for Cold-Headed and Machined Parts.*
14. IS 4170 – 1967: *Specification for Brass Rods for General Engineering Purposes.*
15. IS 4171 – 1983: *Specification for Copper Rods and Bars for General Engineering Purposes.*
16. IS 4412 – 1981: *Specification for Copper Wires for General Engineering Purposes.*
17. IS 4413 – 1981: *Specification for Brass Wires for General Engineering Purposes.*
18. IS 6912 – 1985: *Specification for Copper and Copper Alloy Forging Stock and Forgings.*
19. IS 7608 – 1987: *Specification for Phosphor Bronze Wires for General Engineering Purposes.*
20. IS 7811 – 1985: *Specification for Phosphor Bronze Rods and Bars.*
21. IS 7814 – 1985: *Specification for Phosphor Bronze Sheet and Strip.*
22. IS 8328 – 1977: *Specification for Free Cutting Copper Bars, Rods and Sections.*
23. IS 8364 – 1989: *Free-Cutting Brass Wire - Specification.*
24. IS 10569 – 1983: *Specification for Aluminium Bronze Rods and Bars.*
25. IS 10723 – 1983: *Specification for Aluminium Silicon Bronze Rods and Bars.*
26. IS 10757 – 1983: *Specification for Nickel Silver Rods and Bars.*
27. IS 12443 – 1988: *Specification for Rolled Brass Plates for General Engineering Purposes.*
28. D.T. Peters and K.J.A. Kundig, Selecting Copper and Copper Alloys—Part I: Wrought Products, Advanced Materials & Processes, 145(2), 1994.

5.6

HEAT TREATMENT OF COPPER AND COPPER ALLOYS

1 HOMOGENIZING

Purpose: To eliminate or decrease chemical segregation which occurs as a natural result of solidification. It is applied to some copper alloys, such as copper–nickel alloys, phosphor bronzes and silicon bronzes to improve the hot and cold ductility of cast billets for mill processing, and occasionally to castings to achieve specified mechanical properties.

Recommended heat treatment cycle: Heat to a temperature above the upper annealing range, to within 50 °C of the solidus temperature, in a protective atmosphere. Hold at this temperature for 3–10 h. Cool slowly to ambient temperature.

Note: The temperature and time required for the homogenization process varies with the alloy, the cast grain size and the degree of homogenization desired.

2 STRESS RELIEVING

Purpose: To reduce residual stresses without causing recrystallization. Residual stresses if present in sufficient magnitude can result in stress-corrosion cracking of material in storage or service, unpredictable distortion of material during cutting or machining, and hot cracking of materials during processing, brazing or welding.

Recommended heat treatment cycle: Heat to the recommended stress relieving temperature given in Table 5.59. Hold at this temperature for about 1 h. Cool slowly to ambient temperature to minimize the development of new residual stresses.

3 ANNEALING

3.1 Full annealing of wrought copper and copper alloys

Purpose: To soften the metal or alloy by the removal of strain hardening resulting from cold working, by recrystallization and/or by coalescing precipitates from the solid solution.

Recommended heat treatment cycle: Heat to the recommended full annealing temperature as given in Table 5.59. Hold at this temperature for a sufficient length of time to equalize the temperature throughout the part (minimum holding time = 1 h). Cool slowly to ambient temperature.

Note: Increasing the amount of cold work prior to annealing lowers the recrystallization temperature.

TABLE 5.59 Recommended stress relieving and full annealing temperatures for wrought copper and copper alloys

Material designation	Stress relieving temperature, °C	Full annealing temperature, °C
Coppers		
CuETP	150–200	375–650
CuFRTP	175–225	400–650
CuFRHC	150–200	–
CuDHP	200–250	375–650
CuATP	–	–
CuDPA	225–275	425–650
Cu-Te	225–275	425–650
Cu-Pb	–	425–650
Brasses		
CuZn20	200–300	425–700
CuZn30	250–350	425–750
CuZn35	250–350	425–700
CuZn37	250–350	425–700
CuZn40	250–350	425–600
CuZn20Al2As	250–350	425–600
CuZn28As	–	–
CuZn29Sn1As	250–350	425–600
CuZn30As	–	425–750

(Contd.)

TABLE 5.59 Recommended stress relieving and full annealing temperatures for wrought copper and copper alloys (continued)

Material designation	Stress relieving temperature, °C	Full annealing temperature, °C
Brasses		
CuZn36Sn1	–	425–600
CuZn37Sn1Pb1	–	425–600
CuZn39Sn	225–325	425–600
HT 1	–	425–600
HT 2	–	–
CuZn35Pb1	250–350	425–650
CuZn37Pb1	250–350	425–600
CuZn36Pb3	250–350	425–600
CuZn40Pb	250–350	425–600
CuZn39Pb1	250–350	–
CuZn39Pb3	250–350	425–600
Phosphor bronzes		
I (CuSn4)	200–350	475–675
II (CuSn5)	200–350	475–675
III (CuSn6)	200–350	–
Aluminium bronzes		
CuAl8Fe3	300–400	600–900
CuAl10Fe3	300–400	600–650
CuAl10Fe5Ni5	300–400	600–700

(Contd.)

TABLE 5.59 Recommended stress relieving and full annealing temperatures for wrought copper and copper alloys (continued)

Material designation	Stress relieving temperature, °C	Full annealing temperature, °C
	Aluminium-silicon bronze	
CuAl7Si2	–	600–700
	Nickel silvers	
CuNi10Zn27	250–350	600–750
CuNi10Zn42Pb2	300–400	600–700
CuNi12Zn24	250–350	600–825
CuNi15Zn21	250–350	600–825
CuNi18Zn20	250–350	600–750
CuNi18Zn27	250–350	600–825
	Copper-nickel alloy	
CuNi10Fe1Mn	275–400	600–825
CuNi30Mn1Fe	300–400	–

3.2 Partial or temper annealing of wrought copper and copper alloys

Purpose: To reduce the strength properties of a cold worked metal or alloy to a controlled level.

Recommended heat treatment cycle: Heat to a temperature in the lower annealing range (≈ 400 °C for brasses), with special precautions to avoid any overheating. Hold at this temperature for about 0.5 h. Cool to ambient temperature in still air.

Note: The exact temperature and time for partial annealing should be established by experiment.

3.3 Annealing of cast copper alloys

Purpose: To correct the effect of mould cooling (such as microstructures resulting in high hardness and/or low ductility, and occasionally inferior corrosion resistance) in some duplex alloys such as high tensile brasses and aluminium bronzes.

Recommended heat treatment cycle: Heat to 580–700 °C in a protective atmosphere. Hold at this temperature for about 1 h. Cool slowly to ambient temperature. For aluminium bronzes cool rapidly to ambient temperature in water or in air flowing at high speed.

4 Hardening and tempering of α–β aluminium bronzes

Purpose: To increase strength without unduly sacrificing ductility.

Recommended heat treatment cycle: Heat to 815–1010 °C in a protective atmosphere. Hold at this temperature for a sufficient length of time until a uniform temperature is achieved. Quench in water or oil. Temper at 595–650 °C for about 2 h.

5 Precipitation hardening

Purpose: To strengthen special types of copper alloys, such as heat-treatable high-copper alloys containing small amounts of beryllium, chromium and zirconium, above the levels ordinarily obtained by cold working.

Recommended heat treatment cycle: Solution treat in a protective atmosphere at the appropriate temperature as indicated below:

a) For copper-chromium alloys 980–1010°C;
b) For copper-zirconium alloys 900–980°C; and
c) For copper-beryllium alloys
  i) 98Cu-1.7Be-0.3Co: 775–800°C, and
  ii) 97Cu-0.5Be-2.5Co: 900–925°C.

Quench rapidly to room temperature. Artificially age for 1–4 h at the appropriate temperature as indicated below:

a) For copper-chromium alloys 425–500°C;
b) For copper-zirconium alloys 500–550°C or 370–480°C if cold worked prior to ageing; and
c) For copper-beryllium alloys
  i) 98Cu-1.7Be-0.3Co: 300–330°C, and
  ii) 97Cu-0.5Be-2.5Co: 455–480°C.

6 REFERENCES

Books

1. *ASM Handbook,* vol. 4, ASM International, Materials Park, Ohio, U.S.A., 1991.
2. *Standards Handbook—Wrought Copper and Copper Alloy Mill Products—Part 2: Alloy Data,* 8th ed., Copper Development Association, Greenwich, Connecticut, U.S.A., 1985.

Standard

3. ISO 197 (5) – 1980: *Copper and Copper Alloys—Terms and definitions—Part 5: Methods of Processing and Treatment.*

6

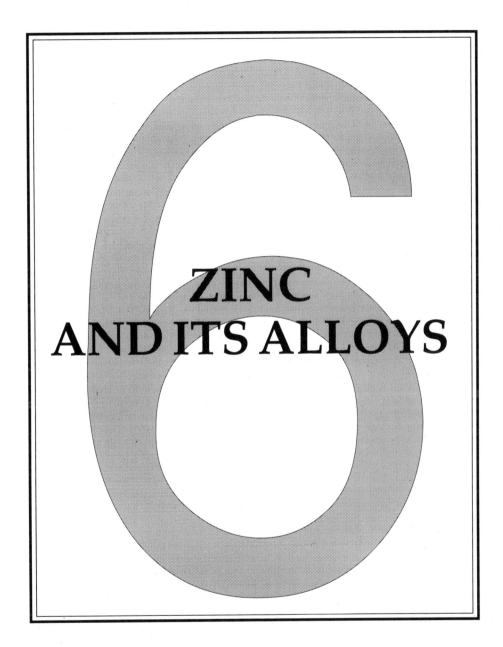

ZINC AND ITS ALLOYS

CHAPTER 6

Zinc and Its Alloys

6.1	Zinc	6.5
6.2	Zinc casting alloys	6.8
6.3	Wrought zinc and zinc alloys	6.31

6.1

ZINC

1 INTRODUCTION

Pure zinc is a bluish-white metal. Except when very pure, zinc is brittle at ordinary temperatures, but soft and ductile above 100 °C. It is a fair conductor of electricity and exhibits excellent atmospheric corrosion resistance. It is non-magnetic. A summary of some of the important physical properties of pure zinc is given in Table 1.1.

Zinc is used extensively:

a) in coatings for corrosion protection of cast iron and steel;
b) in sacrificial anodes for cathodic protection of metals and alloys;
c) in zinc casting alloys;
d) in wrought zinc and zinc alloys;
e) as an alloying element in aluminium-base, copper-base, magnesium-base and tin-base alloys, and in a number of other complex, low-melting point brazing and soldering alloys; and
f) as zinc dust and oxide;
g) in chemicals.

2 SPECIFICATIONS

TABLE 6.1 Chemical composition of grades of zinc metal specified in Indian Standards

Designation	Chemical composition [% (*m*/*m*)]								
	Zn min.	Pb max.	Cd max.	Fe max.	Sn max.	Cu max.	Tl max.	In max.	Total impurities max.
Virgin zinc (IS 209 – 1992)									
Zn 99.99	99.99	0.003	0.003	0.003	0.001	0.002	0.001	0.0005	0.01
Zn 99.95	99.95	0.025	0.02	0.01	0.001	0.002	–	–	0.05

(Contd.)

TABLE 6.1 Chemical composition of grades of zinc metal specified in Indian Standards (continued)

Designation	Chemical composition [% (*m*/*m*)]								
	Zn	Pb	Cd	Fe	Sn	Cu	Tl	In	Total impuri-ties
	min.	max.	max.	max.	max.	max.	max.	max.	max.
Refined secondary zinc (IS 4699–1984)									
SZn 99.5	99.5	0.35	0.15	0.05	0.01	–	–	–	0.5
SZn 98.5	98.5	1.4	0.20	0.07	0.02	–	–	–	1.5

3 TYPICAL APPLICATIONS

TABLE 6.2 Typical applications of grades of zinc metal specified in Indian Standards

Designation	Typical applications
Zn 99.99	Deep drawing brass, deep drawing nickel silver, etching plates, sacrificial anodes for cathodic protection, zinc casting alloys, and zinc sheet strip and wire.
Zn 99.95	Deep drawing brass, and zinc sheet, strip and wire.
SZn 99.5	Alloying purposes, zinc oxide, and zinc sheet and strip.
SZn 98.5	Galvanizing.
Source: References 8–10	

4 EQUIVALENT GRADES

TABLE 6.3 Comparison of grades of zinc metal specified in Indian Standards with nearest equivalent grades specified in other national and international standards

IS 209–92 IS 4699–84	ASTM B 6–95	BS 3436–86	DIN 1706–74	GOST 3640–79	ISO 752–81	JIS H 2107–57
Zn 99.99	Special High Grade (Z13001)	Zn 1	Zn99.99	ЦВ	Zn 99.99	Special Zinc Metal
Zn 99.95	High Grade (Z15001)	Zn 2	Zn99.95	Ц1	Zn 99.95	Ordinary Zinc Metal
SZn 99.5	–	Zn 3	Zn99.5	–	Zn 99.5	Distilled Zinc Metal, Special
SZn 98.5	Prime Western (Z19001)	Zn 4	Zn98.5	Ц3С	Zn 98.5	Distilled Zinc Metal, Class 1

5 REFERENCES

Books

1. H. Morrow III, *Zinc: Properties and Applications, Encyclopedia of Materials Science and Engineering*, vol. 7, M.B. Bever (Ed.), Pergamon Press, Oxford, U.K., 1986.
2. A.L. Ponikvar, *Zinc, McGraw-Hill Encyclopedia of Science & Technology*, vol. 14, 5th ed., McGrawHill, New York, U.S.A., 1982.
3. D.R. Lide (Ed.), *CRC Handbook of Chemistry and Physics*, 73rd ed., CRC Press, Boca Raton, Florida, U.S.A., 1992.
4. *ASM Handbook*, vol. 2, ASM International, Materials Park, Ohio, U.S.A., 1990.
5. *Metals Handbook Desk Edition*, American Society for Metals, Metals Park, Ohio, U.S.A., 1985.

Standards

6. IS 209 – 1992: *Zinc Ingot—Specification.*
7. IS 4699 – 1984: *Specification for Refined Secondary Zinc.*
8. IS 13229 – 1991: *Zinc for Galvanizing—Specification.*
9. BS 3436 – 1986: *Specification for Ingot zinc.*
10. DIN 1706 – 1974: *Zinc.*

6.2

ZINC CASTING ALLOYS

I CONVENTIONAL ZINC DIE CASTING ALLOYS

1 INTRODUCTION

Conventional zinc die casting alloys, popularly known as alloys 2, 3, 5 and 7, are hypoeutectic Zn-4Al alloys which can be hot-chamber die cast to net or near-net shapes. These alloys are characterized by the following properties:

a) Moderate tensile strength: The tensile strength of conventional zinc die casting alloys compares favourably with those of some of the competing alloys;
b) High modulus of elasticity: The modulus of elasticity of conventional zinc die casting alloys is greater than that of aluminium and magnesium alloys, and plastics;
c) High impact strength: The impact strength of conventional zinc die casting alloys is significantly higher than that of most casting alloys;
d) Excellent damping capacity: The damping capacity of conventional zinc die casting alloys is superior to that of steel and some of the cast aluminium alloys at ambient temperature. The damping capacity of these alloys increases with temperature and at 100 °C is greater than that of grey cast iron;
e) Non-sparking characteristics: Conventional zinc die casting alloys are classified as non-sparking and are used in potentially explosive environments such as in coal mines, refineries and tankers;
f) Non-magnetic characteristics: Conventional zinc die casting alloys are non-magnetic;
g) Good corrosion resistance: The corrosion resistance of conventional zinc die casting alloys is similar to that of cast aluminium alloys. The recommended pH range of application is 6.0–11.5. When required the corrosion resistance can be enhanced by chromate and phosphate coatings, and substantially improved by anodizing;
h) Excellent machinability;
i) Superior pressure tightness: Conventional zinc alloy die castings can be successfully used in pressure tight applications without the use of an impregnation treatment;
j) Superior finishing characteristics;
k) Excellent thin-wall castability; and
l) Zero draft angle castability: Conventional zinc die casting alloys can be die cast with less draft angle than that recommended for competing materials. In fact, in some cases castings with zero draft angle can be achieved.

m) Good dimensional accuracy: The casting tolerances for conventional zinc die casting alloys are superior to those of aluminium and magnesium castings produced by the same casting process.

The use of conventional zinc alloy die castings is generally limited to 95 °C

2 SPECIFICATIONS

TABLE 6.4 Chemical composition of zinc alloy ingots for die casting and zinc alloy die castings specified in Indian Standards

Alloy designation	Chemical composition [% (*m*/*m*)]										
	Zn	Al	Cu	Mg	Fe max.	Pb max.	Cd max.	Sn max.	Tl + In max.	Ni max.	Cr max.
Zinc alloy ingots for die casting[1] (IS 713 – 1981)											
Zn Al 4	Rem.	3.9–4.3	≤ 0.03	0.04–0.06	0.03	0.003	0.003	0.001	0.0015	–	–
Zn Al 4 Cu 1	Rem.	3.9–4.3	0.75–1.25	0.04–0.06	0.03	0.003	0.003	0.001	0.0015	–	–
Zinc alloy die castings (IS 742 – 1981)											
Zn Al 4	Rem.	3.8–4.3	≤ 0.10	0.03–0.06	0.10	0.005	0.005	0.002	0.0015	0.006	0.02
Zn Al 4 Cu 1	Rem.	3.8–4.3	0.75–1.25	0.03–0.06	0.10	0.005	0.005	0.002	0.0015	0.006	0.02

1) Zinc of 99.99+ % purity, conforming to IS 209 – Zn 99.99, should be used for the manufacture of these alloy ingots.

3 TYPICAL MECHANICAL PROPERTIES

TABLE 6.5 Typical mechanical properties[1] of zinc die casting alloys, conforming to IS 742 – 1981, at 20–21 °C

Property	Unit	Alloy designation	Original value 5 weeks after casting	After 12 months normal ageing	After 5 years normal ageing	After 8 years normal ageing	After 12 month dry ageing at 95 °C
Tensile strength[2]	MPa	**Zn Al 4**	286	264	260	247	–
		Zn Al 4 (S)	273	264	242	–	232
		Zn Al 4 Cu 1	335	320	295	292	–
		Zn Al 4 Cu 1 (S)	312	290	–	–	245
Percentage elongation after fracture (L_o = 50.8 mm, d = 6.35 mm)	%	**Zn Al 4**	15	25	27	20	–
		Zn Al 4 (S)	17	24	19	–	30
		Zn Al 4 Cu 1	9	12	12	14	–
		Zn Al 4 Cu 1 (S)	10	14	–	–	22
Charpy impact strength[3]	J	**Zn Al 4**	57	58	59	60	–
		Zn Al 4 (S)	61	56	57	–	47
		Zn Al 4 Cu 1	58	57	58	56	–
		Zn Al 4 Cu 1 (S)	60	61	–	–	17
Brinell hardness	HB 10/500/30	**Zn Al 4**	83	67	65	65	–
		Zn Al 4 (S)	69	54	61	–	50
		Zn Al 4 Cu 1	92	74	77	74	–
		Zn Al 4 Cu 1 (S)	83	72	–	–	57

[1] Measured on pressure die cast test pieces.
[2] Strain rate = 6.3 mm/min crosshead speed.
[3] Measured on unnotched, 6.35 mm × 6.35 mm test pieces.

Note: (S) indicates stabilized condition.
Source: References 12 and 14

TABLE 6.6 Other typical mechanical properties[1)] of zinc die casting alloys, conforming to IS 742–1981

Alloy designation	Compressive strength MPa	Shear strength MPa	Modulus of rupture MPa
Zn Al 4	415	215	650
Zn Al 4 Cu 1	600	250	720

1) Measured on pressure die cast test pieces.

Source: Reference 13

TABLE 6.7 Effect of temperature on the mechanical properties[1)] of zinc die casting alloys, conforming to IS 742 – 1981

Alloy designation	Temperature °C	R_m MPa	$A^{2)}_{50.8mm}$ %	Charpy impact strength[3)] J
Zn Al 4	–40	318	4.5	2.8
	–20	–	–	3.4
	–10	–	–	4.7
	0	296	9	10
	10	–	–	42
	20	281	11	57
	40	248	16	57
	95	196	30	54

TABLE 6.7 Effect of temperature on the mechanical properties[1] of zinc die casting alloys, conforming to IS 742 – 1981 (continued)

Alloy designation	Temperature °C	R_m MPa	$A^{2)}_{50.8mm}$ %	Charpy impact strength[3] J
Zn Al 4 Cu 1	–40	375	3	3.2
	–20	–	–	4.8
	–10	–	–	24
	0	375	8	53
	10	–	–	56
	20	348	8	60
	40	294	13	63
	95	240	23	58

[1] Measured on pressure die cast test pieces.
[2] Diameter of the parallel length of the test piece (*d*) = 6.35 mm.
[3] Measured on unnotched, 6.35 mm × 6.35 mm test pieces.

Source: Reference 14

4 TYPICAL PHYSICAL PROPERTIES

TABLE 6.8 Typical physical properties of zinc die casting alloys, conforming to IS 742 – 1981

Property	Unit	Zn Al 4	Zn Al 4 Cu 1
Density	kg/m^3	6700	6700
Casting contraction (mean)	mm/mm	0.006	0.006
Liquidus temperature	°C	387	388
Solidus temperature	°C	382	379

(Contd.)

TABLE 6.8 Typical physical properties of zinc die casting alloys, conforming to IS 742 – 1981 (continued)

Property	Unit	Zn Al 4	Zn Al 4 Cu 1
Coefficient of thermal expansion between 20 °C and 100 °C	µm/(m K)	27	27
Specific heat capacity	J/(kg K)	420	420
Thermal conductivity at 18 °C	W/(m K)	113	109
Electrical conductivity at 20 °C	MS/m	15.7	15.3

Source: Reference 13

5 CASTING TEMPERATURE

TABLE 6.9 Recommended casting temperatures for zinc die casting alloys, conforming to IS 742 – 1981

Alloy designation	Casting temperature range for		Die temperature
	hot chamber machines °C	cold chamber machines °C	°C
Zn Al 4	400–410	410–420	180–260
Zn Al 4 Cu 1	400–410	410–420	180–260

Source: Reference 13

6 DIMENSIONAL CHANGES

TABLE 6.10 Dimensional changes of zinc alloy die castings, conforming to IS 742 – 1981, on ageing at ambient temperature

Alloy designation	Shrinkage, in mm/mm, measured after			
	5 weeks	6 months	5 years	8 years
Zn Al 4				
Air-cooled	0.00032	0.00056	0.00073	0.00079
Quenched	0.0006	0.0012	–	–
Zn Al 4 Cu 1				
Air-cooled	0.00069	0.00103	0.00136	0.00141

Source: References 12 and 14

TABLE 6.11 Dimensional changes of stabilized[1)] zinc alloy die castings, conforming to IS 742 – 1981, on ageing at ambient temperature

Alloy designation	Shrinkage, in mm/mm, measured after		
	5 weeks	3 months	2 years
Zn Al 4 (S)	0.00020	0.00030	0.00030
Zn Al 4 Cu 1 (S)	0.00022	0.00026	0.00037

1) Heat treated at 100 ± 5 °C for 6 h, followed by cooling to ambient temperature in still air.
Note: (S) indicates stabilized condition.
Source: References 12 and 14

7 HEAT TREATMENT (STABILIZING)

Purpose: To achieve a greater degree of dimensional stability.

Recommended heat treatment cycle: Heat to 100 ± 5 °C. Hold at this temperature for 6 h. Cool to ambient temperature in still air.

8 TYPICAL APPLICATIONS

Bathroom fittings, bodies for fuel pumps, body mouldings, car door handles, car radiator grilles, carburettor bodies, control panels, locks, parts for hydraulic parts, speedometer frames, toys and windshield–wiper parts.

9 EQUIVALENT GRADES

TABLE 6.12 Comparison of grades of zinc alloy ingots for die casting specified in IS 713 – 1981 with nearest equivalent grades specified in other national and international standards

IS 713–81	ASTM B 240–95	BS 1004–72	DIN 1743 (1)–78	GOST 19424–74	ISO 301–81	JIS H 2201–57	SAE J468–88	Trade designation
Zn Al 4	Z33521 (AG40A)	Alloy A	GB–ZnAl4	ZnAl4A	Zn Al 4	Class 2	Z33521	Alloy 3
Zn Al 4 Cu1	Z35530 (AC41A)	Alloy B	GB–ZnAl4Cu1	ZnAl4Cu1A	Zn Al 4 Cu 1	Class 1	Z35530	Alloy 5

TABLE 6.13 Comparison of grades of zinc alloy die castings specified in IS 742 – 1981 with nearest equivalent grades specified in other national and international standards

IS 742–81	ASTM B 86–95	BS 1004–72	DIN 1743 (2)–78	GOST	ISO	JIS H 5301–90	SAE J468–88	Trade designation
Zn Al 4	Z33520 (AG40A)	Alloy A	GD–ZnAl4	–	–	ZDC 2	Z33520	Alloy 3
Zn Al 4 Cu1	Z35531 (AC41A)	Alloy B	GD–ZnAl4Cu1	–	–	ZDC 1	Z35531	Alloy 5

II ZINC–ALUMINIUM CASTING ALLOYS

1 INTRODUCTION

Zinc–aluminium (ZA) casting alloys are a family of high performance, hypereutectic alloys which can be cast to net or near net-shape using sand casting, permanent mould casting, pressure die casting, spin casting, investment casting, continuous casting, and centrifugal casting techniques. These alloys are characterized by the following properties:

a) High strength: The tensile and yield strength values of ZA alloys are the highest among the most widely used non-ferrous alloys and match or exceed those of many cast irons;
b) High modulus of elasticity: The modulus of elasticity of ZA alloys is greater than that of aluminium and magnesium alloys, and plastics;
c) High impact strength and good ductility;
d) High hardness: The Brinell hardness values of ZA alloys are much higher than those of competing materials, such as cast aluminium alloys and engineering plastics;
e) Improved creep strength: The creep strength of ZA alloys is higher than that of conventional zinc die casting alloys;
f) Excellent bearing properties: ZA alloy bushes and bearings outperform bronze bushes and bearings in some applications;
g) Excellent damping capacity: The damping capacity of ZA alloys is superior to that of steel and some of the cast aluminium alloys at ambient temperature. The damping capacity of these alloys increases with temperature and at 100 °C is greater than that of grey cast iron;
h) Non-sparking characteristics: ZA alloys, with the exception of ZA–27, are classified as non-sparking and are used in potentially explosive environments such as in coal mines, refineries and tankers;
i) Non-magnetic characteristics: ZA alloys are non-magnetic;
j) Good corrosion resistance: The corrosion resistance of ZA alloys is similar to that of conventional zinc die casting alloys and cast aluminium alloys. The recommended pH range of application is 6.0–11.5. When required, the corrosion resistance can be enhanced by chromate and phosphate coating, and substantially improved by anodizing;
k) Excellent machinability: The machining speeds for ZA alloys are three to four times faster than that for cast iron and often rival those for free-machining brass. The tool wear is also less than with brass;
l) Superior pressure tightness: ZA alloy castings can be successfully used in pressure tight applications without the use of an impregnation treatment;
m) Superior finishing characteristics;
n) Excellent thin-wall castability;
o) Good dimensional accuracy: The casting tolerances for ZA alloys are superior to those of aluminium and magnesium castings produced by the same casting process; and

The outstanding mechanical properties of ZA alloys enables them to successfully replace bronze, brass, aluminium alloys, cast iron or steel in many applications. However, the use of these alloys is generally limited to 120 °C.

2 SPECIFICATIONS

TABLE 6.14 Chemical composition of ZA alloy ingots and castings

Alloy designation	Chemical composition [% (m/m)]							
	Zn	Al	Cu	Mg max.	Fe max.	Pb max.	Cd max.	Sn max.
Zinc-aluminium alloy ingots for foundry and die castings[1),2)] (ASTM B 669–1995)								
Z35635 (ZA–8)	Rem.	8.2–8.8	0.8–1.3	0.020–0.030	0.065	0.005	0.005	0.002
Z35630 (ZA–12)	Rem.	10.8–11.5	0.5–1.2	0.020–0.030	0.065	0.005	0.005	0.002
Z35840 (ZA–27)	Rem.	25.5–28.0	2.0–2.5	0.012–0.020	0.072	0.005	0.005	0.002
Zinc-aluminium alloy foundry and die castings (ASTM B 791–1995)								
Z35636 (ZA–8)	Rem.	8.0–8.8	0.8–1.3	0.015–0.030	0.075	0.006	0.006	0.003
Z35631 (ZA–12)	Rem.	10.5–11.5	0.5–1.2	0.015–0.030	0.075	0.006	0.006	0.003
Z35841 (ZA–27)	Rem.	25.0–28.0	2.0–2.5	0.010–0.020	0.075	0.006	0.006	0.003

1) ZA alloy ingots for foundry and die castings may contain Cr, Mn, or Ni in amounts of up to 0.01 % each or 0.03 % total.

2) Special high grade zinc (Z13001) of 99.990+ % purity, conforming to ASTM B 6 – 1995, should be used for the manufacture of ZA alloy ingots.

3 TYPICAL MECHANICAL PROPERTIES

TABLE 6.15 Typical mechanical properties[1)] of ZA casting alloys at 20 °C

Property	Unit	ZA–8			ZA–12		
		Sand cast	Permanent mould cast	Die cast	Sand cast	Permanent mould cast	Die cast
Tensile strength	MPa	248–276	221–255	374	275–317	310–345	404
0.2 % proof stress	MPa	193–200	200–214	290	207–214	250–275	320

TABLE 6.15 Chemical composition of ZA alloy ingots and castings (continued)

Property	Unit	ZA–8			ZA–12		
		Sand cast	Permanent mould cast	Die cast	Sand cast	Permanent mould cast	Die cast
Percentage elongation after fracture (L_o = 50.8 mm)[4)]	%	1–2	1–2	8	1–2	1.5–2.5	5
Compressive yield strength 0.1 % offset	MPa	193–207	200–221	252	224–234	228–241	269
Shear strength	MPa	–	242	275	248–262	–	296
Modulus of elasticity (Tension)	GPa	87.6	85.5	85.5	82.1	83	82.7
Poisson's ratio	–	0.29–0.30	0.29–0.30	–	0.32–0.33	0.30–0.31	–
Fatigue strength[5)]	MPa	–	≈ 52	103	≈ 103	–	117
Brinell hardness	HB 10/500/30	82–89	85–90	103	92–96	85–95	100
Vickers hardness	HV 5	110–120	105–115	–	115–116	105–115	–
Charpy impact strength	J	17–24[6)]	–	42[7)]	23–30[6)]	–	29[7)]
Creep strength[8)]	MPa	–	≈ 70	–	≈ 70	–	–
Creep rate[9)]	%/1000 h	–	0.2	–	0.2	0.2	–

(Contd.)

TABLE 6.15 Chemical composition of ZA alloy ingots and castings (continued)

Property	Unit	ZA–27				
		Sand cast			Permanent mould cast	Die cast
		As-cast	Stabilized[2)]	Homogeized[3)]		
Tensile strength	MPa	400–441	290–324	310–324	428–455	426
0.2 % proof stress	MPa	365–372	234–241	255–262	358–393	371
Percentage elongation after fracture (L_o = 50.8 mm)[4)]	%	3–6	2–4	8–11	2–3	2.5
Compressive yield strength, 0.1 % offset	MPa	328–332	–	253–262	–	359
Shear strength	MPa	283–297	–	221–228	–	325
Modulus of elasticity (Tension)	GPa	77.9	85	79.3	–	77.9
Poisson's ratio	–	0.30–0.31	–	–	0.32–0.33	–
Fatigue strength[5)]	MPa	≈ 172	–	≈ 103	–	145
Brinell hardness	HB10/500/30	110–120	85–95	90–100	110–117	119
Vickers hardness	HV 5	135–140	105	100–105	144–148	–
Charpy impact strength	J	34–54[6)]	27–38[6)]	47–74[6)]	–	13[7)]
Creep strength[8)]	MPa	≈ 79	≈ 90	≈ 99	–	–
Creep rate[9)]	%/1000 h	0.08	0.38	0.08	–	–

1) Measured on separately cast test samples.
2) Heat treated at 250 °C for 12 h, followed by furnace cooling to ambient temperature.
3) Heat treated at 320 °C for 3 h, followed by furnace cooling to ambient temperature.
4) Diameter of the parallel length of the circular test piece (d) is 12.7 mm for sand cast and permanent mould cast test pieces, and 6.35 mm for die cast test pieces.
5) Rotating bending at 5 × 108 cycles.
6) Measured on unnotched, 10 mm × 10 mm test pieces.
7) Measured on unnotched, 6.35 mm × 6.35 mm test pieces.
8) Stress to produce a steady state creep rate of 1% in 100000 h (11.4 years).
9) Creep rate at a stress of 138 MPa.

Source: References 1, 6 and 7

TABLE 6.16 Effect of temperature on the mechanical properties of ZA casting alloys

Alloy designation	Temperature °C	Tensile strength[1] MPa	Charpy impact strength[1),2)] J	K_{Ic}[3)] Pressure die cast MPa $\sqrt{m}$	K_{Ic}[3)] Sand cast MPa $\sqrt{m}$
ZA–8	–40	409.6	1	10.2	–
	–20	402.7	1	–	–
	–10	–	2	–	–
	0	382.7	2	–	–
	20	373.7	42	–	–
	24	–	–	12.6	–
	40	–	54	–	–
	50	328.2	–	–	–
	60	–	56	–	–
	80	–	65	–	–
	100	224.1	63	27.7	–

TABLE 6.16 Effect of temperature on the mechanical properties of ZA casting alloys (continued)

Alloy designation	Temperature	Tensile strength[1)]	Charpy impact strength [1),2)]	K_{Ic}[3)]	
				Pressure die cast	Sand cast
	°C	MPa	J	MPa $\sqrt{m}$	MPa $\sqrt{m}$
ZA–12	–40	450.2	1.5	11.2	9.8
	–20	–	1.5	–	–
	0	434.4	3	–	–
	20	403.4	29	–	–
	24	–	–	14.4	14.5
	40	–	35	–	–
	50	349.6	–	–	–
	60	–	40	–	–
	80	–	46	–	–
	100	228.9	46	29.0	29.1
	150	119.3	–	–	–

TABLE 6.16 Effect of temperature on the mechanical properties of ZA casting alloys (continued)

Alloy designation	Temperature °C	Tensile strength[1] MPa	Charpy impact strength[1),2)] J	K_{Ic}[3)] Pressure die cast MPa $\sqrt{m}$	K_{Ic}[3)] Sand cast MPa $\sqrt{m}$
ZA–27	–40	520.6	2	11.9	16.4
	–20	500.6	3	–	–
	–10	–	7	–	–
	0	497.1	–	–	–
	20	425.4	13	–	–
	24	–	–	20.2	23.7
	40	–	15	–	–
	50	397.8	–	–	–
	60	–	16	–	–
	80	–	16	–	–
	100	259.3	16	35.2	42.1
	150	129.0	–	–	–

1) Measured on pressure die cast test pieces.
2) Measured on unnotched, 6.35 mm × 6.35 mm test pieces.
3) Average value.

Source: Reference 1

4 TYPICAL PHYSICAL PROPERTIES

TABLE 6.17 Typical physical properties of ZA casting alloys

Property	Unit	ZA–8	ZA–12	ZA–27
Density at 20 °C	kg/m^3	6300	6030	5000
Solid contraction[1)]	%	≈ 1	≈ 1–1.3	≈ 1.3
Liquidus temperature	°C	404	432	484
Solidus temperature	°C	375	377	375
Coefficient of thermal expansion between 20 °C and 100 °C	μm/(m K)	23.2	24	26
Specific heat capacity between 24 °C and 92 °C	J/(kg K)	435	450	525
Latent heat of fusion between 380 °C and 415 °C	kJ/kg	112	118	128
Thermal conductivity at 24 °C	W/(m K)	115	116	125.5
Electrical conductivity at 24 °C	% IACS	27.7	28.3	29.7
Electrical resistivity at 20 °C	μΩ m	0.062	0.061	0.058

1) Equivalent to pattern maker's shrinkage.

Source: References 1, 6 and 7

5 CASTING TEMPERATURE

TABLE 6.18 Recommended casting temperatures for ZA casting alloys

Alloy designation	Casting temperature range °C
ZA–8	435–460
ZA–12	460–490
ZA–27	515–545

Source: Reference 1

6 DESIGN STRESSES

TABLE 6.19 Maximum allowable design stress in tension for ZA casting alloys according to the ASME Boiler Code

Alloy designation	Maximum design stress[1)], in MPa, at a temperature, in °C, of			
	20	60	100	150
ZA–8				
Permanent mould cast	≈ 59	≈ 48	> 21	> 7
ZA–12				
Sand cast	≈ 59	–	> 10	–
Permanent mould cast	≈ 59	–	–	–
ZA–27				
Sand cast	≈ 69	≈ 48	21	10
homogenized[2)]	≈ 90	–	–	–

1) Based upon the stress which results in a creep rate of 0.01 % per 1000 h.
2) Heat treated at 320 °C for 3 h, followed by furnace cooling to ambient temperature.

Source: Reference 9

7 HEAT TREATMENT

Zinc-aluminium alloys are generally used in the as-cast condition. However, ZA-27 may be subjected to a stabilizing or a homogenizing treatment.

7.1 Stabilizing

Purpose: To achieve a greater degree of dimensional stability.

Recommended heat treatment cycle: Heat to 250 °C. Hold at this temperature for 12 h. Furnace cool to ambient temperature.

7.2 Homogenizing

Purpose: To obtain added ductility and creep strength.

Recommended heat treatment cycle: Heat to 320 °C. Hold at this temperature for 3 h. Furnace cool to ambient temperature.

8 TYPICAL APPLICATIONS

Bearings and bushings for high load and low speed applications, electrical conduit fittings, electrical switchgear and transformer parts, electronic chassis hardware, liquid and gas valve housings and fittings, machine and component housings and covers, marine hardware, mine hardware, office accessory components, plated decorative hardware, pressure resistant components, and pulleys and sheaves.

9 EQUIVALENT GRADES

TABLE 6.20 Comparison of grades of ZA alloy ingots for foundry and die castings specified in ASTM B 669 – 1995 nearest equivalent grades specified in other national and international standards

ASTM B 669–95	DD 139–86	ISO 301–81
Z35635 (ZA–8)	ZA8	–
Z35630 (ZA–12)	ZA12	Zn Al 11 Cu 1
Z35840 (ZA–27)	ZA27	–

TABLE 6.21 Comparison of grades of ZA alloy castings specified in ASTM B 791 – 1995 with nearest equivalent grades specified in other national and international standards

ASTM B 791–95	DD 139–86
Z35636 (ZA–8)	ZA8
Z35631 (ZA–12)	ZA12
Z35841 (ZA–27)	ZA27

III OTHER CASTING ALLOYS

1 INTRODUCTION

1.1 Slush casting alloys

Zinc-base slush casting alloys are Zn-4.75Al-0.25Cu (Alloy A) and Zn-5.5Al (Alloy B) alloys , which are used for the production of hollow decorative castings, such as lighting fixtures, lamp bases, small statues and casket hardware

Slush Casting Alloy A is more difficult to cast than Slush Casting Alloy B and, therefore, yields castings with thicker walls, but with improved mechanical properties and having longer service life.

Slush Casting Alloy B , because of its superior castability, yields lower cost thin wall castings.

1.2 Forming die alloys

Forming die alloys are Zn-Al-Cu casting alloys which are characterized by good abrasion resistance, largely because of its self lubricating properties. They are used in the construction of limited life dies and punches for forming sheet metal parts (e.g., airframe skins and prototype automotive parts) and plastic parts. These alloys may be remelted and recast to their original properties.

2 SPECIFICATIONS

TABLE 6.22 Chemical composition of zinc casting alloy ingots for slush casting, and for sheet metal forming dies and plastic injection moulds specified in ASTM Standards

Alloy designation	Chemical composition [% (*m*/*m*)]							
	Zn	Al	Cu	Mg	Fe max.	Pb max.	Cd max.	Sn max.
Zinc alloy ingots for slush casting (ASTM B 792 – 1995)								
Z34510 (Slush Casting Alloy A)	Rem.	4.50–5.00	0.2–0.3	≤ 0.010	0.100	0.007	0.005	0.005
Z30500 (Slush Casting Alloy B)	Rem.	5.25–5.75	≤ 0.1	≤ 0.010	0.100	0.007	0.005	0.005
Zinc alloy ingots for sheet metal forming dies and plastic injection molds (ASTM B 793 – 1995)								
Z35543 (Alloy A)	Rem.	3.5–4.5	2.5–3.5	0.02–0.10	0.100	0.007	0.005	0.005
Z35542 (Alloy B)	Rem.	3.9–4.3	2.5–2.9	0.02–0.05	0.075	0.003	0.003	0.001

Note: Special High Grade zinc (Z13001) of 99.990+ % purity, conforming to ASTM B 6 – 1995, should be used for the manufacture of these alloy ingots.

3 THERMAL PROPERTIES

TABLE 6.23 Liquidus and solidus temperatures of slush casting alloys, conforming to ASTM B 792 – 1995

Alloy designation	Liquidus temperature °C	Solidus temperature °C
Z34510 (Slush Casting Alloy A)	≈ 390	380
Z30500 (Slush Casting Alloy B)	≈ 395	380

Source: Reference 1

IV EFFECT OF IMPURITY ELEMENTS

Zinc casting alloys containing impurity elements, such as cadmium, indium, lead, thallium, tin, and others, in amounts exceeding the permissible limits are susceptible to intergranular corrosion, particularly when exposed to a warm, humid atmosphere, and fail prematurely by warping and cracking. Magnesium when present in sufficient amounts (in the range 0.01–0.08 %, depending upon the alloy) counteracts the ill effects of the impurity elements. However, higher amounts of magnesium will lower fluidity of the melt, cause hot shortness, increase hardness and reduce ductility.

Cadmium

Cadmium when present in amounts greater than 0.005 % can cause hot shortness and reduced resistance to corrosion.

Indium

Indium when present in amounts greater than 0.0005 % can cause intergranular corrosion in zinc die casting alloys.

Iron

Iron when present in amounts greater than 0.10 % can cause problems in casting and, difficulty in plating and machining.

Lead

Lead when present in amounts exceeding the limit given in the specification (≈ 0.007 %) can promote intergranular corrosion.

Thallium

Thallium when present in amounts greater than 0.001 % promotes intergranular corrosion in zinc die casting alloys.

Tin

Tin when present in amounts exceeding the limits given in the specification (≈ 0.005 %) can cause intergranular corrosion, hot shortness and reduced impact strength.

Other Elements

Antimony, arsenic, bismuth and mercury when present in amounts greater than 0.001 % adversely affect the stability of zinc casting alloys.

V COMPARISON RATINGS OF ZINC CASTING ALLOYS

TABLE 6.24 Comparison[1] of the characteristics of zinc casting alloys

Characteristic	Alloy 2	Alloy 3 (Zn Al 4)	Alloy 5 (Zn Al 4 Cu 1)	Alloy 7	ZA–8	ZA–12	ZA–27
Die castability[2]	E	E	E	E	VG	VG	G
Sand castability	G	NR	NR	NR	G	E	F
Permanent mould castability	G	NR	NR	NR	VG	E	F
Strength	VG	G	G	G	VG	E	E
Ductility	VG	E	VG	E	VG	G	F
Impact	G	E	E	E	VG	G	F
Bearing/wear	VG	G	G	G	VG	E	E
Machinability	E	E	E	E	E	VG	G
Pressure tightness	E	E	E	E	VG	E	F
Plating	E	E	E	E	VG	G	NR[3]
Zinc anodizing	E	E	E	E	E	E	VG
Chromating	E	E	E	E	VG	G	F
Painting	E	E	E	E	E	E	E
Dimensional stability	VG	E	E	E	VG	VG	F
Anti-sparking	E	E	E	E	E	E	F[4]

Key: E = excellent; VG = very good; G = good; F = fair; NR = not recommended.

1) General performance ratings can vary depending upon process selection.
2) Alloys 2, 3, 5 and 7, and ZA–8 are hot chamber die cast. ZA–12 and ZA–27 are cold chamber die cast.
3) ZA–27 can be plated using special techniques. However, it is normally not recommended for plating.
4) High aluminium content of ZA–27 reduces anti-sparking rating.

Note: The designation in parenthesis conforms to the IS designation.

Source: Reference 8

VI REFERENCES

Books

1. *ASM Handbook*, vol. 2, ASM International, Materials Park, Ohio, U.S.A, 1990.
2. C.H. Mathew son (Ed.), *Zinc — The Science and Technology of the Metal, Its Alloys and Compounds*, Reinhold Publishing Corp., New York, U.S.A., 1959.
3. S.W.K. Morgan, *Zinc and its Alloys and Compounds*, Ellis Horwood Ltd., Chichester, West Sus sex, U.K., 1985.
4. F.E. Goodwin and A.L. Ponikvar (Eds.), *Engineering Properties of Zinc Alloys*, 3rd ed. – Revised, International Lead Zinc Research Organization, Research Triangle Park, North Carolina, U.S.A., 1989.
5. H. Morrow III and R.F. Lynch, *Zinc Alloys: Engineering Applications, Encyclopedia of Materials Science and Engineering*, vol. 7, M.B. Bever (Ed.), Pergamon Press, Oxford, U.K., 1986.
6. R.J. Barnhurst, *Gravity Casting Manual for Zinc-Aluminum Alloys*, Noranda Technology Centre, Pointe Claire, Quebec, Canada, 1989.
7. *Engineering in Zinc, Today's Answer*, Noranda Sales Corporation, Toronto, Ontario, Canada, 1989.
8. *The Zinc Alloy Guide — The Properties of Zinc Alloys for Use in Casting*, Eastern Alloys Inc., May brook, New York, U.S.A.
9. *Design Stress Considerations for ZA Alloys*, Eastern Alloys Inc., Maybrook, New York, U.S.A.
10. F. Porter, *Zinc Handbook: Properties, Processing, and Use in Design*, Marcel Dekker, New York, U.S.A., 1991.

Standards

11. IS 713 – 1981: *Specification for Zinc Base Alloy Ingots for Die Casting.*
12. IS 742 – 1981: *Specification for Zinc Base Alloy Die Castings.*
13. IS 1655 – 1991: *Metallic Materials – Zinc Alloys – Code of Practice for Manufacture of Pressure Die Castings.*
14. BS 5338 – 1976: *Code of Practice for Zinc Alloy Pressure Die Casting for Engineering.*
15. ASTM B 669 – 1995: *Specification for Zinc–Aluminum Alloys in Ingot Form for Foundry and die Castings.*
16. ASTM B 791 – 1995: *Specification for Zinc–Aluminum (ZA) Alloy Foundry and Die Castings.*
17. ASTM B 792 – 1995: *Specification for Zinc Alloys in Ingot Form for Slush Casting.*
18. ASTM B 793 – 1995: *Specification for Zinc Casting Alloy Ingot for Sheet Metal Forming Dies and Plastic Injection Molds.*
19. JIS H 5301 – 1990: *Zinc Alloy Die Castings.*

6.3

WROUGHT ZINC AND ZINC ALLOYS

1 INTRODUCTION

Wrought zinc and zinc alloys are available in the form of flat-rolled products (rolled zinc), forged and extruded products, and wire drawn products. They possess good forming, machining, joining and finishing characteristics, and exhibit good corrosion resistance in many types of environments. They are also non-magnetic and non-sparking.

Unalloyed zinc is characterized by low strength and low creep resistance. It is not possible to harden unalloyed zinc appreciably by cold working, as it recrystallizes rapidly after deformation at room temperature. Addition of small amounts of copper to pure zinc raises its recrystallization temperature and increases strength, hardness and creep resistance. Titanium has little effect on the tensile strength and hardness of rolled zinc but greatly increases creep resistance, particularly when added with copper. Cadmium is sometimes added to zinc since it raises the recrystallization temperature and improves strength, hardness and creep resistance. For certain applications, such as dry cells, lead reduces the rate of corrosion of zinc and is added to improve shelf life. With the addition of around 22 % aluminium to pure zinc superplasticity can be achieved.

2 SPECIFICATIONS

2.1 Rolled zinc (plate, sheet and strip)

TABLE 6.25 Chemical composition of grades of rolled zinc, conforming to IS 2258 – 1981

Grade	Chemical composition [% (*m*/*m*)]				
	Zn	Pb	Cd	Fe max.	Cu max.
1	Rem.	≤ 0.20	≤ 0.10	0.010	0.005
2[1)]	Rem.	0.15–0.60	0.15–0.60	0.020	0.005

(Contd.)

TABLE 6.25 Chemical composition of grades of rolled zinc, conforming to IS 2258 – 1981 (continued)

Grade	Chemical composition [% (m/m)]				
	Zn	Pb	Cd	Fe max.	Cu max.
3	Rem.	0.80–1.00	≤0.07	0.007	0.001

[1] The Cu content in this grade should be ≤ 0.001 % when used for the manufacture of soldered battery cans.

2.2 *Zinc wire*

2.2.1 *Chemical composition*

TABLE 6.26 Chemical composition of zinc wire for metallizing, conforming to IS 12447 – 1988

Designation	Chemical composition [% (m/m)]						
	Zn min.	Pb max.	Cd max.	Fe max.	Sn max.	Cu max.	Total impurities max.
Zn 99.95	99.95	0.025	0.02	0.01	0.001	0.002	0.05

2.2.2 *Mechanical properties*

TABLE 6.27 Mechanical properties of zinc wire for metallizing, conforming to IS 12447 – 1988

Grade	R_m MPa	$A_{100\,mm}$ %
Zn 99.95	≥100	≥40

3 TYPICAL PHYSICAL PROPERTIES

TABLE 6.28 Typical physical properties of grades of rolled zinc, conforming to IS 2258 – 1981

Property	Unit	Grade 1	Grade 2	Grade 3
Density at 21 °C	kg/m^3	7140	7140	7200
Melting point	°C	419	419	419
Coefficient of thermal expansion between 20 °C and 40 °C	µm/(m k)	32.5–23	39.9–23.4	–
Specific heat capacity between 20 °C and 100 °C	J/(kg k)	395	395	–
Thermal conductivity	W/(m K)	108	108	–
Electrical conductivity at 20 °C	% IACS	28.4	≈ 32	29
Electrical resistivity at 20 °C	µΩ m	0.062	≈ 0.0606	–
Non-sparking	–	Yes	Yes	Yes
Non-magnetic	–	Yes	Yes	Yes

Source: References 1 and 6

4 FABRICATION CHARACTERISTICS

TABLE 6.29 Fabrication characteristics of grades of rolled zinc, conforming to IS 2258 – 1981

Grade	Hot working temperature °C	Hot shortness temperature °C	Annealing temperature °C
1	120–275	300–420	–
2	120–275	300–420	105
3	120–275	300–420	–

Note: Fabrication of wrought zinc should be carried out at a temperature exceeding 21 °C.

Source: Reference 1

5 TYPICAL APPLICATIONS

TABLE 6.30 Typical applications of grades of rolled zinc and zinc wire specified in Indian Standards

Grade/ designation	Typical applications
Rolled zinc	
1	Address plates, drawn articles and photoengraver's plates.
2	Drawn articles, soldered battery cans and weather strips.
3	Extruded battery cans.
Zinc wire	
Zn 99.95	Thermal spraying or metallizing.

Source: References 4 and 5

6 EQUIVALENT GRADES

TABLE 6.31 Comparison of grades of rolled zinc specified in IS 2258 – 1981 with nearest equivalent grades specified in other national standards

IS 2258–81	ASTM B 69–95	BS 6561–85	DIN 17770–90	GOST 598–90 GOST 18327–73	JIS H 4321–53
1	Zn–0.08 Pb (Z21210)	–	–	–	–
2	Zn–0.3 Pb–0.03 Cd (Z21540)	–	–	–	–
3	–	Type B	–	–	–

7 REFERENCES

Books

1. *ASM Handbook*, vol. 2, ASM International, Materials Park, Ohio, U.S.A, 1990.
2. F.E. Goodwin and A.L. Ponikvar (Ed.), *Engineering Properties of Zinc Alloys*, 3rd ed. – Revised, International Lead Zinc Research Organization, Research Triangle Park, North Carolina, U.S.A., 1989.
3. S.W.K. Morgan, *Zinc and its Alloys and Compounds*, Ellis Horwood Ltd., Chichester, West Sussex, U.K., 1985.

Standards

4. IS 2258 – 1981: *Specification for Rolled Zinc Plate, Sheet and Strip.*
5. IS 12447 – 1988: *Specification for Zinc Wire for Sprayed Zinc Coatings.*
6. BS 6561 – 1985: *Specification for Zinc Alloy Sheet and Strip for Building.*

7

POWDER METALLURGY

CHAPTER 7

Powder Metallurgy

7	**Powder Metallurgy**	7.5

7

POWDER METALLURGY

1 INTRODUCTION

1.1 Definition of powder metallurgy

Powder metallurgy (P/M) is that part of metallurgy which relates to the manufacture of metallic powders, or of articles made from such powders, with or without the addition of non-metallic powders, by the application of forming and sintering processes.

1.2 Production of metal powders

Metal powders can be obtained from ores, from salts and other compounds, or from bulk metals or alloys. The processes most commonly employed for the production of metal or alloy powders are:

Atomization: In this process, a thin stream of molten metal or alloy is dispersed into particles by a high pressure jet of gas (usually air, nitrogen or argon) or liquid (usually water) stream or, by other means. This process is extensively used for the manufacture of both elemental (Al, Cd, Cu, Fe, Pb, Sn) and pre-alloyed (brass, bronze, low-alloy steel, stainless steel, tool steel) powders.

Typically, water-atomized powders are irregular in shape and have relatively high surface oxygen contents. Gas-atomized powders, on the other hand, generally are more spherical or rounded in shape and, if atomized by an inert gas, have lower oxygen contents.

Thermal decomposition of metal carbonyls: High purity (> 99.5 %) iron and nickel powders are produced by the thermal decomposition of their respective gaseous carbonyls. The iron powder produced by this technique is usually spherical in shape and very fine, while the nickel powder is usually irregular in shape, porous and fine.

Reduction of metal oxides: Very fine metallic oxides are reduced by a gaseous or solid reducing agent to produce a spongy metallic mass which is later pulverized to the desired size. Powders produced by this technique are usually porous and irregular in shape. Copper, cobalt, iron, molybdenum, nickel, and tungsten powders are commonly produced by this process.

Electrolytic deposition from solutions: In this process, the metal is electrodeposited from a solution of its salt or from a fused salt. By controlling the conditions of electrodeposition a loosely adhering spongy deposit, or a brittle deposit is obtained which is subsequently pulverized to the desired size. Metal powders may also be deposited directly from the electrolyte. Copper, iron, manganese and silver powders are obtained by electrodeposition from their aqueous solution, whereas tantalum, thorium and uranium powders are obtained by electrodeposition from their fused salts. Powders produced by this method are usually very pure but significantly more expensive.

Mechanical comminution: In this method, brittle materials are reduced to angular particles by grinding or milling. Metals such as molybdenum, titanium, tungsten, and many of their alloys and compounds are produced by this method.

Milling of ductile metals produces particles which are flakelike. This method is employed in the production of aluminium and bronze powders for pigments. This type of powder is generally unsuitable for the manufacture of sintered parts. In many cases, embrittlement of ductile metals or alloys facilitates disintegration and prevents the formation of flat platelike particles.

1.3 Conventional method of producing powder metallurgy parts

The production of P/M parts consists of the following steps:

1.3.1 Blending or mixing

Elemental or pre-alloyed metal powders are mixed with admixed lubricants (such as zinc stearate, stearic acid or wax) and other additives to produce a homogeneous blend of ingredients which when sintered will produce the desired chemical composition. Lubricants are added not only to minimize die friction and wear, but also to minimize density gradients in the compact and the force necessary to eject the compact from the die.

1.3.2 Compacting

A measured quantity of the mixed powder is fed into a precision die and is compressed, usually at room temperature at pressures ranging from 70 MPa to 700 MPa, to the desired shape and dimensions. The compact so produced, called the green compact, should have sufficient strength to avoid damage during ejection from the die, and during transfer from the compacting press to the sintering furnace. These green compacts are porous and have a lower density than the cast or wrought products of the same metal.

Compacting of P/M parts is intended to produce parts of reproducible volume and mass to ensure proper dimensional control with stringent tolerances. Also, compaction should achieve uniform density within each part so that uniform shape changes occur during sintering, thus leading to uniform properties in the sintered part.

1.3.3 Sintering

Sintering consists of heating the low strength green compact in a controlled gaseous atmosphere to a temperature below the melting point or liquidus of the main constituent, holding at this temperature for a sufficient length of time, and followed by cooling to ambient temperature in an inert atmosphere. During sintering, the admixed lubricants are burned off and metallurgical bonding occurs between adjacent particles to form a coherent mass with increased density, strength and ductility. In addition, chemical homogeneity is developed by diffusion of alloying elements, metal oxides are reduced and the desired carbon content in ferrous P/M compacts is achieved by diffusion of carbon from the graphite powder. Sintering may also be accompanied by shrinkage or growth, resulting in dimensional changes.

Sintered parts may be subsequently heat treated, repressed, forged, machined, impregnated, or finished to improve mechanical or physical properties, increase dimensional accuracy or enhance decorative appeal. They may also be joined by sinterbonding, brazing or welding, or by mechanical means.

1.3.4 Finishing operations

Sizing: Sizing consists of re-pressing a sintered part to obtain the desired dimensions.

Coining: Coining consists of re-pressing a sintered part to obtain a specific surface configuration.

Impregnation: Impregnation is a process of filling the pores of a sintered part with a non-metallic material, such as oil, wax or resin.

Oil impregnation of sintered parts is carried out to impart a self-lubricating capability (e.g., oil-impregnated bearings) and also to provide corrosion resistance during storage.

Impregnation of sintered parts with wax or resin is carried out for the following reasons:

a) with wax, for moisture exclusion and to improve machinability; and
b) with resin, for making the part gas-tight or liquid-tight, to seal pores so as to prevent the ingress of the plating solution into the pores during electroplating and to improve machinability.

Infiltration: Infiltration is a process of filling the pores of a sintered part with a metal or alloy of lower melting point than that of the sintered material. A prerequisite for infiltration is limited solubility, or complete insolubility, of the materials in each other.

Infiltration increases the density, strength and ductility of sintered parts. It also makes the part gas-tight or liquid-tight, improves machinability and seals pores in preparation for electroplating.

Steam treatment: This process consists of heating the sintered ferrous parts to 540-595 °C and subjecting them to superheated steam under pressure for 1–4 h. Under these conditions, the steam reacts with the iron to form an adherent oxide layer on all external and internal surfaces, forming a lining on the surface of the larger pores and completely filling the smaller ones. This process increases the surface hardness, compressive strength, wear resistance and corrosion resistance of the part. Steam treatment usually results in a decrease in impact strength of the part. Steam treated parts are usually dipped in oil to further increase their corrosion and wear resistance.

Plating: Standard plating techniques can be used for very dense sintered parts of density greater than 95 % of the theoretical density. For parts of lower density, sealing of the pores by resin impregnation is recommended to avoid the ingress of the plating solution into the pores.

Heat treatment: Sintered parts may be heat treated by conventional means to improve mechanical properties.

Heating of sintered parts in a molten salt bath is not recommended even for those parts having maximum density, because of the difficulty of washing the parts free of salt. If the salt remains in the voids, the parts will corrode.

Machining: Sintered objects are sometimes machined to produce special shapes, such as undercuts, re-entrant angles, cross-holes, etc., or to achieve closer dimensional accuracy.

1.4 New consolidation processes

1.4.1 Cold isostatic pressing

In this process the powders are consolidated inside a flexible container or mould by a pressurized liquid such as water or oil.

There are two variations of cold isostatic pressing (CIP):

a) **Drybag isostatic pressing**: In this process, the mould is fixed to the inside of the pressure vessel. After the elastomeric mould is filled with powder, pressure is applied by introducing pressurized oil between the fixed mould and the vessel wall; and

b) **Wetbag isostatic pressing**: In this process, the mould is filled with powder and sealed outside the pressure vessel. After the mould is introduced into the vessel, it is completely immersed in the pressurized medium, usually water containing lubricating and corrosion preventing additions. Pressure is applied isostatically by pressurizing the water around the mould.

After cold isostatic pressing, the unsintered part is sintered to increase part density and improve mechanical properties.

1.4.2 Hot isostatic pressing

In this process, the metal powder compact or sintered object is placed in a mould or can which is subsequently evacuated and sealed. The sealed can is then placed in an autoclave and simultaneously subjected to both high temperature and high isostatic pressure to achieve densification. Materials processed commercially by hot isostatic pressing (HIP) include beryllium, ceramics, hardmetals, stainless steel, superalloys, and tool steels.

1.4.3 Powder forging

Powder forging consists of hot forging of an unsintered, pre-sintered or sintered preform made from powder to its final shape. This process increases the density and thus the strength of the powder compact.

1.4.4 Powder rolling

Powder rolling, also known as roll compaction, is a process in which the metal powder is introduced between a pair of rotating rolls which causes the powder to be compacted into a continuous green (unsintered) strip. The compacted powder strip emerging from the rolls is then sintered and susequently hot rolled or cold rolled to produce an end product with the desired properties. Fully dense materials can be produced by this process. Cobalt, nickel and nickel–cobalt alloy strip are commonly produced by this technique.

1.5 Advantages of the powder metallurgy process

a) Parts can be fabricated to net or near-net shape, thus eliminating or reducing scrap losses, machining and assembly operations. Consequently, the cost of P/M parts is generally lower than that of parts of similar accuracy made by casting, forging, or machining from barstock;

b) Certain metal parts can be produced only by powder metallurgy. Among these are cemented carbide dies and tools, electrical contacts, oxide-dispersion-strengthened alloys, filters, and self-lubricating bearings;

c) Very close dimensional tolerances can be maintained throughout long production runs, thereby reducing inspection of the finished parts.

1.6 Limitations of the powder metallurgy process

The P/M process imposes restrictions on the size, shape and composition of the components that can be economically produced.

1.6.1 Size

The largest size of a part which can be produced by powder metallurgy is limited by:

a) the available press capacity;

b) the difficulty in achieving a uniform density throughout the compact; and

c) the difficulty in uniformly heating the compact during sintering.

1.6.2 Shape

The shape of the powder metallurgy part must be such that the compact can be ejected from the die without damage. Hence, shapes that may be made by powder metallurgy are limited, and undercuts, re-entrant angles or cross-holes cannot be made.

1.6.3 Composition

Addition of alloying elements, such as chromium, manganese or silicon, in their elemental form to the powder mixture should be avoided because their oxides are not reducible during the sintering operation.

1.6.4 Equipment cost

The precision dies in which the parts are compacted are expensive, and unless the high cost can be distributed over large production runs, the cost of the sintered part may be uneconomical.

2 SPECIFICATIONS

2.1 Indian Standard

TABLE 7.1 Chemical composition and, physical and mechanical properties of sintered metal powder, oil-impregnated, bearings, conforming to IS 3980 – 1982

Grade	Class	Sub-class	Chemical composition [% (*m*/*m*)]					Density[1]	Interconnected porosity	Strength constant
			Fe	C	Cu	Sn	Total other elements	kg/m^3	%	MPa
1 (copper-base)	A	–	≤ 1.0	≤ 2.0	Rem.	8–9.5	≤ 2.0	5800–6200 (L)	≥ 25	≥ 100
			≤ 1.0	≤ 2.0	Rem.	8–9.5	≤ 2.0	6200–6600 (M)	≥ 18	≥ 140
			≤ 1.0	≤ 2.0	Rem.	8–9.5	≤ 2.0	6600–7000 (H)	≥ 12	≥ 180
1 (copper-base)	B	B1	28–30	≤ 2.0	Rem.	4–6	≤ 2.0	5700–6100 (L)	≥ 25	≥ 100
			28–30	≤ 2.0	Rem.	4–6	≤ 2.0	6100–6500 (M)	≥ 18	≥ 140
			28–30	≤ 2.0	Rem.	4–6	≤ 2.0	6500–6900 (H)	≥ 12	≥ 180
		B2	38–40	≤ 2.0	Rem.	4–6	≤ 2.0	5600–6000 (L)	≥ 25	≥ 100
			38–40	≤ 2.0	Rem.	4–6	≤ 2.0	6000–6400 (M)	≥ 18	≥ 140
			38–40	≤ 2.0	Rem.	4–6	≤ 2.0	6400–6800 (H)	≥ 12	≥ 180
		B3	48–50	≤ 2.0	Rem.	4–6	≤ 2.0	5500–5900 (L)	≥ 25	≥ 100
			48–50	≤ 2.0	Rem.	4–6	≤ 2.0	5900–6300 (M)	≥ 18	≥ 140
			48–50	≤ 2.0	Rem.	4–6	≤ 2.0	6300–6700 (H)	≥ 12	≥ 180

(Contd.)

TABLE 7.1 **Chemical composition and, physical and mechanical properties of sintered metal powder, oil-impregnated, bearings, conforming to IS 3980 – 1982 (continued)**

Grade	Class	Sub-class	Chemical composition [% (*m*/*m*)]					Density[1)]	Interconnected porosity	Strength constant
			Fe	C	Cu	Sn	Total other elements	kg/m^3	%	MPa
2 (iron-base)	A	–	Rem.	≤ 0.25	≤0.5	–	≤2.0	5400–5800 (L)	≥ 22	≥ 150
			Rem.	≤ 0.25	≤0.5	–	≤2.0	5800–6200 (M)	≥ 18	≥ 170
			Rem.	≤ 0.25	≤0.5	–	≤2.0	6200–6600 (H)	≥ 12	≥ 250
2 (iron-base)	B	B1	Rem.	≤ 0.25	1.5–3.5	–	≤2.0	5400–5800 (L)	≥ 22	≥ 180
			Rem.	≤ 0.25	1.5–3.5	–	≤2.0	5800–6200 (M)	≥ 18	≥ 210
			Rem.	≤ 0.25	1.5–3.5	–	≤2.0	6200–6600 (H)	≥ 12	≥ 250

(Contd.)

TABLE 7.1 Chemical composition and, physical and mechanical properties of sintered metal powder, oil-impregnated, bearings, conforming to IS 3980 – 1982 (continued)

Grade	Class	Sub-class	Chemical composition [% (*m*/*m*)]					Density[1)]	Interconnected porosity	Strength constant
			Fe	C	Cu	Sn	Total other elements	kg/m^3	%	MPa
2 (iron-base)	B	B2	Rem.	≤0.25	4–6	–	≤2.0	5400–5800 (L)	≥22	≥250
			Rem.	≤0.25	4–6	–	≤2.0	5800–6200 (M)	≥18	≥275
			Rem.	≤0.25	4–6	–	≤2.0	6200–6600 (H)	≥12	≥310
		B3	Rem.	≤0.25	9–11	–	≤2.0	5400–5800 (L)	≥22	≥270
			Rem.	≤0.25	9–11	–	≤2.0	5800–6200 (M)	≥18	≥300
			Rem.	≤0.25	9–11	–	≤2.0	6200–6600 (H)	≥12	≥340

1) L = low density; M = medium density; H = high density.

2.3 ASTM Standards

2.3.1 Sintered ferrous materials for structural parts

TABLE 7.2 Chemical composition of sintered ferrous materials for structural parts, conforming to ASTM B 783 – 1991

Material designation	Chemical composition [% (m/m)]											
	Fe	C	Cu	Ni	Mo	Cr	Mn	Si	P	S	N	Total other elements
Iron and carbon steel												
F-0000	97.7–100	≤0.3	–	–	–	–	–	–	–	–	–	≤ 2.0
F-0005	97.4–99.7	0.3–0.6	–	–	–	–	–	–	–	–	–	≤ 2.0
F-0008	97.1–99.4	0.6–0.9	–	–	–	–	–	–	–	–	–	≤ 2.0
Iron-copper and copper steel												
FC-0200	93.8–98.5	≤0.3	1.5–3.9	–	–	–	–	–	–	–	–	≤ 2.0
FC-0205	93.5–98.2	0.3–0.6	1.5–3.9	–	–	–	–	–	–	–	–	≤ 2.0
FC-0208	93.2–97.9	0.6–0.9	1.5–3.9	–	–	–	–	–	–	–	–	≤ 2.0
FC-0505	91.4–95.7	0.3–0.6	4.0–6.0	–	–	–	–	–	–	–	–	≤ 2.0
FC-0508	91.1–95.4	0.6–0.9	4.0–6.0	–	–	–	–	–	–	–	–	≤ 2.0

(Contd.)

TABLE 7.2 Chemical composition of sintered ferrous materials for structural parts, conforming to ASTM B 783 – 1991 (continued)

Material designation	Chemical composition [% (*m*/*m*)]											
	Fe	C	Cu	Ni	Mo	Cr	Mn	Si	P	S	N	Total other elements
Iron-copper and copper steel												
FC-0808	88.1–92.4	0.6–0.9	7.0–9.0	–	–	–	–	–	–	–	–	≤ 2.0
FC-1000	87.2–90.5	≤0.3	9.5–10.5	–	–	–	–	–	–	–	–	≤ 2.0
Iron-nickel and nickel steel												
FN-0200	92.2–99.0	≤0.3	≤2.5	1.0–3.0	–	–	–	–	–	–	–	≤ 2.0
FN-0205	91.9–98.7	0.3–0.6	≤2.5	1.0–3.0	–	–	–	–	–	–	–	≤ 2.0
FN-0208	91.6–98.4	0.6–0.9	≤2.5	1.0–3.0	–	–	–	–	–	–	–	≤ 2.0
FN-0405	89.9–96.7	0.3–0.6	≤2.0	3.0–5.5	–	–	–	–	–	–	–	≤ 2.0
FN-0408	89.6–96.4	0.6–0.9	≤2.0	3.0–5.5	–	–	–	–	–	–	–	≤ 2.0
Low-alloy steel												
FL-4205	95.9–98.7	0.4–0.7	–	0.35–0.55	0.50–0.85	–	–	–	–	–	–	≤ 2.0
FL-4605	94.5–97.5	0.4–0.7	–	1.70–2.00	0.40–0.80	–	–	–	–	–	–	≤ 2.0

(Contd.)

TABLE 7.2 Chemical composition of sintered ferrous materials for structural parts, conforming to ASTM B 783–1991 (continued)

Material designation	Chemical composition [% (*m*/*m*)]										
	Fe	C	Cu	Ni	Mo	Cr	Mn	Si	P	S	N	Total other elements
Copper-infiltrated iron and steel												
FX-1000	82.8–92.0	≤ 0.3	8.0–14.9	–	–	–	–	–	–	–	–	≤ 2.0
FX-1005	82.5–91.7	0.3–0.6	8.0–14.9	–	–	–	–	–	–	–	–	≤ 2.0
FX-1008	82.2–91.4	0.6–0.9	8.0–14.9	–	–	–	–	–	–	–	–	≤ 2.0
FX-2000	72.7–85.0	≤ 0.3	15.0–25.0	–	–	–	–	–	–	–	–	≤ 2.0
FX-2005	72.4–84.7	0.3–0.6	15.0–25.0	–	–	–	–	–	–	–	–	≤ 2.0
FX-2008	72.1–84.4	0.6–0.9	15.0–25.0	–	–	–	–	–	–	–	–	≤ 2.0
Stainless steel												
SS-303N1, N2	Rem.	≤ 0.15	–	8.0–13.0	–	17.0–19.0	≤ 2.0	≤ 1.0	≤ 0.20	0.15–0.30	0.2–0.6	≤ 2.0
SS-303L	Rem.	≤ 0.03	–	8.0–13.0	–	17.0–19.0	≤ 2.0	≤ 1.0	≤ 0.20	0.15–0.30	–	≤ 2.0
SS-304N1, N2	Rem.	≤ 0.08	–	8.0–12.0	–	18.0–20.0	≤ 2.0	≤ 1.0	≤ 0.045	≤ 0.03	0.2–0.6	≤ 2.0
SS-304L	Rem.	≤ 0.03	–	8.0–12.0	–	18.0–20.0	≤ 2.0	≤ 1.0	≤ 0.045	≤ 0.03	–	≤ 2.0

(Contd.)

TABLE 7.2 Chemical composition of sintered ferrous materials for structural parts, conforming to ASTM B 783 – 1991 (continued)

Material designation	Chemical composition [% (*m*/*m*)]											
	Fe	C	Cu	Ni	Mo	Cr	Mn	Si	P	S	N	Total other elements
Stainless steel												
SS-316N1, N2	Rem.	≤0.08	–	10.0–14.0	2.0–3.0	16.0–18.0	≤2.0	≤1.0	≤0.045	≤0.03	0.2–0.6	2.0
SS-316L	Rem.	≤0.03	–	10.0–14.0	2.0–3.0	16.0–18.0	≤2.0	≤1.0	≤0.045	≤0.03	–	2.0
SS-410	Rem.	≤0.25	–	–	–	11.5–13.0	≤1.0	≤1.0	≤0.040	≤0.03	0.2–0.6	2.0

TABLE 7.3 Typical mechanical[1] and physical properties of sintered ferrous materials for structural parts, conforming to ASTM B 783 – 1991

Material designation[2]	R_m	$R_{p0.2}$	$A_{25\,mm}$	E	μ (compressive)	R_{tr}	Unnotched Charpy impact strength	Density	$R_{c0.1}$	Hardness		Fatigue limit[3]
										Apparent (direct)	Matrix (converted)	
	MPa	MPa	%	GPa		MPa	J	kg/m^3	MPa			MPa
Iron and carbon steel												
F-0000-10	124	90	1.5	96.5	0.25	248	4.1	6100	110	40 HRF	N/A	48
F-0000-15	172	124	2.5	117	0.26	345	8.1	6700	124	60 HRF	N/A	69
F-0000-20	262	172	7.0	141	0.28	655	47	7300	131	80 HRF	N/A	97
F-0005-15	165	124	≤1.0	96.5	0.25	331	4.1	6100	200	25 HRB	N/A	62
F-0005-20	221	159	1.0	114	0.26	441	5.4	6600	214	40 HRB	N/A	83
F-0005-25	262	193	1.5	124	0.27	524	6.8	6900	221	55 HRB	N/A	97
F-0005-50HT[4]	414	[5]	≤0.5	114	0.25	724	4.1	6600	296	20 HRC	58 HRC	159
F-0005-60HT[4]	483	[5]	≤0.5	121	0.27	827	4.7	6800	359	22 HRC	58 HRC	186
F-0005-70HT[4]	552	[5]	≤0.5	131	0.30	965	5.4	7000	421	25 HRC	58 HRC	207
F-0008-20	200	172	≤0.5	82.7	0.25	352	3.4	5800	283	35 HRB	N/A	76
F-0008-25	241	207	≤0.5	100	0.25	421	4.1	6200	283	50 HRB	N/A	90
F-0008-30	290	241	≤1.0	114	0.26	510	5.4	6600	290	60 HRB	N/A	110

(Contd.)

TABLE 7.3 Typical mechanical[1)] and physical properties of sintered ferrous materials for structural parts, conforming to ASTM B 783 – 1991 (continued)

Material designation[2)]	R_m	$R_{p0.2}$	$A_{25\,mm}$	E	μ (compressive)	R_{tr}	Unnotched Charpy impact strength	Density	$R_{c0.1}$	Hardness		Fatigue limit[3)]
										Apparent (direct)	Matrix (converted)	
	MPa	MPa	%	GPa		MPa		kg/m^3	MPa			MPa
Iron and carbon steel												
F-0008-35	393	276	1.0	131	0.28	689	6.8	7000	290	70 HRB	N/A	152
F-0008-55HT [4)]	448	5)	≤0.5	103	0.22	689	4.1	6300	290	22 HRC	60 HRC	172
F-0008-65HT [4)]	517	5)	≤0.5	114	0.25	793	5.4	6600	400	28 HRC	60 HRC	200
F-0008-75HT [4)]	586	5)	≤0.5	124	0.28	896	6.1	6900	517	32 HRC	60 HRC	221
F-0008-85HT [4)]	655	5)	≤0.5	134	0.30	1000	6.8	7100	593	35 HRC	60 HRC	248
Iron-copper and copper steel												
FC-0200-15	172	138	1.0	89.6	0.26	310	6.1	6000	124	11 HRB	N/A	69
FC-0200-18	193	159	1.5	103	0.26	352	6.8	6300	145	18 HRB	N/A	76
FC-0200-21	214	179	1.5	114	0.27	386	7.5	6600	159	26 HRB	N/A	83
FC-0200-24	234	200	2.0	124	0.28	434	8.1	6900	179	36 HRB	N/A	90
FC-0205-30	241	241	≤1.0	89.6	0.27	414	≤2.7	6000	345	37 HRB	N/A	90
FC-0205-35	276	276	≤1.0	103	0.27	517	4.1	6300	365	48 HRB	N/A	103
FC-0205-40	345	310	≤1.0	117	0.28	655	6.8	6700	393	60 HRB	N/A	131

(Contd.)

TABLE 7.3 Typical mechanical[1] and physical properties of sintered ferrous materials for structural parts, conforming to ASTM B 783 – 1991 (continued)

Material designation[2]	R_m	$R_{p0.2}$	$A_{25\,mm}$	E	μ (com-pressive)	R_{tr}	Unnotched Charpy impact strength	Density	$R_{c0.1}$	Hardness		Fatigue limit[3]
										Apparent (direct)	Matrix (converted)	
	MPa	MPa	%	GPa		MPa		kg/m³	MPa			MPa
Iron-copper and copper steel												
FC-0205-45	414	345	≤1.0	134	0.29	793	11	7100	414	72 HRB	N/A	159
FC-0205-60HT[4]	483	[5]	≤0.5	100	0.23	655	3.4	6200	393	19 HRC	58 HRC	186
FC-0205-70HT[4]	552	[5]	≤0.5	110	0.26	758	4.7	6500	490	25 HRC	58 HRC	207
FC-0205-80HT[4]	621	[5]	≤0.5	121	0.28	827	6.1	6800	593	31 HRC	58 HRC	234
FC-0205-90HT[4]	689	[5]	≤0.5	131	0.30	931	7.5	7000	655	36 HRC	58 HRC	262
FC-0208-30	241	241	≤1.0	82.7	0.26	414	≤2.7	5800	393	50 HRB	N/A	90
FC-0208-40	345	310	≤1.0	103	0.27	621	2.7	6300	427	61 HRB	N/A	131
FC-0208-50	414	379	≤1.0	117	0.28	862	6.8	6700	455	73 HRB	N/A	159
FC-0208-60	517	448	≤1.0	138	0.29	1069	9.5	7200	490	84 HRB	N/A	200
FC-0208-50HT[4]	448	[5]	≤0.5	96.5	0.23	655	3.4	6100	400	20 HRC	60 HRC	172
FC-0208-65HT[4]	517	[5]	≤0.5	107	0.26	758	4.7	6400	496	27 HRC	60 HRC	200
FC-0208-80HT[4]	621	[5]	≤0.5	121	0.29	896	6.1	6800	627	35 HRC	60 HRC	234
FC-0208-95HT[4]	724	[5]	≤0.5	134	0.30	1034	7.5	7100	724	43 HRC	60 HRC	276
FC-0505-30	303	248	≤0.5	82.7	0.27	531	4.1	5800	345	51 HRB	N/A	117

(Contd.)

TABLE 7.3 Typical mechanical[1)] and physical properties of sintered ferrous materials for structural parts, conforming to ASTM B 783 – 1991 (continued)

Material designation[2)]	R_m	$R_{p0.2}$	$A_{25\,mm}$	E	μ (compressive)	R_{tr}	Unnotched Charpy impact strength	Density	$R_{c0.1}$	Hardness		Fatigue limit[3)]
										Apparent (direct)	Matrix (converted)	
	MPa	MPa	%	GPa		MPa		kg/m³	MPa			MPa
Iron-copper and copper steel												
FC-0505-40	400	324	≤0.5	103	0.28	703	6.1	6300	372	62 HRB	N/A	152
FC-0505-50	490	386	≤1.0	117	0.29	855	6.8	6700	400	72 HRB	N/A	186
FC-0508-40	400	345	≤0.5	86.2	0.27	689	4.1	5900	400	60 HRB	N/A	152
FC-0508-50	469	414	≤0.5	103	0.28	827	4.7	6300	434	68 HRB	N/A	179
FC-0508-60	565	483	≤1.0	121	0.29	1000	6.1	6800	469	80 HRB	N/A	214
FC-0808-45	379	345	≤0.5	89.6	0.27	586	4.1	6000	427	65 HRB	N/A	145
FC-1000-20	207	179	≤1.0	89.6	0.26	365	4.7	6000	228	15 HRB	N/A	76
Iron-nickel and nickel steel												
FN-0200-15	172	117	1.5	107	0.26	–	–	6600	110	–	N/A	69
FN-0200-20	241	172	4.0	134	0.27	552	26	7000	124	75 HRF	N/A	90
FN-0200-25	276	207	6.5	159	0.28	–	–	7300	138	–	N/A	103
FN-0205-20	276	172	1.5	107	0.26	448	8.1	6600	228	44 HRB	N/A	103
FN-0205-25	345	207	2.5	128	0.28	689	16	6900	262	59 HRB	N/A	131
FN-0205-30	414	241	4.0	152	0.30	862	28	7200	290	69 HRB	N/A	159

(Contd.)

TABLE 7.3 Typical mechanical[1] and physical properties of sintered ferrous materials for structural parts, conforming to ASTM B 783 – 1991 (continued)

Material designation[2]	R_m	$R_{p0.2}$	$A_{25\,mm}$	E	μ (compressive)	R_{tr}	Unnotched Charpy impact strength	Density	$R_{c0.1}$	Hardness		Fatigue limit[3]
										Apparent (direct)	Matrix (converted)	
	MPa	MPa	%	GPa		MPa		kg/m³	MPa			MPa
Iron–nickel and nickel steel												
FN-0205-35	483	276	5.5	165	0.30	1034	46	7400	310	78 HRB	N/A	186
FN-0205-80HT[6]	621	[5]	≤0.5	107	0.25	827	4.7	6600	531	23 HRC	55 HRC	234
FN-0205-105HT[6]	827	[5]	≤0.5	128	0.27	1103	6.1	6900	621	29 HRC	55 HRC	317
FN-0205-130HT[6]	1000	[5]	≤0.5	145	0.28	1310	8.1	7100	683	33 HRC	55 HRC	379
FN-0205-155HT[6]	1103	[5]	≤0.5	152	0.29	1482	9.5	7200	710	36 HRC	55 HRC	421
FN-0205-180HT[6]	1276	[5]	≤0.5	165	0.30	1724	13	7400	772	40 HRC	55 HRC	483
FN-0208-30	310	241	1.5	114	0.28	586	7.5	6700	283	63 HRB	N/A	117
FN-0208-35	379	276	1.5	128	0.29	724	11	6900	324	71 HRB	N/A	145
FN-0208-40	483	310	2.0	145	0.29	896	15	7100	365	77 HRB	N/A	186
FN-0208-45	552	345	2.5	159	0.30	1069	22	7300	407	83 HRB	N/A	207
FN-0208-50	621	379	3.0	165	0.30	1172	28	7400	427	88 HRB	N/A	234
FN-0208-80HT[6]	621	[5]	≤0.5	114	0.26	827	5.4	6700	683	26 HRC	57 HRC	234
FN-0208-105HT[6]	827	[5]	≤0.5	128	0.27	1034	6.1	6900	855	31 HRC	57 HRC	317
FN-0208-130HT[6]	1000	[5]	≤0.5	134	0.28	1276	7.5	7000	938	35 HRC	57 HRC	379

(Contd.)

TABLE 7.3 Typical mechanical[1] and physical properties of sintered ferrous materials for structural parts, conforming to ASTM B 783 – 1991 (continued)

Material designation[2]	R_m	$R_{p0.2}$	$A_{25\,mm}$	E	μ (compressive)	R_{tr}	Unnotched Charpy impact strength	Density	$R_{c0.1}$	Hardness		Fatigue limit[3]
										Apparent (direct)	Matrix (converted)	
	MPa	MPa	%	GPa		MPa		kg/m^3	MPa			MPa
Iron–nickel and nickel steel												
FN-0208-155HT[6]	1172	5)	≤0.5	152	0.29	1517	9.5	7200	1117	39 HRC	57 HRC	448
FN-0208-180HT[6]	1344	5)	≤0.5	165	0.30	1724	11	7400	1296	42 HRC	57 HRC	510
FN-0405-25	276	207	≤1.0	96.5	0.25	448	6.1	6500	228	49 HRB	N/A	103
FN-0405-35	414	276	3.0	134	0.28	827	20	7000	276	71 HRB	N/A	159
FN-0405-45	621	345	4.5	165	0.30	1207	45	7400	310	84 HRB	N/A	234
FN-0405-80HT[6]	586	5)	≤0.5	97	0.25	793	5.4	6500	462	19 HRC	55 HRC	221
FN-0405-105HT[6]	758	5)	≤0.5	121	0.26	1000	6.8	6800	614	25 HRC	55 HRC	290
FN-0405-130HT[6]	931	5)	≤0.5	134	0.28	1379	8.8	7000	710	31 HRC	55 HRC	352
FN-0405-155HT[6]	1103	5)	≤0.5	159	0.29	1689	13	7300	855	37 HRC	55 HRC	421
FN-0405-180HT[6]	1276	5)	≤0.5	165	0.30	1931	18	7400	910	40 HRC	55 HRC	483
FN-0408-35	310	276	1.0	100	0.28	517	5.4	6500	255	67 HRB	N/A	117
FN-0408-45	448	345	1.0	128	0.29	793	10	6900	345	78 HRB	N/A	172
FN-0408-55	552	414	1.0	152	0.30	1034	15	7200	407	87 HRB	N/A	207

(Contd.)

TABLE 7.3 Typical mechanical[1] and physical properties of sintered ferrous materials for structural parts, conforming to ASTM B 783 – 1991 (continued)

Material designation[2]	R_m	$R_{p0.2}$	$A_{25\,mm}$	E	μ (compressive)	R_{tr}	Unnotched Charpy impact strength	Density	$R_{c0.1}$	Hardness		Fatigue limit[3]
										Apparent (direct)	Matrix (converted)	
	MPa	MPa	%	GPa		MPa		kg/m^3	MPa			MPa
Low-alloy steel												
FL-4205-80HT[4]	621	5)	≤0.5	117	0.25	931	4.7	6600	552	28 HRC	60 HRC	234
FL-4205-100HT[4]	758	5)	≤0.5	131	0.26	1103	5.4	6800	758	32 HRC	60 HRC	290
FL-4205-120HT[4]	896	5)	≤0.5	152	0.26	1276	5.4	7000	965	36 HRC	60 HRC	338
FL-4205-140HT[4]	1034	5)	≤0.5	172	0.27	1482	6.1	7200	1172	39 HRC	60 HRC	393
FL-4605-80HT[4]	586	5)	≤0.5	114	0.24	896	4.7	6550	627	24 HRC	60 HRC	221
FL-4605-100HT[4]	758	5)	≤0.5	124	0.25	1138	6.1	6750	786	29 HRC	60 HRC	290
FL-4605-120HT[4]	896	5)	≤0.5	138	0.26	1344	8.1	6950	958	34 HRC	60 HRC	338
FL-4605-140HT[4]	1069	5)	≤0.5	148	0.27	1586	9.5	7200	1172	39 HRC	60 HRC	407
Copper infiltrated iron and steel												
FX-1000-25	352	221	7.0	110	0.31	910	34	7300	228	65 HRB	N/A	131
FX-1005-40	531	345	4.0	110	0.31	1089	18	7300	365	82 HRB	N/A	200
FX-1005-110HT[4]	827	5)	≤0.5	110	0.29	1448	9.5	7300	758	38 HRC	55 HRC	317
FX-1008-50	600	414	3.0	110	0.31	1145	14	7300	490	89 HRB	N/A	228
FX-1008-110HT[4]	827	5)	≤0.5	110	0.29	1303	8.8	7300	793	43 HRC	58 HRC	317

(Contd.)

TABLE 7.3 Typical mechanical[1] and physical properties of sintered ferrous materials for structural parts, conforming to ASTM B 783 – 1991 (continued)

Material designation[2]	R_m	$R_{p0.2}$	$A_{25\,mm}$	E	μ (compressive)	R_{tr}	Unnotched Charpy impact strength	Density	$R_{c0.1}$	Hardness		Fatigue limit[3]
										Apparent (direct)	Matrix (converted)	
	MPa	MPa	%	GPa		MPa		kg/m^3	MPa			MPa
Copper infiltrated iron and steel												
FX-2000-25	317	255	3.0	103	0.30	993	20	7300	283	66 HRB	N/A	117
FX-2005-45	517	414	1.5	103	0.31	1020	11	7300	414	85 HRB	N/A	200
FX-2005-90HT[4]	689	5)	≤0.5	103	0.31	1179	9.5	7300	490	36 HRC	55 HRC	262
FX-2008-60	552	483	1.0	103	0.31	1076	9.5	7300	483	90 HRB	N/A	207
FX-2008-90HT[4]	689	5)	≤0.5	103	0.31	1096	6.8	7300	510	36 HRC	58 HRC	262
Stainless steel[7]												
SS-303N1-25	269	221	0.5	–	0.23	593	4.7	6400	262	62 HRB	N/A	–
SS-303N2-35	379	290	5.0		0.24	676	26	6500	317	63 HRB	N/A	–
SS-303L-12	269	117	17.5	–	0.24	565	–	6600	145	21 HRB	N/A	–
SS-304N1-30	296	262	0.5	–	0.24	772	5.4	6400	262	61 HRB	N/A	–
SS-304N2-33	393	276	10.0	–	0.25	876	34	6500	324	62 HRB	N/A	–
SS-304L-13	296	124	23.0	–	0.26	–	–	6600	152	–	N/A	–
SS-316N1-25	283	234	0.5	–	0.24	745	6.8	6400	248	59 HRB	N/A	–
SS-316N2-33	414	269	10.0	–	0.25	862	38	6500	303	62 HRB	N/A	–

(Contd.)

TABLE 7.3 Typical mechanical[1] and physical properties of sintered ferrous materials for structural parts, conforming to ASTM B 783 – 1991 (continued)

Material designation[2]	R_m	$R_{p0.2}$	$A_{25\,mm}$	E	μ (compressive)	R_{tr}	Unnotched Charpy impact strength	Density	$R_{c0.1}$	Hardness		Fatigue limit[3]
										Apparent (direct)	Matrix (converted)	
	MPa	MPa	%	GPa		MPa		kg/m^3	MPa			MPa
Stainless steel[7]												
SS-316L-15	283	138	18.5	–	0.25	552	47	6600	152	20 HRB	N/A	–
SS-410-90HT[4]	724	[5]	≤0.5	–	0.24	779	3.4	6500	641	23 HRC	55 HRC	–

1) Measured on test specimens sintered under commercial manufacturing conditions.
2) The suffix number represents
(a) for the as-sintered condition: the minimum $R_{p0.2}$, in 10^3 psi; and
(b) for the heat treated (HT) condition: the minimum R_m, in 10^3 psi.
3) Fatigue limit (estimated) = $0.38 \times R_m$.
4) Tempering temperature = 177 °C.
5) $R_{p0.2}$ and R_m are approximately the same for heat treated (HT) materials.
6) Tempering temperature = 260 °C.
7) N1 = Nitrogen alloyed, with good strength and low elongation;
N2 = Nitrogen alloyed, with high strength and medium elongation;
L = Low carbon, with lower strength and highest elongation; and
HT = Martensitic grade, heat treated, with highest strength.

Key: N/A = not applicable.

2.3.2 Sintered non-ferrous materials for structural parts

TABLE 7.4 Chemical composition of sintered brass, bronze and nickel silver for structural parts, conforming to ASTM B 823 – 1992

Material designation	Chemical composition [% (*m*/*m*)]					
	Cu	Zn	Pb	Sn	Ni	Total other elements
CZ-1000	88.0–91.0	Rem.	–	–	–	≤ 2.0
CZP-1002	88.0–91.0	Rem.	1.0–2.0	–	–	≤ 2.0
CZP-2002	77.0–80.0	Rem.	1.0–2.0	–	–	≤ 2.0
CZ-3000	68.5–71.5	Rem.	–	–	–	≤ 2.0
CZP-3002	68.5–71.5	Rem.	1.0–2.0	–	–	≤ 2.0
CNZ-1818	62.5–65.5	Rem.	–	–	16.5–19.5	≤ 2.0
CNZP-1816	62.5–65.5	Rem.	1.0–2.0	–	16.5–19.5	≤ 2.0
CT-1000	87.5–90.5	Rem.	–	9.5–10.5	–	≤ 2.0

TABLE 7.5 Typical mechanical[1] and physical properties of sintered brass, bronze and nickel silver for structural parts, conforming to ASTM B 823 – 1992

Material designation[2]	R_m	$R_{p0.2}$	$A_{25\,mm}$	E	μ (compressive)	R_{tr}	Unnotched Charpy impact strength	Density	$R_{c0.1}$	Hardness		Fatigue limit
										Apparent (direct)	Matrix (converted)	
	MPa	MPa	%	GPa		MPa	J	kg/m^3	MPa			MPa
CZ-1000-9	124	66	9.0	51.7	0.29	269	–	7600	83	65 HRH	N/A	–
CZ-1000-10	138	76	10.5	68.9	0.31	317	–	7900	83	72 HRH	N/A	–
CZ-1000-11	159	83	12.0	–	0.33	359	–	8100	83	80 HRH	N/A	–
CZP-1002-7	138	59	10.0	51.7	0.31	310	42	7900	69	66 HRH	N/A	–
CZP-2002-11	159	93	12.0	68.9	0.32	345	38	7600	83	75 HRH	N/A	–
CZP-2002-12	207	110	14.5	82.7	0.32	483	76	8000	97	84 HRH	N/A	–
CZ-3000-14	193	110	14.0	62.1	0.31	427	31	7600	124	84 HRH	N/A	–
CZ-3000-16	234	131	17.0	68.9	0.33	593	52	8000	131	92 HRH	N/A	–
CZP-3002-13	186	103	14.0	62.1	0.31	393	–	7600	83	80 HRH	N/A	–
CZP-3002-14	217	114	16.0	68.9	0.32	490	–	8000	103	88 HRH	N/A	–
CNZ-1818-17	234	138	11.0	75.8	0.30	503	33	7900	165	90 HRH	N/A	–

(Contd.)

TABLE 7.5 Typical mechanical[1] and physical properties of sintered brass, bronze and nickel silver for structural parts, conforming to ASTM B 823 – 1992 (continued)

Material designation[2]	R_m	$R_{p0.2}$	$A_{25\ mm}$	E	μ (compressive)	R_{tr}	Unnotched Charpy impact strength	Density	$R_{c0.1}$	Hardness		Fatigue limit
										Apparent (direct)	Matrix (converted)	
	MPa	MPa	%	GPa		MPa		kg/m³	MPa			MPa
CNZP-1816-13	179	103	10.0	55.2	0.30	345	30	7900	124	86 HRH	N/A	–
CT-1000-13 (repressed)	152	110	4.0	37.9	0.29	310	5.4	7200	138	82 HRH	N/A	–

1) Measured on test specimens sintered under commercial manufacturing conditions.
2) The suffix number represents the minimum $R_{p0.2}$, in 10^3 psi.

Key: N/A = not applicable.

3 TYPICAL APPLICATIONS

3.1 Metal powders

a) For powder metallurgy parts;
b) In coated electrodes;
c) For cutting, lancing, scarfing and welding operations;
d) In pharmaceutical applications;
e) For food enrichment;
f) As copier powders;
g) In paints;
h) In magnetic particle inspection;
i) For magnetic separation of seeds;
j) For fuel propellants, pyrotechnics and explosives; and
k) For hardfacing.

3.2 Powder metallurgy parts

Automotive applications: Timing gears, camshaft thrust plates, connecting rods, oil pump gears, oil pump rotors, rocker arms, valve guides, valve seat inserts.

Business machines: Gear inserts, drive gears, printer hammer guide assembly, printing wheels.

Cutting and forming tools

Electrical applications: Electrical contacts, metal-graphite brushes, resistance welding electrodes, tungsten filaments.

Filters

Magnetic applications: Permanent magnets, soft magnetic materials.

Medical and dental applications: Dental amalgams, porous orthopedic implants, porous coatings for implant fixation.

Nuclear applications: Fuel pellets.

Self-lubricating bearings

Ordnance applications: Frangible ammunition, mortar shell bodies, projectile shells and weapon components.

Friction materials: Clutch plate facings, brake linings.

4 EQUIVALENT SPECIFICATIONS

TABLE 7.6 Comparison of nearest equivalent national and international standards on powder metallurgy

	IS	ASTM	BS	DIN	ISO	JIS
Glossary of terms relating to powder metallurgy						
Vocabulary	5432–1982	B 243–1994	5600: Part 1: Sec. 1.2–1981	30900–1982	3252–1982	Z 2500–1987
Metallic powders						
Bronze	10035–1993	–	–	–	–	–
Cobalt	7505–1985	–	–	–	–	–
Copper	8485–1977	–	–	–	–	H 2114–1983
Copper-lead	11110–1984	–	–	–	–	–
Iron	8370–1977	–	–	–	–	H 2601–1983
Leaded bronze	11111–1984	–	–	–	–	–
Nickel	7506–1987	–	–	–	–	–
Niobium carbide	12217–1987	–	–	–	–	–
Tantalum	7970–1987	–	–	–	–	–
Tantalum carbide	12216–1987	–	–	–	–	–
Tin	8367–1993	–	–	–	–	–
Titanium carbide	8369–1985	–	–	–	–	–

(Contd.)

TABLE 7.6 Comparison of nearest equivalent national and international standards on powder metallurgy (continued)

	IS	ASTM	BS	DIN	ISO	JIS
Metallic powders						
Tungsten	8392–1985	–	–	–	–	H 2116–1979
Tungsten carbide	8368–1985	–	–	–	–	H 2116–1979
Sampling and testing of metallic powders						
Acid-insoluble content in iron, copper, tin and bronze powders	7438–1985	E 194–1990	EN 24496–1993	ISO 4496–1987	4496–1978	–
Apparent density a) Funnel method b) Scott volumeter method	 4848–1981 11627–1986	 B 212–1989 B 329–1990	 EN 23923 (1)–1993 EN 23923 (2)–1993	 ISO 3923 (1)–1980 ISO 3923 (2)–1987	 3923 (1)–1979 3923 (2)–1981	 Z 2504–1979 –
Compactibility (compressibility)	4857–1991	B 331–1985	EN 23927–1993	ISO 3927–1991	3927–1985	–
Dimensional changes associated with compacting and sintering	12570–1988	B 610–1993	EN 24492–1993	ISO 4492–1987	4492–1985	–
Flowability	4840–1984	B 213–1990	5600: Part 2: Sec. 2.6–1985	ISO 4490–1987	4490–1978	Z 2502–1979
Green strength	12571–1988	B 312–1982	EN 23995–1993	ISO 3995–1991	3995–1985	–
Hydrogen loss	5644 (2)–1993	E 159–1992	EN 24491 (2)–1993	ISO 4491 (2)–1990	4491 (2)–1989	–

(Contd.)

TABLE 7.6 Comparison of nearest equivalent national and international standards on powder metallurgy (continued)

	IS	ASTM	BS	DIN	ISO	JIS
Sampling and testing of metallic powders						
Hydrogen-reducible oxygen	5644 (3)–1993	–	–	ISO 4491 (3)–1990	4491 (3)–1989	–
Lubricant content (Soxhlet extraction method)	11506–1985	–	5600: Part 2: Sec. 2.11–1980	ISO 4495 –1987	4495–1978	–
Particle size	7512–1974	B 214–1992	EN 24497–1993	ISO 4497–1991	4497–1983	–
Sampling	–	B 215–1990	EN 23954–1993	ISO 3954–1980	3954–1977	–
Sampling and testing of hardmetal powders using sintered test pieces	–	–	EN 24884–1993	ISO 4884–1991	4884–1978	–
Tap density	8871–1991	B 527–1993	EN 23953–1993	ISO 3953–1991	3953–1993	–
Total oxygen by reduction-extraction	5644 (4) – 1993	–	EN 24491 (4)–1993	ISO 4491 (4)–1990	4491 (4)–1989	–

(Contd.)

TABLE 7.6 Comparison of nearest equivalent national and international standards on powder metallurgy (continued)

	IS	ASTM	BS	DIN	ISO	JIS
Testing of sintered parts, excluding hardmetals						
Apparent hardness						
(a) materials of essentially uniform section hardness	–	–	EN 24498 (1)–1993	ISO 4498 (1)–1993	4498 (1)–1990	–
(b) case-hardened ferrous materials	–	–	5600: Part 3: Sec. 3.11: Subsec. 3.11.2–1984	–	4498 (2)–1981	
Case depth	–	B 721–1991	–	–	4507–1978	–
Density (impermeable materials)	4841–1982	B 311–1993	EN 23369–1993	ISO 3369–1990	3369–1975	–
Density, oil content and open porosity (permeable materials)	5642–1991	B 328–1992	5600: Part 3: Sec. 3.2–1988	–	2738–1987	Z 2501–1989 Z 2505–1989 Z 2506–1989
Fatigue test pieces	–	–	5600: Part 3: Sec. 3.10–1979	ISO 3928–1990	3928–1977	–
Impact test pieces	–	E 23–1994a	–	ISO 5754–1990	5754–1978	–
Modulus of elasticity	13803–1993	–	–	ISO 3312–1990	3312–1987	–
Permeability	13782–1993	–	5600: Part 3: Sec. 3.6–1988	ISO 4022–1990	4022–1987	–
Pore size (bubble test)	13781–1993	–	EN 24003–1993	ISO 4003–1990	4003–1977	–
Radial crushing strength	10385–1982	B 438M–1984 B 439–1983	5600: Part 3: Sec. 3.9–1979	–	2739–1973	Z 2507–1989
Tensile test pieces	–	E 8M–1994a	5600: Part 3: Sec. 3.7–1988	ISO 2740–1991	2740–1991	–
Transverse rupture strength	12279–1988	B 528–1983a	5600: Part 3: Sec. 3.8–1979	ISO 3325–1993	3325–1975	–

(Contd.)

TABLE 7.6 Comparison of nearest equivalent national and international standards on powder metallurgy (continued)

	IS	ASTM	BS	DIN	ISO	JIS
Testing of hardmetals						
Abrasive wear resistance	12286–1988	B 611–1985	–	–	–	–
Compressive strength	13780–1993	–	EN 24506–1993	ISO 4506–1991	4506–1979	–
Coercivity	11518–1985	–	EN 23326–1993	ISO 3326–1990	3326–1975	–
Density	–	B 311–1993	5600: Part 4: Sec. 4.4–1979	ISO 3369–1990	3369–1975	–
Hardness, Rockwell A	5652 (1)–1993	B 294–1992	5600: Part 4: Sec. 4.5–1979	–	3738 (1)–1982	–
Hardness, Vickers	12783–1989	–	EN 23878–1993	–	3878–1983	–
Metallographic determination of microstructure	11959–1987	B 657–1992	EN 24499–1993	ISO 4499–1991	4499–1978	–
Metallographic sample preparation	11520–1985	B 665–1992	–	–	–	–
Modulus of elasticity	13803–1993	–	EN 23312–1993	ISO 3312–1990	3312–1987	–
Porosity	–	B 276–1991	EN 24505–1993	ISO 4505–1991	4505–1978	–
Sampling and testing of sintered hardmetals	–	–	EN 24489–1993	ISO 4489–1991	4489–1978	–
Transverse rupture strength	4842–1982	B 406–1990	EN 23327–1993	–	3327–1982	–

TABLE 7.6 Comparison of nearest equivalent national and international standards on powder metallurgy (continued)

Material	IS	ASTM	BS	DIN	ISO	JIS
Filters, bearings and structural parts						
Sintered metal materials for filters	–	–	–	30910 (2)–1990	–	–
Sintered metal materials for bearing (oil-impregnated)	3980–1982	B 438M–1984 B 439–1983 B 612–1991 B 782–1988	5600: Part 5: Sec. 5.1–1988	30910 (3)–1990	5755 (1)–1987	–
Sintered ferrous materials for structural parts	–	B 783–1993	5600: Part 5: Sec. 5.2–1988 5600: Part 5: Sec. 5.3–1988	30910 (4)–1990	5755 (2)–1987 5755 (3)–1987	Z 2550–1989
Sintered non-ferrous materials for structural parts	–	B 255–1983a B 282–1983a B 458–1983a B 715–1983 B 817–1993 B 823–1993	–	30910 (4)–1990	–	Z 2550–1989
Sintered metal materials for structural parts with soft-magnetic properties	–	–	–	30910 (5)–1990	–	–
Sintered metal materials for hot-forged structural parts	–	B 848–1994	–	30910 (6)–1990	–	–

5 REFERENCES

Books/articles

1. *ASM Handbook*, vol. 7, American Society for Metals, Metals Park, Ohio, U.S.A., 1984.
2. *ASM Handbook*, vol. 1, ASM International, Materials Park, Ohio, U.S.A., 1990.
3. *ASM Handbook*, vol. 4, ASM International, Materials Park, Ohio, U.S.A., 1991.
4. S. Bradbury (Ed.), *Source Book on Powder Metallurgy*, American Society for Metals, Metals Park, Ohio, U.S.A., 1979.
5. R.H.T. Dixon and A. Clayton, *Powder Metallurgy for Engineers*, Machinery Publishing Co., London, U.K., 1971.
6. *P/M Design Guidebook*, Metal Powder Industries Federation, Princeton, New Jersey, U.S.A., 1983.
7. A. G. Dowson, "Designing for Production by Powder Metallurgy", *Engineering Designer*, 14 (1), 14–16, 1988.
8. R. Duggirala and R. Shivpuri: "Effects of Processing Parameters in P/M Steel Forging on Part Properties: A Review — Part I: Powder Preparation, Compaction, and Sintering", *Journal of Materials Engineering and Performance*, 1, (4), 495–503, 1992.
9. W.B. James, "New Shaping Methods for Powder Metallurgy Components", *Materials & Design*, 8, (4), 187–197, 1987.
10. K.M. Kulkarni, "P/M Forging Moves into Volume Production", *Machine Design*, 57(14), 74–79, 1985.

Standards

11. IS 3980 – 1982: *Specification for Porous Metal Powder Oi!- Impregnated Bearings.*
12. ISO 3252 – 1982: *Powder Metallurgy — Vocabulary.*
13. ISO 5755 (1) – 1987: *Sintered metal materials — Specifications — Part 1: Materials, for Bearings, Impregnated with Liquid Lubricant.*
14. ISO 5755 (2) – 1987: *Sintered Metal Materials — Specifications — Part 2: Sintered Iron and Sintered Steel Containing One or Both of the Elements Carbon and Copper, Used for Structural Parts.*
15. ISO 5755 (3) – 1987: *Sintered Metal Materials — Specifications — Part 3: Sintered Alloyed and Sintered Stainless Steels Used for Structural Parts.*
16. ASTM B 783 – 1991: *Specification for Materials for Ferrous Powder Metallurgy Structural Parts.*
17. ASTM B 823 – 1992: *Specification for Materials for Non-Ferrous Powder Metallurgy (P/M) Structural Parts.*

8

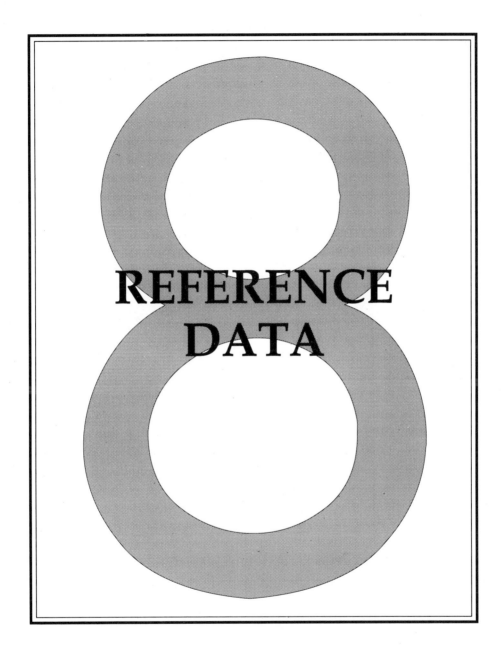

REFERENCE DATA

CHAPTER 8

Reference Data

8 Reference Data 8.5

8

REFERENCE DATA

1 DAMPING CAPACITY

The damping capacity of a material measures its ability to dissipate elastic strain energy. A common measure of the noise-damping capability of a material is the specific damping capacity (SDC)—the percent of strain energy that is dissipated per cycle.

TABLE 8.1 Specific damping capacity of commercial alloys at ambient temperature

Alloy	Treatment condition	Surface shear stress MPa	Specific damping capacity %
Cast irons			
Grey cast iron1)			
• high-carbon inoculated (Ferrite = 5–10 %)	As-cast	34.5	19.3
• uninoculated (Ferrite = 5 %)	As-cast	34.5	8.3
• undercooled graphite (Ferrite = 10–20 %)	As-cast	34.5	4.0
• inoculated (Ferrite = 2 %)	As-cast	34.5	7.2
• alloyed (Ferrite = 5 %)	As-cast	34.5	5.2
• inoculated, pearlitic	As-cast	34.5	3.9
Spheroidal graphite cast iron			
• ferritic	Annealed	34.5	2.7
• pearlitic	As-cast	34.5	1.8
Malleable cast iron			
• pearlitic	–	34.5	1.6
• blackheart	–	34.5	1.5

(Contd.)

TABLE 8.1 Specific damping capacity of commercial alloys at ambient temperature (continued)

Alloy	Treatment condition	Surface shear stress MPa	Specific damping capacity %
Cast irons			
Austenitic flake graphite cast iron			
• 21% Ni	As-cast	34.5	7.1
• Ni–Cu	As-cast	34.5	3.9
Steels			
Low-carbon steel (C = 0.08 %)	–	34.5	1.0
High-carbon steel (C = 0.95 %)	Spheroidized	34.5	0.8
	Water quenched from 800 °C	34.5	0.5
	Water quenched from 800 °C and tempered at 100 °C for 1.5 h	34.5	0.2
Low-alloy steel (42Cr4Mo2)	Quenched and tempered to 269–331 HB	34.5	0.15
Austenitic stainless steel			
• Type 302	Solution treated at 1050 °C and water quenched	34.5	0.3
• Type 321	Solution treated at 1050 °C and water quenched	34.5	1.8
Copper alloys			
Gunmetal (88Cu-10Zn-2Sn)	–	34.5	1.0
High tensile brass	As-cast	34.5	0.25
Aluminium and aluminium alloys			
Aluminium	–	–	0.3
Duralumin[1)]	–	34.5	0.25
Sintered aluminium powder (SAP)	–	34.5	5.0

1) Cu = 3.5–4.5 %; Mg = 0.3–0.6 %; Si ≤ 1.0 %; Fe = 0.4–1.0 %; Mn ≤ 0.8 %; Al = Rem.

Source: References 1, 2 and 8

2 EMISSIVITY

Emissivity is defined as the ratio of the radiation emitted by a surface to the radiation emitted by a perfect blackbody at the same temperature. Emissivity characterizes the radiation or absorption quality of nonblackbodies. Emissivity varies with the wavelength, temperature and angle of emission.

TABLE 8.2 Total normal emissivity of metals and alloys

Metal/Alloy	Total normal emissivity at a temperature, in °C, of									
	20	100	500	1000	1200	1400	1600	2000	2500	3000
Metals										
Aluminium	–	0.038	0.064	–	–	–	–	–	–	–
Chromium	–	0.08	0.11–0.14	–	–	–	–	–	–	–
Copper	–	–	0.02	–	0.12[1)]	–	–	–	–	–
Iron	0.05	0.07	0.14	0.24	–	–	–	–	–	–
Lead	–	0.63	–	–	–	–	–	–	–	–
Molybdenum	0.065	0.08	0.13	0.19	0.22	0.24	0.27	–	–	–
Nickel	–	–	0.09–0.15	0.14–0.22	–	–	–	–	–	–
Tin	–	0.07	–	–	–	–	–	–	–	–
Titanium	–	0.11	–	–	–	–	–	–	–	–
Tungsten	–	–	0.05	0.11	0.14	0.17	0.19	0.23	0.27	0.30
Zinc	–	0.07	–	–	–	–	–	–	–	–
Zinc (galvanized iron)	–	0.21	–	–	–	–	–	–	–	–
Alloys										
Brass	–	0.059	–	–	–	–	–	–	–	–
Cast iron (cleaned)	–	0.21	–	–	–	–	0.29[1)]	–	–	–
Steel (polished)	–	0.13–0.21	0.18–0.26	0.55–0.80	–	–	–	–	–	–
Steel (cleaned)	–	0.21–0.38	0.25–0.42	0.50–0.77	–	–	–	–	–	–

[1)] Value for molten state.

Source: Reference 1

TABLE 8.3 Total normal emissivity of oxidized metals and alloys

Metal/Alloy	Total normal emissivity at a temperature, in °C, of				
	200	400	600	800	1000
Aluminium	0.11	0.15	0.19	–	–
Brass	0.61	0.60	0.59	–	–
Chromium	–	0.09	0.14–0.34	–	–
Copper (red heat for 30 min)	0.15	0.18	0.23	0.24	–
Copper (stably oxidized at 760 °C)	–	0.40–0.50	0.60–0.66	–	–
Copper (extreme oxidation)	–	0.88	0.92	–	–
Cast iron	0.64	0.71	0.78	–	–
Cast iron (strongly oxidized)	0.95	–		–	–
Iron (red heat for 30 min)	0.45	0.52	0.57	–	–
Lead	0.63	–	–	–	–
Molybdenum (oxide volatile in vacuum above 540 °C)	–	0.84	–	–	–
Nickel (stably oxidized at 900 °C)	0.15–0.50	0.33–0.51	0.44–0.57	0.49–0.71	–
Stainless steel (stably oxidized at high temperature)	–	0.80–0.87	0.84–0.91	0.89–0.95	–
Stainless steel (red heat in air for 30 min)	0.12–0.25	0.17–0.30	0.23–0.37	0.30–0.44	–

(Contd.)

TABLE 8.3 Total normal emissivity of oxidized metals and alloys (continued)

Metal/Alloy	Total normal emissivity at a temperature, in °C, of				
	200	400	600	800	1000
Stainless steel (buffed, stably oxidized at 600 °C)	–	0.41	0.44	0.54	–
Stainless steel (polished, oxidized at high temperature)	–	–	0.65–0.70	–	0.73–0.83
Stainless steel (shot blasted, stably oxidized at 600 °C)	–	0.65	0.67	–	–
Zinc	–	0.11	–	–	–

Notes:
1) The values of emissivity depend on the degree of oxidation and the grain size.
2) Unless otherwise stated, the values of emissivity given in this table are for metals and alloys oxidized in general above 600 °C.

Source: Reference 1

3 GALVANIC SERIES

The galvanic series is an arrangement of metals and alloys according to their relative potential measured in a specific electrolyte, such as seawater (see Table 8.4). It allows one to determine which metal or alloy in a galvanic couple is more active. In some cases, the separation between the two metals or alloys in the galvanic series gives an indication of the probable magnitude of the corrosive effect.

TABLE 8.4 Galvanic series in seawater at 25 °C

Corroded end (anodic, or least noble)	Aluminium alloys 5052, 3004, 3003, 1100, 6053, in this order
Magnesium	Cadmium
Magnesium alloys	Aluminium alloys 2117, 2017, 2024, in this order
Zinc	Low-carbon steel
Galvanized steel or galvanized wrought iron	Wrought iron

(Contd.)

TABLE 8.4 Galvanic series in seawater at 25 °C (continued)

Cast iron	Silicon bronze (97Cu–3Si)
Ni-Resist (high-nickel cast iron)	Copper-nickel alloy (70Cu–30Ni)
Type 410 stainless steel (active)	Leaded tin bronze M (88Cu–6Sn–1.5Pb–4.5Zn)
50–50 lead-tin solder	Nickel 200 (passive)
Type 304 stainless steel (active)	Inconel 600 (passive)
Type 316 stainless steel (active)	Monel 400
Lead	Type 410 stainless steel (passive)
Tin	Type 304 stainless stell (passive)
Muntz metal (60Cu–40Zn)	Type 316 stainless steel (passive)
Naval brass (60Cu–39.2Zn–0.8Sn)	Incolony 825
Nickel 200 (active)	Inconel 625
Inconel 600 (active)	Silver
Inconel 600 (active)	Titanium
Yellow brass (65Cu–35Zn)	Graphite
Admiralty brass (71Cu–28Zn–1Sn)	Gold
Aluminium bronze (91Cu–7Al–2Fe)	Platinum
Red brass (85Cu–15Zn)	**Protected end (cathodic, or most noble)**
ETP copper (99.95 Cu–0.04O)	

Source: Reference 3

4 DIMENSIONS AND MASS PER UNIT LENGTH OF STEEL PLATES

TABLE 8.5 Dimensions and mass per unit length of steel plates

Width	Mass per unit length[1), in kg/m, for a thickness, in mm, of																			
mm	5	6	7	8	10	12	14	16	18	20	22	25	28	32	36	40	45	50	56	63
900	35.3	42.4	49.5	56.5	70.6	84.8	98.9	113	127	141	155	177	198	226	254	283	318	353	396	445
950	37.3	44.7	52.2	59.7	74.6	89.5	104	119	134	149	164	186	209	239	268	298	336	373	418	470
1000	39.2	47.1	55.0	62.8	78.5	94.2	110	126	141	157	173	196	220	251	283	314	353	392	440	495
1100	43.2	51.8	60.4	69.1	86.4	104	121	138	155	173	190	216	242	276	311	345	389	432	484	544
1200	47.1	56.5	65.9	75.4	94.2	113	132	151	170	188	207	236	264	301	339	377	424	471	528	593
1250	49.1	58.9	68.7	78.5	98.1	118	137	157	177	196	216	245	275	314	353	392	442	491	550	618
1400	55.0	65.9	76.9	87.9	110	132	154	176	198	220	242	275	308	352	396	440	495	550	615	692
1500	58.9	70.6	82.4	94.2	118	141	165	188	212	236	259	294	330	377	424	471	530	589	659	742
1600	62.8	75.4	87.9	100	126	151	176	201	226	251	276	314	352	402	452	502	565	628	703	791
1800	70.6	84.8	98.9	113	141	170	198	226	254	283	311	353	396	452	509	565	636	706	791	890
2000	78.5	94.2	110	126	157	188	220	251	283	314	345	392	440	502	565	628	706	785	879	989
2200	86.4	104	121	138	173	207	242	276	311	345	380	432	484	553	622	691	777	864	967	1090
2500	98.1	118	137	157	196	236	275	314	353	392	432	491	550	628	706	785	883	981	1100	1240

1) The values of mass per unit length are based on a density of steel of 7850 kg/m^3.

Note: The values of mass per unit length have been rounded to three significant figures.

Source: Dimensions according to IS 1730 – 1989

5 DIMENSIONS, SURFACE AREA AND MASS OF STEEL SHEETS

TABLE 8.6 Dimensions, surface area and mass of steel sheets

Size (length × width)	Surface area	Mass[1)], in kg, for a thickness, in mm, of																				
mm × mm	m^2	0.40	0.50	0.63	0.80	0.90	1.00	1.12	1.25	1.40	1.60	1.80	1.90	2.00	2.24	2.50	2.80	3.15	3.55	4.00	4.30	4.65
1800 × 600	1.08	3.39	4.24	5.34	6.78	7.63	8.48	9.50	10.6	11.9	13.6	15.3	16.1	17.0	19.0	21.2	23.7	26.7	30.1	33.9	36.5	39.4
1800 × 750	1.35	4.24	5.30	6.68	8.48	9.54	10.6	11.9	13.2	14.8	17.0	19.1	20.1	21.2	23.7	26.5	29.7	33.4	37.6	42.4	45.6	49.3
1800 × 900	1.62	5.09	6.36	8.01	10.2	11.4	12.7	14.2	15.9	17.8	20.3	22.9	24.2	25.4	28.5	31.8	35.6	40.1	45.1	50.9	54.7	59.1
1800 × 950	1.71	5.37	6.71	8.46	10.7	12.1	13.4	15.0	16.8	18.8	21.5	24.2	25.5	26.8	30.1	33.6	37.6	42.3	47.7	53.7	57.7	62.4
1800 × 1000	1.80	5.65	7.06	8.90	11.3	12.7	14.1	15.8	17.7	19.8	22.6	25.4	26.8	28.3	31.7	35.3	39.6	44.5	50.2	56.5	60.8	65.7
1800 × 1100	1.98	6.22	7.77	9.79	12.4	14.0	15.5	17.4	19.4	21.8	24.9	28.0	29.5	31.1	34.8	38.9	43.5	49.0	55.2	62.2	66.8	72.3
1800 × 1200	2.16	6.78	8.48	10.7	13.6	15.3	17.0	19.0	21.2	23.7	27.1	30.5	32.2	33.9	38.0	42.4	47.5	53.4	60.2	67.8	72.9	78.8
1800 × 1250	2.25	7.06	8.83	11.1	14.1	15.9	17.7	19.8	22.1	24.7	28.3	31.8	33.6	35.3	39.6	44.2	49.5	55.6	62.7	70.6	75.9	82.1
1800 × 1400	2.52	7.91	9.89	12.5	15.8	17.8	19.8	22.2	24.7	27.7	31.7	35.6	37.6	39.6	44.3	49.5	55.4	62.3	70.2	79.1	85.1	92.0
1800 × 1500	2.70	8.48	10.6	13.4	17.0	19.1	21.2	23.7	26.5	29.7	33.9	38.2	40.3	42.4	47.5	53.0	59.3	66.8	75.2	84.8	91.1	98.6
2000 × 600	1.20	3.77	4.71	5.93	7.54	8.48	9.42	10.6	11.8	13.2	15.1	17.0	17.9	18.8	21.1	23.6	26.4	29.7	33.4	37.7	40.5	43.8
2000 × 750	1.50	4.71	5.89	7.42	9.42	10.6	11.8	13.2	14.7	16.5	18.8	21.2	22.4	23.6	26.4	29.4	33.0	37.1	41.8	47.1	50.6	54.8
2000 × 900	1.80	5.65	7.06	8.90	11.3	12.7	14.1	15.8	17.7	19.8	22.6	25.4	26.8	28.3	31.7	35.3	39.6	44.5	50.2	56.5	60.8	65.7

(Contd.)

TABLE 8.6 Dimensions, surface area and mass of steel sheets (continued)

Size (length × width)	Surface area	Mass[1)], in kg, for a thickness, in mm, of																				
mm × mm	m^2	0.40	0.50	0.63	0.80	0.90	1.00	1.12	1.25	1.40	1.60	1.80	1.90	2.00	2.24	2.50	2.80	3.15	3.55	4.00	4.30	4.65
2000 × 950	1.90	5.97	7.46	9.40	11.9	13.4	14.9	16.7	18.6	20.9	23.9	26.8	28.3	29.8	33.4	37.3	41.8	47.0	52.9	59.7	64.1	69.4
2000 × 1000	2.00	6.28	7.85	9.89	12.6	14.1	15.7	17.6	19.6	22.0	25.1	28.3	29.8	31.4	35.2	39.2	44.0	49.5	55.7	62.8	67.5	73.0
2000 × 1100	2.20	6.91	8.64	10.9	13.8	15.5	17.3	19.3	21.6	24.2	27.6	31.1	32.8	34.5	38.7	43.2	48.4	54.4	61.3	69.1	74.3	80.3
2000 × 1200	2.40	7.54	9.42	11.9	15.1	17.0	18.8	21.1	23.6	26.4	30.1	33.9	35.8	37.7	42.2	47.1	52.8	59.3	66.9	75.4	81.0	87.6
2000 × 1250	2.50	7.85	9.81	12.4	15.7	17.7	19.6	22.0	24.5	27.5	31.4	35.3	37.3	39.2	44.0	49.1	55.0	61.8	69.7	78.5	84.4	91.3
2000 × 1400	2.80	8.79	11.0	13.8	17.6	19.8	22.0	24.6	27.5	30.8	35.2	39.6	41.8	44.0	49.2	55.0	61.5	69.2	78.0	87.9	94.5	102
2000 × 1500	3.00	9.42	11.8	14.8	18.8	21.2	23.6	26.4	29.4	33.0	37.7	42.4	44.7	47.1	52.8	58.9	65.9	74.2	83.6	94.2	101	110
2200 × 600	1.32	4.14	5.18	6.53	8.29	9.33	10.4	11.6	13.0	14.5	16.6	18.7	19.7	20.7	23.2	25.9	29.0	32.6	36.8	41.4	44.6	48.2
2200 × 750	1.65	5.18	6.48	8.16	10.4	11.7	13.0	14.5	16.2	18.1	20.7	23.3	24.6	25.9	29.0	32.4	36.3	40.8	46.0	51.8	55.7	60.2
2200 × 900	1.98	6.22	7.77	9.79	12.4	14.0	15.5	17.4	19.4	21.8	24.9	28.0	29.5	31.1	34.8	38.9	43.5	49.0	55.2	62.2	66.8	72.3
2200 × 950	2.09	6.56	8.20	10.3	13.1	14.8	16.4	18.4	20.5	23.0	26.3	29.5	31.2	32.8	36.8	41.0	45.9	51.7	58.2	65.6	70.5	76.3
2200 × 1000	2.20	6.91	8.64	10.9	13.8	15.5	17.3	19.3	21.6	24.2	27.6	31.1	32.8	34.5	38.7	43.2	48.4	54.4	61.3	69.1	74.3	80.3
2200 × 1100	2.42	7.60	9.50	12.0	15.2	17.1	19.0	21.3	23.7	26.6	30.4	34.2	36.1	38.0	42.6	47.5	53.2	59.8	67.4	76.0	81.7	88.3
2200 × 1200	2.64	8.29	10.4	13.1	16.6	18.7	20.7	23.2	25.9	29.0	33.2	37.3	39.4	41.4	46.4	51.8	58.0	65.3	73.6	82.9	89.1	96.4
2200 × 1250	2.75	8.64	10.8	13.6	17.3	19.4	21.6	24.2	27.0	30.2	34.5	38.9	41.0	43.2	48.4	54.0	60.4	68.0	76.6	86.4	92.8	100

(Contd.)

TABLE 8.6 Dimensions, surface area and mass of steel sheets (continued)

Size (length × width)	Surface area	Mass[1], in kg, for a thickness, in mm, of																				
mm × mm	m^2	0.40	0.50	0.63	0.80	0.90	1.00	1.12	1.25	1.40	1.60	1.80	1.90	2.00	2.24	2.50	2.80	3.15	3.55	4.00	4.30	4.65
2200 × 1400	3.08	9.67	12.1	15.2	19.3	21.8	24.2	27.1	30.2	33.8	38.7	43.5	45.9	48.4	54.2	60.4	67.7	76.2	85.8	96.7	104	112
2200 × 1500	3.30	10.4	13.0	16.3	20.7	23.3	25.9	29.0	32.4	36.3	41.4	46.6	49.2	51.8	58.0	64.8	72.5	81.6	92.0	104	111	120
2500 × 600	1.50	4.71	5.89	7.42	9.42	10.6	11.8	13.2	14.7	16.5	18.8	21.2	22.4	23.6	26.4	29.4	33.0	37.1	41.8	47.1	50.6	54.8
2500 × 750	1.88	5.89	7.36	9.27	11.8	13.2	14.7	16.5	18.4	20.6	23.6	26.5	28.0	29.4	33.0	36.8	41.2	46.4	52.3	58.9	63.3	68.4
2500 × 900	2.25	7.06	8.83	11.1	14.1	15.9	17.7	19.8	22.1	24.7	28.3	31.8	33.6	35.3	39.6	44.2	49.5	55.6	62.7	70.6	75.9	82.1
2500 × 950	2.38	7.46	9.32	11.7	14.9	16.8	18.6	20.9	23.3	26.1	29.8	33.6	35.4	37.3	41.8	46.6	52.2	58.7	66.2	74.6	80.2	86.7
2500 × 1000	2.50	7.85	9.81	12.4	15.7	17.7	19.6	22.0	24.5	27.5	31.4	35.3	37.3	39.2	44.0	49.1	55.0	61.8	69.7	78.5	84.4	91.3
2500 × 1100	2.75	8.64	10.8	13.6	17.3	19.4	21.6	24.2	27.0	30.2	34.5	38.9	41.0	43.2	48.4	54.0	60.4	68.0	76.6	86.4	92.8	100
2500 × 1200	3.00	9.42	11.8	14.8	18.8	21.2	23.6	26.4	29.4	33.0	37.7	42.4	44.7	47.1	52.8	58.9	65.9	74.2	83.6	94.2	101	110
2500 × 1250	3.12	9.81	12.3	15.5	19.6	22.1	24.5	27.5	30.7	34.3	39.2	44.2	46.6	49.1	55.0	61.3	68.7	77.3	87.1	98.1	105	114
2500 × 1400	3.50	11.0	13.7	17.3	22.0	24.7	27.5	30.8	34.3	38.5	44.0	49.5	52.2	55.0	61.5	68.7	76.9	86.5	97.5	110	118	128
2500 × 1500	3.75	11.8	14.7	18.5	23.6	26.5	29.4	33.0	36.8	41.2	47.1	53.0	55.9	58.9	65.9	73.6	82.4	92.7	105	118	127	137
2800 × 600	1.68	5.28	6.59	8.31	10.6	11.9	13.2	14.8	16.5	18.5	21.1	23.7	25.1	26.4	29.5	33.0	36.9	41.5	46.8	52.8	56.7	61.3
2800 × 750	2.10	6.59	8.24	10.4	13.2	14.8	16.5	18.5	20.6	23.1	26.4	29.7	31.3	33.0	36.9	41.2	46.2	51.9	58.5	65.9	70.9	76.7
2800 × 900	2.52	7.91	9.89	12.5	15.8	17.8	19.8	22.2	24.7	27.7	31.7	35.6	37.6	39.6	44.3	49.5	55.4	62.3	70.2	79.1	85.1	92.0

(Contd.)

TABLE 8.6 Dimensions, surface area and mass of steel sheets (continued)

Size (length × width)	Surface area	Mass[1)], in kg, for a thickness, in mm, of																				
mm × mm	m^2	0.40	0.50	0.63	0.80	0.90	1.00	1.12	1.25	1.40	1.60	1.80	1.90	2.00	2.24	2.50	2.80	3.15	3.55	4.00	4.30	4.65
2800 × 950	2.66	8.35	10.4	13.2	16.7	18.8	20.9	23.4	26.1	29.2	33.4	37.6	39.7	41.8	46.8	52.2	58.5	65.8	74.1	83.5	89.8	97.1
2800 × 1000	2.80	8.79	11.0	13.8	17.6	19.8	22.0	24.6	27.5	30.8	35.2	39.6	41.8	44.0	49.2	55.0	61.5	69.2	78.0	87.9	94.5	102
2800 × 1100	3.08	9.67	12.1	15.2	19.3	21.8	24.2	27.1	30.2	33.8	38.7	43.5	45.9	48.4	54.2	60.4	67.7	76.2	85.8	96.7	104	112
2800 × 1200	3.36	10.6	13.2	16.6	21.1	23.7	26.4	29.5	33.0	36.9	42.2	47.5	50.1	52.8	59.1	65.9	73.9	83.1	93.6	106	113	123
2800 × 1250	3.50	11.0	13.7	17.3	22.0	24.7	27.5	30.8	34.3	38.5	44.0	49.5	52.2	55.0	61.5	68.7	76.9	86.5	97.5	110	118	128
2800 × 1400	3.92	12.3	15.4	19.4	24.6	27.7	30.8	34.5	38.5	43.1	49.2	55.4	58.5	61.5	68.9	76.9	86.2	96.9	109	123	132	143
2800 × 1500	4.20	13.2	16.5	20.8	26.4	29.7	33.0	36.9	41.2	46.2	52.8	59.3	62.6	65.9	73.9	82.4	92.3	104	117	132	142	153
3200 × 600	1.92	6.03	7.54	9.50	12.1	13.6	15.1	16.9	18.8	21.1	24.1	27.1	28.6	30.1	33.8	37.7	42.2	47.5	53.5	60.3	64.8	70.1
3200 × 750	2.40	7.54	9.42	11.9	15.1	17.0	18.8	21.1	23.6	26.4	30.1	33.9	35.8	37.7	42.2	47.1	52.8	59.3	66.9	75.4	81.0	87.6
3200 × 900	2.88	9.04	11.3	14.2	18.1	20.3	22.6	25.3	28.3	31.7	36.2	40.7	43.0	45.2	50.6	56.5	63.3	71.2	80.3	90.4	97.2	105
3200 × 950	3.04	9.55	11.9	15.0	19.1	21.5	23.9	26.7	29.8	33.4	38.2	43.0	45.3	47.7	53.5	59.7	66.8	75.2	84.7	95.5	103	111
3200 × 1000	3.20	10.0	12.6	15.8	20.1	22.6	25.1	28.1	31.4	35.2	40.2	45.2	47.7	50.2	56.3	62.8	70.3	79.1	89.2	100	108	117
3200 × 1100	3.52	11.1	13.8	17.4	22.1	24.9	27.6	30.9	34.5	38.7	44.2	49.7	52.5	55.3	61.9	69.1	77.4	87.0	98.1	111	119	128
3200 × 1200	3.84	12.1	15.1	19.0	24.1	27.1	30.1	33.8	37.7	42.2	48.2	54.3	57.3	60.3	67.5	75.4	84.4	95.0	107	121	130	140
3200 × 1250	4.00	12.6	15.7	19.8	25.1	28.3	31.4	35.2	39.2	44.0	50.2	56.5	59.7	62.8	70.3	78.5	87.9	98.9	111	126	135	146

(Contd.)

TABLE 8.6 Dimensions, surface area and mass of steel sheets (continued)

Size (length × width)	Surface area	Mass[1), in kg, for a thickness, in mm, of																				
mm × mm	m^2	0.40	0.50	0.63	0.80	0.90	1.00	1.12	1.25	1.40	1.60	1.80	1.90	2.00	2.24	2.50	2.80	3.15	3.55	4.00	4.30	4.65
3200 × 1400	4.48	14.1	17.6	22.2	28.1	31.7	35.2	39.4	44.0	49.2	56.3	63.3	66.8	70.3	78.8	87.9	98.5	111	125	141	151	164
3200 × 1500	4.80	15.1	18.8	23.7	30.1	33.9	37.7	42.2	47.1	52.8	60.3	67.8	71.6	75.4	84.4	94.2	106	119	134	151	162	175
3600 × 600	2.16	6.78	8.48	10.7	13.6	15.3	17.0	19.0	21.2	23.7	27.1	30.5	32.2	33.9	38.0	42.4	47.5	53.4	60.2	67.8	72.9	78.8
3600 × 750	2.70	8.48	10.6	13.4	17.0	19.1	21.2	23.7	26.5	29.7	33.9	38.2	40.3	42.4	47.5	53.0	59.3	66.8	75.2	84.8	91.1	98.6
3600 × 900	3.24	10.2	12.7	16.0	20.3	22.9	25.4	28.5	31.8	35.6	40.7	45.8	48.3	50.9	57.0	63.6	71.2	80.1	90.3	102	109	118
3600 × 950	3.42	10.7	13.4	16.9	21.5	24.2	26.8	30.1	33.6	37.6	43.0	48.3	51.0	53.7	60.1	67.1	75.2	84.6	95.3	107	115	125
3600 × 1000	3.60	11.3	14.1	17.8	22.6	25.4	28.3	31.7	35.3	39.6	45.2	50.9	53.7	56.5	63.3	70.6	79.1	89.0	100	113	122	131
3600 × 1100	3.96	12.4	15.5	19.6	24.9	28.0	31.1	34.8	38.9	43.5	49.7	56.0	59.1	62.2	69.6	77.7	87.0	97.9	110	124	134	145
3600 × 1200	4.32	13.6	17.0	21.4	27.1	30.5	33.9	38.0	42.4	47.5	54.3	61.0	64.4	67.8	76.0	84.8	95.0	107	120	136	146	158
3600 × 1250	4.50	14.1	17.7	22.3	28.3	31.8	35.3	39.6	44.2	49.5	56.5	63.6	67.1	70.6	79.1	88.3	98.9	111	125	141	152	164
3600 × 1400	5.04	15.8	19.8	24.9	31.7	35.6	39.6	44.3	49.5	55.4	63.3	71.2	75.2	79.1	88.6	98.9	111	125	140	158	170	184
3600 × 1500	5.40	17.0	21.2	26.7	33.9	38.2	42.4	47.5	53.0	59.3	67.8	76.3	80.5	84.8	95.0	106	119	134	150	170	182	197
4000 × 600	2.40	7.54	9.42	11.9	15.1	17.0	18.8	21.1	23.6	26.4	30.1	33.9	35.8	37.7	42.2	47.1	52.8	59.3	66.9	75.4	81.0	87.6
4000 × 750	3.00	9.42	11.8	14.8	18.8	21.2	23.6	26.4	29.4	33.0	37.7	42.4	44.7	47.1	52.8	58.9	65.9	74.2	83.6	94.2	101	110
4000 × 900	3.60	11.3	14.1	17.8	22.6	25.4	28.3	31.7	35.3	39.6	45.2	50.9	53.7	56.5	63.3	70.6	79.1	89.0	100	113	122	131

(Contd.)

TABLE 8.6 Dimensions, surface area and mass of steel sheets (continued)

Size (length × width)	Surface area	Mass[1)], in kg, for a thickness, in mm, of																				
mm × mm	m^2	0.40	0.50	0.63	0.80	0.90	1.00	1.12	1.25	1.40	1.60	1.80	1.90	2.00	2.24	2.50	2.80	3.15	3.55	4.00	4.30	4.65
4000 × 950	3.80	11.9	14.9	18.8	23.9	26.8	29.8	33.4	37.3	41.8	47.7	53.7	56.7	59.7	66.8	74.6	83.5	94.0	106	119	128	139
4000 × 1000	4.00	12.6	15.7	19.8	25.1	28.3	31.4	35.2	39.2	44.0	50.2	56.5	59.7	62.8	70.3	78.5	87.9	98.9	111	126	135	146
4000 × 1100	4.40	13.8	17.3	21.8	27.6	31.1	34.5	38.7	43.2	48.4	55.3	62.2	65.6	69.1	77.4	86.4	96.7	109	123	138	149	161
4000 × 1200	4.80	15.1	18.8	23.7	30.1	33.9	37.7	42.2	47.1	52.8	60.3	67.8	71.6	75.4	84.4	94.2	106	119	134	151	162	175
4000 × 1250	5.00	15.7	19.6	24.7	31.4	35.3	39.2	44.0	49.1	55.0	62.8	70.6	74.6	78.5	87.9	98.1	110	124	139	157	169	183
4000 × 1400	5.60	17.6	22.0	27.7	35.2	39.6	44.0	49.2	55.0	61.5	70.3	79.1	83.5	87.9	98.5	110	123	138	156	176	189	204
4000 × 1500	6.00	18.8	23.6	29.7	37.7	42.4	47.1	52.8	58.9	65.9	75.4	84.8	89.5	94.2	106	118	132	148	167	188	203	219

1) The values of mass are based on a density of steel of 7850 kg/m^3.

Note: The values of surface area and mass have been rounded to three significant figures.

Source: Dimensions according to IS 1730–1989

6 DIMENSIONS AND MASS PER UNIT LENGTH OF STEEL STRIPS

TABLE 8.7 Dimensions and mass per unit length of steel strips

Width	Mass per unit length[1)], in kg/m, for a thickness, in mm, of													
mm	1.60	1.80	2.00	2.24	2.50	2.80	3.15	3.55	4.00	4.50	5.00	6.00	8.00	10.0
100	1.26	1.41	1.57	1.76	1.96	2.20	2.47	2.79	3.14	3.53	3.92	4.71	6.28	7.85
125	1.57	1.77	1.96	2.20	2.45	2.75	3.09	3.48	3.92	4.42	4.91	5.89	7.85	9.81
160	2.01	2.26	2.51	2.81	3.14	3.52	3.96	4.46	5.02	5.65	6.28	7.54	10.0	12.6
200	2.51	2.83	3.14	3.52	3.92	4.40	4.95	5.57	6.28	7.06	7.85	9.42	12.6	15.7
250	3.14	3.53	3.92	4.40	4.91	5.50	6.18	6.97	7.85	8.83	9.81	11.8	15.7	19.6
320	4.02	4.52	5.02	5.63	6.28	7.03	7.91	8.92	10.0	11.3	12.6	15.1	20.1	25.1
400	5.02	5.65	6.28	7.03	7.85	8.79	9.89	11.1	12.6	14.1	15.7	18.8	25.1	31.4
500	6.28	7.06	7.85	8.79	9.81	11.0	12.4	13.9	15.7	17.7	19.6	23.6	31.4	39.2
650	8.16	9.18	10.2	11.4	12.8	14.3	16.1	18.1	20.4	23.0	25.5	30.6	40.8	51.0
800	10.0	11.3	12.6	14.1	15.7	17.6	19.8	22.3	25.1	28.3	31.4	37.7	50.2	62.8
950	–	13.4	14.9	16.7	18.6	20.9	23.5	26.5	29.8	33.6	37.3	44.7	59.7	74.6
1000	–	–	15.7	17.6	19.6	22.0	24.7	27.9	31.4	35.3	39.2	47.1	62.8	78.5
1050	–	–	16.5	18.5	20.6	23.1	26.0	29.3	33.0	37.1	41.2	49.5	65.9	82.4
1150	–	–	–	20.2	22.6	25.3	28.4	32.0	36.1	40.6	45.1	54.2	72.2	90.3
1250	–	–	–	–	24.5	27.5	30.9	34.8	39.2	44.2	49.1	58.9	78.5	98.1
1300	–	–	–	–	–	28.6	32.1	36.2	40.8	45.9	51.0	61.2	81.6	102
1450	–	–	–	–	–	–	35.9	40.4	45.5	51.2	56.9	68.3	91.1	114
1550	–	–	–	–	–	–	38.3	43.2	48.7	54.8	60.8	73.0	97.3	122

1) The values of mass per unit length are based on a density of steel of 7850 kg/m^3.

Note: The values of mass per unit length have been rounded to three significant figures.

Source: Dimensions according to IS 1730 – 1989

7 DIMENSIONS, SECTIONAL AREA AND MASS PER UNIT LENGTH OF STEEL BARS

7.1 Round bars

TABLE 8.8 Dimensions, sectional area and mass per unit length of round steel bars

Diameter (*d*) mm	Sectional area[1)] cm^2	Mass per unit length[2)] kg/m	Diameter (*d*) mm	Sectional area[1)] cm^2	Mass per unit length[2)] kg/m
5	0.196	0.154	23	4.16	3.26
6	0.283	0.222	24	4.53	3.55
7	0.385	0.302	25	4.91	3.85
8	0.503	0.395	26	5.31	4.17
9	0.636	0.500	27	5.73	4.50
10	0.786	0.617	28	6.16	4.84
11	0.951	0.746	29	6.61	5.19
12	1.13	0.888	30	7.07	5.55
13	1.33	1.04	31	7.55	5.93
14	1.54	1.21	32	8.05	6.32
15	1.77	1.39	33	8.56	6.72
16	2.01	1.58	34	9.08	7.13
17	2.27	1.78	35	9.62	7.56
18	2.55	2.00	36	10.2	7.99
19	2.84	2.23	37	10.8	8.44
20	3.14	2.47	38	11.3	8.91
21	3.46	2.72	39	12.0	9.38
22	3.80	2.99	40	12.6	9.87

(Contd.)

TABLE 8.8 Dimensions, sectional area and mass per unit length of round steel bars (continued)

Diameter (*d*) mm	Sectional area[1)] cm²	Mass per unit length[2)] kg/m	Diameter (*d*) mm	Sectional area[1)] cm²	Mass per unit length[2)] kg/m
41	13.2	10.4	61	29.2	23.0
42	13.9	10.9	62	30.2	23.7
43	14.5	11.4	63	31.2	24.5
44	15.2	11.9	64	32.2	25.3
45	15.9	12.5	65	33.2	26.1
46	16.6	13.1	66	34.2	26.9
47	17.4	13.6	67	35.3	27.7
48	18.1	14.2	68	36.3	28.5
49	18.9	14.8	69	37.4	29.4
50	19.6	15.4	70	38.5	30.2
51	20.4	16.0	71	39.6	31.1
52	21.2	16.7	72	40.7	32.0
53	22.1	17.3	73	41.9	32.9
54	22.9	18.0	74	43.0	33.8
55	23.8	18.7	75	44.2	34.7
56	24.6	19.3	76	45.4	35.6
57	25.5	20.0	77	46.6	36.6
58	26.4	20.7	78	47.8	37.5
59	27.4	21.5	79	49.0	38.5
60	28.3	22.2	80	50.3	39.5

(Contd.)

TABLE 8.8 Dimensions, sectional area and mass per unit length of round steel bars (continued)

Diameter (*d*) mm	Sectional area[1)] cm²	Mass per unit length[2)] kg/m	Diameter (*d*) mm	Sectional area[1)] cm²	Mass per unit length[2)] kg/m
81	51.6	40.5	145	165	130
82	52.8	41.5	150	177	139
83	54.1	42.5	155	189	148
84	55.4	43.5	160	201	158
85	56.8	44.6	165	214	168
86	58.1	45.6	170	227	178
87	59.5	46.7	175	241	189
88	60.8	47.8	180	255	200
89	62.2	48.9	185	269	211
90	63.6	50.0	190	284	223
95	70.9	55.7	195	299	235
100	78.6	61.7	200	314	247
105	86.6	68.0	205	330	259
110	95.1	74.6	210	346	272
115	104	81.6	215	363	285
120	113	88.8	220	380	299
125	123	96.4	225	398	312
130	133	104	230	416	326
135	143	112	235	434	341
140	154	121	240	453	355

(Contd.)

TABLE 8.8 Dimensions, sectional area and mass per unit length of round steel bars (continued)

Diameter (*d*) mm	Sectional area[1)] cm^2	Mass per unit length[2)] kg/m	Diameter (*d*) mm	Sectional area[1)] cm^2	Mass per unit length[2)] kg/m
245	472	370	275	594	466
250	491	385	280	616	484
255	511	401	285	638	501
260	531	417	290	661	519
265	552	433	295	684	537
270	573	450	300	707	555

1) Sectional area $= \frac{\pi d^2}{400}$, in cm^2.

2) The values of mass per unit length are based on a density of steel of 7850 kg/m^3.

Note: The values of sectional area and mass per unit length have been rounded to three significant figures.

7.2 Square bars

TABLE 8.9 Dimensions, sectional area and mass per unit length of square steel bars

Width across flats (*b*) mm	Sectional area[1)] cm^2	Mass per unit length[2)] kg/m	Width across flats (*b*) mm	Sectional area[1)] cm^2	Mass per unit length[2)] kg/m
5	0.250	0.196	10	1.00	0.785
6	0.360	0.283	11	1.21	0.950
7	0.490	0.385	12	1.44	1.13
8	0.640	0.502	13	1.69	1.33
9	0.810	0.636	14	1.96	1.54

(Contd.)

TABLE 8.9 Dimensions, sectional area and mass per unit length of square steel bars (continued)

Width across flats (*b*) mm	Sectional area[1)] cm^2	Mass per unit length[2)] kg/m	Width across flats (*b*) mm	Sectional area[1)] cm^2	Mass per unit length[2)] kg/m
15	2.25	1.77	35	12.2	9.62
16	2.56	2.01	36	13.0	10.2
17	2.89	2.27	37	13.7	10.7
18	3.24	2.54	38	14.4	11.3
19	3.61	2.83	39	15.2	11.9
20	4.00	3.14	40	16.0	12.6
21	4.41	3.46	41	16.8	13.2
22	4.84	3.80	42	17.6	13.8
23	5.29	4.15	43	18.5	14.5
24	5.76	4.52	44	19.4	15.2
25	6.25	4.91	45	20.2	15.9
26	6.76	5.31	46	21.2	16.6
27	7.29	5.72	47	22.1	17.3
28	7.84	6.15	48	23.0	18.1
29	8.41	6.60	49	24.0	18.8
30	9.00	7.06	50	25.0	19.6
31	9.61	7.54	51	26.0	20.4
32	10.2	8.04	52	27.0	21.2
33	10.9	8.55	53	28.1	22.1
34	11.6	9.07	54	29.2	22.9

(Contd.)

TABLE 8.9 Dimensions, sectional area and mass per unit length of square steel bars (continued)

Width across flats (b) mm	Sectional area[1)] cm^2	Mass per unit length[2)] kg/m	Width across flats (b) mm	Sectional area[1)] cm^2	Mass per unit length[2)] kg/m
55	30.2	23.7	75	56.2	44.2
56	31.4	24.6	76	57.8	45.3
57	32.5	25.5	77	59.3	46.5
58	33.6	26.4	78	60.8	47.8
59	34.8	27.3	79	62.4	49.0
60	36.0	28.3	80	64.0	50.2
61	37.2	29.2	81	65.6	51.5
62	38.4	30.2	82	67.2	52.8
63	39.7	31.2	83	68.9	54.1
64	41.0	32.2	84	70.6	55.4
65	42.2	33.2	85	72.2	56.7
66	43.6	34.2	86	74.0	58.1
67	44.9	35.2	87	75.7	59.4
68	46.2	36.3	88	77.4	60.8
69	47.6	37.4	89	79.2	62.2
70	49.0	38.5	90	81.0	63.6
71	50.4	39.6	95	90.2	70.8
72	51.8	40.7	100	100	78.5
73	53.3	41.8	105	110	86.5
74	54.8	43.0	110	121	95.0

(Contd.)

TABLE 8.9 Dimensions, sectional area and mass per unit length of square steel bars (continued)

Width across flats (b) mm	Sectional area[1)] cm^2	Mass per unit length[2)] kg/m	Width across flats (b) mm	Sectional area[1)] cm^2	Mass per unit length[2)] kg/m
115	132	104	210	441	346
120	144	113	215	462	363
125	156	123	220	484	380
130	169	133	225	506	397
135	182	143	230	529	415
140	196	154	235	552	434
145	210	165	240	576	452
150	225	177	245	600	471
155	240	189	250	625	491
160	256	201	255	650	510
165	272	214	260	676	531
170	289	227	265	702	551
175	306	240	270	729	572
180	324	254	275	756	594
185	342	269	280	784	615
190	361	283	285	812	638
195	380	298	290	841	660
200	400	314	295	870	683
205	420	330	300	900	706

1) Sectional area = $\frac{b^2}{100}$, in cm^2.

2) The values of mass per unit length are based on a density of steel of 7850 kg/m^3.

Note: The values of sectional area and mass per unit length have been rounded to three significant figures.

7.3 Hexagonal bars

TABLE 8.10 Dimensions, sectional areas and mass per unit length of hexagonal steel bars

Width across flats (s) mm	Sectional area[1)] cm^2	Mass per unit length[2)] kg/m	Width across flats (s) mm	Sectional area[1)] cm^2	Mass per unit length[2)] kg/m
5	0.217	0.170	24	4.99	3.92
6	0.312	0.245	25	5.41	4.25
7	0.424	0.333	26	5.85	4.60
8	0.554	0.435	27	6.31	4.96
9	0.701	0.551	28	6.79	5.33
10	0.866	0.680	29	7.28	5.72
11	1.05	0.823	30	7.79	6.12
12	1.25	0.979	31	8.32	6.53
13	1.46	1.15	32	8.87	6.96
14	1.70	1.33	33	9.43	7.40
15	1.95	1.53	34	10.0	7.86
16	2.22	1.74	35	10.6	8.33
17	2.50	1.96	36	11.2	8.81
18	2.81	2.20	37	11.9	9.31
19	3.13	2.45	38	12.5	9.82
20	3.46	2.72	39	13.2	10.3
21	3.82	3.00	40	13.9	10.9
22	4.19	3.29	41	14.6	11.4
23	4.58	3.60	42	15.3	12.0

(Contd.)

TABLE 8.10 Dimensions, sectional areas and mass per unit length of hexagonal steel bars (continued)

Width across flats (*s*) mm	Sectional area[1)] cm^2	Mass per unit length[2)] kg/m	Width across flats (*s*) mm	Sectional area[1)] cm^2	Mass per unit length[2)] kg/m
43	16.0	12.6	63	34.4	27.0
44	16.8	13.2	64	35.5	27.8
45	17.5	13.8	65	36.6	28.7
46	18.3	14.4	66	37.7	29.6
47	19.1	15.0	67	38.9	30.5
48	20.0	15.7	68	40.0	31.4
49	20.8	16.3	69	41.2	32.4
50	21.7	17.0	70	42.4	33.3
51	22.5	17.7	71	43.7	34.3
52	23.4	18.4	72	44.9	35.2
53	24.3	19.1	73	46.2	36.2
54	25.3	19.8	74	47.4	37.2
55	26.2	20.6	75	48.7	38.2
56	27.2	21.3	76	50.0	39.3
57	28.1	22.1	77	51.3	40.3
58	29.1	22.9	78	52.7	41.4
59	30.1	23.7	79	54.0	42.4
60	31.2	24.5	80	55.4	43.5
61	32.2	25.3	81	56.8	44.6
62	33.3	26.1	82	58.2	45.7

(Contd.)

TABLE 8.10 Dimensions, sectional areas and mass per unit length of hexagonal steel bars (continued)

Width across flats (*s*) mm	Sectional area[1] cm^2	Mass per unit length[2] kg/m	Width across flats (*s*) mm	Sectional area[1] cm^2	Mass per unit length[2] kg/m
83	59.7	46.8	155	208	163
84	61.1	48.0	160	222	174
85	62.6	49.1	165	236	185
86	64.1	50.3	170	250	196
87	65.5	51.5	175	265	208
88	67.1	52.6	180	281	220
89	68.6	53.8	185	296	233
90	70.1	55.1	190	313	245
95	78.2	61.4	195	329	259
100	86.6	68.0	200	346	272
105	95.5	75.0	205	364	286
110	105	82.3	210	382	300
115	115	89.9	215	400	314
120	125	97.9	220	419	329
125	135	106	225	438	344
130	146	115	230	458	360
135	158	124	235	478	375
140	170	133	240	499	392
145	182	143	245	520	408
150	195	153	250	541	425

(Contd.)

TABLE 8.10 Dimensions, sectional areas and mass per unit length of hexagonal steel bars (continued)

Width across flats (*s*) mm	Sectional area[1)] cm^2	Mass per unit length[2)] kg/m	Width across flats (*s*) mm	Sectional area[1)] cm^2	Mass per unit length[2)] kg/m
255	563	442	280	679	533
260	585	460	285	703	552
265	608	477	290	728	572
270	631	496	295	754	592
275	655	514	300	779	612

1) Sectional area = $\frac{\sqrt{3}s^2}{200}$, in cm^2.

2) The values of mass per unit length are based on a density of steel of 7850 kg/m^3.

Note: The values of sectional area and mass per unit length have been rounded to three significant figures.

8 DIMENSIONS AND MASS PER UNIT LENGTH OF STEEL FLATS

TABLE 8.11 Dimensions and mass per unit length of steel flats

Width	Mass per unit length[1], in kg/m, for a thickness, in mm, of												
mm	3	4	5	6	8	10	12	15	20	25	30	40	50
10	0.236	0.314	0.392	0.471	–	–	–	–	–	–	–	–	–
16	0.377	0.502	0.628	0.754	1.00	–	–	–	–	–	–	–	–
20	0.471	0.628	0.785	0.942	1.26	1.57	1.88	2.36	–	–	–	–	–
25	0.589	0.785	0.981	1.18	1.57	1.96	2.36	2.94	–	–	–	–	–
30	0.706	0.942	1.18	1.41	1.88	2.36	2.83	3.53	4.71	–	–	–	–
35	0.824	1.10	1.37	1.65	2.20	2.75	3.30	4.12	5.50	6.87	8.24	–	–
40	0.942	1.26	1.57	1.88	2.51	3.14	3.77	4.71	6.28	7.85	9.42	–	–
45	–	1.41	1.77	2.12	2.83	3.53	4.24	5.30	7.06	8.83	10.6	–	–
50	1.18	1.57	1.96	2.36	3.14	3.92	4.71	5.89	7.85	9.81	11.8	15.7	–
55	–	1.73	2.16	2.59	3.45	4.32	5.18	6.48	8.64	10.8	13.0	17.3	–
60	1.41	1.88	2.36	2.83	3.77	4.71	5.65	7.06	9.42	11.8	14.1	18.8	–
65	–	2.04	2.55	3.06	4.08	5.10	6.12	7.65	10.2	12.8	15.3	20.4	–
70	–	2.20	2.75	3.30	4.40	5.50	6.59	8.24	11.0	13.7	16.5	22.0	27.5
75	–	2.36	2.94	3.53	4.71	5.89	7.06	8.83	11.8	14.7	17.7	23.6	29.4
80	–	2.51	3.14	3.77	5.02	6.28	7.54	9.42	12.6	15.7	18.8	25.1	31.4
90	–	–	3.53	4.24	5.65	7.06	8.48	10.6	14.1	17.7	21.2	28.3	35.3
100	–	–	3.92	4.71	6.28	7.85	9.42	11.8	15.7	19.6	23.6	31.4	39.2
120	–	–	4.71	5.65	7.54	9.42	11.3	14.1	18.8	23.6	28.3	37.7	47.1
130	–	–	–	6.12	8.16	10.2	12.2	15.3	20.4	25.5	30.6	40.8	51.0
140	–	–	–	–	8.79	11.0	13.2	16.5	22.0	27.5	33.0	44.0	55.0

(Contd.)

TABLE 8.11 Dimensions and mass per unit length of steel flats (continued)

Width	Mass per unit length[1), in kg/m, for a thickness, in mm, of												
mm	3	4	5	6	8	10	12	15	20	25	30	40	50
150	–	–	–	–	9.42	11.8	14.1	17.7	23.6	29.4	35.3	47.1	58.9
160	–	–	–	–	10.0	12.6	15.1	18.8	25.1	31.4	37.7	50.2	–
180	–	–	–	–	11.3	14.1	17.0	21.2	28.3	35.3	42.4	56.5	–
200	–	–	–	–	–	15.7	18.8	23.6	31.4	39.2	47.1	62.8	–
250	–	–	–	–	–	19.6	23.6	29.4	39.2	49.1	58.9	78.5	–
300	–	–	–	–	–	–	28.3	35.3	47.1	58.9	70.6	94.2	–
400	–	–	–	–	–	–	–	47.1	62.8	78.5	94.2	126	–

1) The values of mass per unit length are based on a density of steel of 7850 kg/m^3.

Note: The values of mass per unit length have been rounded to three significant figures.

Source: Dimensions according to IS 1730 – 1989

9.9 WIRE AND SHEET-METAL GAUGES

TABLE 8.12 Wire and sheet-metal gauges

Name of Gauge	Steel Wire Gauge or Washburn and Moen Wire Gauge Steel W.G. W. & M.W.G.	Music Wire Gauge M.W.G.	American Wire Gauge or Brown and Sharpe Gauge A.W.G. B. & S.G.	New Birmingham Standard Sheet and Hoop Gauge B.G.	British Imperial or English Legal Standard Wire Gauge S.W.G.	Birmingham or Stubs' Iron Wire Gauge B.W.G.
Principal use	Steel wire except music wire	Steel music wire	Non-ferrous sheets and wire	Iron and steel sheets and hoops	Wire	Flats, plates and wire
Gauge No.	mm	mm	mm	mm	mm	mm
7/0's	12.4	–	–	16.9	12.7	–
6/0's	11.7	0.102	14.7	15.9	11.8	–
5/0's	10.9	0.127	13.1	14.9	11.0	12.7
4/0's	10.0	0.152	11.7	13.8	10.2	11.5
3/0's	9.21	0.178	10.4	12.7	9.45	10.8
2/0's	8.41	0.203	9.27	11.3	8.84	9.65
0	7.79	0.229	8.25	10.1	8.23	8.64
1	7.19	0.254	7.35	8.97	7.62	7.62
2	6.67	0.279	6.54	7.99	7.01	7.21
3	6.19	0.305	5.83	7.12	6.40	6.58
4	5.72	0.330	5.19	6.35	5.89	6.05
5	5.26	0.356	4.62	5.65	5.38	5.59
6	4.88	0.406	4.11	5.03	4.88	5.16

(Contd.)

TABLE 8.12 Wire and sheet-metal gauges (continued)

Name of Gauge	Steel Wire Gauge or Washburn and Moen Wire Gauge Steel W.G. W. & M.W.G.	Music Wire Gauge M.W.G.	American Wire Gauge or Brown and Sharpe Gauge A.W.G. B. & S.G.	New Birmingham Standard Sheet and Hoop Gauge B.G.	British Imperial or English Legal Standard Wire Gauge S.W.G.	Birmingham or Stubs' Iron Wire Gauge B.W.G.
Principal use	Steel wire except music wire	Steel music wire	Non-ferrous sheets and wire	Iron and steel sheets and hoops	Wire	Flats, plates and wire
Gauge No.	mm	mm	mm	mm	mm	mm
7	4.50	0.457	3.67	4.48	4.47	4.57
8	4.11	0.508	3.26	3.99	4.06	4.19
9	3.77	0.559	2.91	3.55	3.66	3.76
10	3.43	0.610	2.59	3.18	3.25	3.40
11	3.06	0.660	2.30	2.83	2.95	3.05
12	2.68	0.737	2.05	2.52	2.64	2.77
13	2.32	0.787	1.83	2.24	2.34	2.41
14	2.03	0.838	1.63	1.99	2.03	2.11
15	1.83	0.889	1.45	1.78	1.83	1.83
16	1.59	0.940	1.29	1.59	1.63	1.65
17	1.37	0.991	1.15	1.41	1.42	1.47
18	1.21	1.04	1.02	1.26	1.22	1.24
19	1.04	1.09	0.912	1.12	1.02	1.07
20	0.884	1.14	0.813	0.996	0.914	0.889

(Contd.)

TABLE 8.12 Wire and sheet-metal gauges (continued)

Name of Gauge	Steel Wire Gauge or Washburn and Moen Wire Gauge Steel W.G. W. & M.W.G.	Music Wire Gauge M.W.G.	American Wire Gauge or Brown and Sharpe Gauge A.W.G. B. & S.G.	New Birmingham Standard Sheet and Hoop Gauge B.G.	British Imperial or English Legal Standard Wire Gauge S.W.G.	Birmingham or Stubs' Iron Wire Gauge B.W.G.
Principal use	Steel wire except music wire	Steel music wire	Non-ferrous sheets and wire	Iron and steel sheets and hoops	Wire	Flats, plates and wire
Gauge No.	mm	mm	mm	mm	mm	mm
21	0.805	1.19	0.724	0.886	0.813	0.813
22	0.726	1.24	0.643	0.795	0.711	0.711
23	0.655	1.30	0.574	0.706	0.610	0.635
24	0.584	1.40	0.511	0.630	0.559	0.559
25	0.518	1.50	0.455	0.559	0.508	0.508
26	0.460	1.60	0.404	0.498	0.457	0.457
27	0.439	1.70	0.361	0.444	0.417	0.406
28	0.411	1.80	0.320	0.396	0.376	0.356
29	0.381	1.90	0.287	0.353	0.345	0.330
30	0.356	2.03	0.254	0.312	0.315	0.305
31	0.335	2.16	0.226	0.279	0.295	0.254
32	0.325	2.29	0.203	0.249	0.274	0.229
33	0.300	2.41	0.180	0.221	0.254	0.203
34	0.264	2.54	0.160	0.196	0.234	0.178

(Contd.)

TABLE 8.12 Wire and sheet-metal gauges (continued)

Name of Gauge	Steel Wire Gauge or Washburn and Moen Wire Gauge Steel W.G. W. & M.W.G.	Music Wire Gauge M.W.G.	American Wire Gauge or Brown and Sharpe Gauge A.W.G. B. & S.G.	New Birmingham Standard Sheet and Hoop Gauge B.G.	British Imperial or English Legal Standard Wire Gauge S.W.G.	Birmingham or Stubs' Iron Wire Gauge B.W.G.
Principal use	Steel wire except music wire	Steel music wire	Non-ferrous sheets and wire	Iron and steel sheets and hoops	Wire	Flats, plates and wire
Gauge No.	mm	mm	mm	mm	mm	mm
35	0.241	2.69	0.142	0.175	0.213	0.127
36	0.229	2.84	0.127	0.155	0.193	0.102
37	0.216	3.00	0.114	0.137	0.173	–
38	0.203	3.15	0.102	0.122	0.152	–
39	0.190	3.30	0.0889	0.109	0.132	–
40	0.178	3.51	0.0787	0.0991	0.122	–

Source: References 4 and 5

10 PREFERRED NUMBERS

Preferred numbers are the conventionally rounded off term values of geometrical series, including the integral powers of 10 and having as ratios the following factors:

$$\sqrt[5]{10} \quad \sqrt[10]{10} \quad \sqrt[20]{10} \quad \sqrt[40]{10} \quad \text{and} \quad \sqrt[80]{10}$$

in accordance with Tables 8.13 and 8.14 set out for the 1 to 10 range. The series of preferred numbers being unlimited in both directions, the values of the terms in other decimal ranges are obtained by multiplying the values in the tables by positive or negative integral powers of 10.

TABLE 8.13 Basic series of preferred numbers

Basic series			
R 5[1]	R 10[2]	R 20[3]	R 40[4]
1.00	1.00	1.00	1.00
			1.06
		1.12	1.12
			1.18
	1.25	1.25	1.25
			1.32
		1.40	1.40
			1.50
1.60	1.60	1.60	1.60
			1.70
		1.80	1.80
			1.90
	2.00	2.00	2.00
			2.12
		2.24	2.24
			2.36
2.50	2.50	2.50	2.50
			2.65
		2.80	2.80
			3.00
	3.15	3.15	3.15
			3.35
		3.55	3.55
			3.75
4.00	4.00	4.00	4.00
			4.25
		4.50	4.50
			4.75
	5.00	5.00	5.00
			5.30
		5.60	5.60
			6.00
6.30	6.30	6.30	6.30
			6.70
		7.10	7.10
			7.50
	8.00	8.00	8.00
			8.50
		9.00	9.00
			9.50
10.00	10.00	10.00	10.00

1) First choice.
2) Second choice.
3) Third choice.
4) Fourth choice.

Source: Reference 11

TABLE 8.14 Exceptional R 80 series[1)]

1.00	1.80	3.15	5.60
1.03	1.85	3.25	5.80
1.06	1.90	3.35	6.00
1.09	1.95	3.45	6.15
1.12	2.00	3.55	6.30
1.15	2.06	3.65	6.50
1.18	2.12	3.75	6.70
1.22	2.18	3.87	6.90
1.25	2.24	4.00	7.10
1.28	2.30	4.12	7.30
1.32	2.36	4.25	7.50
1.36	2.43	4.37	7.75
1.40	2.50	4.50	8.00
1.45	2.58	4.62	8.25
1.50	2.65	4.75	8.50
1.55	2.72	4.87	8.75
1.60	2.80	5.00	9.00
1.65	2.90	5.15	9.25
1.70	3.00	5.30	9.50
1.75	3.07	5.45	9.75

1) The terms of the basic series should be given preference over the terms of the R 80 series.

Source: Reference 11

11 SURFACE ROUGHNESS RANGES FOR COMMON PRODUCTION METHODS

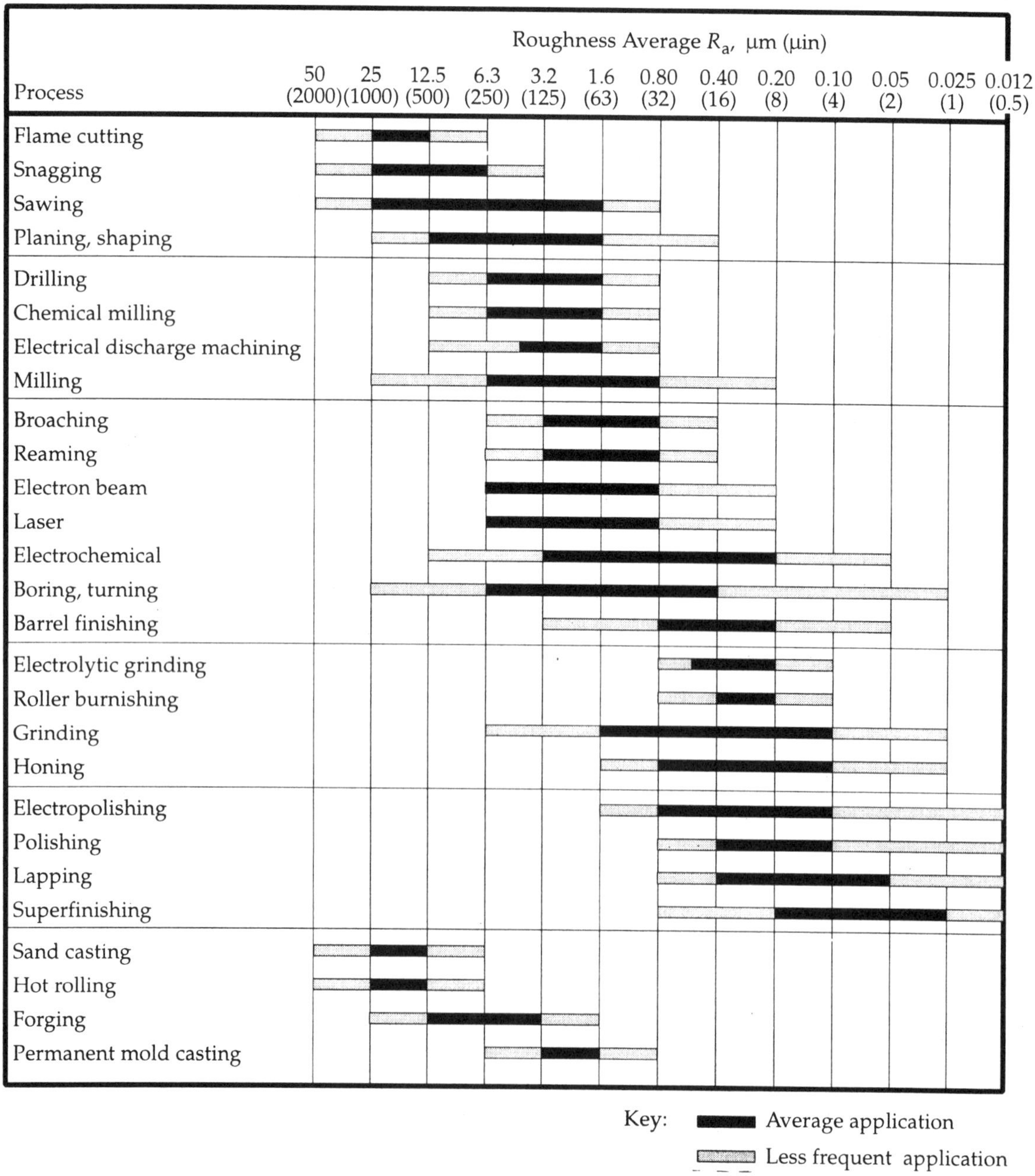

Roughness Average R_a, μm (μin)

Values: 50 (2000), 25 (1000), 12.5 (500), 6.3 (250), 3.2 (125), 1.6 (63), 0.80 (32), 0.40 (16), 0.20 (8), 0.10 (4), 0.05 (2), 0.025 (1), 0.012 (0.5)

Process	50–25	25–12.5	12.5–6.3	6.3–3.2	3.2–1.6	1.6–0.80	0.80–0.40	0.40–0.20	0.20–0.10	0.10–0.05	0.05–0.025	0.025–0.012
Investment casting				□	■	□	□					
Extruding			□	□	■	■	□					
Cold rolling, drawing				□	■	■	□	□				
Die casting					□	■	□					

Key: ■ Average application; □ Less frequent application

Note: The ranges shown above are typical of the processes listed; higher or lower values may be obtained under special conditions.

Source: Reference 9

12 AREAS OF SECTIONS

Triangle

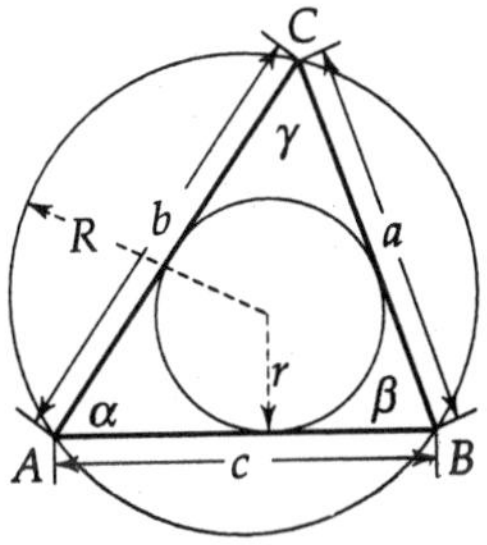

$$\text{Area} = \sqrt{s(s-a)(s-b)(s-c)}$$

$$= rs = \frac{abc}{4R}$$

$$= \frac{ah}{2} = \frac{ab \sin\gamma}{2}$$

$$= \frac{a^2 \sin\beta \sin\gamma}{2\sin\alpha}$$

$$= 2R^2 \sin\alpha \sin\beta \sin\gamma$$

where $2s = a + b + c$

$$\text{Radius of inscribed circle } (r) = \frac{\sqrt{s(s-a)(s-b)(s-c)}}{s}$$

$$\text{Radius of circumscribed circle } (R) = \frac{abc}{4\sqrt{s(s-a)(s-b)(s-c)}}$$

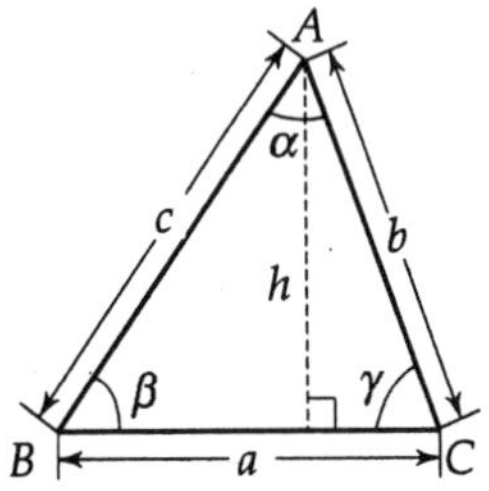

Square

Diagonal $(d) = a\sqrt{2}$

Area $= a^2$

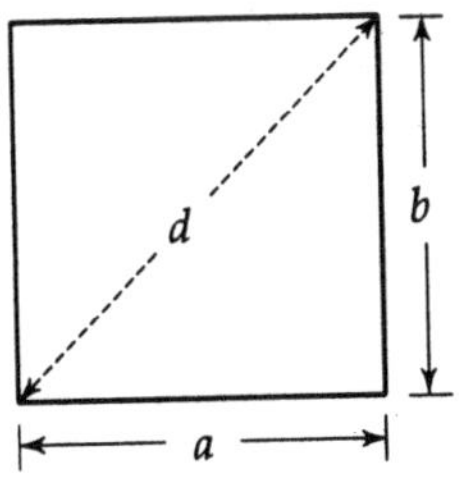

Rectangle

Diagonal $(d) = \sqrt{a^2 + b^2}$

Area $= ab$

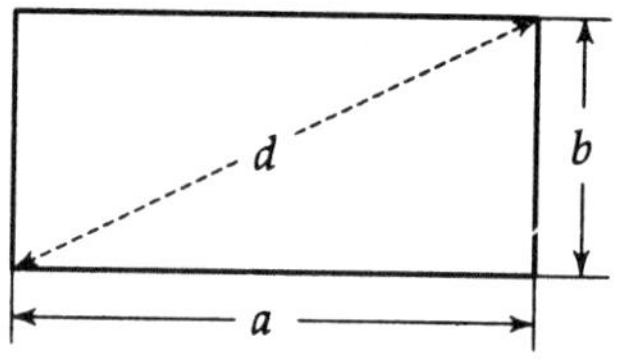

Rhombus

Diagonal $(d_1) = 2a \cos\frac{\alpha}{2}$

Diagonal $(d_2) = 2a\sin\frac{\alpha}{2}$

Area $= ah = a^2\sin\alpha = \frac{d_1 d_2}{2}$

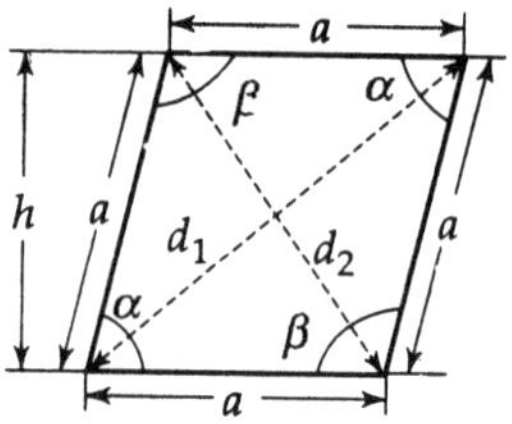

Parallelogram

Diagonal $(d_1) = \sqrt{a^2 + b^2 - 2ab\cos\beta}$

Diagonal (d_2) s= $\sqrt{a^2 + b^2 - 2ab\cos\alpha}$

Area $= ah = ab \sin\alpha$

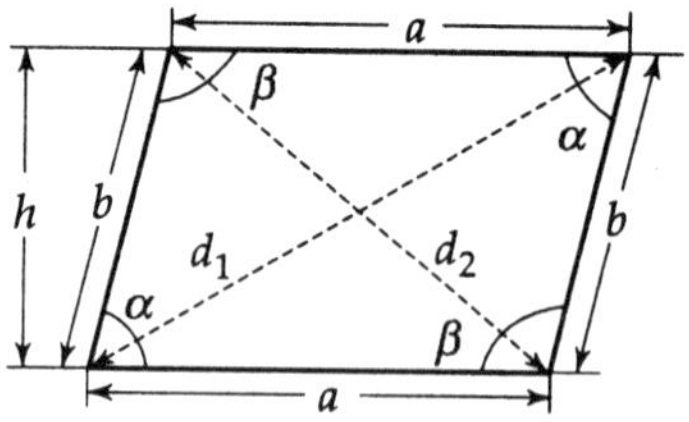

Trapezoid

Area $= \frac{(a + b)\, h}{2}$

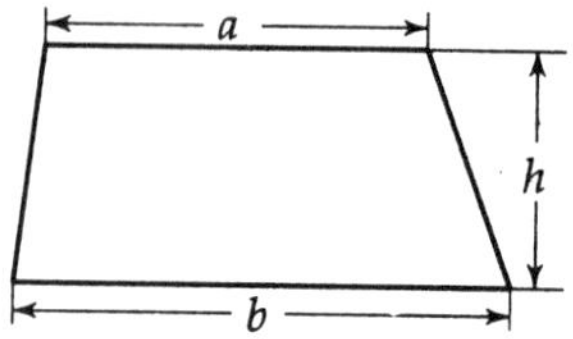

Regular polygon

$$\alpha = \left(\frac{n-2}{n}\right) 180°$$

$$a = 2r \tan\frac{180°}{n} = 2R \sin\frac{180°}{n}$$

$$\text{Area} = \frac{na^2}{4} \cot\frac{180°}{n} = nr^2 \tan\frac{180°}{n}$$

$$= \frac{nR^2}{2} \sin\frac{360°}{n}$$

Radius of inscribed circle (r) $= \frac{a}{2} \cot\frac{180°}{n}$

Radius of circumscribed circle (R) $= \frac{a}{2} \operatorname{cosec}\frac{180°}{n}$

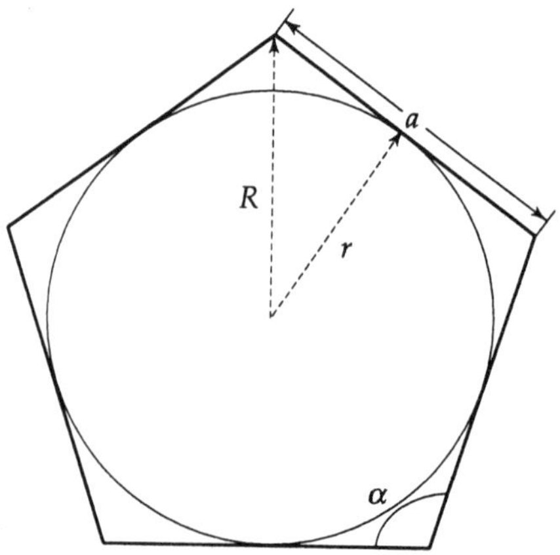

Circle and parts of a circle

Circle

Circumference $= 2\pi r = \pi d$

Area $= \pi r^2 = \dfrac{\pi d^2}{4}$

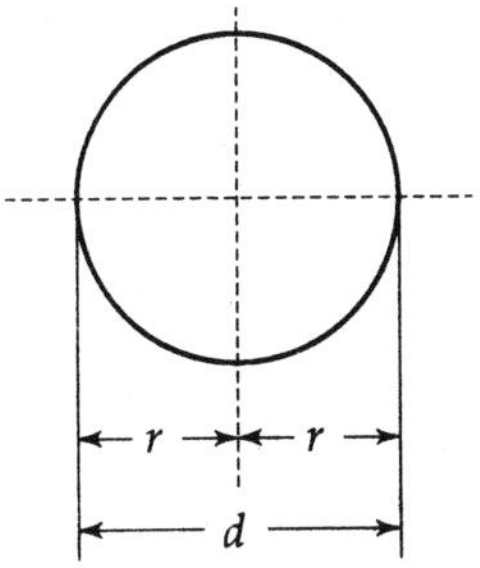

Sector of a circle

Length of arc $(l) = \dfrac{\alpha^\circ}{360^\circ} 2\pi r = \dfrac{\pi r \alpha^\circ}{180^\circ} = r\alpha$

Area $= \dfrac{\alpha^\circ}{360^\circ} \pi r^2 = \dfrac{r^2 \alpha^\circ}{2} = \dfrac{rl}{2}$

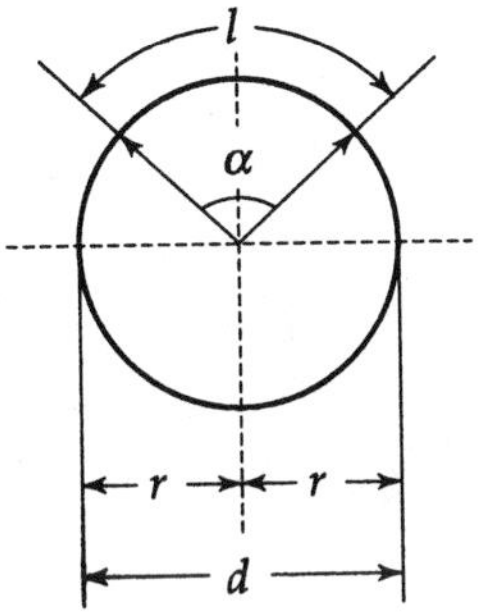

Segment of a circle

Length of chord $(c) = 2r\sin\dfrac{\alpha}{2}$

$h = r\left(1 - \cos\dfrac{\alpha}{2}\right)$

Area $= \dfrac{r^2}{2}\left(\dfrac{\pi\alpha^\circ}{180^\circ} - \sin\alpha\right)$

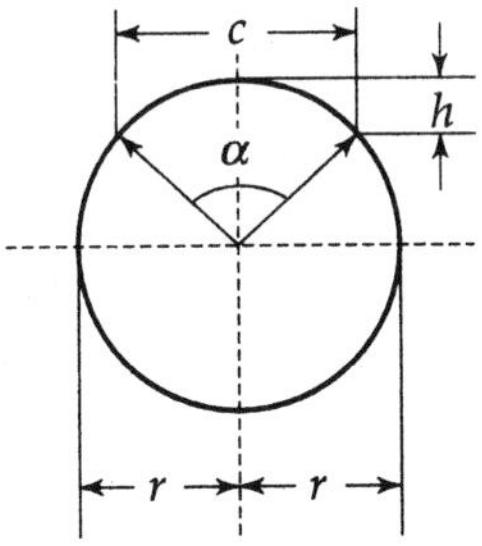

Annulus

Area $= \pi(r_1^2 - r_2^2) = \dfrac{\pi}{4}(d_1^2 - d_2^2) = \dfrac{\pi t}{2}(d_1 + d_2)$

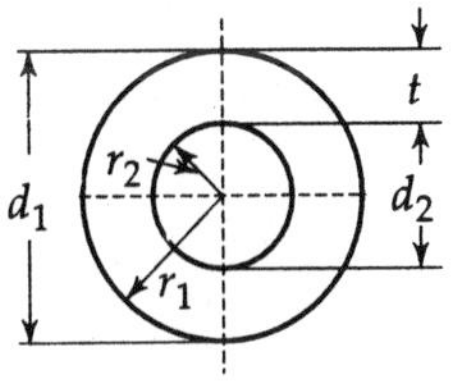

Sector of an annulus

Area $= \dfrac{\alpha}{2}(r_1^2 - r_2^2)$

$= \dfrac{\alpha t}{2}(r_1 + r_2)$

$= \dfrac{t}{2}(l_1 + l_2)$

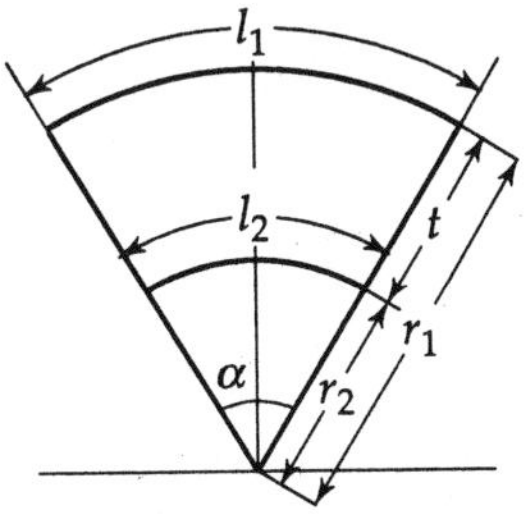

Ellipse

Circumference $\approx 2\pi\sqrt{\dfrac{a^2 + b^2}{2}}$

Area $= \pi ab$

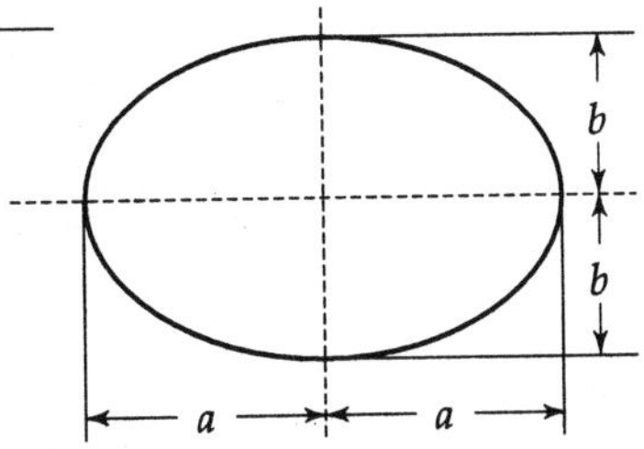

13 AREAS AND VOLUMES OF SOLIDS

Cube

Space-diagonal $(d) = a\sqrt{3}$

Surface area $= 6a^2$

Volume $= a^3$

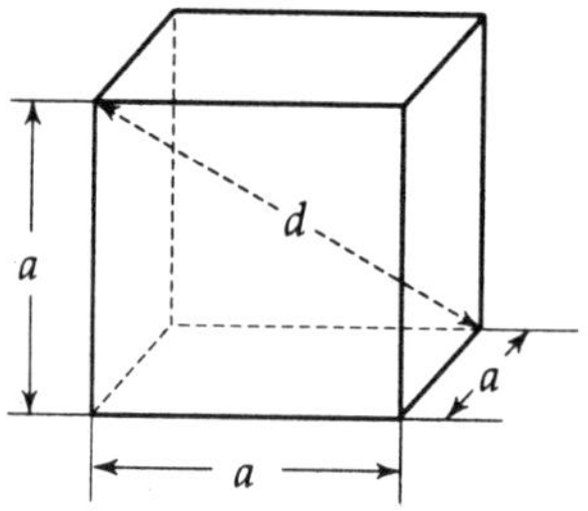

Cuboid

Space-diagonal $d = \sqrt{a^2 + b^2 + c^2}$

Surface area $= 2(ab + bc + ca)$

Volume $= abc$

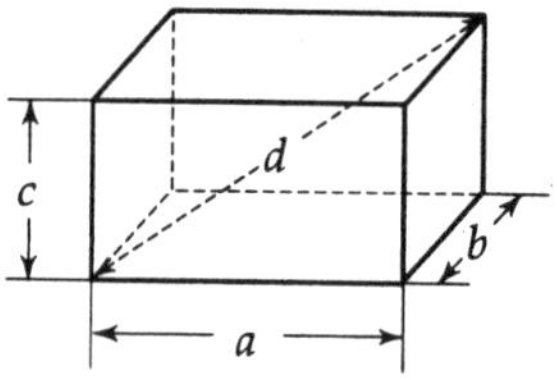

Right pyramid

Lateral area $= a\sqrt{h^2 + (b/2)^2} + b\sqrt{h^2 + (a/2)^2}$

Area of base $= ab$

Volume $= \dfrac{abh}{3}$

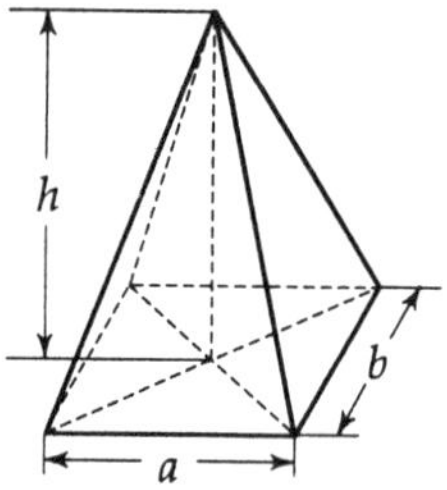

Cylinder

Area of curved surface $= 2\pi rh$

Area of base $= \pi r^2$

Volume $= \pi r^2 h$

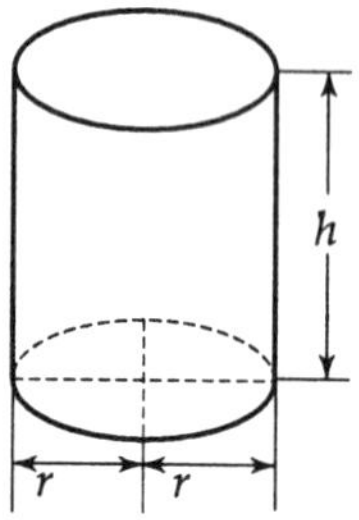

Hollow cylinder

Area of curved surface (outer + inner) $= 2\pi h(r_1 + r_2)$

Area of base $= \pi(r_1^2 - r_2^2)$

Volume $= \pi h(r_1^2 - r_2^2)$

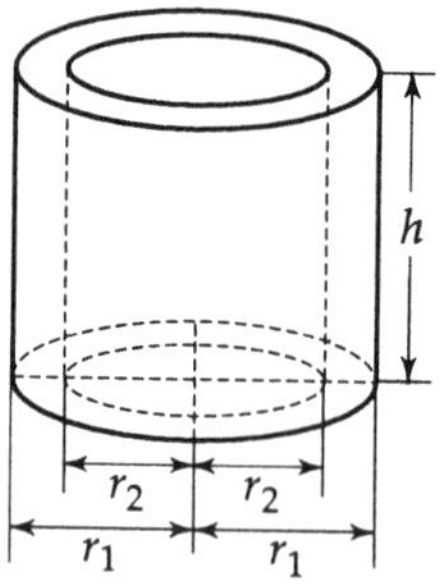

Right circular cone

Area of curved surface $= \pi r\sqrt{r^2 + h^2} = \pi rl$

Area of base $= \pi r^2$

Volume $= \dfrac{\pi r^2 h}{3}$

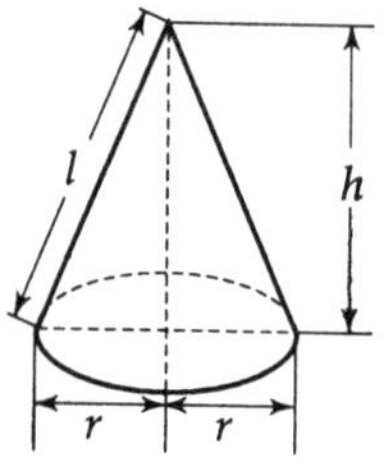

13.7 Sphere and parts of a sphere

Sphere

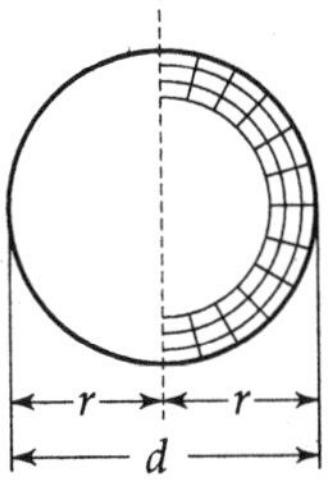

Surface area $= 4\pi r^2 = \pi d^2$

Volume $= \frac{4}{3}\pi r^3 = \frac{\pi d^3}{6}$

Spherical sector

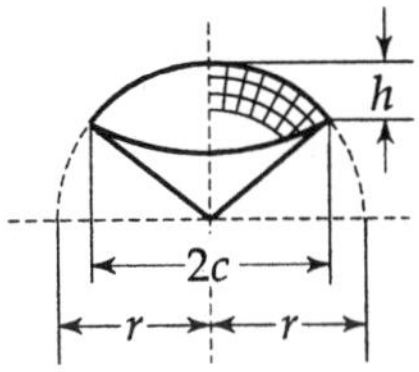

Area of surface $= \pi r(2h + c)$

Volume $= \frac{2\pi r^2 h}{3}$

Spherical segment

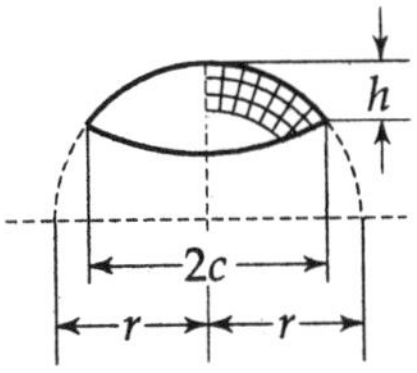

Area of curved surface $= 2\pi rh$

Total surface area $= \pi h(4r - h)$

Volume $= \frac{\pi h^2}{3}(3r - h) = \frac{\pi h}{6}(3c^2 + h^2)$

Spherical zone

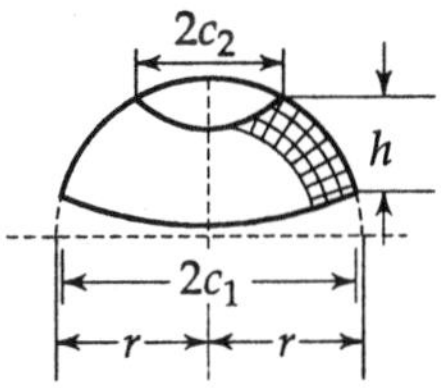

Area of curved surface $= 2\pi rh$

Total surface area $= \pi(2rh + c_1^2 + c_2^2)$

Volume $= \frac{\pi h}{6}(3c_1^2 + 3c_2^2 + h^2)$

Hollow sphere

Volume $= \frac{4\pi}{3}(r_1^3 - r_2^3)$

14 REFERENCES

1. C.J. Smithells (Ed.), *Metals Reference Book*, 5th ed., Butterworths, London, U.K., 1976.
2. H.T. Angus, *Cast Iron: Physical and Engineering Properties*, 2nd ed. Butterworths, London, U.K., 1976.
3. *ASM Handbook*, vol. 13, ASM International, Materials Park, Ohio, U.S.A., 1987.
4. H.E. McGannom (Ed.), *The Making, Shaping and Treating of Steel*, 9th ed., United States Steel, Pittsburgh, Pennsylvania, U.S.A., 1971.

5. *Bethlehem Alloy Steels, Handbook* 301, Bethlehem Steel Company, Bethlehem, Pennsylvania, U.S.A., 1951.
6. J.J. Tuma, *Engineering Mathematics Handbook*, 3rd ed., McGraw-Hill, New York, U.S.A., 1987.
7. W.H. Beyer (Ed.), *CRC Handbook of Mathematical Sciences*, 5th ed., CRC Press, West Palm Beach, Florida, 1978.

Article/Datasheet

8. L.M. Schetky and J. Perkins, "The 'Quiet' Alloys", *Machine Design*, 50(8), 202–206, 1978.
9. "Metal Progress Datasheet: Surface Roughness Averages for Common Production Methods" *Metal Progress*, 118(2), 51, 1980.

Standards

10. IS 1730 – 1989: *Dimensions for Steel Plates, Sheets, Strips and Flats for General Engineering Purposes.*
11. IS 1076 (1) – 1985: Preferred Numbers – Part 1: Series of Preferred Numbers.

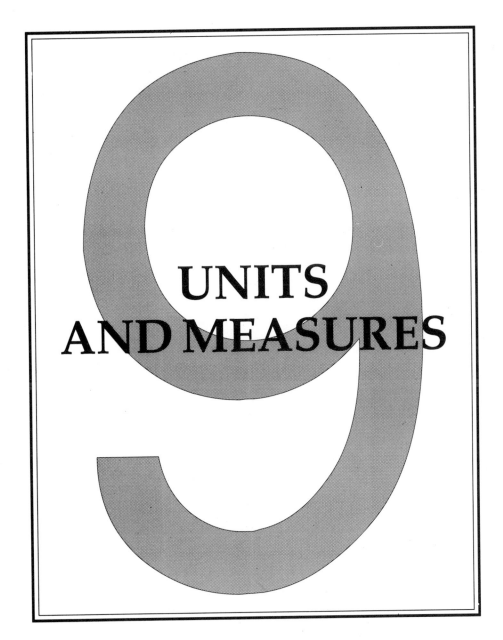

9
UNITS AND MEASURES

CHAPTER **9**

Units and Measures

9	Units and Measures	9.5

9

UNITS AND MEASURES

1 THE INTERNATIONAL SYSTEM OF UNITS (SI)

The International System of Units (SI) is a modernized version of the metric system and is intended as a basis for worldwide standardization of measurement units.

This system includes:

a) Base units, and
b) Derived units including supplementary units,

which together form the coherent system of SI units.

1.1 Base units

The seven base units are listed in Table 9.1.

TABLE 9.1 SI base units

Base quantity	SI base unit	
	Name	Symbol
length	metre	m
mass	kilogram	kg
time	second	s
electric current	ampere	A
thermodynamic temperature	kelvin	K
amount of substance	mole	mol
luminous intensity	candela	cd

1.2 Derived units including supplementary units

Derived units are expressed algebraically in terms of the seven base units, and the supplementary and other derived units in some cases. The symbols for derived units are obtained by means of the mathematical signs of multiplication and division, and the use of exponents. For some of the derived units, special names and symbols exist; those approved by the CGPM are listed in Tables 9.2 and 9.3.

The SI units radian (rad) and steradian (sr) for plane angle and solid angle respectively are classified as "supplementary units". These units are interpreted as dimensionless derived units which may be used or omitted in the expression for derived units.

TABLE 9.2 SI derived units with special names, including SI supplementary units

Derived quantity	SI derived unit		
	Special name	Symbol	Expressed in terms of SI base units and SI derived units
plane angle	radian	rad	1 rad = 1 m/m = 1
solid angle	steradian	sr	1 sr = 1 m^2/m^2 = 1
frequency	hertz	Hz	1 Hz = 1 s^{-1}
force	newton	N	1 N = 1 kg m/s^2
pressure, stress	pascal	Pa	1 Pa = 1 N/m^2
energy, work, quantity of heat	joule	J	1 J = 1 N m
power, radiant flux	watt	W	1 W = 1 J/s
electric charge, quantity of electricity	coulomb	C	1 C = 1 A s
electric potential, potential difference, tension, electromotive force	volt	V	1 V = 1 W/A

(Contd.)

TABLE 9.2 SI derived units with special names, including SI supplementary units (continued)

Derived quantity	SI derived unit		
	Special name	Symbol	Expressed in terms of SI base units and SI derived units
capacitance	farad	F	$1\,F = 1\,C/V$
electric resistance	ohm	Ω	$1\,\Omega = 1\,V/A$
electric conductance	siemens	S	$1\,S = 1\,\Omega^{-1}$
magnetic flux	weber	Wb	$1\,Wb = 1\,V\,s$
magnetic flux density	tesla	T	$1\,T = 1\,Wb/m^2$
inductance	henry	H	$1\,H = 1\,Wb/A$
Celsius temperature	degree Celsius[1)]	°C	1 °C = 1 K
luminous flux	lumen	lm	$1\,lm = 1\,cd\,sr$
illuminance	lux	lx	$1\,lx = 1\,lm/m^2$

1) Degree Celsius is a special name for the unit kelvin for use in stating values of Celsius temperature.

TABLE 9.3 SI derived units with special names admittted for reasons of safeguarding human health

Derived quantity	SI derived unit		
	Special name	Symbol	Expressed in terms of SI base units and SI derived units
activity (of a radionuclide)	becquerel	Bq	$1\,Bq = 1\,s^{-1}$
absorbed dose, specific energy imparted, kerma, absorbed dose index	gray	Gy	$1\,Gy = 1\,J/kg$
dose equivalent, dose equivalent index	sievert	Sv	$1\,Sv = 1\,J/kg$

1.3 SI prefixes

In order to avoid large or small numerical values, decimal multiples and sub-multiples of the SI units are added to the coherent system within the framework of the SI. They are formed by means of the prefixes listed in Table 9.4.

TABLE 9.4 SI prefixes

Factor	Prefix		Factor	Prefix	
	Name	Symbol		Name	Symbol
10^{24}	yotta	Y	10^{-1}	deci	d
10^{21}	zetta	Z	10^{-2}	centi	c
10^{18}	exa	E	10^{-3}	milli	m
10^{15}	peta	P	10^{-6}	micro	µ
10^{12}	tera	T	10^{-9}	nano	n
10^{9}	giga	G	10^{-12}	pico	p
10^{6}	mega	M	10^{-15}	femto	f
10^{3}	kilo	k	10^{-18}	atto	a
10^{2}	hecto	h	10^{-21}	zepto	z
10	deca	da	10^{-24}	yocto	y

2 GUIDE TO THE ROUNDING OF NUMBERS

1. When the figure immediately after the last figure to be retained is less than 5, the last figure to be retained remains unchanged.
2. When the figure immediately after the last figure to be retained is greater than 5, or equal to 5 and followed by at least one figure other than zero, the last figure to be retained is increased by one.
3. When the figure immediately after the last figure to be retained is equal to 5 and followed by zero only, the last figure to be retained:

   a) Rule A: remains unchanged if even and is increased by one if odd; or

   b) Rule B: is increased by one.

Note: Rule A is generally preferred and is of special advantage when treating, for example, series of measurements in such a way that the rounding errors are minimized. Rule B is widely used in computers.

4. Rounding in more than one stage by the application of the rules given above may lead to errors. It is, therefore, recommended always to round in one step.

3 CONVERSION FACTORS AND TABLES

Conversion factors to convert from one unit of measurement to another from the standpoint of the International System of Units (SI) for a number of quantities which are in general use in engineering, industry and trade are covered in Tables 9.5 to 9.41.

Conversion factors are written as a number greater than or equal to one and less than ten with five or fewer decimal places. For numbers greater than or equal to ten the number is followed by a multiplication sign (×) and a powers-of-ten notation.

Note: Where factors are given in bold print it is to show that they are exact.

4 REFERENCES

Books

1. K. Raznjevic, *Handbook of Thermodynamic Tables and Charts*, McGraw-Hill, New York, U.S.A., 1976.
2. P.M. Unterweiser, H.E. Boyer and J.J. Kubbs (Eds.), *Heat Treater's Guide—Standard Practices and Procedures for Steel*, American Society for Metals, Metals Park, Ohio, U.S.A., 1982.
3. G.N.J. Gilbert, *Engineering Data on Grey Cast Irons—SI units*, BCIRA, Alvechurch, Birmingham, U.K., 1977.

Standards

4. IS 10005 – 1994: *Quantities and Units—Part 0: General Principles.*
5. IS 1890 (0) – 1995: *SI Units and Recommendations for the Use of their Multiples and of Certain Other Units.*
6. BS 350 (1) – 1974: *Conversion Factors and Tables—Part 1. Basis of Tables. Conversion Factors.*

3.1 *Length*

TABLE 9.5 Conversion factors for units of length

Unit		metre[1]	millimetre	centimetre	kilometre	inch
Name	Symbol	m	mm	cm	km	in
1 metre[1] =	m	1	1×10^{3}	1×10^{2}	1×10^{-3}	3.93701×10^{1}
1 millimetre =	mm	1×10^{-3}	1	1×10^{-1}	1×10^{-6}	3.93701×10^{-2}
1 centimetre =	cm	1×10^{-2}	1×10^{1}	1	1×10^{-5}	3.93701×10^{-1}
1 kilometre =	km	1×10^{3}	1×10^{6}	1×10^{5}	1	3.93701×10^{4}
1 inch =	in	2.54×10^{-2}	2.54×10^{1}	2.54	2.54×10^{-5}	1
1 foot =	ft	3.048×10^{-1}	3.048×10^{2}	3.048×10^{1}	3.048×10^{-4}	1.2×10^{1}
1 yard =	yd	9.144×10^{-1}	9.144×10^{2}	9.144×10^{1}	9.144×10^{-4}	3.6×10^{1}
1 mile =	–	1.60934×10^{3}	1.60934×10^{6}	1.60934×10^{5}	1.60934	6.336×10^{4}

(Contd.)

TABLE 9.5 Conversion factors for units of length (continued)

Unit			foot	yard	mile
Name		Symbol	ft	yd	
1 metre[1)]	=	m	3.28084	1.09361	6.21371×10^{-4}
1 millimetre	=	mm	3.28084×10^{-3}	1.09361×10^{-3}	6.21371×10^{-7}
1 centimetre	=	cm	3.28084×10^{-2}	1.09361×10^{-2}	6.21371×10^{-6}
1 kilometre	=	km	3.28084×10^{3}	1.09361×10^{3}	6.21371×10^{-1}
1 inch	=	in	8.33333×10^{-2}	2.77778×10^{-2}	1.57828×10^{-5}
1 foot	=	ft	**1**	3.33333×10^{-1}	1.89394×10^{-4}
1 yard	=	yd	**3**	**1**	5.68182×10^{-4}
1 mile	=	–	$\mathbf{5.28 \times 10^{3}}$	$\mathbf{1.76 \times 10^{3}}$	**1**

1) The unit of length in the International System (SI) is the metre [m].
2) 1 International nautical mile = 1 US nautical mile = 1852 m.

TABLE 9.6 Conversion factors for units of length (for precise measurements)

Unit			metre	angstrom	micro-metre	millimetre	microinch	milli-inch[1)]	inch
Name		Symbol	m	Å	μm	mm	μin		in
1 metre	=	m	**1**	**1×10^{10}**	**1×10^{6}**	**1×10^{3}**	3.93701×10^{7}	3.93701×10^{4}	3.93701×10^{1}
1 angstrom	=	Å	**1×10^{-10}**	**1**	**1×10^{-4}**	**1×10^{-7}**	3.93701×10^{-3}	3.93701×10^{-6}	3.93701×10^{-9}
1 micrometre	=	μm	**1×10^{-6}**	**1×10^{4}**	**1**	**1×10^{-3}**	3.93701×10^{1}	3.93701×10^{-2}	3.93701×10^{-5}
1 millimetre	=	mm	**1×10^{-3}**	**1×10^{7}**	**1×10^{3}**	**1**	3.93701×10^{4}	3.93701×10^{1}	3.93701×10^{-2}
1 microinch	=	μin	**2.54×10^{-8}**	**2.54×10^{2}**	**2.54×10^{-2}**	**2.54×10^{-5}**	**1**	**1×10^{-3}**	**1×10^{-6}**
1 milli–inch[1)]	=	–	**2.54×10^{-5}**	**2.54×10^{5}**	**2.54×10^{1}**	**2.54×10^{-2}**	**1×10^{3}**	**1**	**1×10^{-3}**
1 inch	=	in	**2.54×10^{-2}**	**2.54×10^{8}**	**2.54×10^{4}**	**2.54×10^{1}**	**1×10^{6}**	**1×10^{3}**	**1**

[1)] The expressions "mil" or "thou" are sometimes used to denote "milli-inch".

3.2 Area

TABLE 9.7 Conversion factors for units of area

Unit			square metre[1)]	square millimetre	square centimetre	area	hectare
Name		Symbol	m^2	mm^2	cm^2	a	ha
1 square metre[1)]	=	m^2	**1**	**1×10^6**	**1×10^4**	**1×10^{-2}**	**1×10^{-4}**
1 square millimetre	=	mm^2	**1×10^{-6}**	**1**	**1×10^{-2}**	**1×10^{-8}**	**1×10^{-10}**
1 square centimetre	=	cm^2	**1×10^{-4}**	**1×10^2**	**1**	**1×10^{-6}**	**1×10^{-8}**
1 are	=	a	**1×10^2**	**1×10^8**	**1×10^6**	**1**	**1×10^{-2}**
1 hectare	=	ha	**1×10^4**	**1×10^{10}**	**1×10^8**	**1×10^2**	**1**
1 square kilometre	=	km^2	**1×10^6**	**1×10^{12}**	**1×10^{10}**	**1×10^4**	**1×10^2**
1 square inch	=	in^2	**6.4516×10^{-4}**	**6.4516×10^2**	**6.4516**	**6.4516×10^{-6}**	**6.4516×10^{-8}**
1 square foot	=	ft^2	9.29030×10^{-2}	9.29030×10^4	9.29030×10^2	9.29030×10^{-4}	9.29030×10^{-6}
1 square yard	=	yd^2	8.36127×10^{-1}	8.36127×10^5	8.36127×10^3	8.36127×10^{-3}	8.36127×10^{-5}
1 acre	=	–	4.04686×10^3	4.04686×10^9	4.04686×10^7	4.04686×10^1	4.04686×10^{-1}
1 square mile	=	$mile^2$	2.58999×10^6	2.58999×10^{12}	2.58999×10^{10}	2.58999×10^4	2.58999×10^2

(Contd.)

TABLE 9.7 Conversion factors for units of area (continued)

Unit			square kilometre	square inch	square foot	square yard	acre	square mile
Name		Symbol	km^2	in^2	ft^2	yd^2		$mile^2$
1 square metre[1]	=	m^2	**1×10^{-6}**	1.55000×10^{3}	1.07639×10^{1}	1.19599	2.47105×10^{-4}	3.86102×10^{-7}
1 square millimetre	=	mm^2	**1×10^{-12}**	1.55000×10^{-3}	1.07639×10^{-5}	1.19599×10^{-6}	2.47105×10^{-10}	3.86102×10^{-13}
1 square centimetre	=	cm^2	**1×10^{-10}**	1.55000×10^{-1}	1.07639×10^{-3}	1.19599×10^{-4}	2.47105×10^{-8}	3.86102×10^{-11}
1 are	=	a	**1×10^{-4}**	1.55000×10^{5}	1.07639×10^{3}	1.19599×10^{2}	2.47105×10^{-2}	3.86102×10^{-5}
1 hectare	=	ha	**1×10^{-2}**	1.55000×10^{7}	1.07639×10^{5}	1.19599×10^{4}	2.47105	3.86102×10^{-3}
1 square kilometre	=	km^2	**1**	1.55000×10^{9}	1.07639×10^{7}	1.19599×10^{6}	2.47105×10^{2}	3.86102×10^{-1}
1 square inch	=	in^2	**6.4516×10^{-10}**	**1**	6.94444×10^{-3}	7.71605×10^{-4}	1.59423×10^{-7}	2.49098×10^{-10}
1 square foot	=	ft^2	9.29030×10^{-8}	**1.44×10^{2}**	**1**	1.11111×10^{-1}	2.29568×10^{-5}	3.58701×10^{-8}
1 square yard	=	yd^2	8.36127×10^{-7}	**1.296×10^{3}**	**9**	**1**	2.06612×10^{-4}	3.22831×10^{-7}
1 acre	=	–	4.04686×10^{-3}	**6.27264×10^{6}**	**4.356×10^{4}**	**4.84×10^{3}**	**1**	**1.5625×10^{-3}**
1 square mile	=	$mile^2$	2.58999	4.01449×10^{9}	**2.78784×10^{7}**	**3.0976×10^{6}**	**6.4×10^{2}**	**1**

[1] The unit of area in the International System (SI) is the square metre [m^2].

3.3 Volume

TABLE 9.8 Conversion factors for units of volume

Unit			cubic metre[1)]	cubic decimetre, litre	cubic centimetre, millilitre	cubic inch	cubic foot
Name		Symbol	m^3	dm^3 l	cm^3 ml	in^3	ft^3
1 cubic metre[1)]	=	m^3	**1**	**1×10^3**	**1×10^6**	6.10237×10^4	3.53147×10^1
1 cubic decimetre 1 litre	= =	dm^3 l	**1×10^{-3}**	**1**	**1×10^3**	6.10237×10^1	3.53147×10^{-2}
1 cubic centimetre 1 millilitre	= =	cm^3 ml	**1×10^{-6}**	**1×10^{-3}**	**1**	6.10237×10^{-2}	3.53147×10^{-5}
1 cubic inch	=	in^3	1.63871×10^{-5}	1.63871×10^{-2}	1.63871×10^{-1}	**1**	5.78704×10^{-4}
1 cubic foot	=	ft^3	2.83168×10^{-2}	2.83168×10^1	2.83168×10^4	**1.728×10^{-3}**	**1**
1 cubic yard	=	yd^3	7.64555×10^{-1}	7.64555×10^2	7.64555×10^5	**4.6656×10^4**	**2.7×10^1**
1 UK gallon	=	UKgal	4.54609×10^{-3}	4.54609	4.54609×10^3	2.77420×10^2	1.60544×10^{-1}
1 US gallon	=	USgal	3.78541×10^{-3}	3.78541	3.78541×10^3	**2.31×10^2**	1.33681×10^{-1}

(Contd.)

TABLE 9.8 Conversion factors for units of volume (continued)

Unit			cubic yard	UK gallon	US gallon
Name		Symbol	yd^3	UKgal	USgal
1 cubic metre[1]	=	m^3	1.30795	2.19969×10^{2}	2.64172×10^{2}
1 cubic decimetre 1 litre	= =	dm^3 l	1.30795×10^{-3}	2.19969×10^{-1}	2.64172×10^{-1}
1 cubic centimetre 1 millilitre	= =	cm^3 ml	1.30795×10^{-6}	2.19969×10^{-4}	2.64172×10^{-4}
1 cubic inch	=	in^3	2.14335×10^{-5}	3.60465×10^{-3}	4.32900×10^{-3}
1 cubic foot	=	ft^3	3.70370×10^{-2}	6.22883	7.48052
1 cubic yard	=	yd^3	**1**	1.68178×10^{2}	2.01974×10^{2}
1 UK gallon	=	UK gal	5.94607×10^{-3}	1	1.20095
1 US gallon	=	US gal	4.95114×10^{-3}	8.32674×10^{-1}	**1**

1) The unit of volume in the International System (SI) is the cubic metre [m^3].

Notes:
1) 1 UK gallon = 4 UK quart = 8 UK pint = 160 UK fluid ounce.
2) 1 US gallon = 8 US liquid pint = 128 US fluid ounce.
3) 1 barrel (of petroleum) = 42 US gallons = 158.987 dm^3.

3.4 Section modulus

TABLE 9.9 Conversion factors for units of section modulus

Unit			metre cubed[1)]	millimetre cubed	centimetre cubed	inch cubed	foot cubed
Name		Symbol	m^3	mm^3	cm^3	in^3	ft^3
1 metre cubed[1)]	=	m^3	**1**	**1×10^9**	**1×10^6**	6.10237×10^4	3.53147×10^1
1 millimetre cubed	=	mm^3	**1×10^{-9}**	**1**	**1×10^{-3}**	6.10237×10^{-5}	3.53147×10^{-8}
1 centimetre cubed	=	cm^3	**1×10^{-6}**	**1×10^3**	**1**	6.10237×10^{-2}	3.53147×10^{-5}
1 inch cubed	=	in^3	1.63871×10^{-5}	1.63871×10^4	1.63871×10^1	**1**	5.78704×10^{-4}
1 foot cubed	=	ft^3	2.83168×10^{-2}	2.83168×10^7	2.83168×10^4	**1.728×10^3**	**1**

1) The unit of section modulus in the International System (SI) is the metre cubed [m^3].

3.5 Second moment of area

TABLE 9.10 Conversion factors for units of second moment of area

Unit			metre to the fourth power[1)]	centimetre to the fourth power	inch to the fourth power	foot to the fourth power
Name		Symbol	m^4	cm^4	in^4	ft^4
1 metre to the fourth power[1)]	=	m^4	**1**	$\mathbf{1 \times 10^{8}}$	2.40251×10^{6}	1.15862×10^{2}
1 centimetre to the fourth power	=	cm^4	$\mathbf{1 \times 10^{-8}}$	**1**	2.40251×10^{-2}	1.15862×10^{-6}
1 inch to the fourth power	=	in^4	4.16231×10^{-7}	4.16231×10^{1}	**1**	4.82253×10^{-5}
1 foot to the fourth power	=	ft^4	8.63097×10^{-3}	8.63097×10^{5}	$\mathbf{2.0736 \times 10^{4}}$	**1**

1) The unit of second moment of area in the International System (SI) is the metre to the fourth power [m^4].

3.6 *Plane angle*

TABLE 9.11 Conversion factors for units of plane angle

Unit: Name	Unit: Symbol	radian[1)] rad	right angle L	degree °	minute ′	second ″	grade, gon g gon
1 radian[1)] =	rad	1	6.36620×10^{-1}	5.72958×10^{1}	3.43775×10^{3}	2.06265×10^{5}	6.36620×10^{1}
1 right angle =	L	1.57080	**1**	$\mathbf{9 \times 10^{1}}$	$\mathbf{5.4 \times 10^{3}}$	$\mathbf{3.24 \times 10^{5}}$	$\mathbf{1 \times 10^{2}}$
1 degree =	°	1.74533×10^{-2}	1.11111×10^{-2}	**1**	$\mathbf{6 \times 10^{1}}$	$\mathbf{3.6 \times 10^{3}}$	1.11111
1 minute =	′	2.90888×10^{-4}	1.85185×10^{-4}	1.66667×10^{-2}	**1**	$\mathbf{6 \times 10^{1}}$	1.85185×10^{-2}
1 second =	″	4.84814×10^{-6}	3.08642×10^{-6}	2.77778×10^{-4}	1.66667×10^{-2}	**1**	3.08642×10^{-4}
1 grade = 1 gon =	g gon	1.57080×10^{-2}	$\mathbf{1 \times 10^{-2}}$	$\mathbf{9 \times 10^{-1}}$	$\mathbf{5.4 \times 10^{1}}$	$\mathbf{3.24 \times 10^{3}}$	**1**

1) The unit of plane angle in the International System (SI) is the radian [rad].

3.7 Time

TABLE 9.12 Conversion factors for units of time

Unit			second[1)]	minute	hour	day	week
Name		Symbol	s	min	h	d	
1 second	=	s	**1**	–	–	–	–
1 minute	=	min	6×10^1	**1**	–	–	–
1 hour	=	h	3.6×10^3	6×10^1	**1**	–	–
1 day	=	d	8.64×10^4	1.44×10^3	2.4×10^1	**1**	–
1 week	=	–	6.048×10^5	1.008×10^4	1.68×10^2	**7**	**1**

1) The unit of time in the International System (SI) is the second [s].

Notes:
1) 1 month = 28, 29, 30 or 31 days (according to calendar).
2) 1 year = 12 months = 365 or 366 days (according to calendar) = 8760 h or 8784 h (according to calendar).

3.8 *Linear velocity*

TABLE 9.13 Conversion factors for units of linear velocity

Unit: Name		Unit: Symbol	metre per second[1)] m/s	kilometre per hour km/h	inch per second in/s	foot per second ft/s	foot per minute ft/min
1 metre per second[1)]	=	m/s	**–1**	**3.6**	3.93701×10^1	3.28084	1.96850×10^2
1 kilometre per hour	=	km/h	2.77778×10^{-1}	**–1**	1.09361×10^1	9.11344×10^{-1}	5.46807×10^1
1 inch per second	=	in/s	**2.54×10^{-2}**	**9.144×10^{-2}**	**–1**	8.33333×10^{-2}	**5**
1 foot per second	=	ft/s	**3.048×10^{-1}**	**1.09728**	**1.2×10^1**	**–1**	**6×10^1**
1 foot per minute	=	ft/min	**5.08×10^{-3}**	**1.8288×10^{-2}**	**2×10^{-1}**	1.66667×10^{-2}	**–1**
1 mile per hour	=	mile/h	**4.4704×10^{-1}**	1.60934	**1.76×10^1**	1.46667	**8.8×10^1**
1 UK knot	=	–	5.14773×10^{-1}	1.85318	2.02667×10^1	1.68889	1.01333×10^2
1 international knot	=	kn	5.14444×10^{-1}	**1.852**	2.02537×10^1	1.68781	1.01269×10^2

(Contd.)

TABLE 9.13 Conversion factors for units of linear velocity (continued)

Unit			mile per hour	UK knot	international knot
Name		Symbol	mile/h		kn
1 metre per second[1)]	=	m/s	2.23694	1.94260	1.94384
1 kilometre per hour	=	km/h	6.21371×10^{-1}	5.39612×10^{-1}	5.39957×10^{-1}
1 inch per second	=	in/s	5.68182×10^{-2}	4.93421×10^{-2}	4.93737×10^{-2}
1 foot per second	=	ft/s	6.81818×10^{-1}	5.92105×10^{-1}	5.92484×10^{-1}
1 foot per minute	=	ft/min	1.13636×10^{-2}	9.86842×10^{-3}	9.87473×10^{-3}
1 mile per hour	=	mile/h	**1**	8.68421×10^{-1}	8.68976×10^{-1}
1 UK knot	=	–	1.15152	**1**	1.00064
1 international knot	=	kn	1.15078	9.99361×10^{-1}	**1**

1) The unit of linear velocity in the International System (SI) is the metre per second [m/s].

3.9 Angular velocity

TABLE 9.14 Conversion factors for units of angular velocity

Unit: Name		Unit: Symbol	radian per second[1)]	radian per minute	revolution per second	revolution per minute	degree per second	degree per minute
			rad/s	rad/min	rev/s	rev/min	°/s	°/min
1 radian per second[1)]	=	rad/s	**1**	**6×10^1**	1.59155×10^{-1}	9.54930	5.72958×10^1	3.43775×10^3
1 radian per minute	=	rad/min	1.66667×10^{-2}	**1**	2.65258×10^{-3}	1.59155×10^{-1}	9.54930×10^{-1}	5.72958×10^1
1 revolution per second	=	rev/s	6.28319	3.76991×10^2	**1**	**6×10^1**	**3.6×10^2**	**2.16×10^4**
1 revolution per minute	=	rev/min	1.04720×10^{-1}	6.28319	1.66667×10^{-2}	**1**	**6**	**3.6×10^2**
1 degree per second	=	°/s	1.74533×10^{-2}	1.04720	2.77778×10^{-3}	1.66667×10^{-1}	**1**	**6×10^1**
1 degree per minute	=	°/min	2.90888×10^{-4}	1.74533×10^{-2}	4.62963×10^{-5}	2.77778×10^{-3}	1.66667×10^{-2}	**1**

1) The unit of angular velocity in the International System (SI) is the radian per second [rad/s].

3.10 Acceleration

TABLE 9.15 Conversion factors for units of acceleration

Unit: Name		Unit: Symbol	metre per second squared[1)] m/s^2	foot per second squared ft/s^2	standard acceleration due to gravity g_n
1 metre per second squared[1)]	=	m/s^2	**1**	3.28084	1.01972×10^{-1}
1 foot per second squared	=	ft/s^2	$\mathbf{3.048 \times 10^{-1}}$	**1**	3.10810×10^{-2}
standard acceleration due to gravity	=	g_n	**9.80665**	3.21740×10^{1}	**1**

1) The unit of acceleration in the International System (SI) is the metre per second squared [m/s^2].

9.11 Mass

TABLE 9.16 Conversion factors for units of mass

Unit			kilogram[1)]	milligram	gram	tonne, megagram	ounce
Name		Symbol	kg	mg	g	t Mg	oz
1 kilogram[1)]	=	kg	**1**	**1×10^{6}**	**1×10^{3}**	**1×10^{-3}**	3.52740×10^{1}
1 milligram	=	mg	**1×10^{-6}**	**1**	**1×10^{-3}**	**1×10^{-9}**	3.52740×10^{-5}
1 gram	=	g	**1×10^{-3}**	**1×10^{3}**	**1**	**1×10^{-6}**	3.52740×10^{-2}
1 tonne 1 megagram	= =	t Mg	**1×10^{3}**	**1×10^{9}**	**1×10^{6}**	**1**	3.52740×10^{4}
1 ounce	=	oz	2.83495×10^{-2}	2.83495×10^{4}	2.83495×10^{1}	2.83495×10^{-5}	**1**
1 pound	=	lb	4.53592×10^{-1}	4.53592×10^{5}	4.53592×10^{2}	4.53592×10^{-4}	**1.6×10^{1}**
1 UK ton	=	ton	1.01605×10^{3}	1.01605×10^{9}	1.01605×10^{6}	1.01605	**3.584×10^{4}**
1 short ton[2)]	=	sh ton	9.07185×10^{2}	9.07185×10^{8}	9.07185×10^{5}	9.07185×10^{-1}	**3.2×10^{4}**

(Contd.)

TABLE 9.16 Conversion factors for units of mass (continued)

Unit			pound	UK ton	short ton[2]
Name		Symbol	lb	ton	sh ton
1 kilogram[1]	=	kg	2.20462	9.84207×10^{-4}	1.10231×10^{-3}
1 milligram	=	mg	2.20462×10^{-6}	9.84207×10^{-10}	1.10231×10^{-9}
1 gram	=	g	2.20462×10^{-3}	9.84207×10^{-7}	1.10231×10^{-6}
1 tonne 1 megagram	= =	t Mg	2.20462×10^{3}	9.84207×10^{-1}	1.10231
1 ounce	=	oz	$\mathbf{6.25 \times 10^{-2}}$	2.79017×10^{-5}	$\mathbf{3.125 \times 10^{-5}}$
1pound	=	lb	**1**	4.46429×10^{-4}	$\mathbf{5 \times 10^{-4}}$
1 UK ton	=	ton	$\mathbf{2.24 \times 10^{3}}$	**1**	**1.12**
1 short ton[2]	=	sh ton	$\mathbf{2 \times 10^{3}}$	8.92857×10^{-1}	**1**

1) The unit of mass in the International System (SI) is the kilogram [kg].
2) U.S. unit.

3.12 *Linear density*

TABLE 9.17 Conversion factors for units of linear density

Unit			kilogram per metre[1)]	tonne per metre	pound per inch	pound per foot	pound per yard
Name		Symbol	kg/m	t/m	lb/in	lb/ft	lb/yd
1 kilogram per metre[1)]	=	kg/m	**1**	**1×10^{-3}**	5.59974×10^{-2}	6.71969×10^{-1}	2.01591
1 tonne per metre	=	t/m	**1×10^{3}**	**1**	5.59974×10^{1}	6.71969×10^{2}	2.01591×10^{3}
1 pound per inch	=	lb/in	1.78580×10^{1}	1.78580×10^{-2}	**1**	**1.2×10^{1}**	**3.6×10^{1}**
1 pound per foot	=	lb/ft	1.48816	1.48816×10^{-3}	8.33333×10^{-2}	**1**	**3**
1 pound per yard	=	lb/yd	4.96055×10^{-1}	4.96055×10^{-4}	2.77778×10^{-2}	3.33333×10^{-1}	**1**
1 pound per mile	=	lb/mile	2.81849×10^{-4}	2.81849×10^{-7}	1.57828×10^{-5}	1.89394×10^{-4}	5.68182×10^{-4}
1 UK ton per 1000 yards	=	ton/1000 yd	1.11116	1.11116×10^{-3}	6.22222×10^{-2}	7.46667×10^{-1}	**2.24**
1 UK ton per mile	=	ton/mile	6.31342×10^{-1}	6.31342×10^{-4}	3.53535×10^{-2}	4.24242×10^{-1}	1.27273

TABLE 9.17 Conversion factors for units of linear density (continued)

Unit			pound per mile	UK ton per 1000 yards	UK ton per mile
Name		Symbol	lb/mile	ton/1000 yd	ton/mile
1 kilogram per metre[1)]	=	kg/m	3.54800×10^{3}	8.99958×10^{-1}	1.58393
1 tonne per metre	=	t/m	3.54800×10^{6}	8.99958×10^{2}	1.58393×10^{3}
1 pound per inch	=	lb/in	**6.336×10^{4}**	1.60714×10^{1}	2.82857×10^{1}
1 pound per foot	=	lb/ft	**5.28×10^{3}**	1.33929	2.35714
1 pound per yard	=	lb/yd	**1.76×10^{3}**	4.46429×10^{-1}	7.85714×10^{-1}
1 pound per mile	=	lb/mile	**1**	2.53653×10^{-4}	4.46429×10^{-4}
1 UK ton per 1000 yards	=	ton/1000 yd	**3.9424×10^{3}**	**1**	**1.76**
1 UK ton per mile	=	ton/mile	**2.24×10^{3}**	5.68182×10^{-1}	**1**

1) The unit of linear density in the International System (SI) is the kilogram per metre [kg/m].

3.13 *Surface density*

TABLE 9.18 Conversion factors for units of surface density

Unit: Name		Symbol	kilogram per square metre[1)] kg/m^2	kilogram per hectare kg/ha	gram per square metre g/m^2	milligram per square centimetre mg/cm^2	pound per thousand square feet $lb/1000\,ft^2$
1 kilogram per square metre[1)]	=	kg/m^2	**1**	**1×10^4**	**1×10^3**	**1×10^2**	2.04816×10^2
1 kilogram per hectare	=	kg/ha	**1×10^{-4}**	**1**	**1×10^{-1}**	**1×10^{-2}**	2.04816×10^{-2}
1 gram per square metre	=	g/m^2	**1×10^{-3}**	**1×10^1**	**1**	**1×10^{-1}**	2.04816×10^{-1}
1 milligram per square centimetre	=	mg/cm^2	**1×10^{-2}**	**1×10^2**	**1×10^1**	**1**	2.04816
1 pound per thousand square feet	=	$lb/1000\,ft^2$	4.88243×10^{-3}	4.88243×10^1	4.88243	4.88243×10^{-1}	**1**
1 ounce per square yard	=	oz/yd^2	3.39057×10^{-2}	3.39057×10^2	3.39057×10^1	3.39057	6.94444
1 ounce per square foot	=	oz/ft^2	3.05152×10^{-1}	3.05152×10^3	3.05152×10^2	3.05152×10^1	**6.25×10^1**
1 pound per acre	=	lb/acre	1.12085×10^{-4}	1.12085	1.12085×10^{-1}	1.12085×10^{-2}	2.29568×10^{-2}
1 UK ton per square mile	=	$ton/mile^2$	3.92298×10^{-4}	3.92298	3.92298×10^{-1}	3.92298×10^{-2}	8.03489×10^{-2}

(Contd.)

TABLE 9.18 Conversion factors for units of surface density (continued)

Unit			ounce per square yard	ounce per square foot	pound per acre	UK ton per square mile
Name		Symbol	oz/yd^2	oz/ft^2	lb/acre	$ton/mile^2$
1 kilogram per square metre[1)]	=	kg/m^2	2.94935×10^{1}	3.27706	8.92179×10^{3}	2.54908×10^{3}
1 kilogram per hectare	=	kg/ha	2.94935×10^{-3}	3.27706×10^{-4}	8.92179×10^{-1}	2.54908×10^{-1}
1 gram per square metre	=	g/m^2	2.94935×10^{-2}	3.27706×10^{-3}	8.92179	2.54908
1 milligram per square centimetre	=	mg/cm^2	2.94935×10^{-1}	3.27706×10^{-2}	8.92179×10^{1}	2.54908×10^{1}
1 pound per thousand square feet	=	$lb/1000\ ft^2$	**1.44×10^{-1}**	**1.6×10^{-2}**	**4.356×10^{1}**	1.24457×10^{1}
1 ounce per square yard	=	oz/yd^2	**1**	1.11111×10^{-1}	**3.025×10^{2}**	8.64286×10^{1}
1 ounce per square foot	=	oz/ft^2	9	**1**	**2.7225×10^{3}**	7.77857×10^{2}
1 pound per acre	=	lb/acre	3.30579×10^{-3}	3.67309×10^{-4}	**1**	2.85714×10^{-1}
1 UK ton per square mile	=	$ton/mile^2$	1.15702×10^{-2}	1.28558×10^{-3}	**3.5**	**1**

1) The unit of surface density in the International System (SI) is the kilogram per square metre [kg/m^2].

3.14 Density

TABLE 9.19 Conversion factors for units of density

Unit: Name		Symbol	kilogram per cubic metre[1)] kg/m^3	kilogram per cubic decimetre, kilogram per litre kg/dm^3 kg/l	gram per cubic centimetre, gram per millilitre g/cm^3 g/ml	tonne per cubic metre, megagram per cubic metre t/m^3 Mg/m^3	pound per cubic inch lb/in^3
1 kilogram per cubic metre[1)]	=	kg/m^3	**1**	$\mathbf{1 \times 10^{-3}}$	$\mathbf{1 \times 10^{-3}}$	$\mathbf{1 \times 10^{-3}}$	3.61273×10^{-5}
1 kilogram per cubic decimetre 1 kilogram per litre	= =	kg/dm^3 kg/l	$\mathbf{1 \times 10^{3}}$	**1**	**1**	**1**	3.61273×10^{-2}
1 gram per cubic centimetre 1 gram per millilitre	= =	g/cm^3 g/ml	$\mathbf{1 \times 10^{3}}$	**1**	**1**	**1**	3.61273×10^{-2}
1 tonne per cubic metre 1 megagram per cubic metre	= =	t/m^3 Mg/m^3	$\mathbf{1 \times 10^{3}}$	**1**	**1**	**1**	3.61273×10^{-2}
1 pound per cubic inch	=	lb/in^3	2.76799×10^{4}	2.76799×10^{1}	2.76799×10^{1}	2.76799×10^{1}	**1**
1 pound per cubic foot	=	lb/ft^3	1.60185×10^{1}	1.60185×10^{-2}	1.60185×10^{-2}	1.60185×10^{-2}	5.78704×10^{-4}
1 UK ton per cubic yard	=	$UK ton/yd^3$	1.32894×10^{3}	1.32894	1.32894	1.32894	4.80110×10^{-2}
1 pound per UK gallon	=	lb/UKgal	9.97763×10^{1}	9.97763×10^{-2}	9.97763×10^{-2}	9.97763×10^{-2}	3.60465×10^{-3}
1 pound per US gallon	=	lb/USgal	1.19826×10^{2}	1.19826×10^{-1}	1.19826×10^{-1}	1.19826×10^{-1}	4.32900×10^{-3}

(Contd.)

TABLE 9.19 Conversion factors for units of density (continued)

Unit: Name		Symbol	pound per cubic foot lb/ft^3	UK ton per cubic yard $UKton/yd^3$	pound per UK gallon $lb/UKgal$	pound per US gallon $lb/USgal$
1 kilogram per cubic metre[1)]	=	kg/m^3	6.24280×10^{-2}	7.52480×10^{-4}	1.00224×10^{-2}	8.34540×10^{-3}
1 kilogram per cubic decimetre 1 kilogram per litre	= =	kg/dm^3 kg/l	6.24280×10^{1}	7.52480×10^{-1}	1.00224×10^{1}	8.34540
1 gram per cubic centimetre 1 gram per millilitre	= =	g/cm^3 g/ml	6.24280×10^{1}	7.52480×10^{-1}	1.00224×10^{1}	8.34540
1 tonne per cubic metre 1 megagram per cubic metre	= =	t/m^3 Mg/m^3	6.24280×10^{1}	7.52480×10^{-1}	1.00224×10^{1}	8.34540
1 pound per cubic inch	=	lb/in^3	$\mathbf{1.728 \times 10^{3}}$	2.08286×10^{1}	2.77420×10^{2}	$\mathbf{2.31 \times 10^{2}}$
1 pound per cubic foot	=	lb/ft^3	**1**	1.20536×10^{-2}	1.60544×10^{-1}	1.33681×10^{-1}
1 UK ton per cubic yard	=	$UK ton/yd^3$	8.29630×10^{1}	**1**	1.33192×10^{1}	1.10905×10^{1}
1 pound per UK gallon	=	$lb/UKgal$	6.22883	7.50797×10^{-2}	**1**	8.32674×10^{-1}
1 pound per US gallon	=	$lb/USgal$	7.48052	9.01670×10^{-2}	1.20095	**1**

1) The unit of density in the International System (SI) is the kilogram per cubic metre [kg/m^3].

3.15 Moment of inertia

TABLE 9.20 Conversion factors for units of moment of inertia

Unit: Name		Unit: Symbol	kilogram metre squared[1)] $kg\,m^2$	kilogram centimetre squared $kg\,cm^2$	gram centimetre squared $g\,cm^2$	pound foot squared $lb\,ft^2$	pound inch squared $lb\,in^2$	ounce inch squared $oz\,in^2$
1 kilogram metre squared[1)]	=	$kg\,m^2$	**1**	$\mathbf{1 \times 10^{-4}}$	$\mathbf{1 \times 10^{7}}$	2.37304×10^{1}	3.41717×10^{3}	5.46748×10^{4}
1 kilogram centimetre squared	=	$kg\,cm^2$	$\mathbf{1 \times 10^{-4}}$	1	$\mathbf{1 \times 10^{3}}$	2.37304×10^{-3}	3.41717×10^{-1}	5.46748
1 gram centimetre squared	=	$g\,cm^2$	$\mathbf{1 \times 10^{-7}}$	$\mathbf{1 \times 10^{-3}}$	1	2.37304×10^{-6}	3.41717×10^{-4}	5.46748×10^{-3}
1 pound foot squared	=	$lb\,ft^2$	4.21401×10^{-2}	4.21401×10^{2}	4.21401×10^{5}	**1**	$\mathbf{1.44 \times 10^{2}}$	$\mathbf{2.304 \times 10^{3}}$
1 pound inch squared	=	$lb\,in^2$	2.92640×10^{-4}	2.92640	2.92640×10^{3}	6.94444×10^{-3}	**1**	$\mathbf{1.6 \times 10^{1}}$
1 ounce inch squared	=	$oz\,in^2$	1.82900×10^{-5}	1.82900×10^{-1}	1.82900×10^{2}	4.34028×10^{-4}	$\mathbf{6.25 \times 10^{-2}}$	**1**

1) The unit of moment of inertia in the International System (SI) is the kilogram metre squared [$kg\,m^2$].

3.16 Linear momentum

TABLE 9.21 Conversion factors for units of linear momentum

Unit: Name		Unit: Symbol	kilogram metre per second[1] kg m/s	pound foot per second lb ft/s
1 kilogram metre per second[1)]	=	kg m/s	**1**	7.23301
1 pound foot per second	=	lb ft/s	1.38255×10^{-1}	**1**

1) The unit of linear momentum in the International System (SI) is the kilogram metre per second [kg m/s].

3.17 Angular momentum

TABLE 9.22 Conversion factors for units of angular momentum

Unit: Name		Unit: Symbol	kilogram metre squared per second[1)] $kg\,m^2/s$	pound foot squared per second $lb\,ft^2/s$
1 kilogram metre squared per second[1)]	=	$kg\,m^2/s$	**1**	2.37304×10^{1}
1 pound foot squared per second	=	$lb\,ft^2/s$	4.21401×10^{-2}	**1**

1) The unit of angular momentum in the International System (SI) is the kilogram metre squared per second [$kg\,m^2/s$].

3.18 Force

TABLE 9.23 Conversion factors for units of force

Unit			newton[1]	kilonewton	meganewton	dyne	kilogram-force[2]	poundal
Name		Symbol	N	kN	MN	dyn	kgf	pdl
1 newton[1]	=	N	**1**	**1×10^{-3}**	**1×10^{-6}**	**1×10^{5}**	1.01972×10^{-1}	7.23301
1 kilonewton	=	kN	**1×10^{3}**	**1**	**1×10^{-3}**	**1×10^{8}**	1.01972×10^{2}	7.23301×10^{3}
1 meganewton	=	MN	**1×10^{6}**	**1×10^{3}**	**1**	**1×10^{11}**	1.01972×10^{5}	7.23301×10^{6}
1 dyne	=	dyn	**1×10^{-5}**	**1×10^{-8}**	**1×10^{-11}**	**1**	1.01972×10^{-6}	7.23301×10^{-5}
1 kilogram-force[2]	=	kgf	**9.80665**	**9.80665×10^{-3}**	**9.80665×10^{-6}**	**9.80665×10^{5}**	**1**	7.09316×10^{1}
1 poundal	=	pdl	1.38255×10^{-1}	1.38255×10^{-4}	1.38255×10^{-7}	1.38255×10^{4}	1.40981×10^{-2}	**1**
1 pound-force	=	lbf	4.44822	4.44822×10^{-3}	4.44822×10^{-6}	4.44822×10^{5}	4.53592×10^{-1}	3.21740×10^{1}
1 UK ton-force	=	tonf	9.96402×10^{3}	9.96402	9.96042×10^{-3}	9.96402×10^{8}	1.01605×10^{3}	7.20699×10^{4}
1 ounce-force	=	ozf	2.78014×10^{-1}	2.78014×10^{-4}	2.78014×10^{-7}	2.78014×10^{4}	2.83495×10^{-2}	2.01088

(Contd.)

TABLE 9.23 Conversion factors for units of force (continued)

Unit			pound-force	UK ton-force	ounce-force
Name		Symbol	lbf	tonf	ozf
1 newton[1)]	=	N	2.24809×10^{-1}	1.00361×10^{-4}	3.59694
1 kilonewton	=	kN	2.24809×10^{2}	1.00361×10^{-1}	3.59694×10^{3}
1 meganewton	=	MN	2.24809×10^{5}	1.00361×10^{2}	3.59694×10^{6}
1 dyne	=	dyn	2.24809×10^{-6}	1.00361×10^{-9}	3.59694×10^{-5}
1 kilogram-force[2)]	=	kgf	2.20462	9.84207×10^{-4}	3.52740×10^{1}
1 poundal	=	pdl	3.10810×10^{-2}	1.38754×10^{-5}	4.97295×10^{-1}
1 pound-force	=	lbf	**1**	4.46429×10^{-4}	$\mathbf{1.6 \times 10^{1}}$
1 UK ton-force	=	tonf	$\mathbf{2.24 \times 10^{3}}$	**1**	$\mathbf{3.584 \times 10^{4}}$
1 ounce-force	=	ozf	$\mathbf{6.25 \times 10^{-2}}$	2.79018×10^{-5}	**1**

1) The unit of force in the International System (SI) is the newton [N].
2) The kilogram-force (kgf) is known as the kilopond (kp) in Germany.

Notes:
1) 1 kip = 1000 lbf.
2) 1 US ton-force = 2000 lbf.

3.19 Moment of force, or torque

TABLE 9.24 Conversion factors for units of moment of force, or torque

Unit Name		Symbol	newton metre[1)] N m	dyne centimetre dyn cm	kilogram-force metre[2)] kgf m	poundal foot pdl ft	pound-force foot lbf ft
1 newton metre[1)]	=	N m	**1**	**1×10^7**	1.01972×10^{-1}	2.37304×10^1	7.37562×10^{-1}
1 dyne centimetre	=	dyn cm	**1×10^{-7}**	**1**	1.01972×10^{-8}	2.37304×10^{-6}	7.37562×10^{-8}
1 kilogram-force[2)] metre	=	kgf m	**9.80665**	**9.80665×10^7**	**1**	2.32715×10^2	7.23301
1 poundal foot	=	pdl ft	4.21401×10^{-2}	4.21401×10^5	4.29710×10^{-3}	**1**	3.10810×10^{-2}
1 pound-force foot	=	lbf ft	1.35582	1.35582×10^7	1.38255×10^{-1}	3.21740×10^1	**1**
1 pound-force inch	=	lbf in	1.12985×10^{-1}	1.12985×10^6	1.15212×10^{-2}	2.68117	$8..33333 \times 10^{-2}$
1 UK ton-force foot	=	tonf ft	3.03703×10^3	3.03703×10^{10}	3.09691×10^2	7.20699×10^4	**2.24×10^3**
1 ounce-force inch	=	ozf in	7.06155×10^{-3}	7.06155×10^4	7.20078×10^{-4}	1.67573×10^{-1}	5.20833×10^{-3}

(Contd.)

TABLE 9.24 Conversion factors for units of moment of force, or torque (continued)

Unit			pound-force inch	UK ton-force foot	ounce-force inch
Name		Symbol	lbf in	tonf ft	ozf in
1 newton metre[1)]	=	N m	8.85075	3.29269×10^{-4}	1.41612×10^{2}
1 dyne centimetre	=	dyn cm	8.85075×10^{-7}	3.29269×10^{-11}	1.41612×10^{-5}
1 kilogram-force[2)] metre	=	kgf m	8.67962×10^{1}	3.22902×10^{-3}	1.38874×10^{3}
1 poundal foot	=	pdl ft	3.72971×10^{-1}	1.38754×10^{-5}	5.96754
1 pound-force foot	=	lbf ft	$\mathbf{1.2 \times 10^{1}}$	4.46429×10^{-4}	$\mathbf{1.92 \times 10^{2}}$
1 pound-force inch	=	lbf in	**1**	3.72024×10^{-5}	$\mathbf{1.6 \times 10^{1}}$
1 UK ton-force foot	=	tonf ft	$\mathbf{2.688 \times 10^{4}}$	**1**	$\mathbf{4.3008 \times 10^{5}}$
1 ounce-force inch	=	ozf in	$\mathbf{6.25 \times 10^{-2}}$	2.32515×10^{-6}	**1**

1) The unit of moment of force, or torque in the International System (SI) is the newton metre [N m].
2) The kilogram-force is called the kilopond (kp) in Germany.

3.20 Pressure and stress

TABLE 9.25 (a) Conversion factors for units of pressure and stress

Unit			pascal[1)], newton per square metre	megapascal, newton per square millimetre	bar
Name		Symbol	Pa N/m^2	MPa N/mm^2	bar
1 pascal[1)] 1 newton per square metre	= =	Pa N/m^2	**1**	**1× 10^{-6}**	**1× 10^{-5}**
1 megapascal 1 newton per square millimetre	= =	MPa N/mm^2	**1× 10^{-6}**	**1**	**1× 10^{1}**
1 bar	=	bar	**1× 10^{5}**	**1× 10^{-1}**	**1**
1 millibar	=	mbar	**1× 10^{2}**	**1× 10^{-4}**	**1× 10^{-3}**
1 kilogram-force per square metre	=	kgf/m^2	**9.80665**	**9.80665× 10^{-6}**	**9.80665× 10^{-5}**
1 kilogram-force per square centimetre 1 technical atmosphere	= =	kgf/cm^2 at	**9.80665× 10^{4}**	**9.80665× 10^{-2}**	**9.80665× 10^{-1}**
1 kilogram-force per square millimetre	=	kgf/mm^2	**9.80665× 10^{6}**	**9.80665**	**9.80665× 10^{1}**
1 millimetre of water	=	mmH_2O	**9.80665**	**9.80665× 10^{-6}**	**9.80665× 10^{-5}**
1 millimetre of mercury 1 torr	= =	mmHg –	1.33322× 10^{2}	1.33322× 10^{-4}	1.33322× 10^{-3}
1 standard atmosphere	=	atm	**1.01325× 10^{5}**	**1.01325× 10^{-1}**	**1.01325**
1 poundal per square foot	=	pdl/ft^2	1.48816	1.48816× 10^{-6}	1.48816× 10^{-5}
1 pound-force per square inch	=	lbf/in^2	6.89476× 10^{3}	6.89476× 10^{-3}	6.89476× 10^{-2}
1 pound-force per square foot	=	$tonf/ft^2$	4.78803× 10^{1}	4.78803× 10^{-5}	4.78803× 10^{-4}
1 UK ton-force per square inch	=	$tonf/in^2$	1.54443× 10^{7}	1.54443× 10^{1}	1.54443× 10^{2}

(Contd.)

TABLE 9.25 (a) Conversion factors for units of pressure and stress (continued)

Unit			pascal[1)], newton per square metre	megapascal, newton per square millimetre	bar
Name		Symbol	Pa N/m^2	MPa N/mm^2	bar
1 UK ton-force per square foot	=	$tonf/ft^2$	1.07252×10^5	1.07252×10^{-1}	1.07252
1 inch of water	=	inH_2O	2.49089×10^2	2.49089×10^{-4}	2.49089×10^{-3}
1 foot of water	=	ftH_2O	2.98907×10^3	2.98907×10^{-3}	2.98907×10^{-2}
1 inch of mercury	=	inHg	3.38639×10^3	3.38639×10^{-3}	3.38639×10^{-2}

TABLE 9.25 (b) Conversion factors for units of pressure and stress

Unit			millibar	kilogram-force per square metre
Name		Symbol	mbar	kgf/m^2
1 pascal[1)] 1 newton per square metre	= =	Pa N/m^2	1×10^{-2}	1.01972×10^{-1}
1 megapascal 1 newton per square millimetre	= =	MPa N/mm^2	1×10^{-4}	1.01972×10^5
1 bar	=	bar	1×10^3	1.01972×10^4
1 millibar	=	mbar	1	1.01972×10^1
1 kilogram-force per square metre	=	kgf/m^2	9.80665×10^{-2}	1
1 kilogram-force per square centimetre 1 technical atmosphere	= =	kgf/cm^2 at	9.80665×10^2	1×10^4

(Contd.)

TABLE 9.25 (b) Conversion factors for units of pressure and stress (continued)

Unit		millibar	kilogram-force per square metre
Name	Symbol	mbar	kgf/m^2
1 kilogram-force per square millimetre =	kgf/mm^2	**9.80665×10^4**	**1×10^6**
1 millimetre of water =	mmH_2O	**9.80665×10^{-2}**	**1**
1 miilimetre of mercury = 1 torr =	mmHg –	1.33322	1.35951×10^1
1 standard atmosphere =	atm	1.01325×10^3	1.03323×10^4
1 poundal per square foot =	pdl/ft^2	1.48816×10^{-2}	1.51750×10^{-1}
1 pound-force per square inch =	lbf/in^2	6.89476×10^1	7.03070×10^2
1 pound-force per square foot =	$tonf/ft^2$	**4.78803×10^{-1}**	**4.88243**
1 UK ton-force per square inch =	$tonf/in^2$	1.54443×10^5	1.57488×10^6
1 UK ton-force per square foot =	$tonf/ft^2$	1.07252×10^3	1.09366×10^4
1 inch of water =	inH_2O	2.49089	**2.54×10^1**
1 foot of water =	ftH_2O	2.98907×10^1	**3.048×10^2**
1 inch of mercury =	inHg	3.38639×10^1	3.45316×10^2

1) The unit of pressure and stress in the International System (SI) is the pascal (Pa).

TABLE 9.25 (c) Conversion factors for units of pressure and stress

Unit			kilogram-force per square centimetre, technical atmosphere	kilogram-force per square millimetre	millimetre of water
Name		Symbol	kgf/cm^2 at	kgf/mm^2	mmH_2O
1 pascal[1)] 1 newton per square metre	= =	Pa N/m^2	1.01972×10^{-5}	1.01972×10^{-7}	1.01972×10^{-1}
1 megapascal 1 newton per square millimetre	= =	MPa N/mm^2	1.01972×10^{1}	1.01972×10^{-1}	1.01972×10^{5}
1 bar	=	bar	1.01972	1.01972×10^{-2}	1.01972×10^{4}
1 millibar	=	mbar	1.01972×10^{-3}	1.01972×10^{-5}	1.01972×10^{1}
1 kilogram-force per square metre	=	kgf/m^2	**1×10^{-4}**	**1×10^{-6}**	**1**
1 kilogram-force per square centimetre 1 technical atmosphere	= =	kgf/cm^2 at	**1**	**1×10^{-2}**	**1×10^{-4}**
1 kilogram-force per square millimetre	=	kgf/mm^2	**1×10^{2}**	**1**	**1×10^{6}**
1 millimetre of water	=	mmH_2O	**1×10^{-4}**	**1×10^{-6}**	**1**
1 miilimetre of mercury 1 torr	= =	mmHg –	**1.35951×10^{-3}**	**1.35951×10^{-5}**	**1.35951×10^{1}**
1 standard atmosphere	=	atm	1.03323	1.03323×10^{-2}	**1.03323×10^{4}**
1 poundal per square foot	=	pdl/ft^2	1.51750×10^{-5}	1.51750×10^{-7}	1.51750×10^{-1}
1 pound-force per square inch	=	lbf/in^2	7.03070×10^{-2}	7.03070×10^{-4}	7.03070×10^{2}
1 pound-force per square foot	=	$tonf/ft^2$	4.88243×10^{-4}	4.88243×10^{-6}	4.88243

(Contd.)

TABLE 9.25 (c) Conversion factors for units of pressure and stress (continued)

Unit			kilogram-force per square centimetre, technical atmosphere	kilogram-force per square millimetre	millimetre of water
Name		Symbol	kgf/cm^2 at	kgf/mm^2	mmH_2O
1 UK ton-force per square inch	=	$tonf/in^2$	1.57488×10^2	1.57488	1.57488×10^6
1 UK ton-force per square foot	=	$tonf/ft^2$	1.09366	1.09366×10^{-2}	1.09366×10^4
1 inch of water	=	inH_2O	**2.54×10^{-3}**	**2.54×10^{-5}**	**2.54×10^1**
1 foot of water	=	ftH_2O	**3.048×10^{-2}**	**3.048×10^{-4}**	**3.048×10^2**
1 inch of mercury	=	inHg	3.45316×10^{-2}	3.45316×10^{-4}	3.45316×10^2

TABLE 9.25 (d) Conversion factors for units of pressure and stress

Unit			millimetre of mercury, torr	standard atmosphere
Name		Symbol	mmHg	atm
1 pascal[1)] 1 newton per square metre	= =	Pa N/m^2	7.50062×10^{-3}	9.86923×10^{-6}
1 megapascal 1 newton per square millimetre	= =	MPa N/mm^2	7.50062×10^3	9.86923
1 bar	=	bar	7.50062×10^2	9.86923×10^{-1}
1 millibar	=	mbar	7.50062×10^{-1}	9.86923×10^{-4}
1 kilogram-force per square metre	=	kgf/m^2	7.35559×10^{-2}	9.67841×10^{-5}

(Contd.)

TABLE 9.25 (d) Conversion factors for units of pressure and stress (continued)

Unit			millimetre of mercury, torr	standard atmosphere
Name		Symbol	mmHg	atm
1 kilogram-force per square centimetre 1 technical atmosphere	= =	kgf/cm^2 at	7.35559×10^2	9.67841×10^{-1}
1 kilogram-force per square millimetre	=	kgf/mm^2	7.35559×10^4	9.67841×10^1
1 millimetre of water	=	mmH_2O	7.35559×10^{-2}	9.67841×10^{-5}
1 miilimetre of mercury 1 torr	= =	mmHg	**1**	1.31579×10^{-3}
1 standard atmosphere	=	atm	7.6×10^2	**1**
1 poundal per square foot	=	pdl/ft^2	1.11621×10^{-2}	1.46870×10^{-5}
1 pound-force per square inch	=	lbf/in^2	5.17149×10^1	6.80460×10^{-2}
1 pound-force per square foot	=	$tonf/ft^2$	3.59131×10^{-1}	4.72542×10^{-4}
1 UK ton-force per square inch	=	$tonf/in^2$	1.15842×10^5	1.52423×10^2
1 UK ton-force per square foot	=	$tonf/ft^2$	8.04452×10^2	1.05849
1 inch of water	=	inH_2O	1.86832	2.45832×10^{-3}
1 foot of water	=	ftH_2O	2.24198×10^1	2.94998×10^{-2}
1 inch of mercury	=	inHg	**2.54×10^1**	3.34211×10^{-2}

(Contd.)

TABLE 9.25 (e) Conversion factors for units of pressure and stress

Unit (Name)		Unit (Symbol)	poundal per square foot pdl/ft²	pound-force per square inch lbf/in²	pound-force per square foot lbf/ft²
1 pascal[1)] 1 newton per square metre	= =	Pa N/m^2	6.71969×10^{-1}	1.45038×10^{-4}	2.08854×10^{-2}
1 megapascal 1 newton per square millimetre	= =	MPa N/mm^2	6.71969×10^{5}	1.45038×10^{2}	2.08854×10^{4}
1 bar	=	bar	6.71969×10^{4}	1.45038×10^{1}	2.08854×10^{3}
1 millibar	=	mbar	6.71969×10^{1}	1.45038×10^{-2}	2.08854
1 kilogram-force per square metre	=	kgf/m^2	6.58976	1.42233×10^{-3}	2.04816×10^{-1}
1 kilogram-force per square centimetre 1 technical atmosphere	= =	kgf/cm^2 at	6.58976×10^{4}	1.42233×10^{1}	2.04816×10^{3}
1 kilogram-force per square millimetre	=	kgf/mm^2	6.58976×10^{6}	1.42233×10^{3}	2.04816×10^{5}
1 millimetre of water	=	mmH_2O	6.58976	1.42233×10^{-3}	2.04816×10^{-1}
1 miilimetre of mercury 1 torr	= =	mmHg –	8.95885×10^{1}	1.93368×10^{-2}	2.78450
1 standard atmosphere	=	atm	6.80874×10^{4}	1.46959×10^{1}	2.11622×10^{3}
1 poundal per square foot	=	pdl/ft^2	**1**	2.15840×10^{-4}	3.10810×10^{-2}
1 pound-force per square inch	=	lbf/in^2	4.63306×10^{3}	**1**	$\mathbf{1.44 \times 10^{2}}$
1 pound-force per square foot	=	$tonf/ft^2$	3.21740×10^{1}	6.94444×10^{-3}	**1**
1 UK ton-force per square inch	=	$tonf/in^2$	1.03781×10^{7}	$\mathbf{2.24 \times 10^{3}}$	$\mathbf{3.2256 \times 10^{5}}$
1 UK ton-force per square foot	=	$tonf/ft^2$	7.20699×10^{4}	1.55556×10^{1}	$\mathbf{2.24 \times 10^{3}}$
1 inch of water	=	inH_2O	1.67381×10^{2}	3.61273×10^{-2}	5.20233

(Contd.)

TABLE 9.25 (e) Conversion factors for units of pressure and stress (continued)

Unit			poundal per square foot	pound-force per square inch	pound-force per square foot
Name		Symbol	pdl/ft^2	lbf/in^2	lbf/ft^2
1 foot of water	=	ftH$_2$O	2.00857×10^3	4.33527×10^{-1}	6.24280×10^1
1 inch of mercury	=	inHg	2.27556×10^3	4.91154×10^{-1}	7.07262×10^1

TABLE 9.25 (f) Conversion factors for units of pressure and stress

Unit			UK ton-force per square inch	UK ton-force per square foot
Name		Symbol	$tonf/in^2$	$tonf/ft^2$
1 pascal[1)] 1 newton per square metre	= =	Pa N/m^2	6.7490×10^{-8}	9.32385×10^{-6}
1 megapascal 1 newton per square millimetre	= =	MPa N/mm^2	6.47490×10^{-2}	9.32385
1 bar	=	bar	6.47490×10^{-3}	9.32385×10^{-1}
1 millibar	=	mbar	6.47490×10^{-6}	9.32385×10^{-4}
1 kilogram-force per square metre	=	kgf/m^2	6.34971×10^{-7}	9.14358×10^{-5}
1 kilogram-force per square centimetre 1 technical atmosphere	= =	kgf/cm^2 at	6.34971×10^{-3}	9.14358×10^{-1}
1 kilogram-force per square millimetre	=	kgf/mm^2	6.34971×10^{-1}	9.14358×10^1
1 millimetre of water	=	mmH$_2$O	6.34971×10^{-7}	9.14358×10^{-5}

(Contd.)

TABLE 9.25 (f) Conversion factors for units of pressure and stress (continued)

Unit			UK ton-force per square inch	UK ton-force per square foot
Name		Symbol	tonf/in^2	tonf/ft^2
1 miilimetre of mercury 1 torr	= =	mmHg –	8.63247×10^{-6}	1.24308×10^{-3}
1 standard atmosphere	=	atm	6.56067×10^{-3}	9.44738×10^{-1}
1 poundal per square foot	=	pdl/ft^2	9.63571×10^{-8}	1.38754×10^{-5}
1 pound-force per square inch	=	lbf/in^2	4.46429×10^{-4}	6.42857×10^{-2}
1 pound-force per square foot	=	tonf/ft^2	3.10020×10^{-6}	4.46429×10^{-4}
1 UK ton-force per square inch	=	tonf/in^2	**1**	**1.44×10^2**
1 UK ton-force per square foot	=	tonf/ft^2	6.94444×10^{-3}	**1**
1 inch of water	=	inH_2O	1.61282×10^{-5}	2.32248×10^{-3}
1 foot of water	=	ftH_2O	1.93539×10^{-4}	2.78697×10^{-2}
1 inch of mercury	=	inHg	2.19265×10^{-4}	3.15743×10^{-2}

TABLE 9.25 (g) Conversion factors for units of pressure and stress

Unit			inch of water	foot of water	inch of mercury
Name		Symbol	inH_2O	ftH_2O	inHg
1 pascal[1)] 1 newton per square metre	=	Pa N/m^2	4.01463×10^{-3}	3.34553×10^{-4}	2.95300×10^{-4}
1 megapascal 1 newton per square millimetre	= =	MPa N/mm^2	4.01463×10^{3}	3.34553×10^{2}	2.95300×10^{2}
1 bar	=	bar	4.01463×10^{2}	3.34553×10^{1}	2.95300×10^{1}
1 millibar	=	mbar	4.01463×10^{-1}	3.34553×10^{-2}	2.95300×10^{-2}

(Contd.)

TABLE 9.25 (g) Conversion factors for units of pressure and stress (continued)

Unit			inch of water	foot of water	inch of mercury
Name		Symbol	inH_2O	ftH_2O	inHg
1 kilogram-force per square metre	=	kgf/m^2	3.93701×10^{-2}	3.28084×10^{-3}	2.89590×10^{-3}
1 kilogram-force per square centimetre 1 technical atmosphere	= =	kgf/cm^2 at	3.93701×10^{2}	3.28084×10^{1}	2.89590×10^{1}
1 kilogram-force per square millimetre	=	kgf/mm^2	3.93701×10^{4}	3.28084×10^{3}	2.89590×10^{3}
1 millimetre of water	=	mmH_2O	3.93701×10^{-2}	3.28084×10^{-3}	2.89590×10^{-3}
1 millimetre of mercury 1 torr	=	mmHg	5.35240×10^{-1}	4.46033×10^{-2}	3.93701×10^{-2}
1 standard atmosphere	=	atm	4.06782×10^{2}	3.38985×10^{1}	2.99213×10^{1}
1 poundal per square foot	=	pdl/ft^2	5.97441×10^{-3}	4.97867×10^{-4}	4.39453×10^{-4}
1 pound-force per square inch	=	lbf/in^2	2.76799×10^{1}	2.30666	2.03602
1 pound-force per square foot	=	lbf/ft^2	1.92222×10^{-1}	1.60185×10^{-2}	1.41390×10^{-2}
1 UK ton-force per square inch	=	$tonf/in^2$	6.20031×10^{4}	5.16693×10^{3}	4.56069×10^{3}
1 UK ton-force per square foot	=	$tonf/ft^2$	4.30575×10^{2}	3.58812×10^{1}	3.16713×10^{1}
1 inch of water	=	inH_2O	**1**	8.33333×10^{-2}	7.35559×10^{-2}
1 foot of water	=	ftH_2O	**1.2×10^{1}**	1	8.82671×10^{-1}
1 inch of mercury	=	inHg	**1.35951×10^{1}**	1.13292	**1**

Notes:
1) 1 ksi = 10^3 lbf/in^2.
2) 1 hectobar (hbar) = 10 megapascal.
3) In Germany, 1 kp/mm^2 = 1 kgf/mm^2.

3.21 *Dynamic viscosity*

TABLE 9.26 Conversion factors for units of dynamic viscosity

Unit			pascal second[1)], newton second per square metre	centipoise	kilogram-force second per square metre
Name		Symbol	Pa s N s/m^2	cP	kgf s/m^2
1 pascal second[1)] 1 newton second per square metre	=	Pa s N s/m^2	**1**	**1 × 10^3**	1.01972 × 10^{-1}
1 centipoise	=	cP	**1 × 10^{-3}**	**1**	1.01972 × 10^{-4}
1 kilogram-force second per square metre	=	kgf s/m^2	**9.80665**	**9.80665 × 10^3**	**1**
1 poundal second per square foot	=	pdl s/ft^2	1.48816	1.48816 × 10^3	1.51750 × 10^{-1}
1 pound-force second per square foot	=	lbf s/ft^2	4.78803 × 10^1	4.78803 × 10^4	4.88243
1 pound-force hour per square foot	=	lbf h/ft^2	1.72369 × 10^5	1.72369 × 10^8	1.75767 × 10^4

(Contd.)

TABLE 9.26 Conversion factors for units of dynamic viscosity (continued)

Unit: Name		Unit: Symbol	poundal second per square foot pdl s/ft²	pound-force second per square foot lbf s/ft²	pound-force hour per square foot lbf h/ft²
1 pascal second[1)] 1 newton second per square metre	= =	Pa s N s/m²	6.71969×10^{-1}	2.08854×10^{-2}	5.80151×10^{-6}
1 centipoise	=	cP	6.71969×10^{-4}	2.08854×10^{-5}	5.80151×10^{-9}
1 kilogram-force second per square metre	=	kgf s/m²	6.58976	2.04816×10^{-1}	5.68934×10^{-5}
1 poundal second per square foot	=	pdl s/ft²	**1**	3.10810×10^{-2}	8.63360×10^{-6}
1 pound-force second per square foot	=	lbf s/ft²	3.21740×10^{1}	**1**	2.77778×10^{-4}
1 pound-force hour per square foot	=	lbf h/ft²	1.15827×10^{5}	$\mathbf{3.6 \times 10^{3}}$	**1**

1) The unit of dynamic viscosity in the International System (SI) is the pascal second [Pa s].

3.22 *Kinematic viscosity*

TABLE 9.27 Conversion factors for units of kinematic viscosity

Unit: Name		Unit: Symbol	metre squared[1)] per second m^2/s	centistokes, millimetre squared per second cSt mm^2/s	metre squared per hour m^2/h	inch squared per second in^2/s	foot squared per second ft^2/s
1 metre squared per second[1)]	=	m2/s	**1**	**1×10^6**	**3.6×10^3**	1.55000×10^3	1.07639×10^1
1 centistokes 1 millimetre squared per second	= =	cSt mm^2/s	**1×10^{-6}**	**1**	**3.6×10^{-3}**	1.55000×10^{-3}	1.07639×10^{-5}
1 metre squared per hour	=	m^2/h	2.77778×10^{-4}	2.77778×10^2	**1**	4.30556×10^{-1}	2.98998×10^{-3}
1 inch squared per second	=	in^2/s	**6.4516×10^{-4}**	**6.4516×10^2**	2.32258	**1**	6.94444×10^{-3}
1 foot squared per second	=	ft^2/s	9.29030×10^{-2}	9.29030×10^4	3.34451×10^2	**1.44×10^2**	**1**
1 inch squared per hour	=	in^2/h	1.79211×10^{-7}	1.79211×10^{-1}	**6.4516×10^{-4}**	2.77778×10^{-4}	1.92901×10^{-6}
1 foot squared per hour	=	ft^2/h	**2.58064×10^{-5}**	**2.58064×10^1**	9.29030×10^{-2}	**4×10^{-2}**	2.77778×10^{-4}

1) The unit of kinematic viscosity in the International System (SI) is the metre squared per second [m^2/s].

(Contd.)

TABLE 9.27 Conversion factors for units of kinematic viscosity (continued)

Unit			inch squared per hour	foot squared per hour
Name		Symbol	in^2/h	ft^2/h
1 metre squared per second[1)]	=	m^2/s	5.58001×10^6	3.87501×10^4
1 centistokes 1 millimetre squared per second	= =	cSt mm^2/s	5.58001	3.87501×10^{-2}
1 metre squared per hour	=	m^2/h	1.55000×10^3	1.07639×10^1
1 inch squared per second	=	in^2/s	**3.6×10^3**	**2.5×10^1**
1 foot squared per second	=	ft^2/s	**5.184×10^5**	**3.6×10^3**
1 inch squared per hour	=	in^2/h	**1**	6.94444×10^{-3}
1 foot squared per hour	=	ft^2/h	**1.44×10^2**	**1**

3.23 Energy

TABLE 9.28 Conversion factors for units of energy

Unit			joule[1)]	kilojoule	erg	kilowatt hour	kilogram-force metre
Name		Symbol	J	kJ	erg	kW h	kgf m
1 joule[1)]	=	J	**1**	**1×10^{-3}**	**1×10^{7}**	2.77778×10^{-7}	1.01972×10^{-1}
1 kilojoule	=	kJ	**1×10^{3}**	**1**	**1×10^{10}**	2.77778×10^{-4}	1.01972×10^{2}
1 erg	=	erg	**1×10^{-7}**	**1×10^{-10}**	**1**	2.77778×10^{-14}	1.01972×10^{-8}
1 kilowatt hour	=	kW h	**3.6×10^{6}**	**3.6×10^{3}**	**3.6×10^{13}**	**1**	3.67098×10^{5}
1 kilogram-force metre	=	kgf m	**9.80665**	**9.80665×10^{-3}**	**9.80665×10^{7}**	2.72407×10^{-6}	**1**
1 calorie[2)]	=	cal_{IT}	**4.1868**	**4.1868×10^{-3}**	**4.1868×10^{7}**	**1.163×10^{-6}**	4.26935×10^{-1}
1 kilocalorie[2)]	=	$kcal_{IT}$	**4.1868×10^{3}**	**4.1868**	**4.1868×10^{10}**	**1.163×10^{-3}**	4.26935×10^{2}
1 metric horsepower hour	=	KS h	2.64780×10^{6}	2.64780×10^{3}	2.64780×10^{13}	7.35499×10^{-1}	**2.7×10^{5}**
1 foot poundal	=	ft pdl	4.21401×10^{-2}	4.21401×10^{-5}	4.21401×10^{5}	1.17056×10^{-8}	4.29710×10^{-3}
1 foot pound-force	=	ft lbf	1.35582	1.35582×10^{-3}	1.35582×10^{7}	3.76616×10^{-7}	1.38255×10^{-1}
1 horsepower hour	=	hp h	2.68452×10^{6}	2.68452×10^{3}	2.68452×10^{13}	7.45700×10^{-1}	2.73745×10^{5}
1 British thermal unit	=	Btu	1.05506×10^{3}	1.05506	1.05506×10^{10}	2.93071×10^{-4}	1.07586×10^{2}

(Contd.)

TABLE 9.28 Conversion factors for units of energy (continued)

Unit			calorie[2]	kilocalorie[2]	metric horsepower hour	foot poundal	foot pound-force
Name		Symbol	cal_{IT}	$kcal_{IT}$	KS h	ft pdl	ft lbf
1 joule[1]	=	J	2.38846×10^{-1}	2.38846×10^{-4}	3.77673×10^{-7}	2.37304×10^{1}	7.37562×10^{-1}
1 kilojoule	=	kJ	2.38846×10^{2}	2.38846×10^{-1}	3.77673×10^{-4}	2.37304×10^{4}	7.37562×10^{2}
1 erg	=	erg	2.38846×10^{-8}	2.38846×10^{-11}	3.77673×10^{-14}	2.37304×10^{-6}	7.37562×10^{-8}
1 kilowatt hour	=	kW h	8.59845×10^{5}	8.59845×10^{2}	1.35962	8.54293×10^{7}	2.65522×10^{6}
1 kilogram-force metre	=	kgf m	2.34228	2.34228×10^{-3}	3.70370×10^{-6}	2.32715×10^{2}	7.23301
1 calorie[2]	=	cal_{IT}	**1**	**1×10^{-3}**	1.58124×10^{-6}	9.93543×10^{1}	3.08803
1 kilocalorie[2]	=	$kcal_{IT}$	**1×10^{3}**	**1**	1.58124×10^{-3}	9.93543×10^{4}	3.08803×10^{3}
1 metric horsepower hour	=	KS h	6.32415×10^{5}	6.32415×10^{2}	**1**	6.28331×10^{7}	1.95291×10^{5}
1 foot poundal	=	ft pdl	1.00650×10^{-2}	1.00650×10^{-5}	1.59152×10^{-8}	**1**	3.10810×10^{-2}
1 foot pound-force	=	ft lbf	3.23832×10^{-1}	3.23832×10^{-4}	5.12056×10^{-7}	3.21740×10^{1}	**1**
1 horsepower hour	=	hp h	6.41186×10^{5}	6.41186×10^{2}	1.01387	6.37046×10^{7}	**1.98×10^{6}**
1 British thermal unit	=	Btu	2.51996×10^{2}	2.51996×10^{-1}	3.98466×10^{-4}	2.50370×10^{4}	7.78169×10^{2}

TABLE 9.28 Conversion factors for units of energy (continued)

Unit			horsepower hour	British thermal unit
Name		Symbol	hp h	Btu
1 joule[1)]	=	J	3.72506×10^{-7}	9.47817×10^{-4}
1 kilojoule	=	kJ	3.72506×10^{-4}	9.47817×10^{-1}
1 erg	=	erg	3.72506×10^{-14}	9.47817×10^{-11}
1 kilowatt hour	=	kW h	1.34102	3.41214×10^{3}
1 kilogram-force metre	=	kgf m	3.65304×10^{-6}	9.29487×10^{-3}
1 calorie[2)]	=	cal_{IT}	1.55961×10^{-6}	3.96832×10^{-3}
1 kilocalorie[2)]	=	$kcal_{IT}$	1.55961×10^{-3}	3.96832
1 metric horsepower hour	=	KS h	9.86319×10^{-1}	2.50963×10^{3}
1 foot poundal	=	ft pdl	1.56974×10^{-8}	3.99410×10^{-5}
1 foot pound-force	=	ft lbf	5.05051×10^{-7}	1.28507×10^{-3}
1 horsepower hour	=	hp h	**1**	2.54443×10^{3}
1 British thermal unit	=	Btu	3.93015×10^{-4}	**1**

1) The unit of energy in the International System (SI) is the joule [J].
2) This refers to the International Table calorie.

Note: 1 electron volt (eV) = 1.60218×10^{-19} J

3.24 Power

TABLE 9.29 Conversion factors for units of power

Unit			watt[1)]	kilowatt	erg per second	kilogram-force metre per second
Name		Symbol	W	kW	erg/s	kgf m/s
1 watt[1)]	=	W	**1**	**1×10^{-3}**	**1×10^{7}**	1.01972×10^{-1}
1 kilowatt	=	kW	**1×10^{3}**	**1**	**1×10^{10}**	1.01972×10^{2}
1 erg per second	=	erg/s	**1×10^{-7}**	**1×10^{-10}**	**1**	1.01972×10^{-8}
1 kilogram-force metre per second	=	kgf m/s	**9.80665**	**9.80665×10^{-3}**	**9.80665×10^{7}**	**1**
1 metric horsepower	=	KS	7.35499×10^{2}	7.35499×10^{-1}	7.35499×10^{9}	**7.5×10^{1}**
1 calorie[2)] per second	=	cal_{IT}/s	**4.1868**	**4.1868×10^{-3}**	**4.1868×10^{7}**	4.26935×10^{-1}
1 kilocalorie[2)] per hour	=	$kcal_{IT}/h$	**1.163**	**1.163×10^{-3}**	**1.163×10^{7}**	1.18593×10^{-1}
1 foot poundal per second	=	ft pdl/s	4.21401×10^{-2}	4.21401×10^{-5}	4.21401×10^{5}	4.29710×10^{-3}
1 foot pound-force per second	=	ft lbf/s	1.35582	1.35582×10^{-3}	1.35582×10^{7}	1.38255×10^{-1}
1 horsepower	=	hp	7.45700×10^{2}	7.45700×10^{-1}	7.45700×10^{9}	7.60402×10^{1}
1 British thermal unit per hour	=	Btu/h	2.93071×10^{-1}	2.93071×10^{-4}	2.93071×10^{6}	2.98849×10^{-2}

(Contd.)

TABLE 9.29 Conversion factors for units of power (continued)

Unit			metric horsepower	calorie per second	kilocalorie per hour	foot poundal per second
Name		Symbol	KS	cal_{IT} /s	$kcal_{IT}$ /h	ft pdl/s
1 watt[1)]	=	W	1.35962×10^{-3}	2.38846×10^{-1}	8.59845×10^{-1}	2.37304×10^{1}
1 kilowatt	=	kW	1.35962	2.38846×10^{2}	8.59845×10^{2}	2.37304×10^{4}
1 erg per second	=	erg/s	1.35962×10^{-10}	2.38846×10^{-8}	8.59845×10^{-8}	2.37304×10^{-6}
1 kilogram-force metre per second	=	kgf m/s	1.33333×10^{-2}	2.34228	8.43220	2.32715×10^{2}
1 metric horsepower	=	KS	**1**	1.75671×10^{2}	6.32415×10^{2}	1.74536×10^{4}
1 calorie[2)] per second	=	cal_{IT}/s	5.69246×10^{-3}	**1**	**3.6**	9.93543×10^{1}
1 kilocalorie[2)] per hour	=	$kcal_{IT}$/h	1.58124×10^{-3}	2.77778×10^{-1}	**1**	2.75984×10^{1}
1 foot poundal per second	=	ft pdl/s	5.72946×10^{-5}	1.00650×10^{-2}	3.62340×10^{-2}	**1**
1 foot pound-force per second	=	ft lbf/s	1.84340×10^{-3}	3.23832×10^{-1}	1.16579	3.21740×10^{1}
1 horsepower	=	hp	1.01387	1.78107×10^{2}	6.41186×10^{2}	1.76957×10^{4}
1 British thermal unit per hour	=	Btu/h	3.98466×10^{-4}	6.99988×10^{-2}	2.51996×10^{-1}	6.95468

TABLE 9.29 Conversion factors for units of power (continued)

Unit			foot pound-force per second	horsepower	British thermal unit per hour
Name		Symbol	ft lbf/s	hp	Btu/h
1 watt[1)]	=	W	7.37562×10^{-1}	1.34102×10^{-3}	3.41214
1 kilowatt	=	kW	7.37562×10^{2}	1.34102	3.41214×10^{3}
1 erg per second	=	erg/s	7.37562×10^{-8}	1.34102×10^{-10}	3.41214×10^{-7}
1 kilogram-force metre per second	=	kgf m/s	7.23301	1.31509×10^{-2}	3.34617×10^{1}
1 metric horsepower	=	KS	5.42476×10^{2}	9.86320×10^{-1}	2.50963×10^{3}
1 calorie[2)] per second	=	cal_{IT}/s	3.08803	5.61459×10^{-3}	1.42860×10^{1}
1 kilocalorie[2)] per hour	=	$kcal_{IT}$/h	8.57785×10^{-1}	1.55961×10^{-3}	3.96832
1 foot poundal per second	=	ft pdl/s	3.10810×10^{-2}	5.65108×10^{-5}	1.43788×10^{-1}
1 foot pound-force per second	=	ft lbf/s	**1**	1.81818×10^{-3}	4.62624
1 horsepower	=	hp	**5.5×10^{2}**	**1**	2.54443×10^{3}
1 British thermal unit per hour	=	Btu/h	2.16158×10^{-1}	3.93015×10^{-4}	**1**

1) The unit of power in the International System (SI) is the watt [W].

TABLE 9.31 Temperature conversion table (continued)

°C		°F	°C		°F	°C		°F	°C		°F
–40.0	**–40**	–40.0	–15.0	**5**	41.0	–1.1	**30**	86.0	12.8	**55**	131.0
–38.9	**–38**	–36.4	–14.4	**6**	42.8	–0.6	**31**	87.8	13.3	**56**	132.8
–37.8	**–36**	–32.8	–13.9	**7**	44.6	0.0	**32**	89.6	13.9	**57**	134.6
–36.7	**–34**	–29.2	–13.3	**8**	46.4	0.6	**33**	91.4	14.4	**58**	136.4
–35.6	**–32**	–25.6	–12.8	**9**	48.2	1.1	**34**	93.2	15.0	**59**	138.2
–34.4	**–30**	–22.0	–12.2	**10**	50.0	1.7	**35**	95.0	15.6	**60**	140.0
–33.3	**–28**	–18.4	–11.7	**11**	51.8	2.2	**36**	96.8	16.1	**61**	141.8
–32.2	**–26**	–14.8	–11.1	**12**	53.6	2.8	**37**	98.6	16.7	**62**	143.6
–31.1	**–24**	–11.2	–10.6	**13**	55.4	3.3	**38**	100.4	17.2	**63**	145.4
–30.0	**–22**	–7.6	–10.0	**14**	57.2	3.9	**39**	102.2	17.8	**64**	147.2
–28.9	**–20**	–4.0	–9.4	**15**	59.0	4.4	**40**	104.0	18.3	**65**	149.0
–27.8	**–18**	–0.4	–8.9	**16**	60.8	5.0	**41**	105.8	18.9	**66**	150.8
–26.7	**–16**	3.2	–8.3	**17**	62.6	5.6	**42**	107.6	19.4	**67**	152.6
–25.6	**–14**	6.8	–7.8	**18**	64.4	6.1	**43**	109.4	20.0	**68**	154.4
–24.4	**–12**	10.4	–7.2	**19**	66.2	6.7	**44**	111.2	20.6	**69**	156.2
–23.3	**–10**	14.0	–6.7	**20**	68.0	7.2	**45**	113.0	21.1	**70**	158.0
–22.2	**–8**	17.6	–6.1	**21**	69.8	7.8	**46**	114.8	21.7	**71**	159.8
–21.1	**–6**	21.2	–5.6	**22**	71.6	8.3	**47**	116.6	22.2	**72**	161.6
–20.0	**–4**	24.8	–5.0	**23**	73.4	8.9	**48**	118.4	22.8	**73**	163.4
–18.9	**–2**	28.4	–4.4	**24**	75.2	9.4	**49**	120.2	23.3	**74**	165.2
–17.8	**0**	32.0	–3.9	**25**	77.0	10.0	**50**	122.0	23.9	**75**	167.0
–17.2	**1**	33.8	–3.3	**26**	78.8	10.6	**51**	123.8	24.4	**76**	168.8
–16.7	**2**	35.6	–2.8	**27**	80.6	11.1	**52**	125.6	25.0	**77**	170.6
–16.1	**3**	37.4	–2.2	**28**	82.4	11.7	**53**	127.4	25.6	**78**	172.4
–15.6	**4**	39.2	–1.7	**29**	84.2	12.2	**54**	129.2	26.1	**79**	174.2

(Contd.)

TABLE 9.31 Temperature conversion table (continued)

°C		°F	°C		°F	°C		°F	°C		°F
26.7	**80**	176.0	43	**110**	230	71	**160**	320	99	**210**	410
27.2	**81**	177.8	44	**112**	234	72	**162**	324	100	**212**	414
27.8	**82**	179.6	46	**114**	237	73	**164**	327	102	**215**	419
28.3	**83**	181.4	47	**116**	241	74	**166**	331	104	**220**	428
28.9	**84**	183.2	48	**118**	244	76	**168**	334	107	**225**	437
29.4	**85**	185.0	49	**120**	248	77	**170**	338	110	**230**	446
30.0	**86**	186.8	50	**122**	252	78	**172**	342	113	**235**	455
30.6	**87**	188.6	51	**124**	255	79	**174**	345	116	**240**	464
31.1	**88**	190.4	52	**126**	259	80	**176**	349	119	**245**	473
31.7	**89**	192.2	53	**128**	262	81	**178**	352	121	**250**	482
32.2	**90**	194.0	54	**130**	266	82	**180**	356	124	**255**	491
32.8	**91**	195.8	56	**132**	270	83	**182**	360	127	**260**	500
33.3	**92**	197.6	57	**134**	273	84	**184**	363	130	**265**	509
33.9	**93**	199.4	58	**136**	277	86	**186**	367	132	**270**	518
34.4	**94**	201.2	59	**138**	280	87	**188**	370	135	**275**	527
35.0	**95**	203.0	60	**140**	284	88	**190**	374	138	**280**	536
35.6	**96**	204.8	61	**142**	288	89	**192**	378	141	**285**	545
36.1	**97**	206.6	62	**144**	291	90	**194**	381	143	**290**	554
36.7	**98**	208.4	63	**146**	295	91	**196**	385	146	**295**	563
37.2	**99**	210.2	64	**148**	298	92	**198**	388	149	**300**	572
38	**100**	212	66	**150**	302	93	**200**	392	152	**305**	581
39	**102**	216	67	**152**	306	94	**202**	396	154	**310**	590
40	**104**	219	68	**154**	309	96	**204**	399	157	**315**	599
41	**106**	223	69	**156**	313	97	**206**	403	160	**320**	608
42	**108**	226	70	**158**	316	98	**208**	406	163	**325**	617

(Contd.)

TABLE 9.31 Temperature conversion table (continued)

°C		°F	°C		°F	°C		°F	°C		°F
166	**330**	626	235	**455**	851	304	**580**	1076	374	**705**	1301
169	**335**	635	238	**460**	860	307	**585**	1085	377	**710**	1310
171	**340**	644	241	**465**	869	310	**590**	1094	379	**715**	1319
174	**345**	653	243	**470**	878	313	**595**	1103	382	**720**	1328
177	**350**	662	246	**475**	887	316	**600**	1112	385	**725**	1337
179	**355**	671	249	**480**	896	319	**605**	1121	388	**730**	1346
182	**360**	680	252	**485**	905	321	**610**	1130	391	**735**	1355
185	**365**	689	254	**490**	914	324	**615**	1139	393	**740**	1364
188	**370**	698	257	**495**	923	327	**620**	1148	396	**745**	1373
191	**375**	707	260	**500**	932	330	**625**	1157	399	**750**	1382
193	**380**	716	263	**505**	941	332	**630**	1166	402	**755**	1391
196	**385**	725	266	**510**	950	335	**635**	1175	404	**760**	1400
199	**390**	734	268	**515**	959	338	**640**	1184	407	**765**	1409
202	**395**	743	271	**520**	968	341	**645**	1193	410	**770**	1418
204	**400**	752	274	**525**	977	343	**650**	1202	413	**775**	1427
207	**405**	761	277	**530**	986	346	**655**	1211	416	**780**	1436
210	**410**	770	279	**535**	995	349	**660**	1220	418	**785**	1445
213	**415**	779	282	**540**	1004	352	**665**	1229	421	**790**	1454
216	**420**	788	285	**545**	1013	354	**670**	1238	424	**795**	1463
219	**425**	797	288	**550**	1022	357	**675**	1247	427	**800**	1472
221	**430**	806	291	**555**	1031	360	**680**	1256	429	**805**	1481
224	**435**	815	293	**560**	1040	363	**685**	1265	432	**810**	1490
227	**440**	824	296	**565**	1049	366	**690**	1274	435	**815**	1499
229	**445**	833	299	**570**	1058	368	**695**	1283	438	**820**	1508
232	**450**	842	302	**575**	1067	371	**700**	1292	441	**825**	1517

(Contd.)

TABLE 9.31 Temperature conversion table (continued)

°C		°F	°C		°F	°C		°F	°C		°F
443	**830**	1526	571	**1060**	1940	710	**1310**	2390	849	**1560**	2840
446	**835**	1535	577	**1070**	1958	716	**1320**	2408	854	**1570**	2858
449	**840**	1544	582	**1080**	1976	721	**1330**	2426	860	**1580**	2876
454	**850**	1562	588	**1090**	1994	727	**1340**	2444	866	**1590**	2894
460	**860**	1580	593	**1100**	2012	732	**1350**	2462	871	**1600**	2912
466	**870**	1598	599	**1110**	2030	738	**1360**	2480	877	**1610**	2930
471	**880**	1616	604	**1120**	2048	743	**1370**	2498	882	**1620**	2948
477	**890**	1624	610	**1130**	2066	749	**1380**	2516	888	**1630**	2966
482	**900**	1652	616	**1140**	2084	754	**1390**	2534	893	**1640**	2984
488	**910**	1670	621	**1150**	2102	760	**1400**	2552	899	**1650**	3002
493	**920**	1688	627	**1160**	2120	766	**1410**	2570	904	**1660**	3020
499	**930**	1706	632	**1170**	2138	771	**1420**	2588	910	**1670**	3038
504	**940**	1724	638	**1180**	2156	777	**1430**	2606	916	**1680**	3056
510	**950**	1742	643	**1190**	2174	782	**1440**	2624	921	**1690**	3074
516	**960**	1760	649	**1200**	2192	788	**1450**	2642	927	**1700**	3092
521	**970**	1778	654	**1210**	2210	793	**1460**	2660	932	**1710**	3110
527	**980**	1796	660	**1220**	2228	799	**1470**	2678	938	**1720**	3128
532	**990**	1814	666	**1230**	2246	804	**1480**	2696	943	**1730**	3146
538	**1000**	1832	671	**1240**	2264	810	**1490**	2714	949	**1740**	3164
			677	**1250**	2282	816	**1500**	2732	954	**1750**	3182
543	**1010**	1850	682	**1260**	2300	821	**1510**	2750	960	**1760**	3200
549	**1020**	1868	688	**1270**	2318	827	**1520**	2768	966	**1770**	3218
554	**1030**	1886	693	**1280**	2336	832	**1530**	2786	971	**1780**	3236
560	**1040**	1904	699	**1290**	2354	838	**1540**	2804	977	**1790**	3254
566	**1050**	1922	704	**1300**	2372	843	**1550**	2822	982	**1800**	3272

(Contd.)

TABLE 9.31 Temperature conversion table (continued)

°C		°F	°C		°F	°C		°F	°C		°F
988	**1810**	3290	1127	**2060**	3740	1266	**2310**	4190	1404	**2560**	4640
993	**1820**	3308	1132	**2070**	3758	1271	**2320**	4208	1410	**2570**	4658
999	**1830**	3326	1138	**2080**	3776	1277	**2330**	4226	1416	**2580**	4676
1004	**1840**	3344	1143	**2090**	3794	1282	**2340**	4244	1421	**2590**	4694
1010	**1850**	3362	1149	**2100**	3812	1288	**2350**	4262	1427	**2600**	4712
1016	**1860**	3380	1154	**2110**	3830	1293	**2360**	4280	1432	**2610**	4730
1021	**1870**	3398	1160	**2120**	3848	1299	**2370**	4298	1438	**2620**	4748
1027	**1880**	3416	1166	**2130**	3866	1304	**2380**	4316	1443	**2630**	4766
1032	**1890**	3434	1171	**2140**	3884	1310	**2390**	4334	1449	**2640**	4784
1038	**1900**	3452	1177	**2150**	3902	1316	**2400**	4352	1454	**2650**	4802
1043	**1910**	3470	1182	**2160**	3920	1321	**2410**	4370	1460	**2660**	4820
1049	**1920**	3488	1188	**2170**	3938	1327	**2420**	4388	1466	**2670**	4838
1054	**1930**	3506	1193	**2180**	3956	1332	**2430**	4406	1471	**2680**	4856
1060	**1940**	3524	1199	**2190**	3974	1338	**2440**	4424	1477	**2690**	4874
1066	**1950**	3542	1204	**2200**	3992	1343	**2450**	4442	1482	**2700**	4892
1071	**1960**	3560	1210	**2210**	4010	1349	**2460**	4460	1488	**2710**	4910
1077	**1970**	3578	1216	**2220**	4028	1354	**2470**	4478	1493	**2720**	4928
1082	**1980**	3596	1221	**2230**	4046	1360	**2480**	4496	1499	**2730**	4946
1088	**1990**	3614	1227	**2240**	4064	1366	**2490**	4514	1504	**2740**	4964
1093	**2000**	3632	1232	**2250**	4082	1371	**2500**	4532	1510	**2750**	4982
1099	**2010**	3650	1238	**2260**	4100	1377	**2510**	4550	1516	**2760**	5000
1104	**2020**	3668	1243	**2270**	4118	1382	**2520**	4568	1521	**2770**	5018
1110	**2030**	3686	1249	**2280**	4136	1388	**2530**	4586	1527	**2780**	5036
1116	**2040**	3704	1254	**2290**	4154	1393	**2540**	4604	1532	**2790**	5054
1121	**2050**	3722	1260	**2300**	4172	1399	**2550**	4622	1538	**2800**	5072

(Contd.)

TABLE 9.31 Temperature conversion table (continued)

°C		°F	°C		°F	°C		°F	°C		°F
1543	**2810**	5090	1682	**3060**	5540	1821	**3310**	5990	1960	**3560**	6440
1549	**2820**	5108	1688	**3070**	5558	1827	**3320**	6008	1965	**3570**	6458
1554	**2830**	5126	1693	**3080**	5576	1832	**3330**	6026	1971	**3580**	6476
1560	**2840**	5144	1699	**3090**	5594	1838	**3340**	6044	1977	**3590**	6494
1566	**2850**	5162	1704	**3100**	5612	1843	**3350**	6062	1982	**3600**	6512
1571	**2860**	5180	1710	**3110**	5630	1849	**3360**	6080	1988	**3610**	6530
1577	**2870**	5198	1715	**3120**	5648	1854	**3370**	6098	1993	**3620**	6548
1582	**2880**	5216	1721	**3130**	5666	1860	**3380**	6116	1999	**3630**	6566
1588	**2890**	5234	1727	**3140**	5684	1865	**3390**	6134	2004	**3640**	6584
1593	**2900**	5252	1732	**3150**	5702	1871	**3400**	6152	2010	**3650**	6602
1599	**2910**	5270	1738	**3160**	5720	1877	**3410**	6170	2015	**3660**	6620
1604	**2920**	5288	1743	**3170**	5738	1882	**3420**	6188	2021	**3670**	6638
1610	**2930**	5306	1749	**3180**	5756	1888	**3430**	6206	2027	**3680**	6656
1616	**2940**	5324	1754	**3190**	5774	1893	**3440**	6224	2032	**3690**	6674
1621	**2950**	5342	1760	**3200**	5792	1899	**3450**	6242	2038	**3700**	6692
1627	**2960**	5360	1765	**3210**	5810	1904	**3460**	6260	2043	**3710**	6710
1632	**2970**	5378	1771	**3220**	5828	1910	**3470**	6278	2049	**3720**	6728
1638	**2980**	5396	1777	**3230**	5846	1915	**3480**	6296	2054	**3730**	6746
1643	**2990**	5414	1782	**3240**	5864	1921	**3490**	6314	2060	**3740**	6764
1649	**3000**	5432	1788	**3250**	5882	1927	**3500**	6332	2065	**3750**	6782
1654	**3010**	5450	1793	**3260**	5900	1932	**3510**	6350	2071	**3760**	6800
1660	**3020**	5468	1799	**3270**	5918	1938	**3520**	6368	2077	**3770**	6818
1666	**3030**	5486	1804	**3280**	5936	1943	**3530**	6386	2082	**3780**	6836
1671	**3040**	5504	1810	**3290**	5954	1949	**3540**	6404	2088	**3790**	6854
1677	**3050**	5522	1815	**3300**	5972	1954	**3550**	6422	2093	**3800**	6872

(Contd.)

TABLE 9.31 Temperature conversion table (continued)

°C		°F	°C		°F	°C		°F	°C		°F
2099	**3810**	6890	2238	**4060**	7340	2377	**4310**	7790	2515	**4560**	8240
2104	**3820**	6908	2243	**4070**	7358	2382	**4320**	7808	2521	**4570**	8258
2110	**3830**	6926	2249	**4080**	7376	2388	**4330**	7826	2527	**4580**	8276
2115	**3840**	6944	2254	**4090**	7394	2393	**4340**	7844	2532	**4590**	8294
2121	**3850**	6962	2260	**4100**	7412	2399	**4350**	7862	2538	**4600**	8312
2127	**3860**	6980	2265	**4110**	7430	2404	**4360**	7880	2543	**4610**	8330
2132	**3870**	6998	2271	**4120**	7448	2410	**4370**	7898	2549	**4620**	8348
2138	**3880**	7016	2277	**4130**	7466	2415	**4380**	7916	2554	**4630**	8366
2143	**3890**	7034	2282	**4140**	7484	2421	**4390**	7934	2560	**4640**	8384
2149	**3900**	7052	2288	**4150**	7502	2427	**4400**	7952	2565	**4650**	8402
2154	**3910**	7070	2293	**4160**	7520	2432	**4410**	7970	2571	**4660**	8420
2160	**3920**	7088	2299	**4170**	7538	2438	**4420**	7988	2577	**4670**	8438
2165	**3930**	7106	2304	**4180**	7556	2443	**4430**	8006	2582	**4680**	8456
2171	**3940**	7124	2310	**4190**	7574	2449	**4440**	8024	2588	**4690**	8474
2177	**3950**	7142	2315	**4200**	7592	2454	**4450**	8042	2593	**4700**	8492
2182	**3960**	7160	2321	**4210**	7610	2460	**4460**	8060	2599	**4710**	8510
2188	**3970**	7178	2327	**4220**	7628	2465	**4470**	8078	2604	**4720**	8528
2193	**3980**	7196	2332	**4230**	7646	2471	**4480**	8096	2610	**4730**	8546
2199	**3990**	7214	2338	**4240**	7664	2477	**4490**	8114	2615	**4740**	8564
2204	**4000**	7232	2343	**4250**	7682	2482	**4500**	8132	2621	**4750**	8582
2210	**4010**	7250	2349	**4260**	7700	2488	**4510**	8150	2627	**4760**	8600
2215	**4020**	7268	2354	**4270**	7718	2493	**4520**	8168	2632	**4770**	8618
2221	**4030**	7286	2360	**4280**	7736	2499	**4530**	8186	2638	**4780**	8636
2227	**4040**	7304	2365	**4290**	7754	2504	**4540**	8204	2643	**4790**	8654
2232	**4050**	7322	2371	**4300**	7772	2510	**4550**	8222	2649	**4800**	8672

(Contd.)

TABLE 9.31 Temperature conversion table (continued)

°C		°F	°C		°F	°C		°F	°C		°F
2654	**4810**	8690	2682	**4860**	8780	2710	**4910**	8870	2738	**4960**	8960
2660	**4820**	8708	2688	**4870**	8798	2715	**4920**	8888	2743	**4970**	8978
2665	**4830**	8726	2693	**4880**	8816	2721	**4930**	8906	2749	**4980**	8996
2671	**4840**	8744	2699	**4890**	8834	2727	**4940**	8924	2754	**4990**	9014
2677	**4850**	8762	2704	**4900**	8852	2732	**4950**	8942	2760	**5000**	9032

3.26 *Coefficient of thermal expansion*

TABLE 9.32 Conversion factors for units of coefficient of thermal expansion

Unit			kelvin to the power minus one[1)	degree Celsius to the power minus one	degree Fahrenheit to the power minus one
Name		Symbol	K^{-1}	$°C^{-1}$	$°F^{-1}$
1 kelvin to the power minus one[1)	=	K^{-1}	**1**	**1**	5.55556×10^{-1}
1 degree Celsius to the power minus one	=	$°C^{-1}$	**1**	**1**	5.55556×10^{-1}
1 degree Fahrenheit to the power minus one	=	$°F^{-1}$	**1.8**	**1.8**	**1**

1) The unit of coefficient of thermal expansion in the International System (SI) is the kelvin to the power minus one [K^{-1}].

3.27 Specific energy

TABLE 9.33 Conversion factors for units of specific energy

Unit			joule per kilogram[1)]	kilocalorie[2)] per kilogram	thermo-chemical kilocalorie per kilogram
Name		Symbol	J/kg	$kcal_{IT}/kg$	$kcal_{th}/kg$
1 joule per kilogram[1)]	=	J/kg	**1**	2.38846×10^{-4}	2.39006×10^{-4}
1 kilocalorie[2)] per kilogram	=	$kcal_{IT}/kg$	**4.1868×10^{3}**	**1**	1.00067
1 thermochemical kilocalorie Per kilogram	=	$kcal_{th}/kg$	**4.184×10^{3}**	9.99331×10^{-1}	**1**
1 kilogram-force metre per kilogram	=	kgf m/kg	**9.80665**	2.34228×10^{-3}	2.34385×10^{-3}
1 British thermal unit per pound	=	Btu/lb	**2.326×10^{3}**	5.55556×10^{-1}	5.55927×10^{-1}
1 foot pound-force per pound	=	ft lbf/lb	2.98907	7.13926×10^{-4}	7.14404×10^{-4}

(Contd.)

TABLE 9.33 Conversion factors for units of specific energy (continued)

Unit: Name		Unit: Symbol	kilogram-force metre per kilogram	British thermal unit per pound	foot pound-force per pound
			kgf m/kg	Btu/lb	ft lbf/lb
1 joule per kilogram[1)]	=	J/kg	1.01972×10^{-1}	4.29923×10^{-4}	3.34553×10^{-1}
1 kilocalorie[2)] per kilogram	=	$kcal_{IT}/kg$	4.26935×10^{2}	**1.8**	1.40070×10^{3}
1 thermochemical kilocalorie per kilogram	=	$kcal_{th}/kg$	4.26649×10^{2}	1.79880	1.39977×10^{3}
1 kilogram-force metre per kilogram	=	kgf m/kg	**1**	4.21610×10^{-3}	3.28084
1 British thermal unit per pound	=	Btu/lb	2.37186×10^{2}	**1**	7.78169×10^{2}
1 foot pound-force per pound	=	ft lbf/lb	**3.048×10^{-1}**	1.28507×10^{-3}	**1**

1) The unit of specific energy in the International System (SI) is the joule per kilogram [J/kg].
2) This refers to the International Table calorie.

3.28 *Specific heat capacity*

TABLE 9.34 Conversion factors for units of specific heat capacity

Unit			joule per kilogram kelvin[1),2)]	kilocalorie[3)] per kilogram kelvin	thermo-chemical kilocalorie per kilogram kelvin
Name		Symbol	J/(kg K)	$kcal_{IT}$/(kg K)	$kcal_{th}$/(kg K)
1 joule per kilogram kelvin[1),2)]	=	J/(kg K)	**1**	2.38846×10^{-4}	2.39006×10^{-4}
1 kilocalorie[3)] per kilogram kelvin	=	$kcal_{IT}$/(kg K)	**4.1868×10^3**	**1**	1.00067
1 thermochemical kilocalorie per kilogram kelvin	=	$kcal_{th}$/(kg K)	**4.184×10^3**	9.99331×10^{-1}	**1**
1 kilogram-force metre per kilogram kelvin	=	kgf m/(kg K)	**9.80665**	2.34228×10^{-3}	2.34385×10^{-3}
1 British thermal unit per pound degree Fahrenheit	=	Btu/(lb °F)	**4.1868×10^3**	**1**	1.00067
1 foot pound-force per pound degree Fahrenheit	=	ft lbf/(lb °F)	5.38032	1.28507×10^{-3}	1.28593×10^{-3}

(Contd.)

TABLE 9.34 Conversion factors for units of specific heat capacity (continued)

Unit			kilogram-force metre per kilogram kelvin	British thermal unit per pound degree Fahrenheit	foot pound-force per pound degree Fahrenheit
Name		Symbol	kgf m/(kg K)	Btu/(lb °F)	ft lbf/(lb °F)
1 joule per kilogram kelvin[1),2)]	=	J/(kg K)	1.01972×10^{-1}	2.38846×10^{-4}	1.85863×10^{-1}
1 kilocalorie[3)] per kilogram kelvin	=	$kcal_{IT}$/(kg K)	4.26935×10^{2}	**1**	7.78169×10^{2}
1 thermochemical kilocalorie per kilogram kelvin	=	$kcal_{th}$/(kg K)	4.26649×10^{2}	9.99331×10^{-1}	7.77649×10^{2}
1 kilogram-force metre per kilogram kelvin	=	kgf m/(kg K)	**1**	2.34228×10^{-3}	1.82269
1 British thermal unit per pound degree Fahrenheit	=	Btu/(lb °F)	4.26935×10^{2}	**1**	7.78169×10^{2}
1 foot pound-force per pound degree Fahrenheit	=	ft lbf/(lb °F)	5.48640×10^{-1}	1.28507×10^{-3}	**1**

1) The unit of specific heat capacity in the International System (SI) is the joule per kilogram kelvin [J/(kg K)].
2) Wherever the kelvin (K) occurs in this table it may be replaced by the degree Celsius (°C), e.g. J/(kg K) is often shown as J/(kg °C).
3) This refers to the International Table calorie.

3.29 Thermal conductivity

TABLE 9.35 Conversion factors for units of thermal conductivity

Unit			watt per metre kelvin[1),2)]	calorie per centimetre second kelvin	kilocalorie per metre hour kelvin	British thermal unit per foot hour degree Fahrenheit	British thermal unit inch per square foot hour degree Fahrenheit
Name		Symbol	W/(m K)	cal_{IT}/(cm s K)	$kcal_{IT}$/(m h K)	Btu/(ft h °F)	Btu in/(ft^2 h °F)
1 watt per metre kelvin[1),2)]	=	W/(m K)	**1**	2.38846×10^{-3}	8.59845×10^{-1}	5.77789×10^{-1}	6.93347
1 calorie[3)] per centimetre second kelvin	=	cal_{IT}/(cm s K)	**4.1868×10^2**	**1**	**3.6×10^2**	2.41909×10^2	2.90291×10^3
1 kilocalorie per metre hour kelvin	=	$kcal_{IT}$/(m h K)	**1.163**	2.77778×10^{-3}	**1**	6.71969×10^{-1}	8.06363
1 British thermal unit per foot hour degree Fahrenheit	=	Btu/(ft h °F)	1.73073	4.13379×10^{-3}	1.48816	**1**	**1.2×10^1**
1 British thermal unit inch per square foot hour degree Fahrenheit	=	Btu in/(ft^2 h °F)	1.44228×10^{-1}	3.44482×10^{-4}	1.24014×10^{-1}	8.33333×10^{-2}	**1**

1) The unit of thermal conductivity in the International System (SI) is the watt per metre kelvin [W/(m K)].
2) Wherever the kelvin occurs in this table it may be replaced by the degree Celsius (°C), e.g., W/(m K) is often shown as W/(m °C).
3) This refers to the International Table calorie.

3.30 *Electrical resistivity*

TABLE 9.36 Conversion factors for units of electrical resistivity

Unit			ohm metre[1)]	microhm metre	nanohm metre	ohm centimetre	microhm centimetre
Name		Symbol	Ω m	μΩ m	nΩ m	Ω cm	μΩ cm
1 ohm metre[1)]	=	Ω m	1	1×10^{6}	1×10^{9}	1×10^{2}	1×10^{8}
1 microhm metre	=	μΩ m	1×10^{-6}	1	1×10^{3}	1×10^{-4}	1×10^{2}
1 nanohm metre	=	nΩ m	1×10^{-9}	1×10^{-3}	1	1×10^{-7}	1×10^{-1}
1 ohm centimetre	=	Ω cm	1×10^{-2}	1×10^{4}	1×10^{7}	1	1×10^{6}
1 microhm centimetre	=	μΩ cm	1×10^{-8}	1×10^{-2}	1×10^{1}	1×10^{-6}	1

1) The unit of electrical resistivity in the International System (SI) is the ohm metre [Ω m].

3.31 *Magnetic permeability*

TABLE 9.37 Conversion factors for units of magnetic permeability

Unit			microhenry per metre[1)]	CGS unit (μ max. value)
Name		Symbol	μH/m	
1 microhenry per metre[1)]	=	μH/m	1	7.95775×10^{-1}
1 CGS unit (μ max. value)	=	–	1.25664	1

1) The unit of magnetic permeability in the International System (SI) is the microhenry per metre [μH/m].

3.32 Remanent magnetism

TABLE 9.38 Conversion factors for units of remanent magnetism

Unit			tesla[1)]	gauss
Name		Symbol	T	
1 tesla[1)]	=	T	**1**	**1×10^4**
1 gauss	=	–	**1×10^{-4}**	**1**

[1)] The unit of remanent magnetism in the International System (SI) is the tesla [T].

3.33 Coercive force

TABLE 9.39 Conversion factors for units of coercive force

Unit			ampere per metre[1)]	oersted
Name		Symbol	A/m	
1 ampere per metre[1)]	=	A/m	**1**	1.25664×10^{-2}
1 oersted	=	–	**7.95775×10^1**	**1**

[1)] The unit of coercive force in the International System (SI) is the ampere per metre [A/m].

3.34 Hysteresis loss

TABLE 9.40 Conversion factors for unit of hysteresis loss (volume basis)

Unit			Joule per cubic metre[1)]	erg per cubic centimetre
Name		Symbol	J/m^3	erg/cm^3
1 joule per cubic metre[1)]	=	J/m^3	**1**	**1×10^{-1}**
1 erg per cubic centimetre	=	erg/cm^3	**1×10^{-1}**	**1**

[1)] The unit of hysteresis loss (volume basis) in the International System (SI) is the joule per cubic metre [J/m^3].

TABLE 9.41 Conversion factors for unit of hysteresis loss (mass basis)

Unit			watt per kilogram[1)]	watt per pound
Name		Symbol	W/kg	W/lb
1 watt per kilogram[1)]	=	W/kg	**1**	2.20462
1 watt per pound	=	W/lb	4.53592×10^{-1}	**1**

1) The unit of hysteresis loss (mass basis) in the International System (SI) is the watt per kilogram [W/kg].

4 REFERENCES

Books

1. K. Raznjevic., *Handbook of Thermodynamic Tables and Charts*, McGraw-Hill, New York, U.S.A., 1976.
2. P.M. Unterweiser, H.E. Boyer and J.J. Kubbs (Eds.), *Heat Treater's Guide—Standard Practices and Procedures for Steel*, American Society for Metals, Metals Park, Ohio, U.S.A., 1982.
3. G.N.J. Gilbert, *Engineering Data on Grey Cast Irons—SI units*, BCIRA, Alvechurch, Birmingham, U.K., 1977.

Standards

4. ISO 31 (0) – 1992: *Quantities and Units—Part 0: General Principles.*
5. ISO 1000 – 1992: *SI Units and Recommendations for the Use of their Multiples and of Certain Other Units.*
6. BS 350 (1) – 1974: *Conversion Factors and Tables—Part 1. Basis of Tables. Conversion Factors.*

ANNEXURE

CONVERSION TABLE FOR VICKERS HARDNESS, BRINELL HARDNESS, ROCKWELL HARDNESS AND TENSILE STRENGTH

TABLE A.1 Approximate relationship of Vickers, Brinell and Rockwell hardness values, and corresponding approximate tensile strengths of unalloyed and low-alloy steels in the as-cast, hot worked and heat treated condition

Tensile strength	Vickers hardness	Brinell hardness[1)]	Rockwell hardness							
MPa	(F ≥98 N)	(0.102 F/D^2 = 30 MPa)	HRB	HRF	HRC	HRA	HRD	HR 15 N	HR 30 N	HR 45 N
255	80	76.0								
270	85	80.7	41.0							
285	90	85.5	48.0	82 6						
305	95	90.2	52.0							
320	100	95.0	56.2	87.0						
335	105	99.8								
350	110	105	62.3	90.5						
370	115	109								

(Contd.)

TABLE A.1 Approximate relationship of Vickers, Brinell and Rockwell hardness values, and corresponding approximate tensile strengths of unalloyed and low-alloy steels in the as-cast, hot worked and heat treated condition (continued)

Tensile strength MPa	Vickers hardness ($F \geq 98$ N)	Brinell hardness[1)] ($0.102 F/D^2 = 30$ MPa)	Rockwell hardness							
			HRB	HRF	HRC	HRA	HRD	HR15N	HR30N	HR45N
385	120	114	66.7	93.6						
400	125	119								
415	130	124	71.2	96.4						
430	135	128								
450	140	133	75.0	99.0						
465	145	138								
480	150	143	78.7	101.4						
495	155	147								
510	160	152	81.7	103.6						
530	165	156								
545	170	162	85.0	105.5						
560	175	166								
575	180	171	87.1	107.2						
595	185	176								
610	190	181	89.5	108.7						
625	195	185								
640	200	190	91.5	110.1						
660	205	195	92.5							
675	210	199	93.5	111.3						
690	215	204	94.0							

(Contd.)

TABLE A.1 Approximate relationship of Vickers, Brinell and Rockwell hardness values, and corresponding approximate tensile strengths of unalloyed and low-alloy steels in the as-cast, hot worked and heat treated condition (continued)

Tensile strength MPa	Vickers hardness ($F \geq 98$ N)	Brinell hardness[1)] ($0.102\,F/D^2 =$ 30 MPa)	Rockwell hardness							
			HRB	HRF	HRC	HRA	HRD	HR15N	HR30N	HR45N
705	220	209	95.0	112.4						
720	225	214	96.0							
740	230	219	96.7	113.4						
755	235	223								
770	240	228	98.1	114.3	20.3	60.7	40.3	69.6	41.7	19.9
785	245	233			21.3	61.2	41.1	70.1	42.5	21.1
800	250	238	99.5	115.1	22.2	61.6	41.7	70.6	43.4	22.2
820	255	242			23.1	62.0	42.2	71.1	44.2	23.2
835	260	247	(101)		24.0	62.4	43.1	71.6	45.0	24.3
850	265	252			24.8	62.7	43.7	72.1	45.7	25.2
865	270	257	(102)		25.6	63.1	44.3	72.6	46.4	26.2
880	275	261			26.4	63.5	44.9	73.0	47.2	27.1
900	280	266	(104)		27.1	63.8	45.3	73.4	47.8	27.9
915	285	271			27.8	64.2	46.0	73.8	48.4	28.7
930	290	276	(105)		28.5	64.5	46.5	74.2	49.0	29.5
950	295	280			29.2	64.8	47.1	74.6	49.7	30.4
965	300	285			29.8	65.2	47.5	74.9	50.2	31.1
995	310	295			31.0	65.8	48.4	75.6	51.3	32.5
1030	320	304			32.2	66.4	49.4	76.2	52.3	33.9
1060	330	314			33.3	67.0	50.2	76.8	53.6	35.2

(Contd.)

TABLE A.1 Approximate relationship of Vickers, Brinell and Rockwell hardness values, and corresponding approximate tensile strengths of unalloyed and low-alloy steels in the as-cast, hot worked and heat treated condition (continued)

Tensile strength MPa	Vickers hardness ($F \geq 98$ N)	Brinell hardness[1)] ($0.102 F/D^2 = 30$ MPa)	Rockwell hardness							
			HRB	HRF	HRC	HRA	HRD	HR15N	HR30N	HR45N
1095	340	323			34.4	67.6	51.1	77.4	54.4	36.5
1125	350	333			35.5	68.1	51.9	78.0	55.4	37.8
1155	360	342			36.6	68.7	52.8	78.6	56.4	39.1
1190	370	352			37.7	69.2	53.6	79.2	57.4	40.4
1220	380	361			38.8	69.8	54.4	79.8	58.4	41.7
1255	390	371			39.8	70.3	55.3	80.3	59.3	42.9
1290	400	380			40.8	70.8	56.0	80.8	60.2	44.1
1320	410	390			41.8	71.4	56.8	81.4	61.1	45.3
1350	420	399			42.7	71.8	57.5	81.8	61.9	46.4
1385	430	409			43.6	72.3	58.2	82.3	62.7	47.4
1420	440	418			44.5	72.8	58.8	82.8	63.5	48.4
1455	450	428			45.3	73.3	59.4	83.2	64.3	49.4
1485	460	437			46.1	73.6	60.1	83.6	64.9	50.4
1520	470	447			46.9	74.1	60.7	83.9	65.7	51.3
1555	480	(456)			47.7	74.5	61.3	84.3	66.4	52.2
1595	490	(466)			48.4	74.9	61.6	84.7	67.1	53.1
1630	500	(475)			49.1	75.3	62.2	85.0	67.7	53.9
1665	510	(485)			49.8	75.7	62.9	85.4	68.3	54.7
1700	520	(494)			50.5	76.1	63.5	85.7	69.0	55.6
1740	530	(504)			51.1	76.4	63.9	86.0	69.5	56.2

(Contd.)

TABLE A.1 Approximate relationship of Vickers, Brinell and Rockwell hardness values, and corresponding approximate tensile strengths of unalloyed and low-alloy steels in the as-cast, hot worked and heat treated condition (continued)

Tensile strength MPa	Vickers hardness ($F \geq 98$ N)	Brinell hardness[1)] ($0.102 F/D^2 =$ 30 MPa)	Rockwell hardness							
			HRB	HRF	HRC	HRA	HRD	HR15N	HR30N	HR45N
1775	540	(513)			51.7	76.7	64.4	86.3	70.0	57.0
1810	550	(523)			52.3	77.0	64.8	86.6	70.5	57.8
1845	560	(532)			53.0	77.4	65.4	86.9	71.2	58.6
1880	570	(542)			53.6	77.8	65.8	87.2	71.7	59.3
1920	580	(551)			54.1	78.0	66.2	87.5	72.1	59.9
1955	590	(561)			54.7	78.4	66.7	87.8	72.7	60.5
1995	600	(570)			55.2	78.6	67.0	88.0	73.2	61.2
2030	610	(580)			55.7	78.9	67.5	88.2	73.7	61.7
2070	620	(589)			56.3	79.2	67.9	88.5	74.2	62.4
2105	630	(599)			56.8	79.5	68.3	88.8	74.6	63.0
2145	640	(608)			57.3	79.8	68.7	89.0	75.1	63.5
2180	650	(618)			57.8	80.0	69.0	89.2	75.5	64.1
	660				58.3	80.3	69.4	89.5	75.9	64.7
	670				58.8	80.6	69.8	89.7	76.4	65.3
	680				59.2	80.8	70.1	89.8	76.8	65.7
	690				59.7	81.1	70.5	90.1	77.2	66.2
	700				60.1	81.3	70.8	90.3	77.6	66.7
	720				61.0	81.8	71.5	90.7	78.4	67.7
	740				61.8	82.2	72.1	91.0	79.1	68.6
	760				62.5	82.6	72.6	91.2	79.7	69.4

(Contd.)

TABLE A.1 Approximate relationship of Vickers, Brinell and Rockwell hardness values, and corresponding approximate tensile strengths of unalloyed and low-alloy steels in the as-cast, hot worked and heat treated condition (continued)

Tensile strength MPa	Vickers hardness ($F \geq 98$ N)	Brinell hardness[1] ($0.102 F/D^2 = 30$ MPa)	Rockwell hardness							
			HRB	HRF	HRC	HRA	HRD	HR15N	HR30N	HR45N
	780				63.3	83.0	73.3	91.5	80.4	70.2
	800				64.0	83.4	73.8	91.8	81.1	71.0
	820				64.7	83.8	74.3	92.1	81.7	71.8
	840				65.3	84.1	74.8	92.3	82.2	72.2
	860				65.9	84.4	75.3	92.5	82.7	73.1
	880				66.4	84.7	75.7	92.7	83.1	73.6
	900				67.0	85.0	76.1	92.9	83.6	74.2
	920				67.5	85.3	76.5	93.0	84.0	74.8
	940				68.0	85.6	76.9	93.2	84.4	75.4

Notes: Calculated from: HB = 0.95 × HV.

1) The figures in brackets are hardness values that lie outside the range of definition of the standardized hardness test methods but which in practice are frequently used as approximate values. Apart from this, the Brinell hardness values in brackets apply only if the measurement is made with a carbide ball.

2) In the case of high-alloy and/or work-hardened steels, substantial deviations are likely to occur.

Source: Reference 1

2 REFERENCE

Standard

1 DIN 50150 – 1976: *Testing of Steel and Cast Steel—Conversion Table for Vickers Hardness, Brinell Hardness, Rockwell Hardness and Tensile Strength.*

INDEX

A

Aluminium and its alloys

Aluminium, primary ingots 4.6–4.7
  Equivalent grades 4.7
  Specifications 4.6

Aluminium, pure 1.6, 1.7, 4.5

Cast aluminium and aluminium alloys 4.16–4.34
  Aluminium–copper alloys 4.16
  Aluminium–magnesium alloys 4.17
  Aluminium–silicon alloys 4.16–4.17
  Aluminium–zinc–magnesium alloys 4.17
  Comparison ratings of cast aluminium and aluminium alloys 4.26–4.27
  Equivalent grades 4.30–4.33
  Recommended casting temperature 4.25
  Specification 4.18–4.22
  Typical applications 4.28, 4.29
  Typical physical and mechanical properties 4.23, 4.24
  Unalloyed aluminium 4.16

Designation system for aluminium and aluminium alloys 4.8–4.11

Heat treatment of aluminium and aluminium alloys 4.77–4.86
  Full annealing 4.77
  Partial annealing 4.77–4.78
  Precipitation hardening
    Ageing 4.84–4.85
    Quenching 4.83–4.84
    Solution heat treatment 4.78–4.83
  Reheat-treatment 4.85

Temper designation system for aluminium and aluminium alloys 4.12–4.15
  Equivalent temper designations 4.14–4.15

Wrought aluminium and aluminium alloys 4.35–4.76
  Aluminium–copper alloys 4.35
  Aluminium–magnesium alloys 4.36
  Aluminium–magnesium–silicon alloys 4.36
  Aluminium–manganese alloys 4.35
  Aluminium–manganese–magnesium alloys 4.35
  Aluminium–silicon alloys 4.35
  Aluminium–zinc–magnesium alloys 4.36
  Aluminium–zinc–magnesium–copper alloys 4.36
  Comparison ratings of wrought aluminium and aluminium alloys 4.68–4.70
  Equivalent grades 4.73–4.75
  Specifications 4.36–4.64
  Typical applications 4.71, 4.72
  Typical physical and mechanical properties 4.65–4.67
  Unalloyed aluminium 4.35

Areas and volumes of solids 8.42–8.43
  Cone 8.42
  Cube 8.42
  Cuboid 8.42
  Cylinder 8.42
  Pyramid 8.42
  Sphere and parts of a sphere 8.43

Areas of sections 8.39–8.41
  Circle and parts of a circle 8.41
  Ellipse 8.41
  Parallelogram 8.40
  Polygon 8.40
  Rectangle 8.40
  Rhombus 8.40
  Square 8.39
  Trapezoid 8.40
  Triangle 8.39

C

Cast irons

Abrasion-resistant cast iron 2.60–2.66
  Equivalent grades 2.65
  Heat treatment 2.63, 2.64

Specifications 2.60–2.62
Typical applications 2.64, 2.65
Typical physical properties 2.63

Austempered ductile iron 2.44–2.48
Austempering heat treatment 2.46
Specifications 2.46, 2.47
Typical applications 2.47

Austenitic cast iron 2.67–2.81
Design stresses 2.76, 2.77
Equivalent grades 2.79, 2.80
Heat treatment 2.77, 2.78
Specification 2.67–2.70
Typical applications 2.72, 2.74, 2.75
Typical mechanical properties 2.70, 2.71, 2.73, 2.76
Typical physical properties 2.71, 2.73, 2.74

Classification of cast irons 2.5, 2.6

Comparison ratings of unalloyed cast irons 2.10, 2.50

Compacted graphite cast iron 2.49–2.52
Specification 2.51
Typical applications 2.52
Typical mechanical and physical properties 2.51, 2.52

Corrosion-resistant high-silicon cast iron 2.82–2.84
Equivalent grades 2.84
Heat treatment 2.83
Specification 2.82, 2.83
Typical applications 2.84
Typical mechanical and physical properties 2.83

Designation of microstructure of graphite in cast irons 2.15–2.21

Designation system for cast irons 2.12–2.14

Ductile iron, (see Spheroidal graphite cast iron)

Effect of elements on the structure of cast irons 2.7–2.10

Flake graphite cast iron, (see Grey cast iron)

Grey cast iron 2.22, 2.32
Design stresses 2.27
Equivalent grades 2.31
Heat treatment 2.27–2.30
Specification 2.23
Typical applications 2.30, 2.31
Typical mechanical properties 2.23–2.25
Typical physical properties 2.25, 2.26

Malleable cast iron 2.53–2.59
Equivalent grades 2.58, 2.59
Heat treatment 2.57, 2.58
Specifications 2.54, 2.55
Typical applications 2.58
Typical mechanical properties 2.55, 2.56
Typical physical properties 2.57

Spheroidal graphite cast iron 2.33–2.43
Design stresses 2.40
Equivalent grades 2.42, 2.43
Heat treatment 2.40–2.42
Specification 2.34–2.36
Typical applications 2.42
Typical mechanical properties 2.37, 2.38
Typical physical properties 2.38, 2.39

Conversion factors and tables 9.10–9.77
Acceleration 9.24
Angular momentum 9.34
Angular velocity 9.23
Area 9.13, 9.14
Coefficient of thermal expansion 9.69
Coercive force 9.76
Density 9.31, 9.32
Dynamic viscosity 9.49, 9.50
Electrical resistivity 9.75
Energy 9.53–9.55
Force 9.35, 9.36
Hysteresis loss 9.76, 9.77
Kinematic viscosity 9.51, 9.52
Length 9.10–9.12
Linear density 9.27, 9.28
Linear momentum 9.34
Linear velocity 9.21, 9.22
Magnetic permeability 9.75
Mass 9.25, 9.26
Moment of force 9.37, 9.38
Moment of inertia 9.33
Plane angle 9.19
Power 9.56–9.58
Pressure 9.39–9.50
Remanent magnetism 9.76
Second moment of area 9.18
Section modulus 9.17
Specific energy 9.70, 9.71
Specific heat capacity 9.72, 9.73
Stress 9.39–9.50
Surface density 9.29, 9.30
Temperature 9.59–9.68
Thermal conductivity 9.74
Time 9.20
Torque 9.37, 9.38
Volume 9.15, 9.16

Copper and its alloys
Cast copper and copper alloys 5.15–5.40
  Aluminium bronzes 5.16
  Brasses 5.15
    Gunmetals 5.15
    High conductivity coppers 5.15
    Leaded tin bronzes 5.16
    Phosphor bronzes 5.16
    Silicon bronzes 5.16
  Comparison ratings of cast copper and copper alloys 5.30–5.32
    Equivalent grades 5.35–5.39
    Recommended casting temperature 5.29
    Specifications 5.17–5.25
    Typical applications 5.32–5.34
    Typical physical properties 5.26–5.28

Copper, commercially pure 5.5–5.9
  Arsenical tough-pitch copper (CuATP) 5.6
  Cathode copper (CuCATH) 5.5
  Copper, pure 1.8, 1.9, 5.5
  Electrolytic tough-pitch copper (CuETP) 5.5
  Electrical properties 5.8
  Equivalent grades 5.8, 5.9
  Fire-refined tough-pitch high-conductivity (CuFRHC) 5.5
  Fire-refined tough-pitch copper (CuFRTP) 5.6
  Oxygen-free copper (CuOF) 5.6
  Phosphorus-deoxidized copper, high residual phosphorus (CuDHP) 5.6
  Phosphorus-deoxidized arsenical copper (CuDPA) 5.6
  Specification 5.7

Designation system for copper and copper alloys 5.10, 5.11

Heat treatment 5.98–5.102
  Annealing of cast copper alloys 5.101
  Full annealing of wrought copper and copper alloys 5.98–5.101
  Hardening and tempering 5.102
  Homogenizing 5.99
  Partial annealing 5.101
  Precipitation hardening 5.102
  Stress-relieving 5.98

Temper designation system for copper and copper alloys 5.12–5.14

Wrought copper and copper alloys 5.41–5.97
  Aluminium bronzes 5.42
  Aluminium silicon bronzes 5.42
  Brasses 5.41–5.42
  Copper-nickel alloys 5.42
  Equivalent grades 5.89–5.96
  High-copper alloys 5.41
  Nickel silvers 5.42
  Phosphor bronzes 5.42
  Specifications 5.43–5.80
  Typical applications 5.86–5.88
  Typical physical and mechanical properties 5.81–5.85
  Unalloyed coppers 5.41

D

Damping capacity 8.5, 8.6

E

Elements

Emissivity 8.7–8.9

Periodic table 1.5

Physical properties 1.6–1.20

G

Galvanic series 8.9, 8.10

H

Hardness – tensile strength conversion table, steel A.1–A.6

P

Periodic table of the elements 1.5

Powder metallurgy (P/M) 7.1–7.36
  Advantages 7.9
  Atomization 7.5
  Blending 7.6
  Coining 7.7
  Cold isostatic pressing (CIP) 7.8
  Compacting 7.6
  Definition 7.5
  Electrolytic deposition 7.5, 7.6
  Equivalent standards 7.30–7.35
  Heat treatment 7.7
  Hot isostatic pressing (HTP) 7.8
  Impregnation 7.7
  Infiltration 7.7
  Limitations 7.9
  Mechanical comminution 7.6
  Mixing, (see Blending)
  Powder forging 7.8

Powder rolling 7.8
Plating 7.7
Reduction of metal oxides 7.5
Sintering 7.6
Sizing 7.7
Specifications 7.10–7.28
Steam treatment 7.7
Thermal decomposition 7.5
Typical applications 7.29

Preferred numbers 8.35–8.37

R

Rounding of numbers 9.8

S

Sheet metal gauges 8.32–8.35

SI system 9.5–9.10

Steels
Bearing steels 3.213–3.220
Equivalent grades 3.218, 3.219
Hot-working and heat treatment temperatures 3.216, 3.217
Specifications 3.214, 3.215

Bright bars 3.86–3.93
Equivalent grades 3.90–3.92
Specifications 3.87–3.89

Case-hardening steels 3.132–3.140
Equivalent grades 3.139, 3.140
Hot-working and heat treatment temperatures 3.137, 3.138
Specification 3.133–3.136

Cast analysis 3.18

Cast steels 3.238–3.261

Classification of steels 3.5–3.12

Coarse-grain steels 3.18

Designation system for wrought steels 3.13–3.16

Effect of elements 3.19–3.21

Equivalent diameter 3.19

Fine-grain steels 3.18

Free-cutting steels 3.81–3.85
Equivalent grades 3.85
Hot-working and heat treatment temperatures 3.84
Specifications 3.82–3.84

Fully killed steels 3.18

Heat treatment
Annealing 3.29, 3.30
Full annealing 3.29
Isothermal annealing 3.29, 3.30
Process annealing 3.30
Soft annealing, (see Softening)
Softening 3.30
Spheroidizing 3.30
Austempering 3.32
Austenitizing 3.36
Carbonitriding 3.33
Carburizing 3.32, 3.33
Cyaniding 3.33, 3.34
Flame hardening 3.35, 3.36
Gaseous ferritic nitrocarburizing 3.34
Hardening and tempering 3.31
Homogenizing 3.30
Induction hardening 3.35
Liquid carbonitriding, (see Cyaniding)
Martempering 3.31
Nitriding 3.34
Normalizing 3.31
Soaking 3.36
Stress-relieving 3.29
Tempering 3.32

Iron-carbon equilibrium diagram 3.17

Killed steels, (see Fully killed steels) 3.18

Limiting ruling section 3.19

Low-carbon steel sheet and strip 3.37–3.45
Equivalent grades 3.44, 3.45
Specifications 3.38–3.44

Machinability rating 3.26, 3.27

Metal coated steel sheet and strip 3.50–3.60
Equivalent grades 3.59
Specifications 3.51–3.58

Microalloyed ferritic-pearlitic forging steels 3.112–3.114
Specifications 3.113

Nitriding steels 3.141–3.145
Equivalent grades 3.145
Hot-working and heat treatment temperature 3.144
Specification 3.142–3.145

Physical properties 3.22–3.25

Product analysis 3.18

Residual elements 3.18

Rimmed steels 3.18

Ruling section 3.19

Semi-killed steels 3.18

Spring steels 3.159–3.184
- Equivalent standards/grades 3.180–3.185
- Hot-working and heat treatment temperatures 3.179, 3.180
- Specifications 3.160–3.179

Stainless steels, wrought 3.185–3.201
- Equivalent grades 3.198–3.200
- Heat treatment temperatures 3.195, 3.197
- Specification 3.187–3.195

Steel sheet and strip for porcelain enamelling 3.46–3.49
- Equivalent grades 3.49
- Specifications 3.47, 3.48

Steel tubes for structural and mechanical purposes 3.71–3.80
- Specifications 3.72–3.79

Steels for cold heading and cold extruding 3.94–3.111
- Equivalent grades 3.105–3.110
- Specifications 3.95–3.105

Steels for flame and induction hardening 3.145–3.153
- Equivalent grades 3.155–3.157
- Hot-working and heat treatment temperatures 3.153–3.154
- Specification 3.147–3.151

Steels for hardening and tempering 3.115–3.131
- Equivalent grades 3.127–3.131
- Hot-working and heat treatment temperatures 3.125–3.126
- Specification 3.115–3.125

Structural steels 3.61–3.70
- Equivalent grades 3.69, 3.70
- Specification 3.62–3.68

Sections
- Bars 8.19–8.29
- Flats 8.30, 8.31
- Plates 8.11
- Sheets 8.12–8.17
- Strips 8.18

Tool steels 3.221–3.237
- Equivalent grades 3.232–3.236
- Hot-working and heat treatment temperatures 3.231
- Specifications 3.222–3.230

Valve steels for internal combustion engines 3.202, 3.212
- Equivalent grades 3.210, 3.211
- Hot-working and heat treatment temperatures 3.209
- Specification 3.203–3.208

Surface roughness ranges of common production processes 8.38–8.39

W

Wire gauges 8.32–8.35

Z

Zinc and its alloys

Zinc, pure 1.18, 1.19, 6.5

Zinc casting alloys 6.8–6.30
- Comparative ratings of zinc casting alloys 6.29
- Conventional zinc die casting alloys 6.8–6.15
  - Casting temperature 6.13
  - Dimensional changes 6.14
  - Equivalent grades 6.15
  - Heat treatment 6.14
  - Specifications 6.9
  - Typical applications 6.14
  - Typical mechanical properties 6.10–6.12
  - Typical physical properties 6.12, 6.13
- Effect of impurity elements 6.28
- Other casting alloys 6.26, 6.27
  - Forming die alloys 6.26
  - Slush casting alloys 6.26
  - Specifications 6.27
  - Thermal properties 6.27
- Zinc-aluminium (ZA) casting alloys 6.16–6.26
  - Casting temperature 6.24
  - Design stresses 6.24
  - Equivalent grades 6.25, 6.26
  - Heat treatment 6.25
  - Specifications 6.17
  - Typical applications 6.25
  - Typical mechanical properties 6.17–6.22
  - Typical physical properties 6.23

Wrought zinc and zinc alloys 6.31–6.35
- Equivalent grades 6.35
- Fabrication characteristics 6.33, 6.34
- Specifications 6.31, 6.32
- Typical applications 6.34
- Typical physical properties 6.33

Zinc, commercially pure 6.5–6.7
- Equivalent grades 6.7
- Specifications 6.5, 6.6
- Typical applications 6.6